Spatial Behavior:
A Geographic Perspective

Spatial Behavior:
A Geographic Perspective

REGINALD G. GOLLEDGE
Department of Geography,
University of California, Santa Barbara

ROBERT J. STIMSON
Australian Housing and Urban Research Institute and
Faculty of Built Environment and Engineering,
Queensland University of Technology

THE GUILFORD PRESS
New York   London

© 1997 The Guilford Press
A Division of Guilford Publications, Inc.
72 Spring Street, New York, NY 10012

Printed in the United States of America

This book is printed on acid-free paper.

Last digit is print number: 9 8 7 6 5 4 3 2 1

Library of Congress Cataloging-in-Publication Data

Golledge, Reginald G., 1937–
  Spatial behavior : a geographic perspective / by Reginald G. Golledge
    and Robert J. Stimson
      p. cm.
    Includes bibliographical references and index.
    ISBN 1-57230-049-3. — ISBN 1-57230-050-7 (pbk.)
    1. Human geography.   2. Spatial behavior.   3. Decision-making.
  4. Geographical perception.   I. Stimson, R. J. (Robert John)
  II. Title.
  GF95.G65   1997
  304.8—dc21   96-47144
                                          CIP

To Ellis J. Thorpe,
whom we both acknowledge as
our mentor in Geography

Preface

This is not a book oriented toward a single narrow viewpoint; it is certainly not Marxist or materialist; it is somewhat realist; it is partly postmodern; it has elements of positivism; it draws in part on the tradition of spatial analysis; but above all, it focuses on *decision making* and *choice behavior*. We have no ideological axe to grind in this book. We think it reflects a lot of geography as it is done today.

This book is an update and revision of *Analytical Behavioural Geography*. In the course of this process, we added new material and introduced new topics. Many will argue that it is now a different book. Some elements of our original book are maintained however. We hope these are the ones that are the most useful in terms of understanding the text as it now appears and that they assist in our discussions of some fundamental principles and strategies for examining spatial problems.

In a real sense, this book is about spatial problems. It provides some indication of how different people over time have solved spatial problems. It examines problems at widely different scales from individual behavior in imaginary settings or in the confines of a laboratory to the activities of corporations and governments.

Topics covered include human decision making and choice behavior at scales ranging from the individual and the private and corporate capitalist to the state. In the first chapter, we look at some social and behavioral constraints on human behavior. This is done initially in the context of tracing the origins of behavioral work in geography. We stress that, as interest in human behavior at all scales increased, geographers began to explore new models of the environment, new models of decision making and choice, new models of spatial action and activity, new sources of data, new methods for collecting and analyzing data, and new ways of looking at the relationships between persons and the societies in which they live.

In Chapter 2 we introduce a conceptual model of the individual decision-making process. We later expand this to the context of private and corporate capitalists and state decision making for planning and policy purposes. In this chapter we introduce concepts of information search, behavior–space perception, cognitive mapping, and movement imagery. We also examine in detail how normative assumptions have constrained geographic theory—emphasizing classical models of agricultural and industrial location theory. We trace the emergence of interest in the built environment and the cognitive environment and introduce the constraints of ecological and psychological fallacy. We also discuss decision making and choice procedures and the way these are conceptualized and modeled. And we introduce current geographic information systems (GIS) via the medium of decision support systems as a tool to aid in planning and policy making at all scales.

The two following chapters discuss the changing operational milieu. Initially, we examine the process and impact of economic, social, and technological change along with the practices of globalization and internationalization of economies and societies. The enrichment of locational decision making by moving to the transnational arena is emphasized and is then related to ongoing trends in the flow of people and capital. The next chapter complements this by emphasizing the demographic and economic changes that have been instrumental in producing the decline of regions at various scales. Emphasis is placed on changes in age structure (such as the graying of the population), changes in household structure, the emerging significance of female participation in the labor force, and problems of ethnic diversity and multiculturalism. These concepts are elaborated within the context of urban space. To tie both national and urban space together, we then consider institutional and private planning processes ranging from the smaller-scale "not in my backyard" (NIMBY) syndrome to the strategic planning practices of institutions and governments.

The next three chapters discuss disaggregate behavioral processes. Initially, we examine theories of the acquisition of spatial knowledge, including psychologically based and geographically based conceptualizations. Landmark, route, and survey knowledge are examined along with notions of hierarchical structure, regionalization, and anchorpoints. Case studies are given of children's wayfinding in unfamiliar environments, and we introduce the idea of sets of interacting computer programs (called computational process models). These are sometimes used in the context of artificial intelligence modeling to operationalize the activity of wayfinding. Chapter 6 examines perceptions, attitudes, and risk. Here we evaluate factors influencing the structure of natural and perceived environments and elaborate on people's attitudes and reactions toward natural and technological hazards. This is followed by a discussion of the dimensions of risk and risk aversion strategies. Chapter 7 emphasizes spatial cognition, cognitive mapping, and cognitive maps. It begins with an overview of the processes of spatial cognition, and after presenting a variety of different metaphors that are used in the discussion of cognitive maps, it examines processes of externally representing such maps, analyzing, and using them.

Chapter 8 links the disaggregate behavioral approach emphasized in the previous three chapters with individual and group scheduling of activities in space and time. Activity scheduling, time–space constraints on behavior, and empirical studies of action space and activity spaces are presented in the balance of the chapter.

Activity analysis in travel and transportation modeling continues the spatial–behavioral emphasis of the previous chapter by focusing on decision making and choice behavior in a transportation context. Examples are given of some types of transportation models in general, but most emphasis is placed on individual and household scheduling of trips. We also add a discussion of recent trends to incorporate cognitive factors into activity-scheduling behavior by presenting a computational process model that uses GIS technology for simulating movement behavior of household members in real world environments.

In Chapter 10 we discuss the related topics of consumer behavior and retail center location. Classical descriptive models of consumer behavior (such as Reilly's Law and Huff's probabilistic model) are initially explained along with the more recently emphasized disaggregate discrete choice and preference models. The role of imagery in consumer decision making is stressed as are other topics such as search and learning and feedback. Store and center images are presented as factors influencing location by retailers, and this leads to a historical overview of the development of retail centers in the United States, their locational characteristics and attributes, and the physical and perceptual factors that influence consumer patronage. Retail location is then examined within the context of national, state, and local expectations with respect to conforming to legislation on the conduct of economic, environmental, and cultural impact studies prior to location and development.

The affective components of place and space are discussed in Chapter 11. The roles that feelings and emotions play in perception of environmental quality and evaluating and assessing landscapes are presented. Particular case studies involve recreation and leisure behavior, and the links among linguistics, environment, and behavior are also examined.

Chapters 12 and 13 focus on a different type of spatial behavior. In Chapter 12, the more or less permanent moves across the landscape (known as migration) are discussed in both the original deterministic and later probabilistic forms. Types of migration and the beliefs, values, aspirations, and stresses underlying the decision to move are evaluated. In Chapter 13 we return to a disaggregate approach and examine residential site selection processes in the context of short-term intracity moves, a process usually referred to as mobility. Differences between the factors influencing mobility and migration are elaborated, and examples of movement patterns in urban space are presented.

The final few chapters cover relatively new topics in geography—the spatial characteristics and problems of special populations. They are often grouped under the term "The Other"; these are populations that have been implicitly and explicitly discriminated against, disenfranchised, and differentiated from the general population. Some are differentiated because of physical disabilities and functioning (such as the wheelchair bound, the mentally retarded, and the blind). Others are differentiated because of traditional roles and attitudes in societies and cultures. From among this latter group, we elected in the final chapter to discuss characteristics of the elderly and issues of societal gendering in the context of work, movement, and empowerment. This chapter completes the book, however, by emphasizing the breaking of new ground in this treatment of special populations. Thus, we present the geography of society, space, and behavior as having much unfinished business.

This book is intended to be a teaching text. We hope that it will be used in a number of contexts from introductory courses to the advanced undergraduate level. It discusses topics to be found in many introductory economic, social, or general human geography texts. It also provides some discussions for more advanced treatment of topics. We hope that it is versatile enough for those who wish to teach courses in which human behavior is a focus to adopt as a text.

As usual, we have many people to thank for assistance in compiling this volume. First we acknowledge the cooperation of many editors of journals and book publishers who have allowed us to reproduce tables, charts, figures, and maps. Next we are indebted to Mary MacDonald, Howard Pommerening, Anne Haque, and Michelle Dixon for their work in translating material from tapes and hastily scrawled scripts as well as for compiling tables and drawing and redrawing figures used in the text. We acknowledge The Guilford Press for being willing to take the risk of publishing what was intended to be a revision of a behavioral text. We thank *La España* of Santa Barbara for closing down its restaurant and bar so that we could devote extra time to working on this manuscript instead of indulging in "Lotusland" activities related to the consumption of salsa, chips, and margaritas.

We hope this book meets the needs of many individuals for the teaching of undergraduate and possibly graduate courses. We are certain that geographers in many countries will find equal benefit in the use of the text.

Contents

1

Society, Space, and Behavior

1.1 Spatial Behavior in a Changing World

We live in a rapidly changing world. Most of us live in cities, which are exceedingly complex in terms of their functions, form, and structure. The built environment is the outcome of the myriad decisions that have been made by individuals, households, firms, corporations, and institutions, including governments and public agencies. Decisions are made under constraints, and they reflect the attitudes, values, and beliefs of people and of society. And the basis of making decisions may change over time.

This book is about the decision-making process of individuals, groups, and institutions in a spatial context. It examines choices and behaviors within small- to large-scale environments, at various scales, ranging from the individual to the societal level.

Our concern is to focus on what is embodied in the behavioral approach in the study of human geography in the context of modern urban society. This provides a particular perspective to the analysis and understanding of how men and women, households, firms, and institutions make decisions in conducting their lives and in performing their functions in spatial environments. These decisions include deciding where to locate activities, where to perform tasks like shopping, where to live, and how people develop knowledge about their operational environment that enables decision making to occur in geographic space. We are concerned also with exploring contemporary economic, demographic, and social trends that are important in shaping the macro environment within which decisions and choices are made.

1.2 The Origin of Behavioral Analysis in Geography

For much of this century, geographic research was dominated by a search for definitive structures in both the human and physical environments. Research was undertaken to develop inventories and to describe and map the spatial and temporal facts of existence. An important component of this research was oriented toward the human and built environments, and, in such cases, the search was for patterns of human activities and the economic, social, and political dimensions of the system within which those patterns developed. It was assumed that human behavior could be given substance and stability by regarding it as consisting of relatively invariant and repetitive events within any time–space context. Mostly the concern was with aggregate patterns of spatial behavior, focusing on the locations of economic activities, the flows of people and goods, and on spatial variations in the density or intensity of particular phenomena.

At times, isolated researchers in human geography pointed to the need for expanding the geographer's horizons. They talked of perceived spatial orientation (Gulliver, 1908), imaginary maps (Trowbridge, 1913), the world of the imagination (Wright, 1947; Kirk, 1951), and the perception of natural hazards (White, 1945). The message emanating from these early pioneers was that the *subjective* component of human spatial existence was as important as the *objective* component. For example, cities were not only the structures in which urban functions were located and where people performed their daily activities, but also the city *was* the people who lived in the structures and gave life to this objective form by means of daily interactions and activities. It was realized also that interactions were limited by social, economic, cultural, political, and other constraints, each of which was a more or less "hidden" structure and which had to be elaborated and examined as a contextual dimension of behavior.

By the early 1960s, a growing number of geographers realized that, in order to exist in and comprehend a given environment, people had to learn about critical subsets of information from the mass of experiences open to them. They had to sense, store, record, organize, and use bits of information for the ultimate purpose of coping with the everyday task of living. In doing this, they created knowledge structures based on information selected from the mass of "to whom it may concern" messages emanating from the world in which we live. Different sociocultural groups gave different meanings to these messages. It was the explicit recognition of the relationship among personal cognition, objective environment, societal and cultural constraints, and human behavior that helped to broaden the base of research and reading in human geography.

In the 1960s and early 1970s, there was a particular surge of interest in disaggregate behavioral research. One stream of thought, now identified as the *humanist approach*, investigated the relevance of imagination (Lowenthal, 1961), values and beliefs (Buttimer, 1969, 1972), environmental meanings (Tuan, 1971) and, in a variety of literary contexts, examined the varied meanings associated with place and space (Relph, 1976). Another stream, one we emphasize in this book, followed a more generally *scientific/analytic approach,* measuring and examining attitudes and expectations (White, 1966–1986; Saarinen, 1966), risks and uncertainty (Wolpert, 1965), learning and habit formation (Golledge & Brown, 1967), decisions and choice (Pred, 1964; Burnett, 1973), preferences for places (Gould, 1966; Rushton, 1969a), cognitive maps (Downs, 1970b; Stea, 1969), and the general process of acquiring spatial knowledge (Golledge & Zannaras, 1973). This behaviorally oriented research also tied in strongly to earlier work on migration and innovation diffusion

(Hägerstrand, 1952, 1957) and to complementary research that emphasized the critical role of space in the processes of information diffusion and adoption. Unfortunately, this work emphasizing the cognitive behavioral domain was misunderstood by some critics who mistakenly identified it with Skinnerian behaviorism and behavioral modification—a misunderstanding that is perpetuated even today by critics who rely only on these misrepresenting sources and who do not understand the essential difference between the two approaches (Cloke, Philo, & Sadler, 1991).

In retrospect, it can be seen that many of these new research interests represented a desire to increase the geographer's level of understanding of particular types of problems. This intention appeared to be part of a general and widespread academic desire to gain a deeper level of understanding of problem situations than was possible within the context of narrow disciplinary boundaries. The consequent *cross-disciplinary* interactions and sharing of epistemologies, theories, methods, concepts, models, and even data, exposed many academic areas to different points of view. Part of this involved the formation of cross-disciplinary academic institutes and associations (e.g., urban studies, human–environment relations, natural hazard and disaster research centers dedicated to equitable distribution of human services) that proved to be fertile grounds for the widening of the knowledge horizon of geographers.

1.3   Process Orientation

Traditionally, geographers had achieved a facility to *observe, record, describe, classify,* and *represent* both physical and human components of environmental phenomena. Perhaps the single most important feature of the 1960s for the human geographer was the change in emphasis from *form* to *process*.

1.3.1   Form

The emphasis on *form* had long been a geographic tradition, and this emphasis was strengthened by the powerful theoretical and quantitative revolutions of the 1950s and early 1960s. An emphasis on form or structure allowed researchers to focus on objectively identified characteristics and patterns and to use the many and varied tools and languages of mathematics, statistics, and scientific explanation in their research. In the context of human geography, however, it soon became obvious that the geographer's usual laboratory—the world of *objective reality*—was far too complex to be incorporated into a comprehensible mathematical or statistical model of human activity. A process of simplification began so that problems of human thought and action in different environmental settings could be made tractable. In doing this, several critical generalizations were made. Some involved subdividing the objective environment (e.g., "assume three regions, A, B, C"; "assume a uniform plain"; or "assume a city with a single and centrally located business district"). Others involved removing much of the individual variability of human-kind (e.g., "assume a uniform population with similar tastes, preferences, and utilities") replacing it with *normative beings* who were economically and spatially rational.

1.3.2 Process

As a wider range of alternative human choice and behavior models became available (Simon, 1957), probabilistic methods, innovative statistical techniques, topological languages, nonparametric measurement techniques, and analytical measures were used more widely by geographers. Thus there emerged a message that superficial and general representations of the natural, human, or built environments (whether verbal, cartographic, or mathematical) were no longer enough. What was required for both understanding and explanation was insight into "how and why things are where they are." This theme became known as a *process-driven* search for explanation and understanding.

Thus, within a decade, through a theoretical and quantitative revolution that sought to build normative models of where things ought to be and to define ideal patterns that were abstractions from reality, a large segment of human geography changed from being interested primarily in classifying and categorizing phenomena on the earth's surface to embracing a process-driven search for knowledge of the various aspects of our spatial existence.

At first, the processes on which researchers focused were *individual* or *disaggregate* behavior and included things such as learning, thinking, forming attitudes, perceiving, sensing, feeling, giving meaning and value, imaging, representing, and using spatial knowledge.

Making the leap from an individual behavioral approach to a more *aggregate societal-based* approach followed quickly. It was realized that people usually did *not* have complete freedom of choice or freedom of behavior. A simple look at the oppressive environments found in societies dominated by Communist and Marxist thought quickly showed the importance of state policies and practices in providing constraining templates within which daily living took place. But, as Harvey (1972), Cox (1973), N. Smith (1984), and an increasingly vociferous band of Marxist social theorists quickly pointed out, there were "hidden" constraints in capitalist-dominated and other democratic societies that also limited individual freedoms.

Thus, the emerging emphasis on individual behavioral processes was soon complemented by an emphasis on processes relating to social theory. As pointed out by Couclelis and Golledge (1983), since Marxist social theory really held no place for the individual acting independently to pursue a desired life-style and quality of life, these two different approaches were often at loggerheads. But in both cases the transition to process-driven explanation remained.

1.4 Behavior in Space versus Spatial Behavior

Geographers are, and always have been, interested in the *overt* behavior of individuals, households, firms, corporations, and institutions. They have mapped, classified, and described these behaviors in a spatial context, often ignoring or drastically simplifying the complicated decision-making processes underlying those overt activities. Behaviors were described in terms of movement frequencies, distances between origins and destinations, decline of movement frequency with increasing distance, and other physically observable properties of spatial acts. The geographer

thus produced abundant descriptive data and normative theories about the properties of distributions, interactions, network connections, patterns, nodes, surface
properties, and hierarchical elements of spatial systems. Figure 1.1 illustrates some
of these basic elements of spatial systems.

While geographers became experts in describing *what* was there, when seeking to explain *why* or how things were there, usually they would search for explanation among the physical correlates within the spatial structure in which phenomena
were distributed or where the spatial behavior occurred (e.g., the length of transportation network segments over which freight moved, the number of parking
spaces near a shopping center, the average distance between shops or towns). Thus,
spatial behavior acquired a very loose meaning. As noted by Amedeo and Golledge
(1975):

> It was used to describe on the one hand the physical manifestations of directed acts of
> human beings, and on the other it was used to describe fluctuations of systems or sys-
> tems elements, such as regional growth, commodity flows, population growth, and so

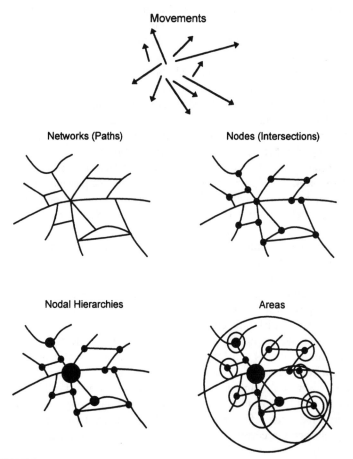

FIGURE 1.1. The elements of spatial systems. *Source*: Haggett, 1965, p. 18.

on. Such uses lumped together human spatial behavior (i.e., those behaviors that are caused, have directedness, motivation, action and achievement), and fluctuations of nonsensate system elements. (p. 348)

An important advantage of adopting a *process-oriented* behavioral approach was to distinguish between *goal-directed* human behaviors and the movement of *nonsensate* system elements that are generated by *exogenous* forces. Thus, the objective was to find process-oriented explanations concerning spatial phenomena rather than cross-sectional explanations that correlate elements of the physical structure of systems with human actions in those systems.

It is important to understand the above proposition. It is the basis on which the process-oriented behavioral approach in geography is differentiated from other approaches. It incorporates the belief that the physical elements of existing and past spatial systems represent the manifestations of a myriad of past and present decision-making behaviors by individuals, groups, and institutions in society. The outcomes are observable landscapes and built environments, and the spatial behaviors that occur within them. The total objective environment thus was seen to encompass not only the *physical* (or natural) environmental phenomena, but also the *built* environment that derived from individual and collective decisions and activities. This differs little from the goals of today's so-called postmodernist research.

If spatial behaviors deduced by mechanistic examination of the properties of spatial systems do not match the empirical reality of the behavior of people acting in those systems, then it must be presumed that factors other than the physical characteristics of those systems need to be considered in developing explanations. There is then a need to incorporate variables derived from analysis of *behavior within the system*. Consider an example such as the movement of passengers by air in a nation. Aggregate flows (or movements) are not strictly goal-directed but are the results of individual sensate behavior occurring in response to a multitude of needs. Although it is possible to seek description of the movement of airline passengers over the air route network in terms of gravity model postulates about interaction between cities of specific sizes and their intervening distances (i.e., the physics of the structure in which the flows occur), to achieve explanations, we need to look to other variables that are relevant and that might include behavioral processes, such as the goal-directedness underlying the movements.

Thus, even though we may use a range of theories to describe and even predict with considerable accuracy the aggregate flows and movements of goods, people, and messages over networks connecting hierarchically arranged nodes, nonetheless, we can furnish little explanation of why those behaviors occurred. The problem is that regularities apparent at aggregate levels of scale may lead us to believe that spatial behavior is equally predictable at finer scales. But, at disaggregated levels, all we can manage is to assign ranges of probabilities to individual decision choices. Even at the aggregate level, in the 1950s and 1960s there was a tendency for geographers to fall into the trap known as the *ecological fallacy*. In so doing they inferred cause–effect relationships between area-based spatial variables and human behavior via correlation and regression models. However, all one can reasonably and validly conclude is that there may be a degree of spatial association between the component variables. To infer that there is causal explanation of human behaviors in terms of those factors, as portrayed by surrogate ratio-type data variables, is invalid.

It may be that the behavior of the structures and elements of spatial system is related directly to behavior occurring over time within those structures. Thus, we need to ask the question: "Are there processes that operate regardless of the spatial structures in which given acts take place?" If this is so, then changes in the *system* characteristics would change the forces operating within it, but they would *not* change the *behavioral processes*. An example will clarify this proposition. Imagine an economically rational person placed first in a spatial system that is compact and then in one that is diffuse. If we examine the mechanics of the behavior of such a "Marshallian person," we may find that "spatial behavior" (e.g., distances traveled) is different in the two systems, and we may conclude that this is due to the different arrangements of the opportunity set in each structure. However, decision-making goals might remain the same—those of minimizing effort, time, or cost. We would need to explain apparently different spatial behaviors by careful specification of the rules of operation and to assess separately the behavior-directing effects of the surrounding structures.

This leads to a definition of the *behavioral process* as "a mechanism for inducing a temporal unfolding of a behavioral system" (Amedeo & Golledge, 1975: 351). It enables us to use the distinction between *sensate* and other types of behavior as the basis for distinguishing between the types of processes that are relevant in searching for explanations of them. Thus, the human behavioral process induces a *temporal sequence of directed acts* by individuals, groups, and institutions in a society. The mechanistic processes that describe the operations of the spatial system over time are not able to explain the behaviors themselves but only the geometric or physical properties of the system. It is the examination of goal-directed spatial behavior, as one manifestation of sensate behavior per se, that may provide the comprehensive explanations of behaviors.

1.5 Dominant Characteristics of Spatial Behavior

Apart from being process-driven, human spatial behavior research has a number of distinguishing characteristics. Some of these were evident in the earliest behavioral research, while others have emerged in response to research barriers to hostile criticism. Since not all of these characteristics have been generally recognized, we now outline them briefly.

1.5.1 New Models of Behavior

Over the centuries, either implicitly or explicitly, writers, thinkers, and scientists have proposed models of human behavior that have incorporated psychological factors, plus the effects of interpersonal influences, social interactions and constraints, motivations for group membership, and decision-making logic. In human geography, traditionally the focus was on the model of *economic behavior* proposed by classical and neoclassical economic theorists. They posed the human decision maker as being totally rational, influenced by objective external factors that were perfectly known. When acting as a consumer, a decision maker sought to *maximize utility;* as an entrepreneur, to *maximize* profit. Not surprisingly, models of economic behavior failed to account for much of the variation generally found in human behavior, particularly in its spatial manifestation.

An alternative model was introduced in the late 1950s by Simon (1957). He saw humans as behaving within the bounds of rationality in a complex world. It was the objective weighting of alternative criteria that led to the principle of *bounded rationality*. Behavior thus generated may appear to be economically or spatially irrational, but it merely reflects the outcomes of variations in individual ability to cope with and store information that is fragmented and incomplete while operating under severe time constraints. The result is that humans *satisfice,* taking a course that allows achievement of limited goals and is "O.K. for me at that time." Simon (1957: 261) suggested, "However adaptive the behavior of organizers in learning and choice situations, this adaptiveness falls far short of the ideal maximizing postulated in economic theory. Evidently organizers adapt well enough to satisfice; they do not, in general, optimize."

It was largely out of this model of a *satisficer* that the behavioral approach in human geography developed. It was a solution to the search by geographers for a model that was a viable alternative to the omniscient economically or spatially rational human being who was assumed within many predictive and explanatory models during the theoretical and quantitative revolution. Such human beings were found to be an inadequate base for those interested in the variability of behavior, as well as its uniformity or its societally constrained nature. In their articles on game theory and decision theory, Gould (1963), Wolpert (1964), and later Webber (1972) exposed geographers to a variety of behavioral criteria other than that of complete rationality. Thus, notions of conservative low-risk behavior, minimum regret, satisficing, bounded rationality, and risky and uncertain behavior, appeared with increasing frequency as the human characteristics that ought to be included in explanatory schemas proposed in human geography.

1.5.2   New Models of Environment

Along with this search for a new model of behavior came an understanding that there existed environments other than the observable external environment. Increased interdisciplinary interaction clearly indicated to geographers that there were *multiple constraints* imposed by "hidden" dimensions of economic, cultural, social, political, legal, moral, and other environments, and that those were equally as important as the *physical* constraints imposed by the *objective* environment. Thus, there developed an interest in the *perceptual, cognitive, ideological, philosophical, sociological,* and other environments that were all part of the dialectical relationship between humanity and the realities in which they live.

1.5.3 Micro Level Focus

For many researchers, the idea of dealing with artificially aggregated masses of humanity was unappealing. While it was recognized that *macro* (or aggregate) levels of analysis could produce some interesting general trends—and, in fact, were a necessary part of the geographer's search for general explanation—it was felt also that a focusing of attention at the *micro* level would give higher levels of understanding and explanation than otherwise could be achieved by the macro level approach.

It must be remembered, however, that while some researchers pursued a disaggregate or behavioral approach in their attempts to obtain more satisfactory levels of understanding and explanation, it was implicit that a major aim of their research was to search for new and improved *generalizations*. Such generalizations could be tied to aggregates based on ideological or societal dimensions of behavior rather than on arbitrary criteria, such as location, demography, or socioeconomic indices of uniformity.

1.5.4 Macro Level Focus

Some geographer's retained their *macro* level focus but emphasized decision making and choice behavior of groups or leaders rather than the mass of ordinary people. Thus decision-making profiles were developed to comprehend how private and corporate capitalists located and arranged their business activities. In governmental decision making at various scales, the power structures that defined decision paths were highlighted. Attempts were made to discern the different decision-making processes inherent in different economic systems, such as those with capitalist, communist, social-welfare, or authoritarian administrations. Goal-directed behaviors that differed from those existing at the individual level emerged, and the fundamental conflict of *individual* versus *group* well-being was emphasized.

Of critical importance in this macro level approach was the definition of plans and policies. These represent strategies by which societal and governmental decisions are made. Unlike the *micro* level emphasis on planning at the personal level and at a reduced spatial and temporal scale (e.g., within the daily activity space), *macro* level researchers examined strategies and plans for intermediate and long-term spatiotemporal contexts. For example, it was found that individual developers (e.g., fast food franchises) used the same set of criteria to locate all of their outlets. It was found that the long-term plans and objectives of Western capitalist societies often benefited small, elite groups of capitalists more than they benefited people with low incomes and that planning for personal welfare diverged widely among capitalist and socialist and social-welfare economies. Decision making at the *macro* level was thus defined as a *constrained* process, undertaken and implemented by a powerful *elite*, often encompassing goals that differentially considered segments of society or economy. But in so doing, this research opened a path for the examination of conditions of empowerment of the many and varied subgroups found within today's economies and societies. Thus it helped focus attention on practices of *discrimination, disenfranchisement,* and *differentiation*—processes that have become a significant part of geographic investigation over the past three decades.

1.5.5 The Basis for Generalization

The search for a new basis for generalization was undertaken by the attempt to find sets of human actions—internal or external—about which researchers could make legitimate, general, realistic, and appropriate statements. It was felt that many of the analyses of existing aggregated objective data sets represented the methods of collecting the data (e.g., the census structure) more than they represented human behaviors and actions per se, and that research based on secondary data analysis has severe limitations. Behavioral researchers, therefore, reorganized their thought processes from the base up, starting with the *individual*, searching for *commonalties* across *groups* of individuals, *aggregating* on the basis of those commonalties, and then searching for potential *generalizations* that might be used to modify existing theories or to develop new ones. As opposed to this "bottom-up" procedure, social theorists adopted a "top-down" approach in which the welfare of individuals was subordinated to *societal* well-being. Much effort was expended on exposing the ills and oppressions (often defined in value-laden, Marxist terms) that "were hidden" within the fabric of democratic and capitalistic societies.

1.6 Issues in Research Design: Data Collection and Analysis

Research into social, economic, and behavioral phenomena involves issues of research design, data collection, and analysis. It is important to consider these issues and, as a result, to address as well a number of philosophical questions about research methodologies.

Objectivity is central to scientific inquiry. Research designs, whether they employ *quantitative* or *qualitative* methods, are supposed to generate *valid* and *reliable* data, and the researcher is supposed to report the outcomes in terms of meaningful variables, justifiable in terms of relevant theories. Objectivity is not the province of positivism; it is essential as the basis for *all* good research.

1.6.1 Critical Questions for Researchers to Ask

A whole range of questions need to be asked by researchers in choosing among methodologies, design, and techniques for collecting data. There is not always a *best* design; rather, it is a matter of considering *alternatives*, and adopting a strategy that will provide *useful* data to answer the research question at hand. Social science methodologists, such as Patton (1990:195–198), have posed a number of key questions that the researcher should ask:

1. What is the *primary purpose* of the study? Is it basic, applied, summative, formative, or action research?

2. What is the *focus* of the study? How does it trade off breadth versus depth?
3. What are the *units of analysis*? Are they individuals, groups, programs, organizations, or communities? Are there critical time periods?
4. What will be the *sampling* strategy or strategies? Will it involve probability random sampling in order to be able to make inferences from sample data to estimate population parameters? Does one need to stratify the sample target population into categories or groups? Will purposive (non-probability) sampling be used? What should be the sample size?
5. What *types* of data will be collected? Will data be qualitative, quantitative, or both?
6. What *controls* will be exercised? What should be the methodology—naturalistic inquiry, experimental design, quasiexperimental design?
7. What *analytical* approach or approaches will be used? Will it (they) involve inductive generalization or deductive reasoning? Will it (they) involve content analysis, statistical analysis, or a combination of the two?
8. How will *validity* of, and *confidence* in, the findings be addressed? Which options of design, methods, and data will be used—triangulation, multiple data, multiple methods, multiple perspectives, or multiple investigators?
9. Are *time* issues important? When will the study occur? How will it involve long-term field work, rapid reconnaissance, an explanatory phase and a confirmatory phase, fixed times versus open time lines?
10. How will *logistics* and *practicalities* be handled? Are there issues in gaining access to the setting, access to people and records, contacts, training, endurance, and so forth?
11. How will *ethical* issues and matters of *confidentiality* be handled? Will it be necessary to obtain informed consent? Is protection of human subjects an issue? How should the researcher and data collectors present themselves?
12. What *resources* will be available? What will the study cost? What are the requirements for personnel, supplies, data collection, materials, analysis time and costs, reporting and publishing costs?

"What is certain is that different methods produce quite different information. The challenge is to find out which information is most needed and most useful in a given situation and then to employ those methods best suited to producing the needed information" (Patton, 1990:196).

1.6.2 Some Methodological Considerations and Choices

There are a number of general methodological approaches that are widely used to investigate the behavior of people in space and society. Four key approaches are:

1. *Naturalistic inquiry*, which involves studying real-world situations without the researcher attempting to manipulate the research setting, through non-manipulative, unobtrusive observation.

2. *Experimental design*, where the researcher attempts to completely (or partially in quasiexperimental design) control conditions of the phenomenon being studied through manipulating, changing, or holding constant external influences to measure a limited set of outcome variables. Often this will involve a *pre*test, a *post*test, and a *control* group. Quantitative or qualitative data are analyzed.

3. *Inductive analysis*, which uses qualitative methods to explore, discover, and inductively derive generalizations about a phenomenon. It involves observation, categorization, searching for dimensions, and building general patterns that exist empirically. Inductive analysis does not presuppose in advance what the important dimensions will be.

4. *Hypothetical deductive analysis*, which requires specification of the main data variables and the statement of specified research hypotheses before data collection begins, based on an explicit theoretical framework consisting of general contents that provide a framework for understanding specific observations or cases.

Hypothesis testing is an integral component of the hypothetical deductive approach to theorizing and testing of theories. It depends largely on *quantitative* specification of variables, and it relies mainly on the use of *statistical* techniques of analysis. *Qualitative* methods permit the researcher to study an issue in depth and in detail, and the researcher is not constrained by predetermined categories (Patton, 1990:14).

Hypothesis testing is not the only basis of research activity, especially in the social sciences, where many of the most dramatic discoveries have come about in other ways. The logical testing of hypotheses using quantitative methods is appropriate only in specific situations and to answer only a limited number of research questions. *Qualitative* methods are more exploratory, departing from the strict, hypothetical, deductive model approach.

Quantitative methods require the use of standardized measures so that varying perspectives and experiences of people can be fitted into a number of predetermined response categories to which numbers are assigned (Patton, 1990:14).

An important way of strengthening the research design for a study is through what is known as *triangulation*. The notion in triangulation is that no single method ever adequately solves the problem of rival factors. "Because each method reveals different aspects of empirical reality, multiple methods of observations must be employed" (Denzin, 1978:28). Triangulation mixes methodologies in four ways (Patton, 1990:187):

1. *Data triangulation*, where a variety of data sources are used.

2. *Investigator triangulation*, where a number of different researchers are used.

3. *Theory triangulation*, where multiple perspectives are used to interpret a single set of data.

4. *Methodological triangulation*, where multiple methods are used to study a single problem.

Six mixes of measurement, design, and analysis are common in research in the social sciences:

1. *Pure hypothetical–deductive approach*, using experimental design, quantitative data, and statistical analysis.
2. *Pure qualitative approach*, using naturalistic inquiry, qualitative data, and content analysis.
3. *Mixed approach*, using *experimental design*, *qualitative data*, and *content analysis*.
4. *Mixed approach*, using *experimental design*, *qualitative data*, and *statistical* analysis.
5. *Mixed approach*, using *naturalistic inquiry*, *qualitative* data, and *statistical* analysis.
6. *Mixed approach*, using *naturalistic inquiry*, *quantitative* data, and *statistical* analysis.

1.6.3 Validity and Reliability

A major concern over many decades in the development and the use of both *quantitative* and *qualitative* research methods has been to ensure greater *validity* and improved *reliability* of data.

Validity is about accuracy in measurement, whether it be quantitative or qualitative. It is a fundamental issue in the development and testing of any theory. Much of the focus of attention has been on the validity of an observation or an instrument, whether a measurement has accuracy, and whether a phenomenon is properly labeled. Thus, there are: (1) *apparent* validity, (2) *instrumental* validity, and (3) *theoretical* or *construct* validity. The theoretical validity underlies apparent and instrumental validity. Believing a principle to be true when it is not is referred to as *type one error*. A *type two error* is about rejecting a principle when it is true. A *type three error* is asking the wrong question.

Reliability is about comparisons of findings of research and the drawing of conclusions. It depends on the explicit description of observed phenomena and procedures. There are three kinds of reliability: (1) *quixotic* reliability, where a single method of observation continually yields an unvarying measurement; (2) *diachronic* reliability, which refers to the stability of an observation through time; and (3) *synchronic* reliability, which refers to the similarity of observations within the same time period.

It is important to keep in mind the issue of objectivity in developing theories and the issues of validity and reliability of measurement in the generating of data. It is also important to recognize that these are crucial considerations in research design and in the use of both quantitative and qualitative methods of data collection and analysis.

The focus of much research in recent decades on the individual and the subjective evaluation of phenomena in various environments at various levels of scale meant that the majority of existing data sets used by geographers were no longer adequate or even appropriate. Researchers have had to create new data sets by *survey research* or other *interactive* methods. The creation of new data sets was paralleled by an emphasis on different methods of data analysis, mainly quantitative but also qualitative, including:

1. *Parametric* analytic measures, such as the test and analysis of variance, and correlation and regression analysis.
2. *Nonparametric* analytic measures, such as χ^2, Kolmogrov–Smirnov, and Mann–Whitney U tests.
3. *Multidimensional* and *multivariate* methods for the representation and analysis of phenomena, such as multidimensional scaling, factor analysis, multidiscriminant analysis, and cluster analysis.
4. The use of complex *experimental designs* and *incomplete data sets.*
5. The search for new *graphic* and *cartographic* modes for presenting mixed metric data.
6. Experimentation with panels, profiles, verbal protocols, and other *transactional* data.

To a large extent, the search for new variables, new data, and new measurement methods has led to the development of strong ties between behavioral research in human geography and the discipline of psychology. This complemented existing ties with other social and behavioral sciences such as sociology, anthropology, political science, economics, and communications.

1.6.4 Qualitative Methods

In their introduction to the Sage *Qualitative Research Methods Series*, Kirk and Miller state that in distinguishing between *quantitative* and *qualitative* research, it is best to "think of qualitative methods as procedures for counting to one. Deciding what is to count as a unit of analysis is fundamentally an interpretative issue requiring judgment and choice." In qualitative methods, "meanings rather than frequencies assume permanent significance" (1986:5). Technically, a qualitative observation identifies the presence or absence of something.

Basically, qualitative methodologies are concerned with (1) *invention* (research design), (2) *discovering* (observation and measurement), (3) *interpretation* (evaluation or analysis), and (4) *explanation* (communication or packaging).

All four phases must be carried out in qualitative research; they are the hallmark of *ethnographic* and *case study* processes. As Yin (1989:21) has said, the case study, like the experiment, does not represent a "sample." The investigator's goal is to expand and generalize theories (analytical generalizations) not to enumerate frequencies (statistical inferences). The researcher seeks to generalize from a particular study to a broader theory, which needs to be tested through replication of the findings in further experiments or case studies where the theory has specified that the same results should occur.

Geographers have a long history of using qualitative methodologies. Their traditional concern has been to *record, describe,* and *classify* phenomena, such as land use in field studies. Methodologies such as *participant observation* involve the same process of recording and categorization, usually of human subjects and their behavior. In fact, the *inductive* approach to developing a theory inevitably incorporates qualitative methodologies in order to identify, specify, and describe the elements of the theory. It involves what has been referred to as "intrepid subjectivity" (Zonabend, 1992) in explicitly defining objectively the object of a study. Within the social sciences, qualitative research is a tradition that "fundamentally depends on watching people in their own territory and interacting with them in their own language, on their own terms" (Kirk & Miller, 1986:9).

It is important to realize that qualitative research methods have a long pedigree, and they can and should be as refined and sophisticated as quantitative methodologies, such as survey research. In recent times, it has become far too common among some human geographers operating under the guise of *postmodernist* and *humanist* approaches to equate the use of qualitative research methods with a break from *positivist* analysis. This represents an ignorance of the historical evolution of the use of both qualitative and quantitative research methods in the social sciences. In addition, too often the recent champions of qualitative methodologies fail to recognize the rigor that is implicit in the proper use of such methodologies.

Too often adherents of qualitative research claim it to be "descriptive," which, according to Kirk and Miller, has caused qualitative research to get a "bad press" for the wrong reasons. Qualitative research is an "empirical, socially located phenomenon, defined by its own history, not simply a residual rag-bag comprising all things that are not quantitative'" (1986:71). It ranges across analytical induction, content analysis, semiotics, hermeneutics, elite interviewing, the study of life histories, and some archival, computer, and statistical manipulation.

In qualitative research, it is important to know as much as possible about the "cognitive idiosyncrasies of the observer—which is to say about his or her theories" (Kirk & Miller, 1986:51). They include academic commitment, values, behavioral style, and experiences. A major issue is what some researchers consider to be the pretense of the social scientist as a "neutral observer."

Qualitative research is intended to be objective. As Kirk and Miller (1986) say:

> There is a world of experienced reality out there. The way we perceive and understand that world is largely up to us, but the world does not tolerate all understandings of it equally. There is a long-standing intellectual community for which it seems worthwhile to figure out collectively how best to talk about the empirical world, by means of incremental, partial improvements in understanding. Often, the improvements come about by identifying ambiguity in prior, apparently clear, views, or by showing that there are cases in which some alternative view works better.

Above all, qualitative research and its objectivity must be measured in terms of its validity and reliability.

The theoretical traditions of qualitative inquiry are many and varied, as set out in Table 1.1. These approaches all use qualitative empirical observation techniques, but they vary in their conceptualization of what it is that is important to ask and to consider in understanding the real world. They use techniques that include

TABLE 1.1.

Theoretical Traditions and Approaches in Qualitative Inquiry

Ethnography	Understanding the culture of a group of people through participant observation
Phenomenology	Seeking to understand the structure or essence of experience of phenomena or people through participant observation in a shared experience
Heuristic inquiry	The inquirer shares the intensity of the experience with those being observed through systematic observation leading to depictions of essential meanings about the structure of experiences
Ethnomethodology	Seeks to ask how people make sure of their everyday activities so as to behave in socially acceptable ways, employing qualitative experiments
Symbolic interactionism	Seeks to identify common sets of symbols and understandings to give meaning to people's interactions using group discussions, interviews with key informants or panel of experts.
Ecological relationships	Seeks to explore the relationship between human behavior and the environment through nonparticipant observation and quantitative coding of behavior
Systems theory	Asks how and why a system functions as a whole, employing quantitative and/or qualitative methods through holistic thinking about the system and its component parts
Hermeneutics	Asks what are the conditions under which a human act takes place or something is produced that makes it possible to interpret its meaning, with investigators reporting from their perspective or standpoints and of the people being studied through a praxis

Source: Derived from Patton (1990), pp. 66–89.

field studies, participant observation, nonparticipant observation, oral histories, and case studies.

Wolcott (1990:64) has warned of the dangers in affixing the label *ethnography* to qualitative studies and draws attention to the need to be aware of the "wide range of alternative terms and approaches." In fact, ethnographic and natural history type of methodologies represent but a small spectrum of the range of qualitative research methodologies in the social sciences. Although research methodologies in anthropology have relied largely on ethnographic case studies, many social scientists regard them as having limited scientific or disciplinary value, thinking of them as exploratory forays into unexplored situations or areas of research, which are useful as systematic means of developing definitions, analysis, and understanding through the use of other methodologies. However, case studies also are regarded by many social scientists as having a great deal of scientific or disciplinary value, standing or falling on their merit. This is particularly so among sociologists.

1.6.5   Case Studies

Case studies have been widely used in social and behavioral research. Like a laboratory test, a case study requires detailed attention to conceptualization and design. As Hamel, Dufow, and Fortin (1993) point out,

The methodological value of the experimental devices in the case study is essentially based on:

(a) the quality of the strategies selected in defining the object of study and in the selection of the social unit that makes up the ideal vantage point from which to understand it,

as well as

(b) the methodological rigor displayed in the description of this subject in the form of a [sociological] analysis that can be understood in action. (p. 40)

The case study approach was influenced greatly by the Chicago School of Sociology from the late 19th century and beginning of the 20th century. Researchers investigated small local communities and urban neighborhoods where rural and immigrant populations had settled, and they included studies of unemployment, poverty, delinquency, and violence. Robert Park was particularly influential in developing applications of the case study methodology to such urban social issues, believing that they "demonstrated a meticulousness and a systematic approach that was lacking in journalistic accounts" (Hamel et al., 1993:14). Park encouraged his students to go beyond official documents and to use sources such as letters and other personal documents, newspaper articles, and open-ended interviews with people. The development of the case study approach was strongly linked to the emergence of the *urban ecology* theoretical perspective on the transformation of small and simple communities to large cities with complex social structures. The case study was used to help people develop an understanding of the processes and outcomes involved in the evolution of the city and its social characteristics and problems. Field studies were developed, using on-site techniques such as observations, open-ended interviews, and the collection of various documentary evidence (Hamel et al., 1993:16). The case studies were used to develop *inductive theory* about the form and structure of the city.

Inherent in the approach of the Chicago School has been the notion that the case study could provide an understanding of the personal experiences of social actors. Becker (1970) expressed the concern as follows:

To understand an individual's behavior, we must know how he perceives the situation, the obstacles he believed he had to face, the alternatives he saw opening up to him. We cannot understand the effects of the range of possibilities, delinquent subfractures, social norms and other explanations of behavior which are commonly invoked, unless we consider them from the actor's point of view. (p. 64)

The main issues to emerge from the use of the case study in developing urban social theory were the degree to which the individual actor's point of view could be incorporated, the status assigned to it, and the "process of transformation defined as part of a rigorous objective process that does not result from the social, intellectual and psychological makeup of the researcher" (Hamel et al., 1993:17). To address such issues, in the 1930s and 1940s the *statistical survey*, developed largely from the work of researchers at Columbia University in New York, became more widely used both to provide explanations and to provide a basis for forecasts. In addition, there was considerable debate that surveys were required to validate a theoretical idea and that this required researchers to hold their subjective attitudes or feelings in check while conducting their studies. *Deductive* approaches to the development of theory began to incorporate those statistical procedures appropriate to test for accuracy and validity and to generalize, and qualitative methodologies, such as participant observation, were deemed inappropriate for such approaches. As a result, the case study tended to become used more as an exploratory investigative tool, as "a preliminary survey giving rise to a statistical study that could validate or eliminate a theory or a general model" (Hamel et al., 1993). Thus, the limitations of the case study (1993:21) were seen to be its (1) lack of representativeness, (2) lack of rigor in the collection, construction and analysis of empirical data, (3) bias.

However, there has been a considerable revival of interest in the case study, reflecting the views such as that expressed by Jacob (1989:10) that "there is no scientific method,' recipe or algorithm known that will permit scientific discovery. We know of no mechanical means that will generate a hypothesis or theory based on certain facts observed in a finite series of steps," with all theories being based initially on a particular case or object. Views such as this may be disputed, and we suggest that they are a distorted interpretation of what constitutes scientific method of inquiry. But, what is evident is that an *in-depth* case study can elicit theories, the elements of which might or might not be validated by further case studies and/or by population-wide surveys. As suggested by Godelier (1982), it is "the empirical description of such elements which yields good theoretical questions. If they are based on a few factual elements, . . . they can pave the way to a discovery of the general structure of some sphere of reality" (25).

If one of the purposes of social behavioral research is to explain the experiences of actors from the perspectives of their beliefs, attitudes, perceptions, cognitions, and experiences, then the case study approach has potential as a tool to help develop explanation, particularly through the use of repetitive case studies that confirm or refute verification of postulated explanatory factors. The issue is not the actual number of case studies but, rather, the suitability of the case for the actual aims of a study and what the cases intend to explain. The methodological qualities of the case study or the case studies are what determine the representative value of the study, and this determination requires careful and explicit construction of the case study within a theoretical framework. Thus, if it is determined that the case being studied is representative of the selected phenomenon, then the scope of the study thus proves to be *macroscopic* through the methodological virtues of the case and the method used. This may permit the in-depth case study to be regarded as explaining characteristics or properties that are inherent in society as a whole—that is, on a global scale.

> In other words, what is an experimental prototype in these fields if not a device, instruments or machines whose theoretical, methodological, or even technical definitions of their use represents the living or inanimate material that constitutes the object of study? Singularity is thus characterized as a concentration of the global in the local. Singularity is not perceived as a particular feature of a fact, a species, or a thing. It is seen, rather, as characterizing a fact, a species or a thing." (Hamel et al., 1993:37–38)

In recent years there has been a rush of interest among some researchers in human geography, and among many current graduate students, in the product of research or the written account in its own right. This concern with the *narrative* raises serious issues of a methodological kind about writing up the results of qualitative research and, in particular, about ensuring that the researchers' textual claims and interpretative perspectives are illustrated well by the data collected. A useful overview of the issues involved in writing up qualitative research is provided by Wolcott (1990). A crucial point he makes is that a researcher cannot create an ethnography unless one has done intensive fieldwork based on rigorous conceptualization of the research problem and the application of the method(s) employed to conduct a study.

1.6.6 Survey Research Methods

Much social and behavioral research requires the collection of data using *survey research* methods, a field in which many geographers often have little if any training. Too often it is the case that human geographers (along with many other social scientists) have an arrogant regard to their capacities and capabilities for using survey research methods to collect the empirical data needed to test their theories and to operationalize their models. Often researchers and students have taken the attitude that it takes no expertise to choose a sample and to design a survey data-collection instrument, collecting data themselves or using their friends or associates. All too often they assume that, by using arbitrary methods, they can generate data that are valid and reliable. Unfortunately, this is far from the truth, as survey research is a diverse and highly complex and specialized field of legitimate theoretical and methodological concern in its own right.

Conducting surveys must involve an extraordinary commitment to the rigors of research design, procedures for data collection, and data coding, analysis, and interpretation. Surveys also require substantial monetary resources, often on a large scale, as well as access to specialized staff (such as interviewers) and equipment (such as computer-assisted telephone interviewing systems, or CATI).

There is a range of tasks involved in collecting data using survey research methods, and these are summarized in Table 1.2. In conducting a survey, a minimal set of decision choices need to be made including:

1. The selection of *probability* or *nonprobability sample design*; stratification and clustering of the sample; methods of selecting sampling units using randomized procedures.
2. The *mode of data collection*; self-completion questionnaires; mail surveys; home-based face-to-face interviews; on-site face-to-face interviewing; telephone interviewing.'
3. The *design of the survey instrument*; the wording of questions to collect data on facts, opinions, and attitudes; the design of scales; the use of precoded, categorical, and open-ended questions; question ordering; use of funnel sequences, flash cards, sketches.
4. The *training of interviewers*; monitoring of interviewer performance; pro-

TABLE 1.2.
Tasks Involved in Conducting a Survey

1. Definition of the research question: formulating hypothesis for testing
2. Budget and time constraints
3. Target population: primary sample units
4. Parameter definition
5. Data variables specification
6. Sample design issues: type of sample, sample frame, error estimates
7. Instrument design issues: mode, sample questions, validity and variability data
8. Administrative procedures: field and office arrangements, training materials and programs, monitoring procedures
9. Pilot survey or pretesting
10. Data processing: coding manuals, establishing a clean data set
11. Data analysis: dealing with nonresponse and missing data, cross-tabulation of data variables, statistical testing of research hypotheses
12. Report writing and data presentation

cedures to standardize interviewer behavior; interviewer feed-back, probing and reinforcement.
5. How to deal with and measure *error* and *bias*; the use of imputation procedures to deal with missing data.

It is essential to use *probability random sampling designs* where the intention is to use *sample statistics* to make inferences about the parameters of the population from which the sample is drawn. For example, public opinion surveys tend to employ multistage, cluster random sampling designs, and it is common to do so when investigating a behavioral phenomenon such as the travel behavior of people in a metropolitan area in transportation studies.

However, *non-probability sampling designs* are most appropriate in many research situations where the concern is to test hypotheses about the determinants of behavior using *quasi-experimental designs.* In such cases it might be appropriate to select a *quota sample*, whereby individuals are chosen so as to ensure that they have specific characteristics that are common and that represent independent variables, which it is hypothesized influence the dependent variable that is being investigated.

It is important to recognize that there are *errors* and *biases* in all surveys and that survey design and data-collection procedures play a vital role in minimizing the various elements that make up *total survey error*, which arise from *sampling* and *non-sampling* (i.e., question design, interviewer behavior, respondent, and misreporting) sources. In fact, even allowing for good sampling procedures, about 70% of total survey error is likely to emanate from non-sampling sources.

In using survey research methods, it is useful to conceptualize the *question and answer process*, which in itself includes complex behavioral interactions between the respondent to the survey and the interviewer asking questions through the medium of the survey instrument. This process involves cognitive processing of information on the part of the respondent, retrieval of information, and its organization in response to the question wording and the coding category to give an answer to a question. Figure 1.2 identifies the complexity of the factors involved in this process.

It is important to recognize that, without due and rigorous attention to the issues outlined above, the data generated using survey methods almost certainly will be of poor quality. Every effort is needed to ensure that survey design, data-collection procedures and their management, and data analysis are undertaken in such a way as to minimize errors and bias in order to produce reliable and valid data from the answers respondents give to questions.

There exist many standard textbooks on survey research methods (e.g., Moser & Kalton, 1971). Sage Publishing produces an excellent series in both quantitative and qualitative research covering a wide range of concerns in research design and data analysis.

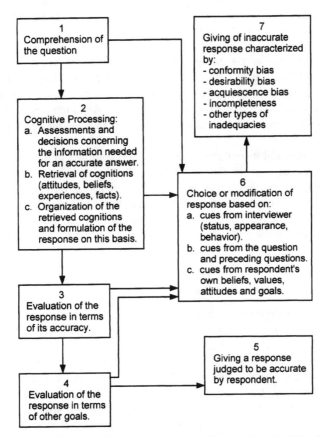

FIGURE 1.2. Model of the question–answer process. *Source*: Cannell, Miller & Oskenberg, 1981.

1.6.7   Gender Roles in Field Work and Data Collection

The investigation of many aspects of human spatial behavior usually involves inter-action between the researcher and/or an intermediary, such as an interviewer, and the subject or subjects being studied. The role of *gender* in field-work situations, not only in research employing qualitative methods, as in ethnographic studies, but also in survey research situations, has been identified by Warren (1988) in the fol-lowing terms:

> The web of gender denotes not only one's place in the social structure but also the deep structures of human experience: body, sexuality, feelings, self. What is presented to the host culture is a body: a size and shape, hair and skin, clothing and movement, sexual invitation and untouchability. The embodied characteristics of the male or female field worker affect not only the place in the social order to which he or she is assigned, but also the fieldworker's and the informants' feelings about attractiveness and sexuality, body functions and display. Some of the way in which the self is presented—such as hair style and clothing—can be altered; some—such as skin color and hand size—can-not. Thus the process of developing relationships in the field involves the monitoring and (perhaps) modification not only of behavior but also of the researcher's self: his or her body and its uses in the field.

There are some stereotypes about the role of men and women in collecting data in the field. Douglas (1979:214) wrote that "in most settings, the ultimate so-ciability specialists are women. These low-key women do not threaten either the women or the men. They are liked by and commonly share intimacies with both sex-es. Men are simply more threatening to both sexes, even when they are the most so-ciable."

Overall, it would seem that methodological researchers in the social sciences investigating gender roles in field- and data-collection situations have identified that the advantages and disadvantages of women versus men reflect women's place in their own social order. "Women fieldworkers are portrayed as more accessible and less threatening than men; coupled with their superior communicative abilities, this makes the interactions of fieldwork generally easier" (Warren, 1988:45).

1.7   Society, Economy, and Space

Paralleling the development of behavioral concerns in human geography, there emerged a concern for the analysis of the social structuring of human actions within the everyday frameworks of society. Social theorists—especially Marxists—criti-cized conservative retreat within discipline boundaries, often creating protected specialty groups and journals within the geography profession. Such criticism, how-ever, did not stop them from doing likewise, and journals such as *Antipode* and *So-ciety and Space* emerged as hosts for "radical" research on the structure and evolu-tion of society in space and time. Emphasis was placed on the social context of many of today's most serious urban problems (e.g., the poor, the homeless, the ethnic subgroups), as well as on ideologically based reinterpretations of classical geographic problems (e.g., urban structure). As stated by the founding editors, the literature in these journals emphasized questions about reproduction and control systems in everyday life, including: gender and patriarchy; theory of the state; power and empowerment; urban social movements; language and cultural ideology relating to social and spatial systems; and industrial restructuring.

Like macro behavioral research, these approaches favored *societal* scales and processes and searched for common ground on which to link both analytical and ideological concerns. Just as micro behavioral research was often mistakenly associated with Skinnerian behaviorism, so too has social theory often been portrayed as being dominated by a single subgroup—Marxists and neo-Marxists. But this movement also encompassed phenomenologists, postmodernists, structuralists, realists, and others interested in humanist modes of investigation.

Although some similarities existed between *behavioral* and *social* theorists, there were also differences. Much behavioral research was disaggregate and oriented to individuals or households; much societally based work focused on social and political groupings. Much behavioral investigation turned for inspiration to psychology; much of the work on social structure was tied to sociology, social theory, and political economy. But, the end product was that behavioral researchers now generally acknowledge the importance of societal, economic, and political constraints (authority constraints) on individual activity, although they are not necessarily tied to any ideological interpretation of these constraints. The individual decision maker is acknowledged to work in a *constrained choice*, not a free choice, situation. Perceptions, cognitions, and learning, take place through *societal filters*, thus adding to the power of the *boundedly rational* assumptions underlying the investigation of human–environment relationships. But, whereas behavioral researchers have acknowledged the importance of society and economy, often the reverse is not the case (i.e., social theorists often ignored individual behavior).

1.8 Epistemological Bases

1.8.1 What Remains of the Positivist Tradition

Analytical research in human geography, including a good part of research on human spatial behavior, originally was based on the positivist tradition. *Positivism* was a critical part of the quantitative and theoretical revolution that swept the discipline in the late 1950s and early 1960s. It combined with the use of quantitative methods and the desire for generalization, via normative theory, to push the discipline into an era of scientific thought. However, it soon became obvious that, although some of the fundamental principles of positivism are ingrained in much of the analytical or scientific work in human geography, the limitations and constraints in the philosophy itself were also quite confining. By detailing some critical characteristics of positivism, it should at once be obvious that this philosophy in its entirety was not readily adaptable to human behavioral research.

The critical elements of positivism were:

1. A physicalist view of the flux of existence;
2. A characterization of reality as a collection of atomistic facts;
3. An emphasis on the objective and the observable;
4. The critical importance of hypotheses testing within that mode of reasoning called "scientific method";
5. An empiricist base;
6. The dominant theme of searching for generalizations and the ultimate aim of producing process theory;
7. A need for public verification of results;
8. The necessity of logical thought;
9. The separation of fact and value;
10. The assumption of the scientist as a passive observer of an objective reality;
11. The principle that value judgments must be excluded from science.

Part of the behavioral research undertaken in the past two decades was born in the positivist tradition, and it still reflects some of the underlying principles of that tradition, including those relating to verification, reliability, and evaluation. However, the bulk of this research has progressed well beyond its original positivist base, and it is increasingly difficult to identify today's research with traditional positivism (Couclelis & Golledge, 1983). It is surprising, therefore, that the positivist bogey-man is still the favorite straw man for social theorists looking for help to advance their own ideas at the expense of behavioral research.

1.8.2 Alternative Epistemologies

Both analytical and more descriptive and transactional research on human behavior have evolved by shedding many of the classical positivist tenants. For example, as the importance of *cognition* as a mediating factor between humanity and environment was recognized, the critical positivist tenant of the *non-sense* of the unobservable was quickly perceived to be untenable. With this loss came a weakening of the principle of reductionism and the physicalist interpretation of human behavior. In behavioral geographic terms, *spatial behavior* became differentiated from *behavior in space* (Cox & Golledge, 1969).

The often criticized tenet of the scientist as a passive observer of an objective reality also came under criticism. It was recognized that humans and environment interact constantly in all time–space contexts, and it was these ongoing transactions that added the essential structure and meaning to the relationship between human and environment. In turn, this meant that alternative epistemologies, such as the *transactional* or *constructivist* positions, based as they are on the dynamics of interaction between humanity and environment, were strengthened as alternatives to positivism. As alternative epistemological positions were taken, the classical positivist separation of *value* and *fact* became less tenable, and the attempt to interpret values and beliefs in a scientific or analytical manner increased. Much research on human spatial behavior began to accept that values must indeed be facts since they help determine overt spatial behaviors. This ongoing epistemological debate also laid to rest the original positivist position of an *a priori given world*, and provided its replacement with an assumption that mind and world are in constant dynamic interaction.

For the last two decades, behavioral research has made major advances by leaping the barriers imposed by positivist philosophy while retaining fundamental and important principles of scientific research—principles that have endured the rise and fall of many philosophies—or by searching for reasonable alternative positions in modes of thought such as postmodernism and realism.

What is contained in current analytical behavioral geography is what was truly positive in positivist thought. These tenet are:

1. The need for logical (and often mathematical) thinking;
2. The need for public verifiability of results;
3. The search for generalization;
4. The emphasis on objective languages for researching and expressing knowledge structures;
6. The importance of hypothesis testing and the importance of selecting the most appropriate bases for generalization or theorizing.

These tenets allow behavioral research to be conducted in a rigorous manner but are far from constraining such research to a "scientism" paradigm—another misused and misunderstood pejorative used by some critics of behavioral analysis.

1.8.3 Cross-Disciplinary Enrichment

Adoption of an analytical mode of discourse has brought geographic researchers to a stage where they can interact freely with, and comprehend researchers in a large number of other disciplines. In turn, many of those researchers now appreciate the unique contributions made from the particular viewpoint of the geographer. This *cross-disciplinary interaction* has been fertile in the search for solutions to many individual and societal problems. Thus the geographer's knowledge and skills have contributed significantly to the current state of knowledge in environmental cognition, environmental learning, and the general process of modeling human decision processes. This latter area, sometimes known as *artificial intelligence modeling*, is one of the several active growth areas for the analytical behavioral geographer. By focusing on the processes of acquisition, storage, recall, and representation of spatial knowledge, geographers are able to use their particular skills–cartographic and graphic representational modes, and their particular interrogative methods—in a context comparable to researchers in disciplines such as computer science, psychology, electrical engineering, and cognitive science.

Since the geographer's entry into the general field of behavioral research has been relatively recent, significant problems persist that require continuing research. Those include creating appropriate experimental designs, selecting appropriate survey research procedures for data collection, selecting appropriate methods for data analysis, and ensuring that research undertaken is verifiable and reliable.

Like other researchers, behavioral geographers need to make a long-term commitment to research with no guarantee of successful results. However, we are so convinced that significant progress is being made that we devote most of this book to discussing the evolution of, and illustrating the state of the art of, a selection of human *behavioral* research areas in the *context of society and space*.

1.9 A Paradigm for Understanding Human–Environment Relationships

A *paradigm* for examining human–environment settings needs to encompass a complex set of relevant variables and their functional relationships. It includes the *physical* and the *built* aspects of environment; it allows for roles of *culture* and its related *social* and *political systems* and *institutions*; it identifies the evolution of culture over time through technology; and it recognizes *intervening psychological processes* as filtering mechanisms in how humanity *perceives* the environment and *acts* within it. The complexity of the interrelationships between and among these variables within the operational milieu of modern western society is demonstrated in Figure 1.3.

However, we have argued earlier in this chapter that the psychological variables intervening between humans and their environment are all-important in expanding the behavioral outcomes of this interaction. They provide a paradigm for investigating the behavioral bases per se of those relationships that manifest themselves as spatial movements and location decisions. Thus, we propose a person–environment behavioral interface model in the simple form of that shown in Figure 1.4.

The *behavioral interface* is the "black-box" within which humans form the image of their world. The *schemata*, or the basic framework, within which past and present environmental experiences are organized and given locational meaning constitute the *cognitive mapping* process. The key psychological variables intervening between environment and human behavior within it are a mixture of (1) cognitive and affective *attitudes*, *values*, *emotions* or other affective responses, (2)*perception* and *cognition*, and (3)*learning*. And all occur within socially and culturally defined spheres. As Figure 1.4 demonstrates, these are linked. It is with the understanding of these relationships that the process-oriented approach to the study of human spatial behavior is concerned.

The paradigm is applicable to the analysis of everyday behavior of humans in their environment. It is based on what Burnett (1976: 25–26) listed as eight beliefs about the mind as a mediator between the environment and behavior in it (Table 1.3). The paradigm postulates both *causal* and *noncausal connections* between the *overt* behavioral process and the *external* world of changing *objective* spatial structures. As suggested in transactional and postmodernist thoughts, the individual is simultaneously part of both the *objective* and the *behavioral* (or subjective) *environments*, receiving *locational* and *attributive information* from the latter, and acting *individually* or as *a group member* within such environments.

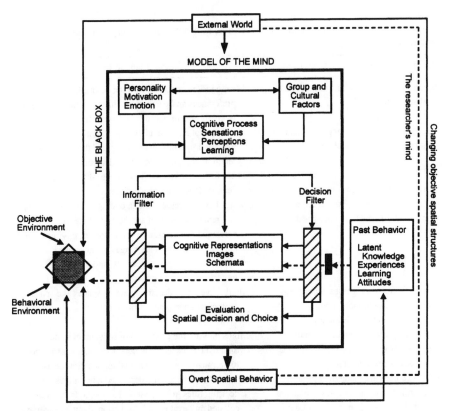

FIGURE 1.3. A paradigm of individual behavior, spatial cognition, and overt spatial behavior. Derived from Gold, 1980, p. 42; Burnett, 1976, p. 27.

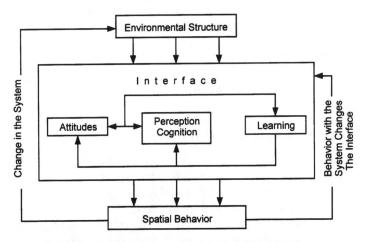

FIGURE 1.4. The human–environment behavioral interface.

TABLE 1.3.

Beliefs Inherent in the Proposition of the Mind as Mediator between Environment and Behavior

Belief	Elaboration
Minds exist and constitute valid objects of scientific enquiry	We are more concerned with the description of preferences and perceptions than with the descriptions of conditions of neurons and nerve fibers
Minds are described in psychological and not neurophysiological language	Minds do not have peculiarly mental, nonmaterial, or ghostly properties that would place them outside the realm of acceptable scientific discourse
There is an external world of spatial stimuli with objective places outside the mind	These include things such as industrial agglomerations, central places, and residential sites
Minds observe, select, and structure information about the real world	Minds have processes corresponding to spatial learning and remembering and are the seat of mental maps, perceived distances, awareness spaces, environmental cues, multidimensional images of shopping, residential, and other locations, and more or less imperfect spatial knowledge
Mental events or processes occur that correspond to thinking about or evaluating patial information	Minds have states describing action spaces and space preferences and place utility functions
Minds are the seat of emotions and sensations and are the seat of attitudes, needs, desires, and motives	Minds thus are the producers of satisfactions and dissatisfactions, environmental stress, and aspirations to optimize or satisfice in making location decisions
Spatial choices are made by thinking according to decision rules, which are made in the mind and result from prior mental states, events, and processes	Choices are made among perceived alternatives; the decision rules may be viewed as methods of relating collated and evaluated information about alternatives on the one hand to motives on the other
Spatial decisions are the cause of an overt act, and over time sequences of spatial choices by individuals and groups cause overt behavioral processes, which in turn cause changes in spatial structures in the external world. Thus, ultimately, location processes are explained (caused) by mental states, events, and processes	An overt act is one such as a search for a new residence, purchase of a new industrial site, or a shopping trip; sequences of choice over time are such as intra- and interurban migrations; changes in spatial structures are like transitions in urban land use

Source: Tabulated presentation of material from Burnett (1976), pp. 25–26.

1.10   Spatial Behavioral Research Today

The trend away from classical positivism has accelerated over the last decade. This has occurred for several reasons including the advantages that alternate epistemologies provide for undertaking human behavioral research and their suitability for exploring new areas of the subjective and unobservable.

To some human geographers, this move from positivism has signaled a flight to humanist alternatives. We interpret it not in this way but, rather, as a search for the development of an appropriate mode of discourse that serves well the original aims of behavioral researchers in geography.

Despite a variety of vigorous attacks on behavioral research from some British human geographers (Johnston, 1972; Thrift, 1980; Cloke et al., 1991), there has been an ever-increasing trend to incorporate behavioral research into the day-by-day practice of geography. There is no basis for recent suggestions that behavioral research has slid into a "quiet backwater." For example, Golledge and Timmermans (1990) and Timmermans and Golledge (1990) provided hundreds of references to recent theoretical and applied behavioral research either conducted by geographers or undertaken by researchers from other disciplines but firmly placed in the spatial domain. It is obvious that defeatist criticisms either misrepresent the significance of behavioral research or are based on a complete misunderstanding of what this research is all about. Increasingly, introductory textbooks in human geography have incorporated behavioral concepts into a variety of conventional geographic subareas and have stressed that the understanding of spatial problems is incomplete without the inclusion of the behavioral dimension.

A significant development since the early 1970s has been the forging of stronger and stronger links between human geography and psychology. Although many of the earliest links between these disciplines were conceptual, of late these have been strengthened by common use of measurement procedures, model-building activities, experimental designs useful for handling large and incomplete data sets, and by the more widespread use of segments of the real world as a laboratory. One indirect result of this tie has been an increased interest by geographers in the spatial problems of populations other than those comprised of fully enabled adults, including those of children, the poor, the elderly, the terminally ill, and other physically and socially disadvantaged subgroups (such as the blind, the deaf, the mentally ill, the mentally retarded, and the physically handicapped or infirm) (C.J. Smith, 1983; Dear, 1987; Matthews, 1992; Golledge, 1993). Thus, the research undertaken by behavioral geographers spans many traditional and nontraditional subareas of their discipline and can provide strong links to hitherto poorly connected subsets of other disciplines.

Current behavioral geography, therefore, can be characterized by a concern both for *scientific rigor* or *experimental realism* and for *phenomenological* and *anthropomorphic understanding* of *human–environment systems* and relationships. This concern covers topics such as:

1. Experimental design.
2. Data collection procedures.
3. The search for validity and reliability.
4. The selection of appropriate analytical methods.
5. The choice of a modern, analytical, epistemological basis.
6. The innovative use of models of behavior and models of environment.
7. Recognition of societally constraining dimensions.
8. Differing viewpoints based on bottom-up or top-down approaches.

Research areas dominated by *scientific approaches* include:

1. Studies of cognitive mapping and spatial behavior.
2. Attitudes, utility, choice, preference, search, learning.
3. Consumer behavior.
4. Location decision-making.
5. Wayfinding, mode choice, and travel behavior.
6. Mobility and migration behavior.

Typical areas dominated by *other* approaches include:

1. Societal, materialist, or feminist interpretations of urban structure.
2. The social reality of housing markets.
3. Environmental ethics.
4. Planning and policy making.
5. Social problems of the homeless and other disadvantaged groups.

It appears that some areas lend themselves more to the process of measurement, model building, abstraction, generalization, and theory building than do other areas of human geography, and we will focus more on the former in this text. We will, however, provide a representative coverage of different behavioral and social approaches.

2

Decision Making and
Choice Behaviors

2.1 Conceptualizing the Decision Process

Decision making and choice behavior represent two of the most significant human behavioral processes emphasized in geography. Whether undertaken by individuals, groups, institutions, or societies, an essential component needed to understand how and why decisions are made is elaboration of the structures and rules used in the decision-making process. In this chapter, we summarize a range of approaches that geographers and others have developed to investigate models of decision making and choice in spatial contexts in both macro and micro environments.

Amedeo and Golledge (1975) provided a schema describing the *decision-making process*. The schema was modified by Golledge (1991)and is further developed in Figure 2.1 and explained as follows.

For any person or population at any given time, there is an existing*knowledge base*. At the individual level, this knowledge structure is often called a *cognitive representation* or a *cognitive map*. For groups it is often called *culture*—that is, the shared habits and rules of a society. For business firms and for government it is referred to as the *institutional knowledge base*. Institutional memory is often presumed to exist in a subjective manner—as in the "institutional memories" of past events and activities contained in the knowledge structure of long-term employees. At times those knowledge bases are described as latent—that is, existing but obscured in some way that requires conscious effort and stimulation to be accessed. We can imagine that a person, group, or an institution exists in a quiescent state when they lack motivation to draw on the knowledge structure to solve a problem or perform a behavior.

It is assumed that the decision-making unit exists in some type of *environmental*, *social*, *political*, and *cultural* context. Evidence of the existence of these contexts is often made available in terms of personal functional and personal structural variables (such as occupations, movement propensity, religion, age, marital status, and other data often collected in a population census), social/cultural/political variables (such as income, social class, ethnic origins, state of birth, and citizenship), and spa-

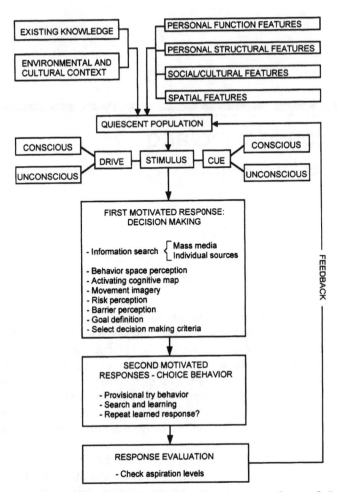

Figure 2.1. Conceptual structure of an individual's decision process. *Source*: Golledge, 1993.

tial variables (including location, density, and distribution characteristics). These are but some of the many obvious and hidden dimensions that can affect decision-making processes. Others include the affective components of our *feelings*, *emotions*, *beliefs*, and *values*. Traditionally, geographers examined these variables as static, descriptive phenomena that could be classified, manipulated, and displayed. There was no incentive to assume the population was anything but quiescent.

2.2　First Motivated Response: Activating the Decision-Making Process

Let us assume that such a quiescent population, however so defined, is stimulated by being faced with a problem that requires solution. The stimulation can be conscious or unconscious and can occur in the form of a *drive* or a *cue*. For example, a conscious drive may be hunger; an unconscious cue may be subliminal advertising. Whatever the nature of the stimulus, the decision-making unit is motivated to act, which we refer to as the *first motivated response*.

2.2.1 Information Search

This first motivated act involves an *information search*. Information may be sought about the nature of the problem itself, the environmental setting, the societal or cultural constraints, or the prior experience of the *decision-making unit* (DMU). This search for information may involve recalling stored information into a working memory at the individual level. It could also involve a search for external sources of information such as might be found in the mass media or via interpersonal contact or communication (e.g., by telephone, fax, or e-mail). The extent to which information about the problem situation may be obtained might depend on processes such as the rate at which relevant information has been diffused and adopted. Search may take place by actually moving through an environment, or it may take place without overt behavior by accessing information stored in convenient forms (e.g., in telephone directories, newspapers, or computer files). Search continues until a sufficient amount of information has been collected to allow activation of other components of the decision process.

2.2.2 Behavior-Space Perception

Another component of this first motivated response is the development of *behavior-space perception*. This requires tapping information contained in one's *cognitive map*. This component can occur in either a spatial or a nonspatial context. In a *nonspatial* context, information may be mapped into multidimensional spaces relating to, say, a product attribute (e.g., if motivated to search for food, relevant attributes might be chewiness, astringency of taste, protein content, carbohydrate content, color, etc.). In the *spatial* domain, the behavior space may consist of a subset of the cognitive map structured in working memory as an image or survey type representation of a segment of the task environment. Thus, representation may consist of a list of places at which a potential goal could be reached, or conceivably, it may consist of an image that could be (but certainly is not always) represented in a manner similar to a cartographic map. Whether in list or spatial form, potential objectives can be located and ordered, and the process of comparing one with another can be initiated. The output could be the selection of one or more potential destinations.

In addition, within the behavior space, specific paths that could be followed to the ultimate destination are reviewed. Criteria such as minimizing distance, minimizing time or effort, maximizing aesthetics, minimizing the number of left turns, and so on (i.e., rules for path selection of a given destination choice) are accessed. Also at this stage expectations are formalized in terms of constraining criteria, such as economic or spatial rationality, bounded rationality, satisficing, or minimizing regret. These criteria are the ones used to help solve conflict situations among possible alternatives identified in the behavior space. If one uses optimality assumptions, a single place at which a goal can be achieved will usually be designated. If one of the other many criteria are adopted, the *feasible opportunity set* will be defined.

2.2.3 Activating the Cognitive Map

Next comes the task of *activating the cognitive map.* Specific components of the cognitive map may be transferred to working memory when one is trying to identify the dimensions and attributes of the behavior space. The appropriate components of the cognitive map are assembled in working memory and selected decision-making criteria are applied to the alternate strategies that could be initiated. For example, members of a *feasible opportunity set* may be ordered according to a selected criterion. Alternate paths that could provide access to a goal object may be evaluated in terms of *time* or *distance* or even *authority constraints* on travel. In other words, working memory becomes a repository for bits of stored knowledge that are purposely activated and assembled as part of the process of making a decision.

2.2.4 Movement Imagery

Having activated the appropriate segment of one's cognitive map, the decision-making unit then engages in an elaboration of *movement imagery.* This is sometimes called the development of a *travel plan* (Gärling & Golledge, 1989). At this stage, barriers to movement may be imaged. These may include distance (as a measure of separation of self from goal object), cost, time, preference, attitude, or desirability. Each of these may be constrained in turn by societal or cultural factors, such as acceptability by a particular nationalistic or ethnic group.

Next, the physical process of moving from an *origin* to a *destination* must be imaged. In large-scale environments this inevitably involves selecting a *mode of travel.* The selection of travel mode depends on the availability of modes within the economic and social system in which the decision-making unit is located. *Preference* for travel mode in any given task situation might also be constrained by personal factors, such as the need to rationalize travel plans with general household requirements and the activity schedules of other members of the decision-making unit. *Conflict resolution* becomes a significant part of movement imagery (Recker, McNally, & Root, 1986a, 1986b). *Barriers* that are perceived to occur as a result of physical, societal, or cultural dimensions of the setting in which decisions have to be made must then be reconciled with the emerging travel plans. Once this has been accomplished, *goal-directed behavior* can be implemented.

2.3 The Second Motivated Act

2.3.1 The Choice Act

The result of this long internal process is the designation of a *goal object,* usually specified by a location at which the goal object can be achieved. This begins the second motivated response. Selection of a specific location represents the *choice act.* Choices are the results of decision-making processes. *Overt behavior,* or that which can be freely observed by someone other than the behaving unit, has traditionally provided the basic data for geographic analysis. Overt behavior is often recorded at the individual level in terms of a *link* between an origin and a destination. In aggregate terms, it is often recorded as the *frequency* with which links occur between origins and destinations. These choices often are characterized to be the result of a boundedly rational decision-making process. In this formulation we stress *bounded rationality* because we do not assume that the decision-making unit has perfect knowledge, a perfect cognitive map, and perfect awareness of the rules of reasoning and inference that allow the decision-making procedure to be integrated and implemented.

The overt evidence or manifestation of a choice act is, as we specified in Chapter 1, referred to as *behavior in space.* It is seen to have measurable characteristics of *linkage, direction, frequency,* and *episodic interval.* This information can be collected at the *individual* level or it can be *aggregated.* When aggregated, it is often used as the basis for making generalizations about different types of spatial activities.

2.3.2 Provisional Try Behavior

The type of behavior initiated by a novice decision maker is sometimes called *provisional try behavior.* If information collected for the first try is quite incomplete (as is usually the case when the decision maker is required to make a choice in a novel environment), the spatial manifestation of the choice act is often considered to be "irrational." On successive trials, variability of the spatial or overt act usually tends to diminish and, after a significant number of trials, converges on *repetitive learned behaviors.*

2.3.3 Feedback and Evaluation

After each trial, *feedback* is provided, and the existing knowledge structure is thereby altered. Evaluations, such as whether levels of satisfaction were achieved on specific trials are encoded with the objective spatial information attached to movement to different members of the feasible opportunity set. As more information about the environment mounts, aspiration levels may be refined. With this refinement, more and more alternative paths, destinations, and actions may be deleted from the compendium of potential travel plans. Once a suitable level of satisfaction is achieved, a tendency toward habit formation develops. The complex sets of rules, reasoning, logic, and inference used in the early stages of decision making become codified and stored in long-term memory. Thus, given a particular stimulus situation, a habitual or almost automatic response (or behavioral template) might be elicited. Over time, this means that the response time between exposure to a stimulus and choice is consistently reduced. As *habits* emerge, response times should stabilize, and behaviors emerge that are relatively invariant, repetitive, and difficult to extinguish.

Whereas it is difficult to model the uncertain behaviors involved in the early stages of decision making, it is a much simpler matter to model behaviors represented by habitual acts. In the former circumstances, uncertainty requires a greater knowledge and interest in the process of decision making itself to achieve levels of understanding; in habitual behavior, one may be able to predict the outcome of a problem situation without detailed formulation of the decision process involved in producing the final choice act. In other words, the spatial manifestation of behavior becomes a predictive tool for future occurrences of the behavior. Achieving this state of affairs has been the goal of many geographic researchers in the economic, social, political, cultural, transportation, and other domains.

2.4   Normative Assumptions about Decisions and Behavior

The above conceptualization discusses spatial decision making, spatial choices, and spatial behaviors of individuals by comprehending the decision-making processes that underlie those choices and behaviors. This is the approach we will follow throughout much of this book. However, a substantial part of geographic research has adopted an alternative strategy. This requires articulating assumptions about how people *might* or *should* behave rather than attempting to understand and state how they *do* behave. Such assumptions are often called *normative* assumptions. In much of classical economic geography, the decision-making process most frequently stressed was that involved in using normative principles to select locations for activities.

Location theory was, in fact, a prime example of the development of spatial theories from normative assumptions. The principal theories imported to the discipline at this time had their origins in economics (see Isard, 1956, for a temporal overview.) In these theories and the models representing them, sets of constraining assumptions were adopted. The environment was assumed to be an infinite uniform surface. No variability was allowed in soil quality, temperature, precipitation or climate, or surface gradient. But there were exceptions. The classical economist Ricardo (1817), for example, speculated about what would happen to land rents if soil fertility changed (see Figure 2.2). He was thus able to show that the changing quality of the environment contributed to the desirability and productivity of land, and this could be translated into a measurable quantity called *rent*. McCarty and Lindberg (1966) provided an alternate view of the interrelationship of physical environmental constraints and rent-yielding capability. For example, they assumed that only two physical constraints existed (precipitation and temperature). For any given crop, an ideal point (or optimum) could be defined in terms of centrality in the physically constrained environment. Productivity would grade equally from this optimum to the marginal boundaries where one or the other environmental extreme would be reached. By extrapolation, they then produced a graph of potential yields (or rents). By defining the break-even point (where returns would equal increasing costs), they then defined an area of feasible production for a land use (see Figure 2.3).

But, for the most part, producers and consumers alike were assumed to be economically and spatially rational beings. Economic rationality required perfect knowledge, complete and immediate communication with all segments of the population throughout the total environment, immediate access to market conditions, and

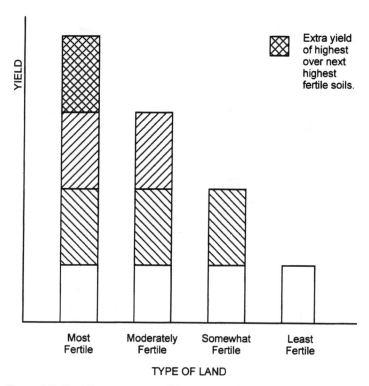

Figure 2.2. Rent theory constructed from arguments made in Ricardo, 1817.

the ability to make optimal decisions that maximized utility. Any variability among producers was thereby eliminated: all thought the same and acted according to the same set of rules. Given a constant and unchanging situation, all the decisions would be the same. Within this context, manipulation of the specific variables by relaxing different assumptions provided insights into how behavior might change.

The simple clear and powerful location models of von Thünen and Weber that are presented in most introductory geography texts owed their success to their simplifying assumptions that removed most sources of variability from the production and consumption process and allowed a focus on a single critical variable—transportation costs—as an indicator of the influence of distance and spatial layout. As one changed these assumptions, the complexity of the location problem became greater. For example, locational shifts might be required because of things such as the exhaustion of a basic raw material at a given site. As technology changed, there might be changes in the proportion of materials used, or indeed, changes in the mix of fundamental materials required. As transportation technology evolved, changes in freight rate structure occurred that caused an imbalance in the relative availability of raw materials. Increasing technology can also change the actual production process itself, making it more efficient and requiring less weight loss or weight gain.

In addition to these generally recognized *regional* locational factors, *local* factors may begin to emerge as having significance. For example, growing urban

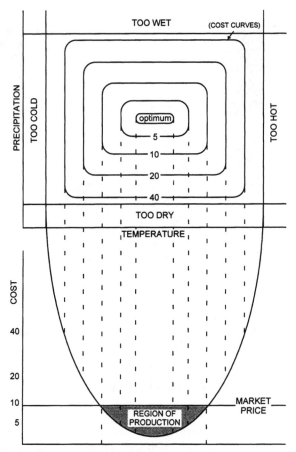

Figure 2.3. Physically constrained agricultural location model. *Source*: McCarty & Lindberg, 1966, p. 62.

areas may increase the tax loads on industries thereby reducing a city's locational desirability. Zoning laws may be enforced in such a way as to restrict sites for expansion. No-growth laws in the Santa Barbara area in California since the 1970s, for example, discouraged local firms from expanding and thus contributed to firm closures or out-migration. In addition, changing market tastes and preferences can cause a lower or increased demand for particular products.

Thus, the locational problem today is envisaged as a far more complex one than can be handled using the normative simplifying assumptions of von Thünen and Weber. Locational decision making is recognized as an error-prone activity, and its nature depends on whether one is a private capitalist, a corporate capitalist, or a state institution. As can be seen in Figure 2.4, depending on which of the above circumstances hold, personal considerations, corporate interests, strategic needs, or social costs and planning goals may become the dominant driving force in the locational process. Nevertheless, one still has to examine the characteristics of materials

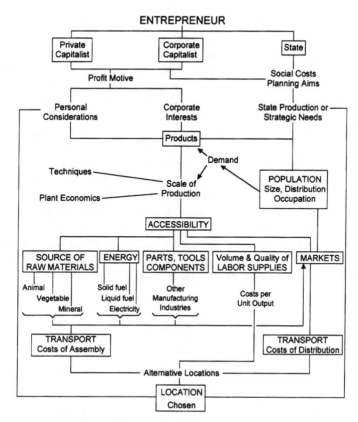

Figure 2.4. Location factors for corporations and institutions.

and products, the scale of a production process, the accessibility of factors of production, the influence of demographic features of the population, the societal and cultural constraints, and the type and cost of transportation as integral parts of the location process.

As more information was collected about the built environment, interest developed in building models of city systems rather than individual cities. From the 1930s, the German theorists Christaller (1933/1963) and later Lösch (1954) offered models of settlement patterns labeled *central place theory*. These played a significant part in the emergence of spatial theory and mathematical and statistical (or quantitative) procedures in geography.

Like classical models of agricultural and industrial location, models of the location of cities in urban systems required strict normative assumptions. Like von Thünen and Weber before him, Christaller assumed an unbounded homogeneous plain. Distributed uniformly over this plain was a population homogeneous with respect to income or purchasing power. Individuals had a uniform demand for goods and uniform patterns of consumption. There was a transportation system that allowed urban places of the same size to be equally accessible. Transportation costs were directly proportional to distance. Producers and consumers were economically and spatially rational. Economically they behaved in a simple *utility-maximizing* fashion. Spatially, consumers were assumed to patronize the nearest centers in which goods and services were available. And given the ultimate aim of distributing sets of urban places across the landscape, Christaller (1933/1963) assumed demands of the rural population were fully met by the minimum number of necessary central places.

Central places were defined as settlements at the center of market areas. In the most widely referred to case (that is, the market system), Christaller assumed that the dominant function of each urban place was to act as a point of contact, exchange, and provision of goods and services both for its own resident population and for the rural population within the surrounding region.

Different levels or orders of central places were acknowledged as existing; the *order* of a place depended on the types of functions performed. The highest level of central places served entire regions. The next-level places were nested within the tributary area or service area of a higher-order place, and in turn they dominated a number of levels of lower-order places (see Figure 2.5). Both Christaller

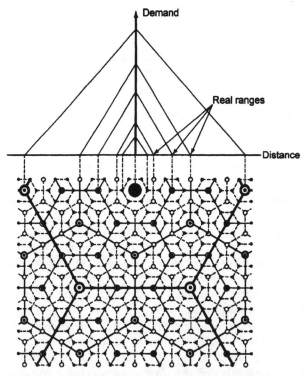

Diagrammatic representation of the nesting effect. The vertical axis in the top part of the diagram is not drawn to any scale.

Figure 2.5. Hierarchical urban system (Christaller *K* = 3).

(1933/1963) and Lösch (1954) suggested that different types of systems would emerge depending on the dominant functions performed by the urban places in a given system. The most commonly recognized were: (1) the *market system* ($K = 3$); (2) the *transportation system* ($K = 4$); and (3) the *administrative system* ($K = 7$). Figure 2.6 describes how each of these three systems is arranged with respect to the number of lower-order centers dominated by the center of the next highest level. In a $K = 3$ system, each center dominates the equivalent of three complete lower-order centers; in the transportation system, each center dominates the equivalent of four lower-order places; and in the administrative system, each center dominates the equivalent of seven lower-order centers. Thus the number of places at each level is a simple progression, in the above cases based on 3, 4, and 7, respectively.

Using sets of normative assumptions, Christaller and Lösch provided mathematical models of how centers with certain dominant functions would be distributed across the landscape. In these systems, people were economically and spatially rational, and uniform hierarchical arrangement of places in the system could allow one to quickly calculate the total number of places in the entire system or at any level. Given also the concept of *threshold*—which was the minimum size population

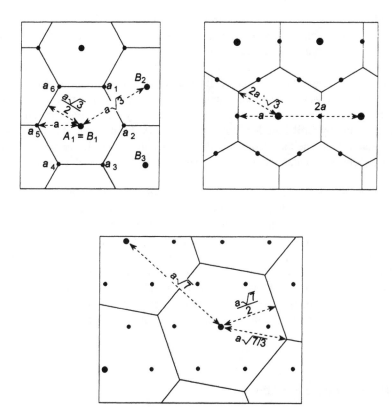

Figure 2.6. Basic arrangement of central places for $K = 3$, $K = 4$, and $K = 7$ levels. *Source:* Lösch, 1954.

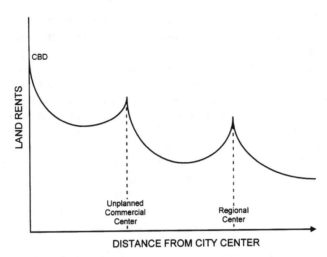

Figure 2.7. Effects of regional shopping center on rent gradients.

that could generate sufficient income to support a given function—it was a simple matter to calculate the number of functions needed in the system and the number of people that a system of any particular order could support.

As was found to be the case with many other normative models, however, casting doubt on the validity of the basic assumptions threw doubt on the validity and explanatory power of the entire system. In particular, extensive research on consumer behavior in both rural and urban environments by Berry, Barnum, and Tennant (1962) and by Golledge, Rushton, and Clark (1966) indicated that much consumer behavior was neither economically nor spatially rational. We shall explore the consequences of boundedly rational behavior in the consumer domain further in Chapter 10.

The simple decay functions in which rents declined exponentially away from city centers through high-density inner suburbs to low-density outer suburbs are exemplified in another normative city model—that offered by Alonso (1960) to account for rent gradients in cities. These negative exponential functions, however, were disrupted by the empirical reality of regional shopping centers and decentralized development (see Figure 2.7). Other disruptive factors included environmentally sectorialized highway systems and efforts to break down the ghetto boundaries imprisoning racial groups. Each of these processes can also be viewed as being antithetical to the simplifying mélange of normative assumptions of embryonic spatial theory.

2.5 Population Assumptions

In the preceding discussions of normative location models, people were seen as rational beings with *uniform tastes and preferences*. Populations were assumed to be *demographically identical*; there were no societally or culturally distinct ethnic groups with different values, beliefs, passions, feelings, emotions, preferences, tastes, and needs.

Each person consumed *a uniform amount of product*, and arranged his or her portfolio of tastes according to similar principles. There was no uncertainty in the system, and *maximizing satisfaction* was a single acceptable goal. Consumers were, in effect, turned into robots, automatons, or creatures of *habit*, devoid of feelings, emotions, beliefs, and values and exhibiting no individual differences in their ability to process information and make decisions.

Given this tightly constrained normative being, behavioral variability could only be allowed in the system by relaxing specific assumptions, generally one at a time. Fundamentally, however, behavior was *habitual* and even *stereotyped*. Even a minute change in economic circumstances would produce a change in behavior of these rational beings.

2.6   The Environments in Which Decisions Are Made

2.6.1   The Physical Environment

Behavioral researchers (Gould, 1963; Wolpert, 1964; Golledge & Brown, 1967) interested in locational decision making or its component parts (such as search, learning, preference, and choice) followed the lead of Simon (1957) in adopting a wider variety of behavioral assumptions that were relevant to decision-making processes of all kinds. This same trend had been pioneered earlier in research on people's attitudes toward floods and other hazards (White, 1945; White et al., 1958) and studies of the variability of physical environments and their impacts on human thought and action (Kates, 1966; Lowenthal, 1967; Saarinen, 1966; Burton & Kates, 1964).

These researchers led the way in adoption of significant changes in the way environments were represented. No longer was the physical environment a benign and harmless setting; rather, it was filled with hazards and dangers, with variable strategies (such as drought and flooding) that induced widespread human suffering, death, and loss of property. Physical environments were recognized as varying in their productive capacity, and although capable of being modified by human action and human resources, they sometimes proved to be equally as important as economic constraints expressed via transportation costs in terms of influencing the choice of human activity undertaken therein.

2.6.2   The Built Environment: Examples of Descriptive Theory

At first a surge of interest developed in building models of city form and the structure of large cities. These were the environments in which the majority of human behaviors took place, whether at the individual or the group level. In particular, theories about the internal structure of cities were developed dating from the 1920s.

These models are illustrated in Figure 2.8. Assumptions concerning the class structure of populations were imported from sociology (Park, Burgess, & McKenzie,

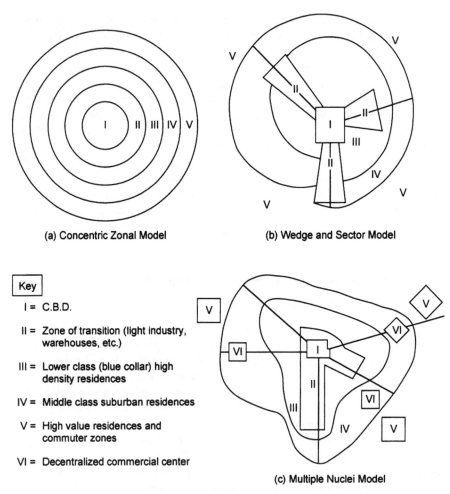

(a) Concentric Zonal Model

(b) Wedge and Sector Model

Key

I = C.B.D.

II = Zone of transition (light industry, warehouses, etc.)

III = Lower class (blue collar) high density residences

IV = Middle class suburban residences

V = High value residences and commuter zones

VI = Decentralized commercial center

(c) Multiple Nuclei Model

Figure 2.8. Descriptive models of city structure.

1925) and supported the notion of *concentric zones* of social classes arranged around a centrally located business district (Figure 2.8a). These zones were modified somewhat in the *wedge and sector hypothesis* of Hoyt (1939), who argued that the built environment would work through pressure on land rents to develop sectors and wedges of dominant land use, often extending along major arteries of transportation (Figure 2.8b). They would be attracted to or repelled from each other by factors of negative or positive valence (such as noise, pollution, and congestion, or serenity, environmental quality, and social class homogeneity). Later Harris and Ullman (1945) proposed a *multiple nuclei model of land uses* largely around a spatially distributed set of commercial nodes in cities (Figure 2.8c). All these theories were descriptive in that they relied on empirical evidence of *what was there* rather than being inferred from an axiomatic or simplified assumption basis.

2.6.3 Economic and Societal Environments

In normative theories, for the most part, physical environments and economic or social environments were tightly constrained. In fact, until the mid- to late 1960s, social environments and the ideologies under which they were organized were considered irrelevant in much of the geographic literature in Western capitalistic societies. The emergence of the Marxist/materialist position (as epitomized in the early work of Harvey, 1973 and Cox, 1973) showed that this significant dimension could no longer be ignored.

By the early 1970s then, the assumptions of a uniform society were being attacked. A clear distinction was made between what might be expected to occur in capitalist and noncapitalist societies. Activities taken for granted in westernized, capitalistic societies were reinterpreted in terms of harsh Marxist principles. *Labor theories* of economic and social structure were offered as alternatives to the more traditional capitalistic economic theories. And at the same time, a rising interest in the disparity between developed and lesser-developed social and economic systems brought home the principle of variability at all scales from the individual to the institution to the economic and social system.

Expansion of knowledge of developing economies and societies was a direct result of economic imperialism. While there is continuing controversy over whether this was a deliberate destructive strategy on the part of capitalists or whether it was a meaningful and serious attempt to use resources of developed economies to improve the well-being of those in lesser-developed areas, the result was that many inappropriate decisions were made. In particular, theories and models developed for the Western capitalist societies often were used as the yardstick for controlling the way that foreign aid funds were allocated in the lesser developed systems. Unfortunately, often there was a complete mismatch among the environmental, behavioral, economic, and social structures of the aid-granting agencies and the realities of the environment, the behaviors, the societies, and the cultures of the systems to which aid was being directed. This mismatch was often seen to introduce massive disharmony into the aid-receiving economic system.

2.6.4 Cognitive Environments

As assumptions and knowledge about the physical, built, economic, and social environments altered, yet another environment was being investigated. Forging links via behavioral research to psychology, geographers began investigating *perceptual* and *cognitive* environments.

It was suggested (Cox & Golledge, 1969) that in many cases the perceived or *cognized environment* was the most important one in which decision-making processes took place. Often it was not what was physically or objectively there but what was *perceived, remembered,* or *recalled to be there* that was the critical factor in invoking decision rules and making choices. It became recognized that cognized environments were schematized, simplified, abstracted, incomplete, and distorted or fuzzy representations of objective reality and that sometimes hidden societal and economic dimensions constraining that reality were significant factors.

Thus, by the beginning of the 1970s, the concept of environment itself had changed dramatically. Within the space of 10 years, in our theories and models it was transformed from being a simple constrained uniform plain to being the widely varying structures, both real and imagined, in which we live our daily lives.

2.7 Types of Systems and Societies

So far, the principal DMU were embedded in one or another distinctly organized system. Perhaps the most versatile was that assuming *perfect competition*. Different economic systems varied in terms of the number of producers and consumers they allowed, the freedom (or lack thereof) to enter the productive process (i.e., to become an entrepreneur), and so on (see Table 2.1). As one varied the system from monopoly, to monopsony, to monopolistic competition, duopoly, oligopoly, and to perfect competition, the seeds of behavioral variability were sown. However, the environments remained the same. Space was epitomized by barrier-free linkages. Even regions did not exist (despite several decades of intense geographic interest in the question of regionalization), until the theories and models that dominated thought at this time were made robust enough to "assume there are two regions A and B." This was achieved in the context of a new discipline, that of regional science, which complemented theoretical and quantitative geographic research.

Table 2.1 summarizes some of the conditions of different types of economic situations. For example, a *monopoly* has one seller and many buyers. Entry to the market is prohibited. A monopoly has a small and controlled set of goods, often a single good or a single product. In *duopoly*, there is an increase in the number of sellers; there are still many buyers, and there is no full access to the market. The sellers can act in competition or collusion. In an *oligopoly*, there are three or more sellers—this is reflective of the range of markets that normally exist in most economies and societies today. There are many buyers, and there are many sellers

TABLE 2.1.

Types of Economic Systems

Characteristics	Monopoly	Monopsony	Monopolistic competition	Duopoly	Perfect competition	Oligopoly
Number of producers	1	Many	Limited	2	Many	Limited
Number of consumers	Many	1	Many	Many	Many	Many
Freedom of entry into market	Nil	Open	Limited	Nil	Open	Limited
Knowledge	Perfect	Perfect	Perfect	Perfect	Perfect	Perfect
Competition	None	Much	Constrained	Competitive/ collusion	Open competition	Constrained competition

competing for a share of the market. The products are often differentiated deliberately, often through advertising (e.g., Post™ corn flakes or Kellogg's™ corn flakes). Product differentiation allows for partitioning of the market using whatever special attributes that an advertising campaign can dream up in order to sell the product.

Occasionally we will refer to a monopolistic or an oligopolistic situation, but for the most part we will concentrate on free market situations. However, many of the models are not free market models; they tend to be strictly constrained.

2.8   Risk and Uncertainty

Research on people's reactions to extreme environmental events or hazards indicated that many choices were made as a result of attitudes developed toward problem situations. White (1945) wondered why people continued to place themselves at risk year after year by inhabiting floodplains. Obviously, floodplains are a source of highly fertile soil that lend themselves to agricultural practices. But flooding, in some places, was known to be an annual event. However, what was not well understood was the variation in the severity and frequency of flood events. This lack of knowledge (or uncertainty) made the extreme flood a major hazard, for it caught people unaware, often unprotected, and frequently unwilling to take precautions. White and many of his co-workers argued that what kept people living in hazardous areas was the attitude they developed. Attitudes were seen to develop as the result of information processing. When precise data on the magnitude and severity of events were not available, subjective probabilities were substituted for concise knowledge.

Edwards (1954) demonstrated that people in general are not very good at developing *subjective probabilities* that closely reflect the *real probabilities* with which events occur. Acting on the basis of inadequate and frequently incorrect information, people develop attitudes about the probability that extreme events of a certain magnitude will occur during their tenure at a place or within their life-span.

A frequent position that underlay formation of these attitudes was the "It can't happen to me" syndrome. With such an attitude, processes of adaptation, alteration, or movement were given reduced urgency. Even following floods or other extreme events (e.g., coastal storms), the "Out of sight, out of mind" syndrome often became reflected in the attitudes people developed with respect to how they anticipated future occurrences of an event. As the number of previous experiences increased, it was found that attitudes and expectations were modified in the direction of comprehending reality, but the gap was often still large.

The *attitudes* so developed were seen as being the result of how people *perceived* their environment. If an event was perceived to be something that had a great interepisodic interval, it had a greater likelihood of being accepted as being part of the future and not having any influence on the present. Dramatic changes were described in people's perceptions of event severity during and immediately after an event as opposed to perceptions measured at some later time. Attitudes were seen to be transitory with their strength depending on the recency of exposure to harmful events. Perceptions of danger were likewise found to be temporally dependent on the size, frequency, and experience of the perceived event.

In circumstances where event uncertainty is high, *risk* becomes a major factor in choosing a *decision path*. For example, if it is perceived that a coastal storm hazard is not a frequent event, a decision path may be followed that involves not consulting authorities or experts (e.g., the Army Corps. of Engineers) or not taking action such as acquiring substantial insurance, weatherproofing buildings, or erecting barriers to wave action. Under these circumstances, *perceived risk* is not great. This is apparently why people continue to live at the foot of active volcanoes that erupt with some regularity; why people persist in living on flood-plains that are inundated on a regular basis with consequent loss of life and property; why some of the world's most densely settled populations exist on coastal areas that are subject to cyclonic disturbances; or why people and developers build houses across the top of geologic faults or live in valleys filled with dense brush that are subject to wildfire intrusion. We examine the notions of risk and uncertainty in more detail in Chapter 6.

2.9 Typologies of Decision-Making Theories and Models

2.9.1 Classical Normative Theories and Models

Broad classes of models involving decision making and choice behavior can be defined using discriminatory factors of *scale* (aggregate vs. disaggregate) and *context category* (normative, descriptive, behavioral, and applied). Even though such a simple cross-tabulation does not exhaust all the theories and models introduced into (or developed by) geography, it does provide a frame of reference that will help in understanding why certain positions are taken (and examples are used) in the later chapters. Such a classification is presented in Figure 2.9.

So far we have discussed some features of *aggregate* and *disaggregate* normative and *descriptive* models. Since social theory by definition discounts the role of the individual and concerns only the well-being of social groups, we plan to exclude it from our coverage. Both disaggregate and aggregate behavioral models will, however, be presented in detail in later chapters (Chapters 3, 4, 7, 8, 9, 10, 12). In this section, we give a brief overview of some "classical" models and theories of decision making and choice, and we follow that discussion with a more focused view of the decision, choice, preference, and attitude approaches that lay the foundation for much of the material discussed later in this book.

Decision theory and the models that represent that theory are diverse. Decision theory ranges from the most basic, normative, and precisely mathematical to the most descriptive, applied, subjective, and fuzzy in terms of the nature of the theories and models developed. Decision theory varies from largely verbal to largely mathematical. The literature on decision theory ranges from prescriptions of *how* decisions could be made, to specification of decisions that *are* made, to those that focus more on how a decision *should* be made.

Mathematically a *decision rule* for a given experiment is a mapping. In other words, it embodies a function that assigns a value to each alternative and lays out what will happen when a particular strategy is adopted. An *outcome* is a result of applying a decision rule in an empirical or task situation. The decision-making *cri-*

	Model Type		
	Normative	Descriptive	Behavioral
Disaggregate (Individual, single firm, or household based)	Hotelling: Duopolistic competition	Saarinen: Attitudes & perception Downs: Cognition of shopping centers Golledge: Cognitive distance & cognitive maps Hart & Moore: Developmental theory of spatial knowledge	Golledge: Market decision process Rushton: Revealed spatial preferences Timmermans: Consumer behavior/shopping center location Patricios: Shopping behavior
Aggregate	von Thünen: Agricultural location Weber: Industrial location Christaller/Lösch: Location structure of urban systems Alonso: Urban rent theory Gould: Game theory Church & Revelle: location/allocation Dorrigo & Tobler: Gravity models of migration Ravenstein: Migration laws	Park, Burgess & McKenzie: Concentric zone Hoyt: Wedge & Sector Harris & Ullman: Multiple nuclei Shevky & Bell: Social area analysis Berry: factorial ecology White: Populations at risk from hazards	Wolpert: Boundedly rational farmer populations McDermott & Taylor: Decision traces in firms

Figure 2.9. Examples of types of models involving decision making and choice.

teria are the set of procedural rules that govern the evaluation or benefit obtained from an outcome or set of outcomes. A *strategy* is a set of decision rules that govern all possible ways of making relevant decisions and seeking an outcome.

Classical decision-making models often can be classified into four categories: (1) *riskless* theories; (2) theories of *risky* decision making; (3) *transitivity* in decision making; and(4) the theory of *games* and *statistical* decision functions.

A principal characteristic of the *theory of riskless decision making* and consequent choice involves an assumption of *economic rationality*. It has been said these decision-making theories are basically armchair methods in which an axiomatic and assumption base is examined deductively for theorems that are proven and combined to produce the theory as an end product. Economic rationality, of course, includes access to complete or perfect information, infinite sensitivity to the infinite divisibility of functions so that they can be regarded as continuous and differentiable, and an ability to order outcomes as part of a process of maximizing some criteria (usually called a *utility*). This implies that all preferences must be *transitive*. Thus the economically rational being must always be able to choose the best alternative. Being able to maximize something lends itself to specification as a procedure designed to find a unique solution from among those available. All these points have been disputed by behaviorally based decision theorists who argue that such a person as described above is unlike any real person.

The *theory of risky decision making*, however, allowed the distinction between concepts of *risk* and *uncertainty*. As we elaborate in Chapter 6 (sections 6.5–6.8), risk is a concept related to knowledge of the probabilities of outcomes, whereas uncertainty implies that it may be impossible for a decision maker to find out what the probability of an outcome might be. The traditional economic and mathematical way to deal with uncertainty and with risky decisions is to use concepts of *expected value*. This procedure has become quite widely used in geographic studies of decision making including those based on game theory.

Psychologists interested in decision making took somewhat different approaches to many economists and mathematicians by following a *probability preference* approach. This allowed the introduction of subjective probabilities into the decision-making process and the idea of weighted preferences, multiattribute utilities, and part-worth attributes. It also provided the incentive for a new set of psychologically based decision theories to develop. These were based on concepts such as potential surprise (Shackle, 1949), regret (Hurwicz, 1945), and cognitive dissonance (Festinger, 1964).

The *theory of transitivity* in decision making gives one view of what happens when preferences diverge slightly from indifference. Psychologists showed that it was highly probable that one would begin getting intransitive choices when four or more objects were arrayed in a paired comparison judgment task. In such a situation, as least one intransitivity involving three objects is likely to occur. Throughout this book we show how behavioral researchers have rejected *utility maximization theory*—which cannot work except with the transitivity of preferences. *Intransitivities* are likely to occur when the decision-making situation contains objects with multiple dimensions such that different things can be evaluated using different dominant dimensions when set up in a paired or triadic comparison situation.

And finally, the *theory of games and decision functions* provided a springboard for behavioral decision theory by allowing many different decision-making criteria to be used in preference and judgment tasks. The development of game theory introduced decision makers to a variety of critical terms and concepts that remain with us today. Because of the significance of this development to behavioral decision theory generally, we shall spend a little more time on it.

As an example of a decision-making problem consider the following situation. Here we have four alternative strategies and four alternative scenarios (Table 2.2). Using a simple riskless criterion (The Wald criterion or MIN–MAX/MAX–MIN criterion) ,one would search for a *pure* strategy as a solution to the task situation. A *mixed* strategy would involve defining a subset of possible alternatives and using each alternative in the subset according to a set proportion or probability. A *pure* strategy requires continuous use of a given rule.

As an example of the variety of decision theories and criteria that can be used, consider the following. In Table 2.2, if we use the Wald (or MIN–MAX) criteria and require a pure strategy to be selected, one would search the payoff matrix for a saddle point, and if one exists, then the mix of strategies intersecting at that saddle point should be used to the exclusion of all others. In essence one would choose the strategy that *maximized the minimum expected outcome*. Solution A-2 would be selected. Considering the Savage *minimum regret* criterion, a mix of strategies A-1, A-

TABLE 2.2.
Strategies and Four Alternative Scenarios

Alternatives	States			
	B_1	B_2	B_3	B_4
A_1	2,500	3,500	0	1,500
A_2	1,500	2,000	500	1,000
A_3	0	6,000	0	0
A_4	1,500	4,500	0	0

Source: Amedeo and Golledge (1986), p. 355.

3, and A-4 would be selected and played with the proportions of 9 : 23 : 68 respectively. Using a *pessimism–optimism* criterion as suggested by Hurwicz, alternatives A-1 and A3 would be eliminated, and choice would always lie between A-2 and A-4. Given no a priori ideas or considerations about the probabilities of each alternative providing successful outcomes, then a criterion of *insufficient reason* (offered by LaPlace) would suggest that equal probabilities should be given to each of the states. Consequently, calculating the expected value of each state would produce A-1 as the chosen alternative. However, if we replaced maximizing theory with Simon's *satisficing theory*, and if a zero level of aspiration were set, any of the different alternatives could be used with consequent achievement of this goal. With non-zero levels of aspiration, A-2 would be selected. As one moves more into psychological theory in which beliefs substitute to some extent for utilities, Shackle's *theory of potential surprise and focus outcome* would suggest that either A-2 or A-3 would be selected with the final choice being related to the risk-taking potential or gambling propensity of the decision maker.

Bayesian models of decision making are essentially *conditional probability* models. King and Golledge (1969) show that the basic theorem can be stated as follows (Savage, 1962):

$$\text{Prob(Hypothesis|Datum)} = \frac{\text{Prob(D|H)·Prob(H)}}{\text{Prob(D)}}$$

This theory follows from the product axiom of probability and requires an a priori probability for the hypothesis in question and a conditional probability for the datum given the hypothesis. Bayes theorem involves the use of subjective or personal probabilities that are indicative of the opinions that someone has about an event. Usually uncertainty is present. The model requires definition of a number of *states* (S) of the world consisting of the *objects* and *events* about which the person is concerned. States are *mutually exclusive* and *exhaustive*. A potential decision maker is uncertain as to which state is the true one but has *prior personal probabilities* concerned with the possible occurrence of each state. After choosing among a number of *alternative possible actions* (A)—which choice may be the result of conducting a set of *experiments* (E) in order to obtain more information about the likelihood of different actions in different states—then the decision maker is presumed to be able to give a numerical value or utility from an evaluation of all this information. The Bayesian model assumes that the decision maker is *rational* and will choose the action for which *utility* is greatest. Examples of the use of Bayesian decision models in geography include prediction of a choice of markets by hog farmers in Iowa (King & Golledge, 1969), description of various processes of subsistence cropping in uncertain natural environments (Gould, 1963), and prediction of climatic variation (Curry, 1966).

Where decisions are made by humans rather than mathematical computer algorithms, it is difficult to assume that one is likely to work in a riskless environment. Consequently accepting that risk and/or uncertainty is a critical part of most decision-making environments encourages us to be more versatile in terms of the criteria we use, the theories we develop, and the models deemed acceptable for studying such decision processes and their outcomes. Note also that there are no strong theoretical grounds for selecting any one of the classical alternative theories and models over another. The algorithm one selects depends on the decision maker's psychological makeup. It also depends on that decision maker's risk-taking propensity, his or her problem-solving ability, and his or her judgmental ability. There does not appear to be a single "best" mathematical procedure for solving decision problems that include uncertainty. However, it is possible to pursue models for which one can give plausible arguments and explanations.

We also identified the fact that decision making is *conscious* and *active* (generally associated with problem solving behavior) and that it may become virtually *automatic* (as when on successive trials or problem-solving scenarios the same act or sequence of behaviors is used). Thus decision making may represent the implementation of a well-learned, repetitive behavior that is extremely difficult to extinguish. But most of the time we look at decision making in the problem-solving context.

Decision models frequently are used in examining human activity patterns. The latter are daily, weekly, or other episodically regular behaviors undertaken by individuals or households. They often involve multipurpose behavior and trip chaining. Models used to examine these types of decision-making situations include Markov chains, modifications of discrete choice models, and simulation models. Others include stochastic models of buyer behavior, purchase incidence models, and models of variety-seeking behavior.

2.9.2 Behavioral Decision Models

It is obvious from the above discussion of traditional decision theories that perception, preference, attitude, cognition, and risk all appear to be relevant in decision-making processes. Recognition of this fact was a major consideration in the departure of behavioral decision theory from classical economic constraints. As we shall see in later chapters, there are still circumstances where maximizing utility or optimizing some other function are perfectly acceptable and adequate decision rules. This is often the case with respect to management decisions concerning production activities in a firm or industry or in terms of the selection of a policy to be pursued by a governing authority or institution. However, it also seems to imply that as one drops below these aggregate levels, the traditional maximizing utility or optimizing framework, whether it comes from an economic or a mathematical origin, appears to be much less acceptable or representative of human decision-making processes. As a result, we emphasize in much of the rest of this book *behavioral* decision theory rather than classical decision theory.

2.9.3 Applied Decision Models

Timmermans (1991) suggests a *typology of decision-making processes* tailored to some of the basic characteristics of choice behavior. Examples include:

1. Decision models that account for *variety-seeking behavior* (e.g., recreational choice [Van der Heijden & Timmermans, 1987]).
2. Decision models that account for *uncomplicated choice among limited alternatives* (e.g., travel mode choice).
3. Decision models that account for *complex choice* situations involving preference and attitude.
4. Decision models that account for *temporal choice* (stochastic models).
5. Decision models that help *simulate complicated choice outcomes* (hybrid models).

Timmermans has offered a typology of decision-making processes based in part on their typical application or use. He argues that this is possible because the same basic theories have been applied over and over again in a wide variety of spatial choice situations. Many of these situations are discussed further in Chapter 10, Sections 10.3 and 10.4, and include activities such as shopping behavior, route choice behavior, housing market choice behavior, and recreational choice behavior. Each of these activities has sets of distinctive characteristics that encourage the use of a particular type of decision-making model.

For example, models of *consumer shopping behavior* emphasize its repetitive nature and almost always assume that behavior is habitual. Under such circumstances, the feasible opportunity set (choice set) is relatively small. Individuals are usually not aware of all possible opportunities at which their goal may be achieved but may have undertaken some preparatory search and exploration behavior in their attempt to define members of the feasible set. Many of the attributes of the choice set in this behavior domain change rather slowly, and consequently, decisions are often made in an environment in which little risk or uncertainty is involved. An incorrect decision will therefore have only little negative impact.

The *housing choice decision* is a more robust and significant one. Only a few of these decisions are likely to be made during the life cycle of an individual. Although some individuals extensively search alternatives available at a particular time in the housing market, the majority view only a small number. Information used in the decision-making process is most often obtained through a mediating agent—the real estate broker. However, in this environment, the feasible alternative set may change rapidly, and the impact of a wrong decision can be dramatic (e.g., buying a high-priced house and then watching its value drop rapidly in a recession).

Recreational choice is often influenced more by a conscious drive or by variety-seeking behavior then by a habit (Timmermans, 1985). Dynamic process models assume a decision maker seeks and evaluates information sequentially. Types of these models include *decision net* or *decision plan* models and *computational process* models (CPM). The former involve a detailed exploration of the entire decision process by an individual. Decision net models are tested for relevance by describing the choice alternative in terms of its attribute level and then searching along the decision net to see whether the chosen alternative is acceptable. CPM are sets of interacting computer programs. Problems with CPM involve determining the relevance of different types of information, deciding on the coding and organization of the information, defining the activation spreading mechanism, and selecting the cognitive processes that act upon the knowledge structure to generate behavior.

2.9.4   Choice Models

We defined a *decision-making process* as a set of *strategies* that guide *decision-making behaviors* such that they appear to cover many possible scenarios. The decision-making process, on the one hand, is the act of tying together all of the components involved in making a decision. The *choice act*, on the other hand, is the outcome of the decision-making process and is the end result or *spatial manifestation* of what has gone on during decision making. Each choice requires indicating the alternative chosen. Choice is revealed by some type of *overt act* (e.g., pushing a button or making a trip). *Preference* is an *activity* that enters into the decision-making process to express or to identify what is desirable. Preference and choice are not the same thing. At times they may coincide, but usually preference and choice are different.

We can regard choice as being *constrained* and showing *revealed preference* (Rushton, 1969a). Revealed preference states that by making choices we reveal the preferences that we have among sets of alternatives. Preference itself, however, is a much stronger concept than that—what we prefer is not always what we are able to achieve because of different constraints. We may prefer to eat at an expensive restaurant five nights a week, but our choice turns out to be Taco Bell™ or McDonalds™ on four of the five nights! Generally intervening between preference and choice there are constraints that can be *socially* based, *income* based, *cognitively* based, or can be *attitudinal* in nature. And there can be *environmental constraints*. Each one of these will represent in part a *barrier* that has to be overcome as part of the decision-making process.

From the early 1970s, different types of decision making and choice models have been advanced, but most are based on a simple conceptual premise that relates choice behavior to an environment via the processes of *perceptions*, *preference formation*, and *cognitive mapping*.

The theories and models assume that individuals develop *cognitive representations* of an environment and that while there are many commonalties in these representations, they all incorporate some idiosyncratic characteristics. Even if perceived in the same way, attributes of places may be given different *saliences*, or weights. Thus, one individual may rate a store highest because of the quality of its products; another because of its price structure; another because of the speed of service; and yet another because of the available parking.

Thus, it is assumed that individual choice sets differ one from another and that all contain only a subset of the total attribute set. Individuals attach *subjective utility* to the attributes they deem important and combine these *part utilities* into some composite utility according to a combination rule or decision-making heuristic. This results in a *preference structure*, which defines the positioning of the different alternatives in the preference space. Preferences are ordered in this space and evaluated for potential choice in terms of individual, institutional, and societal constraints. Within this context, most models have been concerned with *single choice behavior*, but some are made to extend to *multiple choice* of products or choice of multiple locations through concepts such as trip chaining and activity scheduling.

Two of the most used models of single choice behavior are *discrete choice models* and *decompositional multiattribute preference models*. Discrete choice models are derived from two formal theories: Luce's *choice axiom* (Luce, 1959) and Thurstone's *random utility theory* (Thurstone, 1927). Luce's model (developed later in Chapter 10) is a strict utility model and is usually considered as an extension of constant utility theory developed by psychologists to account for the occurrence of intransitivities in making choices among multiple alternatives. This theory assumes that the probability of choosing an alternative is equal to the ratio of the utility associated with that alternative to the sum of the utilities for all the other alternatives in a choice set. His "law" is essentially a constant ratio decision rule.

Random utility theory was developed as part of the basis of measurement scales for comparative judgments. Whereas Luce assumed deterministic preference structures, random utility theory assumes a stochastic preference structure. Thus, an individual is presumed to develop a different utility function for each choice occasion. Random utility models now assume that there are unknown factors influencing a choice and that the utility component consists of a deterministic component and an error component due to the effect of the unobserved factors that influence choice. This type of model also assumes a utility-maximizing decision rule. Furthermore, it is assumed that the random utility components are "independently and identically normally distributed" (the IID property).

Timmermans and Golledge (1990), in discussing preference and choice models, generally state that, under strict utility theory, different functions may be allowed if they are unique except for multiplication by a positive constant. However, random utility theory also allows for different specifications of utility functions. A variety of mathematical and statistical procedures for establishing parameters and values have been identified in Wrigley and Dunn (1984a,1984b) and Wrigley (1984). Estimation of *discrete choice model* parameters involves establishing a functional relationship between the choice alternative's attributes and observed choice behavior. Estimation techniques include conditioned logit analysis, weighted least squares, maximum likelihood, and normal linear least squares (Horowitz, 1982, 1983). Alternatively, robust estimation techniques (Wrigley & Dunn, 1984c) and generalized linear models have been suggested as viable alternatives. In addition, adjusted likelihood functions and alternative estimates have been proposed when choice-based sampling is considered (Manski & Lerman, 1977). Moreover, utility-based models generally do not make explicit assumptions about the preference-forming mechanism but, rather, assume that preferences satisfy conditions of transitivity, continuity, and completeness. Satisfying these conditions assures that a utility function can be determined. Discrete choice models are among the most widely used in geography, and they occur not only in consumer choice behavior but also with respect to residential choice, retailing, recreational choice, transportation-mode choice, and even at a more macro level for models of the urban economy.

Other preference and choice models are detailed by Timmermans and Golledge (1990) and include the following:

1. Decompositional multiattribute preference models.
2. Compositional multiattribute attitude models.
3. Hybrid evaluation models.

Most of the models discussed or listed above assume:

1. Utility of a choice alternative is independent from the attributes of other alternatives in the choice set (independence from irrelevant alternatives property or IIA).
2. Choice behavior is compensatory.
3. The models typically have invariant parameters that do not depend on variation and attribute levels of members of the choice set.
4. Introducing a new alternative into the choice set does not increase the choice probabilities associated with existing alternatives.
5. The active choice is not influenced by indifference between other choice alternatives.
6. In many of these models, the geographic location of members of the feasible choice set does not influence the choice probabilities.

Traditionally, the main argument against the use of decompositional models is that their use has been impeded by a number of unresolved problems including the fact that they lack an integrated theoretical framework to link preferences to choice behavior. General compositional and decompositional preference models are different from choice models in that attributes are calculated rather than being explicitly measured. And, strictly speaking, they are not choice models since they do not include an explicit decision rule.

Numerous varieties of choice models now exist; many of them have been developed in an attempt to avoid the IIA property. One model that has been given favorable reception in geography but also avoids the IIA property is the *elimination by aspects model* (Tversky, 1972). Here each choice alternative is assumed to consist of a set of aspects, and sequential elimination of alternatives occurs at different stages of the choice process. Aspects are chosen with a probability proportional to their importance, and choice alternatives that do not possess a considered aspect are eliminated. The process continues until a single choice alternative remains.

Most discrete choice models do not allow indifference between choice alternatives despite psychological evidence that such indifferences might exist, particularly when there are only small differences in the utility associated with the alternatives. Spatial choice models are often based on the assumption of independence of the spatial structure. Fotheringham (1983, 1985) modified traditional spatial interaction models to correct this misspecification. He did this by adding an accessibility variable representing the accessibility of a given destination compared to all possible destinations.

2.9.5   Choosing between Decision Models

The decision maker may benefit from achieving a solution to a decision-making and choice process in a variety of ways such as:

1. Benefit may accrue from the mere fact that a search for a solution was undertaken even if no single acceptable solution is found.
2. Benefit may accrue by discovering that some factor or set of factors that have been considered important are not so, or a set of factors considered unimportant were more important than assumed.
3. Benefit may accrue as a result of a more precise specification of a problem as a decision process is invoked.
4. Benefit may accrue as a result of becoming acquainted with a number of different strategies that may produce a desirable result.
5. Benefit may accrue by changing the criteria for selecting a solution (e.g., from economic efficiency to equality or equity).
6. Benefit may accrue by obtaining good estimates of the parameters and significance of the variables in the decision-making process.
7. Benefit may accrue by achieving a satisfactory solution and implementing it.
8. Benefit may accrue as a result of spread effects from undertaking a decision-making and choice process in that unanticipated insights may be gained into the solution of other problems.
9. Benefit may accrue by achieving an exact solution that meets expectations or a priori levels of aspiration.

The achievement of a solution after going through a decision-making and choice process brings with it both benefits and problems. In order to assess relative worth of each of these scenarios, we first need to understand what criteria are commonly used in assessing the output from decision making. Significant criteria include:

1. A solution must be considered at least satisfactory by the decision-making unit or must conform to the exactness criteria established by the decision-making unit.
2. A solution must be acceptable to those who must implement it.
3. A solution must be feasible or within the abilities of a decision-making unit.

4. A solution should be judged in terms of the anticipated responses to its implementation.
5. The risks associated with potential solutions should be known.
6. Multiple solutions should be arranged in sequence according to clearly established criteria.
7. A solution should be assessed on the basis of quality and acceptability; quality should be related to the degree to which a solution can contribute to achieving short- and/or long-run goals and objectives; acceptability is primarily a human resource criterion relating the degree to which the implementation of a solution achieves some useful purpose.
8. Solutions may be evaluated in terms of anticipated reactions to them (i.e., consequent behaviors must be anticipated and judged as acceptable).

In all problem-solving disciplines where more than one alternative approach can be used to achieve an answer or solution to a problem, the ultimate question is how to choose among the several treatments. A soft answer is always to choose the "best" of them, but significant variation occurs in what one may consider to be the "best." In many cases, the selection of "best" is a matter of personal preference and reflects the weight or salience an individual gives to particular dimensions of the problem or the obtained solution. For example, a government agency in a city faced with the task of school redistricting may come up with different "best" solutions depending on whether they are trying to minimize cost, maximize educational quality, ensure appropriate conditions of diversity or ethnic mix were obtained, or simply to minimize the distance that students travel to a school. And even within each of these alternatives, it may be possible to produce many different solutions that give approximately the same product and could qualify for being a considered "best" solution.

Considering the disciplinary origins of many decision and choice theories before they were applied to the discipline of geography, it would not be surprising that the most commonly used criterion in searching for the best solution has been expressed in terms of differential utility. Inevitably using such a criterion involves determining an expected value associated with different solutions and ordering the expected values to highlight the largest values.

In some cases it may be desirable to go through a preprocessing operation to select a "best" theory or model before coming up with a solution. Again in the school redistricting situation, practice and economy may dictate that a particular type of model be selected a priori and that only outcomes associated with the application of that model be considered. For example, the school district might decide that it is driven entirely by cost considerations and search only among those models that provide a minimum cost solution. On the other hand, some decisions are made in an a posteriori manner after the outputs of several different models are examined, and the one that appears to have the greatest acceptability is selected. In many practical planning situations, this requires looking at a simulation, for it may not be feasible to run all possible viable models to determine a set of outcomes that would provide the data for a meta-decision-making process.

A second criterion that could be used in the selection of models or theories includes the value of information obtained from applying the theory or model. There is often a difference between whether a decision-making unit wants confirmatory or exploratory results. Confirmatory results may just provide objective justification for an a priori or subjectively stated hypothesis. Exploratory results may involve the decision-making unit searching areas or decision paths not previously considered.

In a less formal domain, one may simply examine the legitimacy of the axioms or assumptions on which the decision-making theory or model is based in order to evaluate it. For example, as we showed earlier in this chapter, closer examination of classical location theories of agriculture and industry clearly shows sets of untenable assumptions about the nature of the decision-making population, the environment, and even the society and economy in which decisions may be made.

Another way of evaluating the worth of different theories or models is to examine in detail their inferential structure. Thus, any logical inconsistency or violation of rules of inference and reasoning may be used as a mechanism for discarding one theory or model or retaining another.

One rigorous and more objective way of evaluating the worth of different theories or models is to assess the degree to which they are capable of reproducing an empirical data set related to a problem-solving or task situation. Such a method has been developed by Hubert and Schultz (1976) and applied in the context of evaluating different movement models by Golledge, Hubert, and Costanzo (1982). This is known as the *quadratic assignment procedure* (QAP); it uses combinatorial procedures in simultaneously mixing rows and columns of a matrix (or two of them) thus developing all possible outcomes. Each single output is compared to the original data matrix, and a cross-product measure is calculated. The cross-products form a reference distribution consisting of a sample of the range of possible outcomes and provide a means for estimating significance of a single theory or model. In essence, the model that produces an outcome that correlates most highly with the data on which it is based and which it is designed to replicate is chosen as the best.

Horowitz (1982, 1983) has indicated that many decision-making and choice models can be evaluated in terms of their *robustness*, *reliability*, and *validity*. Mathematical measures of robustness using maximum likelihood procedures are suggested as being preferable to the randomization procedures offered by Golledge, Hubert, and Costanzo (1982). The relative value of each approach is debatable.

Other criteria used to select among decision-making and choice models include *transparency* (i.e., simplicity and user-friendliness), *generalizability*, and *transferability* (i.e., the degree to which the model developed in one area can successfully be used again in another area), and *acceptability* (i.e., the degree to which the final decision-making unit can comprehend the process by which an outcome is reached).

2.10 Decision Support Systems

A *decision support system* (DSS) is often based on a comprehensive data base that is usually spatial in nature. It includes a variety of things ranging from specifics about the physical environment to characteristics of the social, economic, and governmental environments, all of which are digitally coded and stored on computer tape or disk. This support system is accessed when the decision maker or management team needs to solve a particular problem.

A *geographic information system* (GIS) is a set of operating computer programs (e.g., ARCINFO®), that allow decision makers to combine different levels of a data base in helpful ways. One example occurs when a decision-making team wants to locate a manufacturing plant. They may have specific criteria such that it has to be located within 1½ miles of a major freeway, within 1 mile of a perennial stream, within an urban local government area for access to sewage disposal and water, and so on. Each one of these constraints is expressed as a layer of information that can be superimposed on a computer screen. As each feature is overlain, it defines a smaller and smaller area of opportunity or potential location. The decision-making team can then look at the final set of possible alternative sites and make a decision based on both the objective criteria and the GIS representation.

2.10.1 Characteristics of a Decision Support System

A DSS consists of a set of computer programs designed to bring a knowledge base to bear on the search for a problem solution. According to Densham and Rushton (1988), this is done by providing a flexible and adaptive system that is constructed in such a way as to make explicit use of the analyst's models and the decision maker's expert knowledge system. Its objective is to *support* rather than *replace* human judgment. It aims at combining the strengths of both human thought and machine processes. The DSS is, in effect, a supportive tool under the control of a user. It does not consist of a completely automated decision process but involves human input at critical parts of the process. The DSS is often brought into play when it is thought not desirable to have decisions made by a fully automated system (e.g., human takeover from an automatic pilot system when landing an airliner).

Usually people differentiate between a DSS, which is used as input for the activities of a decision-making team, and an *artificial intelligence system* (AIS), which actually carries out and controls the entire problem-solving process. An AIS as embodied in some types of robots, for example, would simply be activated, given a set of instructions, and allowed to follow the instructions according to preprogrammed sets of strategies. It may select from among the strategies according to some formal decision rule, or it may select among the strategies at random depending on how the operating system was originally set up. Some AIS, such as autopilots on airplanes or autonavigational systems on boats and ocean liners, actually perform the functions of steering, guiding, and navigating the aircraft or the boat. They are not always perfect, obviously, as we have seen in episodes such as the Exxon™ oil spill off the Alaskan coast. Often human intervention is required in support of the AIS, particularly at critical points such as entrance or exit from a harbor or navigation through areas where information is incomplete.

A DSS assists managers in their decision processes, supports managerial judgments, and improves the effectiveness rather than the efficiency of decision making. Densham and Rushton (1988) suggest that specific characteristics of a DSS include the following:

1. They can be used to tackle semistructured problems such as when the decision maker's goals and objectives or the problem itself cannot be coherently defined.
2. They are designed to be easy to use, though they often rely on quite sophisticated computer technology. An essential part of a DSS is a user-friendly interface.
3. They are enabling technologies. They allow the decision maker to make full use of as much data as possible and as many relevant models as are available.
4. They can be used to display and evaluate different outcomes. To do this, DSS usually are interactive, often requiring a decision maker to select an option or a path before proceeding to the next step.

A DSS is thus used when no automatic algorithm exists to solve a problem or is deemed appropriate to solve a problem. Under these circumstances, problem solving usually requires intermittent evaluation of alternatives, and the final solution is often a good or satisficing solution as opposed to the optimum solutions of many automated processes. This occurs because the many important aspects of problem-solving behavior still cannot be adequately captured in mathematical formulae or in terms of computer software.

2.10.2 DSS Subsystems

Most DSS provide three subsystems. These consist of: (1) a *knowledge* system that contains data and data manipulation procedures; (2) a *language* system that represents the user interface; and (3) a *problem processing* system that processes models and data in the search for solutions (Densham & Rushton, 1988).

The question arises under which circumstances should we use DSS? So far in this chapter, we have examined a number of decision-making situations that conceivably can be represented by an objective function (e.g., to maximize or minimize cost, distance, etc.) and a set of constraints on the solution space (e.g., subject to a budget constraint, maximal distance constraint, or maximal service constraint). An automated system would provide as exact a solution as possible to the problem being investigated. Typical decision-making aids include mathematical models, mathematical programming, combinatorial techniques, and dynamic programming. As a set of structured computer programs, the DSS provides the analyst with a variety of ways of manipulating data and representing it for visual inspection. To operate effectively, Densham and Rushton (1988) claim that a DSS must have four working elements:

1. One or more individuals who are users and whose job is to execute some task in a competent way.
2. A task environment that may include historical, physical, or social dimen-

sions and attributes that are relatively permanent and upon which decisions are supported.

3. A procedure for ordering the use of the support system by the users.
4. Some criteria for judging competence or adequacy of the output.

2.10.3 Use of DSS

Often DSS are used for locational planning. As we move from conditions of perfect knowledge and perfect markets, multiple locations can serve as sites for specific industries. Within a locational planning context, a DSS can determine a variety of alternate locations for a given activity, providing estimates of factor–cost combinations at each site. The decision-making agency thus would have at its fingertips information about a variety of alternative locations and could then make decisions according to the particular mix of input that agency wished to incorporate into the production and distribution processes. Decisions can thus be made by comparing alternative outcomes to some a priori ideal state, sometimes produced as a result of applying an optimization model. Alternatively no a priori objective need be specified; then the decision-making agency simply looks at alternative outcomes and selects one of them on the basis of criteria deemed most significant at the time.

2.11 Macro Behavioral Assumptions: Institutions, Strategies, and Policies

Much of the discussion to date has focused on the micro level. We now undertake a change in scale and examine decision making and choice behavior at a macro or aggregate level rather than at a disaggregate level.

Decision-making activities of private and corporate capitalists and government agencies are more complex in that there are *multiple stakeholders* involved. However, within the operational environment of the real world, such as a city system, which is dynamic, a multitude of past, present, and future decisions at both the *macro* and *micro* levels create and share that dynamic environment. And those decisions are taken in the context of *multiple goals and objectives*, which may be complimentary or in conflict at different levels and between or among decision makers at any one level. In particular, there will be differences in the objectives of the public sector, which incorporates values concerned with the public good and equity, in comparison to the private sector organizations and firms, whose objectives relate to efficiency, returns, and profit.

2.11.1 A Framework for Planning and Decision Making in an Urban Setting

Cities and regions are dynamic organisms. Often regional scientists analyze them, and planners develop plans for them, using a static rather than a dynamic approach. Also, often they use a sectorally oriented approach rather than focusing on *inter sectoral linkages*; and they use physicalist strategies rather than comprehensive and integrative strategies that are oriented to understanding the interrelationships among the physical, economic, social, and political environments and potentials of a city or region.

There is a complex set of trends, concerns, issues, and processes that need to be addressed in planning for the future development, growth, and management of a city or a city region. Figure 2.10 illustrates the overall milieu in which planning is undertaken within a spatial analytical and decision-making framework that is appropriate to investigate a city and to evaluate its potential evolution over time. It identifies the elements that could be incorporated. It suggests that the *urban setting* (the built environment, structure, form, and processes of change in the city) can be described and analyzed in a *sectoral, spatial*, and *temporal* framework, or in a combination of these. The urban setting evolves over time through *processes* that include planning and involve decision-making actions at the *public, corporate, group*, and *individual* levels within a broad *political, social* and *economic* domain. We can evaluate the past and current performance of a city and evaluate potential future scenarios for the city against criteria relating, for example, to effectiveness, efficiency, equity, and quality-of-life issues that reflect societal values. *Objectives* and *policies* for shaping the future of a city reflect *values* of society that are manifest through the decisions of elected representatives and their judgments as to what the myriad of stakeholders in the city would like to see happen.

From Figure 2.10, it is evident that we are dealing with an exceedingly complex and multifaceted phenomenon in understanding and analyzing the operational milieu of a city or region and in attempting to plan for its possible future and evaluate its performance.

2.11.2 Strategic Approaches to Decision Making

Institutions in the private and public sectors have developed corporate strategies to plan their operations and to perform their functions. *Strategic planning processes* developed in the private sector have focused on both the *internal* and the *external* environments of an organization; the issues it needs to address; how it serves its *stakeholders* and *customers*; and how it formulates *goals* and *objectives* and puts into place *plans, programs* and *actions* to achieve those goals and objectives.

There are a number of reasons why institutions engage in strategic planning. It can help them to: (1) think strategically and develop effective strategies and plans; (2) clarify future direction; (3) establish priorities; (4) develop a coherent and definable basis for decision making; (5) exercise maximum discretion in areas under organizational control; (6) make decisions across levels and functions in the organization; (7) improve organizational performance; (8) deal effectively with rapidly changing circumstances; (9) build teamwork and expertise.

Increasingly governments, public agencies, nonprofit organizations, and communities face considerable difficulties and challenges in their various environments in an increasingly dynamic, turbulent, and competitive world. They are being held more and more accountable for their actions. As a result, they have been turning to strategic planning and management approaches to provide a "disciplined effort to produce fundamental decisions and actions that shape and guide what an or-

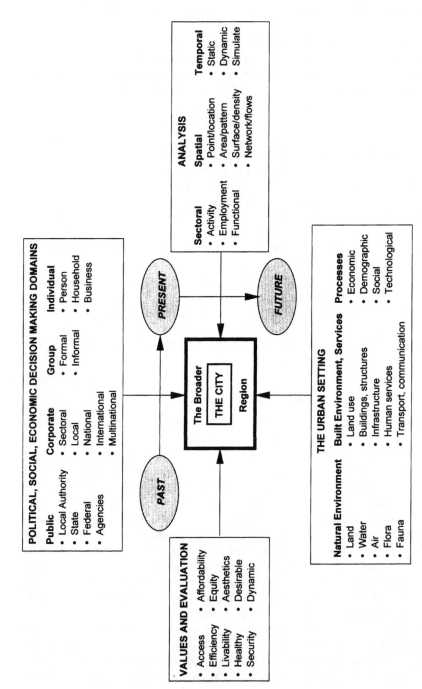

Figure 2.10. A spatial-analytic framework for evaluating the evolution and performance of a city.

ganization is, or what it does, and why it does it" (Bryson & Einsweiler, 1988:1). As a result, since the late 1970s there has been a rush of activity, particularly in the United States, by city administrations and state government agencies to use strategic planning methodologies to assist with the development, growth, and management of cities, regions, and states.

A distinction can be made between the private sector, corporate strategic planning approach and conventional public planning on five grounds (Bryson & Roering, 1988:38):

1. Corporate planning promotes broader and more diverse participation in the planning process.
2. Corporate planning places more emphasis on the community in its external context, determining the opportunities and threats to a community via the environmental scan.
3. Corporate planning embraces competitive behavior on the part of communities.
4. Corporate planning assesses a community's strengths and weaknesses in the context of opportunities and threats.

The key distinction in strategic planning is that strategies are action- and results-oriented, and as such, are *proactive* rather than *reactive*. Strategic planning embraces skills in facilitation, communication, analysis of secondary data, collection of new data, evaluation, and forecasting.

2.11.3   The Harvard SWOT Analysis

Since the 1920s, the Harvard Business School has developed models of corporate management and planning that evolved to help a firm to develop the best fit between itself and its environment. The *SWOT model* evolved which seeks to develop strategies by:

- Analyzing the *strengths* and *weaknesses* of the firm.
- Identifying the external *threats* and *opportunities* in the environment in which the firm operates.
- Identifying the *social* objectives of the firm.

It involves giving attention to the values of key stakeholders, including management and employees, suppliers, distributors, and customers. And it requires a mandate from management to be initiated and carried through.

The strategic planning process involves an organization in addressing four fundamental questions:

- Where are we going?: that is, the *mission*.
- How do we get there?: that is, the *strategies*.
- What is our blueprint for action?: that is, *budgets* and *programs*.
- How do we know if we are on track?: that is, *control* and *evaluation*.

The framework for the SWOT analysis method in strategic planning is illustrated in Figure 2.11. In the context of a public sector agency, such as the administration of a city, Bryson (1988) has shown how it can help decision makers to:

1. Think strategically and develop effective strategies.
2. Clarify future direction; establish priorities.
3. Make today's decisions in light of their future consequences.
4. Develop a coherent and defensible basis for decision-making.
5. Service maximum discretion in the areas under organizational control.
6. Make decisions across levels and functions.
7. Solve major organizational problems.
8. Improve organizational performance.
9. Deal effectively with rapidly changing circumstances.
10. Build teamwork and expertise.

Operationally, typically there are seven basic steps in strategic planning for a city (or a community) that relate to this SWOT model (Sorkin, Ferris, & Hudak, 1984). These are:

1. Scan the environment.
2. Select the issues.
3. Set mission statements or broad goals.
4. Undertake external and internal analysis.
5. Develop goals, objectives and strategies with respect to each issue.
6. Develop an implementation plan and carry out strategic citations.
7. Monitor, update and scan.

Figure 2.12 sets out the key phases and their sequence in the strategic planning process. It is important to explain that it is an on-going process.

There is an absolute requirement for both public and private involvement in the process of strategic planning. It is a technique that creates the opportunity for the development of powerful and lasting public–private partnerships. It can produce better coordination and concerted action within a community by enabling many different groups and programs to act on the same information. A shared vision can link individual efforts in the pursuit of consensus goals.

In the end, *implementation* is the key to strategic planning. It requires institutions, including governments, to focus on the allocation of scarce resources to critical issues. Relatively few cities have established a continuing strategic planning process, and instead the development of urban strategies have tended to be one of projects that are often considered complete after initial implementation without monitoring. Also, all too often governments do not even proceed into the implementation phase, and the strategic plans sit on the shelf collecting dust. In part this is because, by the very nature of the processes involved, they will bring into the open issues that involve potentially difficult political situations.

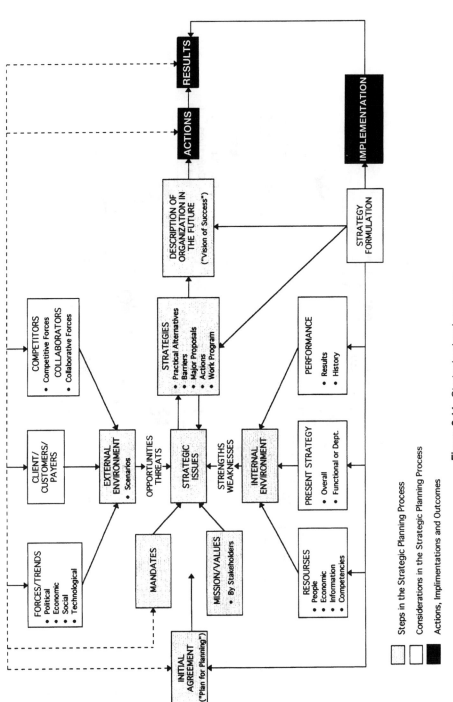

Figure 2.11. Strategic planning process.

Steps in the Strategic Planning Process

Considerations in the Strategic Planning Process

Actions, Implimentations and Outcomes

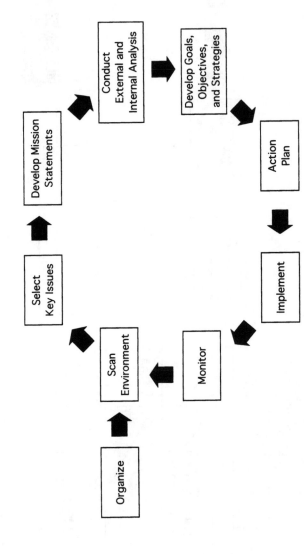

Figure 2.12. Strategic planning process: typical sequence of task.

2.11.4 Plans, Policies, and Projects

State and local governments as well as corporations tend to make three broad categories of decisions: those relating to *plans*, *policies*, and *projects*. In each category it is important to make decisions on the basis of the best information available. Today, environmental, economic, social, and cultural impact assessments are a critical part of this set of information. To elaborate on the decision-making activities of government, *plans* can vary from comprehensive state or community profiles to the specific functions of a governmental body (e.g., a water district). *Policies* are adopted to achieve definite goals. They range from broad guiding principles to detailed regulatory systems, zoning ordinances or policies limiting the acceptable land use in different areas of a community. Policies that regulate air or water quality constitute attempts to achieve standards that are said to influence the welfare of a community. *Projects* are special land use or development proposals initiated usually by private developers but sometimes by a government agency or a combination of both. Projects evolve in the context of local, state, or national policy and plans.

Plans do not exist as useful entities by themselves. Their role is to act as an instrument of policy and to help an administrative system cope with change. Both are needed for the guidance of executive decisions. Change often takes place in the context of controlled policy. Often, however, it takes place as part of a natural process of change in the economic or social system in which growth is occurring. For example, one of the larger cities in the United States, Houston, grew without a city plan. In general, however, when change reaches a critical threshold, communities and administrators no longer tolerate its unbridled action. Even in the absence of projects, specific plans, and policies, decision making can be effective and less capricious if it is consistent with a set of principles established for dealing with different classes of problems. These sets of principles are embedded in land use plans and governmental policies. They do not necessarily control executive decisions, but they can act as guidelines for making such decisions.

Executive decisions are based on facts and values. *Facts* are the objectively verified statements about the real world. For example, economic impact analyses are supposed to provide facts to decision makers about the costs and benefits of potential developments. *Values*, on the other hand, are the statements of preference concerning the nature of the real world and are the currency of politics. But decision makers' values usually have to be brought to bear on the facts so that an appropriate decision can be made.

Thus, policy incorporates factors such as complements and incompatibilities, thresholds and tradeoffs, and facts and beliefs:

1. *Complements* express essential relationships between variables in the system (e.g., capital borrowed must be repaid or buildings must be ventilated).
2. *Incompatibilities* are the converse of complements such as: there must be no frontage access to freeways ,or there must be no industrial development in residential areas.
3. *Thresholds* define a minimum of acceptable level of an occurrence such as a maximum building height or a minimum lane width on roads.

4. *Tradeoffs* define acceptable levels of exchange rates between variables such as the number of cash dollars in lieu of parking spaces in a planned commercial development.

In many situations, thresholds and tradeoffs are left to local governments or some local unit of decision making. Compatibilities and incompatibilities, facts and beliefs and values are often involved in *institutional* or government decision making. This division of labor provides great flexibility in allowing a policy to operate either as a stimulant or as a depressant to growth depending on what values are held by the appropriate decision-making units. Thus, a given area may use policy to stimulate growth by allowing new shopping centers to exist, whereas another area might use the same policy to protect its local environment and deny such a development plan.

2.12 Conclusion

In this chapter we have provided a broad overview of the nature of decision making related to spatial behavior. If one examines individual decision-making activity, then it is obvious that some decisions are made primarily for personal convenience, some for economic efficiency, some for equity considerations, some for taking advantage of specific production or distribution factors, and some may even appear to have no specifically designed unique rationale for their selection. Institutions tend to operate in a more complex milieu concerning the range of values and operational circumstances in which they make decisions, and this becomes more and more complex in an era of globalization and the multinational corporation.

Throughout this book, we will see examples of decision making at both the macro level and micro level scales. At the *macro* level we will consider the decision-making activities of national and state governments, institutions such as corporate and private capitalists, and institutions that are political or welfare oriented in their behavior. This latter area involves the development of policy and planning. The majority of the book, however, focuses on the *micro* level decision making of individuals and households when going about their daily lives. In both macro and micro level decision making, it is important to realize that this takes place within an operational environment that is dynamic. The operational milieu will be influenced by changes of an economic, social, environmental, and technological nature. It is these processes of change to which we turn in Chapter 3. In Chapter 4, the focus is directed toward the most typical operational environment for spatial behavior, that is, the large-scale urban environment known as the city.

3

The Big Picture: Processes of Economic, Technological, and Social Change

3.1 A Rapidly Changing World

We live in a rapidly changing and increasingly complex and competitive world. Our operational environment is the product of decisions made in the past and the present that are influenced by macro processes of economic, technological, and social change and that are the outcomes of a myriad of individual and institutional decisions. The processes of change, and the challenges they present, are of a nature and magnitude that are not always fully appreciated by policy makers, and our institutions of government often appear to be struggling because they are largely inappropriate to manage change in our urban-dominated society. Yet, the changes are pervasive, and over time, they have major impacts on the daily lives of many individuals and households and on the mode of operation of firms.

Processes of *globalization* and *internationalization* have had dramatic impacts on the nature of production, capital flows, trade, and passenger travel. When coupled with changes in *technology*, that have revolutionized transportation and communications systems, led to the development of new manufacturing processes through the application of microchip and new materials technology, and resulted in the diffusion of just-in-time inventory and delivery systems, the outcome has been a fundamental restructuring of economic systems over the last two to three decades. *Product cycles* have shortened; manufacturing processes and producer services functions have become more *knowledge based* and *information* and *capital intensive* and less *energy* and *labor intensive*.

Furthermore, we are experiencing increased mobility, new waves of international migration, changing patterns of internal migration, declining fertility, increased life expectancy, the aging of the population, the changing structure of households, the increasing participation of women in the labor force, changing gender relations, increasing mobility, and increasing community participation in public decision making. These represent fundamental *demographic* and *social changes* affecting decisions and the nature of behavior in modern urban society. In addition, deregulation of labor markets, financial markets, trading arrangements, and communications creates significant changes in *institutional environments* within which decisions are made.

These processes of change have created, and will continue to create, a dynamic operational milieu within which human behaviors occur and location decision making takes place at the institutional, corporate, small business, household, and individual levels.

Studies by geographers of trends and patterns of the distribution of population and economic activities have highlighted both processes of *concentration* and of *dispersal* and have identified the evolution of *hierarchies* across and within urban and regional spatial systems. In a recent review of population geography, White *et al.* (1989:266) noted that "the relative concentration or dispersal of population in settlements is the most obvious outcome of human behavior that population geographic theory should be able to explain." An empirical survey of the evolution of human settlements by Berry (1981) has revealed a continuous and inevitable concentration of population—initially in the form of the burgeoning population of agricultural civilizations having favored physical environments and effective organizational structures and, later, in the form of never-ending *urbanization* and *metropolization*, associated first with the industrial revolution and later with the rise of the services sectors.

As a *metaprinciple*, "whether humans are pursuing goals of greater production/consumption or social interaction, the larger the concentration, the greater the value of output and interaction" (White et al., 1989:266). The incredible innovations that have occurred in technologies in transportation and communications continue to "shrink the world" and make it possible for ever-greater concentrations of populations to occur on a global scale through the process of *urbanization*. Processes of an economic, social, and political nature also create dispersal of populations. The major forces of dispersal appear to be within major metropolitan regions, as evidenced by *suburbanization*. Directional flows of population through internal-migration movements also are bolstering existing concentrations of populations at various levels in the human settlement hierarchy of nations. New urbanized concentrations of less structured form are emerging in some regions and especially in high-amenity, "sun-belt" locations that attract people for reasons to do with *lifestyle* and *retirement*.

Thus, there are complex and sometimes seemingly conflicting processes at work generating both concentration and dispersal of populations. It is the net balance between in-flows and out-flows for a region—at whatever spatial scale it is defined—that is an increasingly significant component determining its patterns of growth and decline in population. And in this context, patterns of immigration have an additional crucial impact on the geography of population.

New and dynamic metropolitan city regional spatial forms and structures are emerging in which the suburbs are increasingly dominant in modern service-based, information-intensive urban economies. Within cities, there are marked patterns of differentiation in the economic and social fortunes of citizens marked by dualism and segregation.

It is the cities that are now the nexus of the economic system. They are the places that are impacted most by economic, demographic, and technological change. And it is the cities that provide the operational milieu for most human decision making and spatial behavior. The majority of people in the developed world are urban, and a rapidly increasing proportion of people in the developing world also live in cities.

This chapter is about the big picture. It outlines many of the processes of change that are restructuring economies and societies at a *macro* level and the changes these processes of change are producing in the nature and structure of society. It is important to have an appreciation of these processes of change and their impacts as they affect the social frameworks and the built environments in which human spatial behaviors take place. These processes of change pose important questions as to the relevance and validity of some of the theories and assumptions in the models that were outlined in Chapter 2. The latter have formed the basis for specialized theories about the economic and social structure and functioning of cities and regions within which behavior and decision making occur. Such environments are themselves the outcomes of human behaviors and decisions.

The chapter also looks at the major societal processes of change that are occurring that are predominantly of a demographic nature. Chapter 4 looks further at these changes specifically in the context of cities.

3.2 Globalization and the Processes of Internationalization

The last few decades have witnessed incredible changes in the nature of the world's economy. Post–World-War-II reconstruction led to the emergence of West Germany and Japan as new economic giants in an era of a cold-war strategic balance dominated by the United States and the former USSR. *Transnational corporations* (TNCs) rose to positions of dominance. *Production processes* increasingly became internationalized. The focus and concentration of trade shifted from the Atlantic to the Pacific. Within the Pacific Rim region, the newly industrialized countries (NICs) emerged with high rates of economic growth and positive balance of payments—especially Taiwan, South Korea, and the city states of Singapore and Hong Kong—the so called "little tigers." The international *financial system* has been deregulated, particularly following the collapse of the Bretton Woods agreement in the early 1970s. Commercial banks have created huge financial empires spanning the world, and capital markets have become truly international. The OPEC-induced oil crises of the 1970s spurred the development, adoption, and diffusion of energy-saving technologies, processes, and policies. Technological revolutions have occurred in transportation and telecommunications, and knowledge-based, information-intensive activities, especially in the *elaborately transformed manufacturers* (ETMs) and *producer services* industries, have become the new growth and high value-added activities. *Tourism* has boomed in the age of cheap intercontinental air travel made possible by the jumbo jet. Many of the old industrial nations, including the United Kingdom and the United States, have lost their competitive edge to become debtor nations. And more recently the Eastern European, communist-bloc nations have fragmented, abandoning socialism and central planning in favor of a market economy. Meanwhile, vast areas of Africa and parts of Latin America remain ravaged by famine and disease and are making little headway in economic and social development, often under oppressive, dictatorial, and brutal political/military regimes.

Globalization and processes of internationalization have altered not only the structure of economies, but they have also changed the relationships between cities and regions by increasing the economic strengths of metropolitan systems globally, and in so doing, they have uncoupled cities from their domestic base. No longer do manufacturing functions require control by one firm located in a single geographic area, and the demand for central nodal production space is not as strong as it once was. Decentralization of activities has been permitted by new technologies that have enabled production processes to be globalized. Now it is the information technologies that are crucial, and, as a result, the agglomeration demands of central nodes are less strong for many activities. The inner-city core regions are readapting and readopting to new uses in response to both internal and external demands for retail, cultural, and government services required to serve the needs of rapidly changing metropolitan demographic structures. In addition, metropolitan regions, other cities and towns, and even rural areas, are reshaping their economic options by using land assets to develop new enterprises that are not necessarily associated with their natural resource base.

According to Knight (1989), we now live in an era in which cities and city regions are: "the nexus of the emerging global society . . . now that development is being drawn more by globalization than by nationalization, the role of cities is increasing. Power comes from global economies that are realized by integrating national economies into the global economy, and cities provide the strategic linkage functions" (p. 327).

This represents a fundamental change in the context within which decision making occurs in locating activities in space. This contextual change also affects the flows of goods, people, information, and ideas.

3.2.1 The Emerging Multipolar Global Economy

At the macro level, the patterns of production, capital flows, and trade are much more complicated today than was the case a few decades ago. The processes of globalization and internationalization have resulted in a much greater degree of *interconnectedness* between economies (Dicken, 1992:16). This has occurred largely through the growth in exports, which have accelerated since 1960 at a much faster rate than the growth in output, which itself has increased rapidly. By the late 1980s, exports were over four times greater than they were in 1960, whereas output was only three times greater. This widening gap has expressed itself in a large increase in the volume of world trade, and the growth in output has been financed by a massive increase in the volume of capital flows between nations. This booming international trade is in both manufactured goods and producer services and is a "reflection of the increasingly close interconnections at a global scale between production and circulation activities" (Dicken, 1992:44).

Three major regions dominate global production and trade, as shown in Figure 3.1 These are North America, the European Community, and East and Southeast

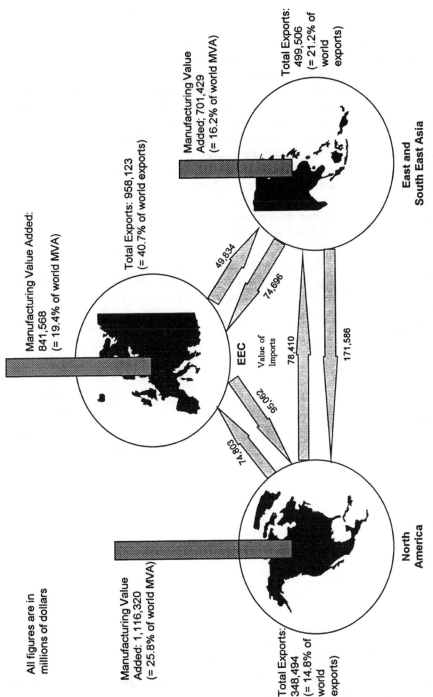

All figures are in millions of dollars

Manufacturing Value Added: 841,568 (= 19.4% of world MVA)

Total Exports: 958,123 (= 40.7% of world exports)

Manufacturing Value Added; 701,429 (= 16.2% of world MVA)

Total Exports: 499,506 (= 21.2% of world exports)

East and South East Asia

EEC

Value of Imports

49,834

74,696

78,410

171,586

95,062

74,803

Manufacturing Value Added: 1,116,320 (= 25.8% of world MVA)

Total Exports: 348,494 (= 14.8% of world exports)

North America

Figure 3.1. A multipolar global economy: the "triad" of economic power. *Source:* Dicken, 1992:45.

Asia, which generate almost 80% of the world's exports and over 60% of the world's manufacturing. These are "the 'mega markets' of today's global economy" (Dicken, 1992:45).

This triad of economic power in the world emphasizes the unevenness of the extent of shifts that have been occurring in the world's economy, and only a relatively small number of advanced and developing countries have experienced the benefits of economic growth. Although it is common to talk of trickle-down effects, they are relatively limited, with this triad of regional blocks dominating trade and capital flows to the exclusion of large components of the developing world.

It is evident also that the growth in world trade has been associated with a phenomenal increase in the volume of *foreign direct investment* (FDI). As illustrated in Figure 3.2, this has been dominated by the United States, Canada, the United Kingdom, France, and West Germany, with Japan having emerged as a major player since the 1970s. By the mid-1980s, the combined share of the United States and the United Kingdom had been reduced to under half compared to two-thirds in 1960. In the same period, Japan had captured almost 12% of total FDI. It was also significant that the NICs of Asia (South Korea, Taiwan, Singapore, and Hong Kong) had emerged as significant providers of FDI, and in particular to nations in the Pacific Rim region. In recent years, much of the FDI has been directed to the property and services sectors in the advanced economies, as well as to the manufacturing, property, and business sectors in developing economies. By way of example, Figure 3.3 illustrates the magnitude of Japanese FDI in the property sector of some U.S. cities in the late 1980s.

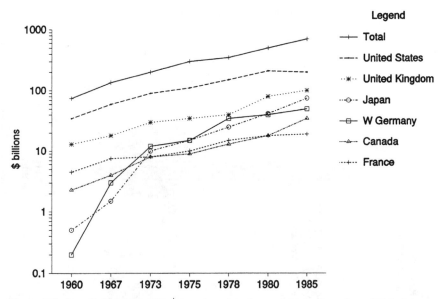

Figure 3.2. Growth in foreign direct investment in the world economy since 1960. *Source:* UNCTC publications; Dicken, 1992:53.

Black bars show the cumulative percentages of Japanese investment through 1989. Screened bars show the relative percentages for the year 1989 alone.

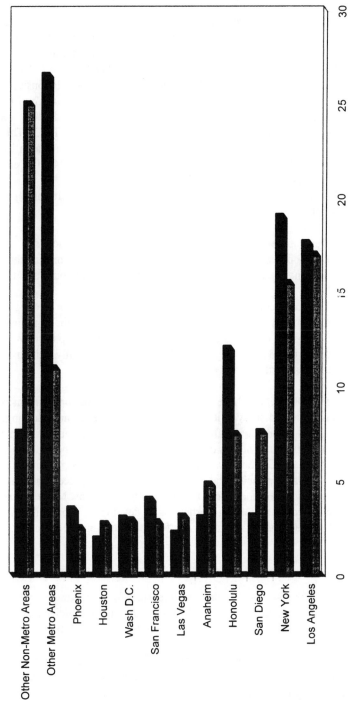

Figure 3.3. Japanese direct foreign investment in property in selected United States cities. *Source:* Kenneth Levanthel & Co.

3.2.2 The New International Division of Labor

The shift in production from the industrialized economies of what used to be referred to as the *core* (North America and Western Europe) to the economies of the so-called *global periphery* has been occurring for many decades. In 1980, Fröbel, Heinrichs, and Kreye's book, *The New International Division of Labor* (NIDL), was published in English, and researchers began to discuss the relationships among the internationalization of circuits of capital, the interconnected nature of finance, production, and commodity trade, and the shift in industrial production from the *core* to the *periphery*, all in the context of the emerging role of the TNCs. The NIDL thesis has several variants, but "all assign the major role to the search by TNCs for cheap, controllable labor at a global scale" (Dicken, 1992:124). The emphasis was seen to be on the minimization of labor costs.

However, as outlined by Fröbel et al. (1980), there were three preconditions for the NIDL:

1. Developments in transport and communication technology.
2. Developments in production process technology permitting the fragmentation and the standardization of specific tasks that could use nonskilled labor.
3. The emergence of a world reservoir of potential labor power in the nations of Asia and Latin America in particular. TNCs sought to minimize their labor cost by off-shoring production functions to cheap third-world locations in a drive to increase profits. This occurred especially in industries such as textiles, clothing, and electronics.

The NIDL theories have been seen to be somewhat one-dimensional. Elson (1988) pointed out that, even in industries such as textiles and clothing, the availability of cheap labor was not always the overwhelming consideration in off-shoring production. Rather it is necessary to view this process as having to do with strategies by TNCs to develop new markets. In addition, governments in developing economies initiated policies that provided considerable enticements and incentives for TNCs to establish production processes in their nations. Many *enterprise zones* were proclaimed by national governments in developing economies as vehicles to attract off-shore activities of enterprises from the developed economies.

3.2.3 The Transnational Corporation

It is the TNCs that are the "single most important force creating global shifts in economic activity" (Dicken, 1992:47). TNCs are the major vehicles that channel FDI. In addition, an increasing proportion of world trade is *intrafirm* rather than *international* in nature.

The largest TNCs control economic activities in many countries; they take advantage of geographic differences between nations and regions in factor endowment, including labor and government policies; and they have considerable geographic flexibility in their activities to shift resources and operations between locations at a global scale. It is the operations of TNCs that are reshaping much of the new global economic systems through their use of advanced telecommunications and transportation technologies and *just-in-time* (JIT) delivery systems. These innovations enable TNCs to disperse and decentralize functions such as design, production, and marketing across many locations through the use of complex organizational systems.

The rise of the TNCs was considerable in the first half of the 20th century, but their growth has been spectacular since the 1950s, and even more since the 1970s. By the mid-1980s, there were over 600 TNCs in the mining and manufacturing sectors, each with annual sales in excess of U.S. $1 billion, and only a little more than 70 of them accounted for over 50% of total sales. Although they had penetrated most areas of economic activity, TNCs seemed to be concentrated largely in four broad types of manufacturing industries and services: (1) technologically more advanced sectors (e.g., pharmaceuticals, computers, scientific investments, electronics, synthetic fibers); (2) large volume, medium technology consumer goods industries (e.g., motor vehicles, tires, televisions, refrigerators); (3) mass-production consumer goods industries supplying branded products (e.g., cigarettes, soft drinks, toilet preparations, breakfast foods); and (4) producer services involved in trading (e.g., banking, finance, insurance, accounting, legal services, real estate) (Dicken, 1992:58).

Once it was companies from the United States, the United Kingdom, Germany, the Netherlands, France, Italy, and Switzerland that dominated the TNCs, but by the late 1980s, Japanese TNCs were among the leading TNCs, and South Korean companies also have now entered the fold.

In 1950 only three of the 315 largest TNCs had manufacturing branches or subsidiaries in more than 20 countries; but by 1975, 44 TNCs from the United States alone had this level of geographic spread. In 1950, 138 out of the 180 largest TNCs in the United States, and 29% of TNCs in the United Kingdom had branches or subsidiaries in only six countries, but by the 1970s, in excess of 60% of TNCs operated in more than six countries. However, "much of the increase in transnationality during the 1980s seems to have been due to the rapid expansion of smaller and medium-sized TNCs, whereas those giant corporations which had already developed globally extensive networks were inclined to consolidate their existing operations" (Dicken, 1992:51). TNCs tend to develop because it is beneficial for them to utilize those major circuits of money, production, and commodity capital by operating them concurrently.

Dunning (1980) proposed that three fundamental principles drawn from the theory of the firm in economics—organization theory, trade theory, and location theory—are crucial for a firm deciding to operate as a TNC. He saw firms engaging in international production when each of the following conditions are met:

1. A firm has action-specific advantages not possessed by competing firms of other nationalities—an *ownership-specific advantage*.
2. The advantages are most suitably exploited by the firm itself rather than by selling or leaving them to other firms—the firm *internalizes the use of its ownership-specific advantage*.

3. It must be profitable for the firm to exploit its assets overseas, rather than in domestic locations—*location-specific factors* play an important part in combination with the internationalization of ownership-specific advantages in determining whether or not, and where, production overseas occurs.

Dicken (1992:192–193) has shown that TNCs tend to display four common organizational structures, and the one chosen depends on the age and experience of the enterprise, the nature of its operations, the degree of product and geographic diversity. These structures are shown in Figure 3.4.

3.3 Technological Change

Technology is an *enabling* agent as well as *imperative* for many economic activities in the global economy. Choices and application of technology are influenced by the

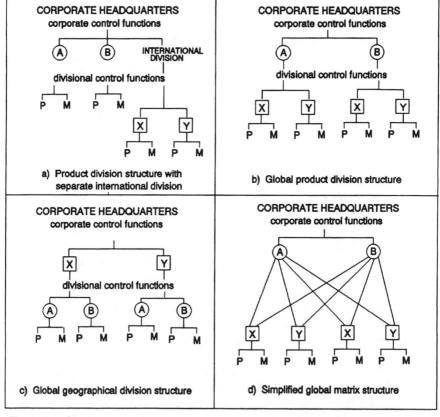

Figure 3.4. Some common types of organizational structures in transnational corporations. *Source*: Dicken, 1992:193.

drive by enterprises for profit, capital accumulation, investment, and market share. And technology makes possible new organizational structures, new geographic arrangements of economic activities, new products, and new processes (Dicken, 1992:98).

According to Cohen and Zysman (1987), technology change and new technologies provide a "frontier of new possibilities." Specific choices within the frontier of technological possibilities "are not the product of technological change; they are, rather, the product of those who make the choices within the frontier of possibilities. Technology does not drive choice; choice drives technology" (p. 183).

Much technology change passes unnoticed as it consists of small-scale, progressive modification of products and production inputs. This is *incremental innovation*. However, single radical innovations in technology have profound and widespread impacts on economic systems.

Freeman (1987) has referred to changes of the *technology system* as being the most crucial, and they generate totally new industries. He has proposed four generic technologies that have created new technological systems that are significant in the current context of globalization and internationalization of economic activity: (1) information technology; (2) biotechnology; (3) energy technology; and (4) space technology. In addition, Freeman has referred to changes in the *techno-economic paradigm*, that are the truly large-scale and revolutionary changes. Schumpeter's long-wave theory encompassed this concept. The changes are so pervasive in their impacts that they revolutionize the whole style of production and management throughout the economic system.

3.3.1 Long Waves of Innovation and Economic Growth

The notion of long waves of creativity in global economic growth has been developed following the ideas in the 1920s of the Russian economist Kondratiev. Since 1780, four completed *K-waves* have been identified, each lasting about 50 years; and each has within it phases of prosperity, recession, depression, and recovery. Each of the four K-waves has been associated with significant technological innovations and their diffusion, which have affected the nature of production, distribution, and organization. The four K-waves are illustrated in Figure 3.5. They are:

1. *First K*: Steam power, cotton textiles and iron; the leaders were Britain, France, and Belgium.
2. *Second K*: Railways, iron, and steel; the leaders were Britain, France, Belgium, Germany, and the United States.
3. *Third K*: Electricity, chemicals, automobiles; the leaders were Germany and the United States, plus other European nations and Britain.
4. *Fourth K*: electronics, synthetic materials, petrochemicals; the leaders were North America, Western Europe and Britain, and Japan.

The *fifth K*-wave, of which we are currently in the early phase, is being labeled the *information technology* (IT) wave, and it is especially prominent in North America, Western Europe, Japan, and some of the NICs of East and South East

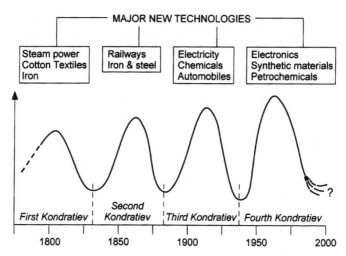

Figure 3.5. Long waves of economic activity and their associated new technologies. *Source*: Dicken, 1992:99.

Asia. It is not IT per se that is new; what is new is IT using microelectronic technologies in computing, robotics, office equipment, telecommunications, and virtual reality. Hall and Preston (1988) and Freeman (1987) have coined the term *convergent IT*, whereby the distinct technologies of computers and telecommunications are integrated or converged into a single system of information processing and exchange. This convergence has been integrated progressively since the 1960s.

3.3.2   The Shrinking of Distance

In a geographic context, the impact of technological advances may be illustrated dramatically by how *innovations in transportation* have *shrunk distances* (McHale, 1969; Dicken, 1992).

Figure 3.6 shows the effect of changing transportation technology on *real* distance over the period from 1500 to the 1960s. Until the modern use of steam locomotives and steamships, movement was slow, and transportation speed and efficiency were low. Steam power provided the crucial breakthrough from 1850, resulting in a telescoping of distance at the global scale, enabling a much enlarged scale of human activity, enhancing dramatic increase in the flow of materials and goods, and stimulating geographic specialization. In the 1950s, it was the propeller aircraft, and later in the 1960s and since it has been the jet aircraft, container ships, and superfreighters, that have provided an accelerated global shrinkage in terms of distance and cost of trade and travel. This was "the era of rapid take-off in the role and pervasive diffusion of the transnational corporation" (Dicken, 1992:103–105).

Since the 1960s, innovations in computer and telecommunications technology have shrunk time-distance to virtually zero, as the world has been linked through satellites and fiber-optic cable networks enabling information and capital to be

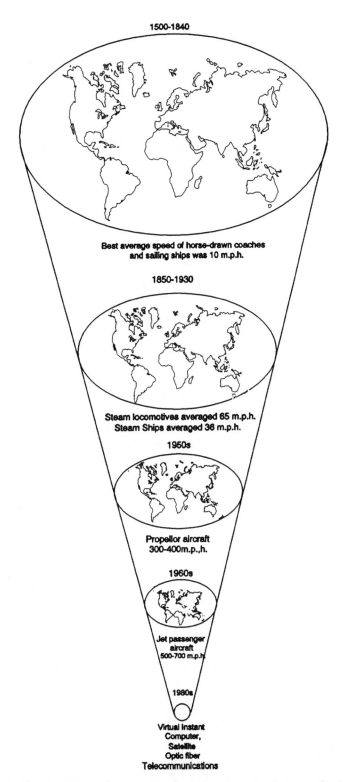

Figure 3.6. The shrinking of distance: the effect of major technologies on "real distance."
Source: McHale, 1969; Dicken, 1992.

transmitted instantaneously around the world. "The new telecommunications technologies are the electronic highways of the information age, equivalent to the role played by railway systems in the process of industrialization" (Henderson & Castells, 1987:6). As we embark on the age of the *information superhighway*, providing multimedia modes on the one network and potentially to locations anywhere in the world, we enter a new era where these technologies are increasingly essential to the survival of existing economic activities and are rapidly developing new activities that are concerned primarily to collect, transform, and transmit information, which is the "lifeblood of the burgeoning producer services sectors of economies" (Dicken, 1992:105).

3.4   Implications of Globalization for Location Theory

The new global economy, and the role of the TNCs, have implications for traditional theories about the location of economic activities. In the discussion on classic location theory in Chapter 2, we saw how the concern was to incorporate space into economic theory to explain the location of economic activities by considering the costs of factor inputs to production and the firm's market area. In terms of spatial location, the focus was on agglomeration and deconcentration or dispersal.

The internationalization of capital has meant that it is no longer appropriate to equate differences in returns on investment in an economic activity to geographic variations in factors such as interest rates on financial investments. This holds even when such variations have been theorized to have an important affect on portfolio investment. But FDI in an era of increasingly deregulated finance means that capital can flow around the globe largely unrestricted.

While traditional explanations of patterns and trade and the location of production and investment are "clearly inadequate to explain the complexities of the modern global economy . . . nevertheless . . . they should not yet be rejected out of hand" (Dicken, 1992:95). This is because: (1) *Comparative advantage* based on different factor endowments remains important; (2) *geographic distance* reflected in movement cost does influence the spatial patterns of economic activities at the global scale; (3) *differential labor costs;* and (4) *economies of spatial agglomeration,* all of which have significant influences on global shifts in economic activity.

These were all central considerations in classical location theories. Although they continue to have relevance, they are static. As a result, they have limitations in a dynamic and volatile real world that is characterized by the new global economic order and the modern era of rapid technological innovation in which decision making by firms takes place. This should involve a consideration of quite complex interactions between the domestic and the foreign components of a firm as well as a recognition of the reality that corporations have developed wide networks of often locationally dispersed suppliers and that they serve locationally dispersed markets. In addition, there are differences in government regulations, taxation policies and the like, at the national, state, and local levels that have to be taken into account. They can create discontinuities across administrative boundaries. Increasingly, the institutional and organizational factors are the significant considerations in the loca-

tion decisions of firms. "Comparative advantage is not simply 'given'; it is created
and re-created by human action" (Dicken, 1992:95).

In the context of globalization and internationalization, we are faced with a
highly complex and interconnected set of processes in which there are many possible solutions for firms making a locational choice. For the TNC, the interaction of
these factors and its corporate competitive strategy have to be placed in a total context of a networked organization. It involves a consideration of:

- Aspects of the technological environment relevant to globalization and the
  international operational milieu of the TNC.
- Why TNCs are involved in the international production of goods and services.
- The role of the nation state in the global economy, and in particular those
  aspects of political decision making that impinge most directly on the
  processes of globalization.
- The nature of network relationships between firms.

3.4.1 Basic Elements in the Production Process

Smith (1981) has conceptualized the elements in the production process of a firm as
shown in Figure 3.7. These key decision areas concern the *technique, scale*, and *location* of *production. Technology* is a critical factor in how the *factor inputs* in production are combined in *linking* operations across locations and in the allocation inputs and outputs between locations.

A further factor that is important to firms is the tendency for *product cycles*
to compress, requiring flexibility and rapid response to changing fashions, markets,

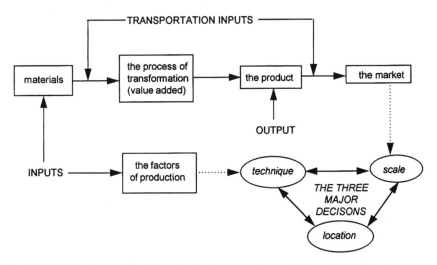

Figure 3.7. Basic elements in the production process. *Source*: Dicken, 1992:113 after Smith,
1981, Fig. 2.1.

and technology. Capital intensity is crucial to the firm, especially at the design, innovation and growth stages of product development and supply. Administrative and marketing skills, research and development (R&D), and production factor efficiency all are crucial in the *globally linked firm.*

Increasingly firms are adopting what is known as flexible production in the current *post-Fordist* era of manufacturing. *Information* and *organization* have become the key elements in the operation of both the unit of production (a factory) and in mobilizing the relationships between customers and suppliers of the firm through the use of JIT inventory and delivery systems. Bessant and Haywood (1988) have referred to increasing flexibility in the production process as altering the relationship between scale and cost of production, allowing smaller production runs, increasing product variety, and changing the way production and labor inputs are organized. This flexibility may be more organizational in nature than technological for most firms.

3.4.2 Strategic Alliances

The emergence of linked production systems in the global economy has made it extremely important for firms to develop what are referred to as *strategic alliances* both locally and nationally, as well as internationally. Such alliances enable an individual firm to do what it could not achieve by itself. They do not have to involve mergers or vertical integration, which are used to create many TNCs; rather, networks of independent firms are formed for the purposes of design, R&D, supply, production, marketing, and distribution. Strategic alliances are increasingly common for firms in both the manufacturing and the producer services sectors of the economy. Often they cut across both these economic functions for the purpose of undertaking joint ventures and to form *functionally specific competitive alliances.* Of course TNCs are active in forming strategic alliances, and increasingly, we see them created in activities such as international airlines, the motor vehicle industry, computing, telecommunications, electronics, and in the legal, accounting, and business services.

Strategic alliances have also developed through public–private *partnerships,* including private enterprises, public trading enterprises, and government agencies. Often strategic alliances are formed to circumvent government regulations that prohibit cartels and seek to protect domestic markets for local firms. These strategic alliances go beyond international subcontracting, which has become increasingly common in the global economy.

A special form of behavior networking has emerged as an important operational component of today's TNCs. Miles and Snow (1986) have referred to this as the *dynamic network.* This involves designers, producers, distributors, and suppliers linked through brokers. Companies such as Nike, the footwear manufacturer, have developed large-scale vertically disintegrated functions and a high level of subcontracting by firms performing specialist roles (Donaghu & Barff, 1990):

> Nike's "in-house" production may be thought of as production from its exclusive partners. Nike develops and produces all high-end products with exclusive partners, while

volume producers manufacture more standardized footwear that experience larger fluctuations in demand . . . Nike acts as the production coordinator, and three categories of primary production alliance form the first tier of subcontractors. A second tier of material and component subcontractors supports production in the first tier . . . [all] . . . production takes place in South East Asia while the headquarters in Beaverton, Oregon, houses Nike's research facilities. (p. 544)

3.4.3 Global Networks and Global Concentrations

As a result of IT, container sea freight, and jet transportation, which have facilitated the internationalized production processes in the new global economy, increasingly firms are dependent on *networks* to operate.

In the transportation and telecommunications industries in particular, global and regional hub and spoke networks predominate. This is also the case for most TNCs and large national corporations. Work by Langdale (1989) shows how TNCs are major users of international-based telecommunications networks that permit them to transmit information instantaneously to all parts of their corporate structure and to their suppliers and markets. Figure 3.8 illustrates the types of global networks that TNCs might evolve, ranging from truly global networks with numerous national hubs, to regional hub and spoke networks, to centralized networks.

In the international shipping industry, major container services are provided by consortia to serve both global and regionally linked markets. The case of the NYK Line from Japan is illustrated in Table 3.1.

The hub and spoke network models are typical of the operations of national airlines. Internationally, airlines and shipping lines are developing global networks that link major *mega cities* that are focal points in what are rapidly emerging urban development corridors. Rimmer (1989) has illustrated this with respect to international air linkages in the Pacific Rim region and the growth of urban mega-agglomerations, as shown in Figure 3.9.

TABLE 3.1.
Major Container Shipping Services Provided by Consortia Involving the NYK Line of Japan, 1993

Service	Consortium	Loops	Vessels
Semi-round the world	NYK/Hapag-Lloyd/Neptune Orient Line	2	12
Asia–Europe	NYK/Hapag-Lloyd/Neptune Orient Line	2	18
Asia–Mediterranean	NYK/Nitsui/Lloyd Triestino	1	9
U.S. east coast/Gulf/Europe	NYK/Hapag-Lloyd/Neptune Orient Line + Atlantic Container Line	1	9
Asia–U.S. west coast	NYK/Neptune Orient Line	4	21
Asia–U.S. east coast	NYK/Neptune Orient Line	1	6

Source: Tanaka, 1993:24.

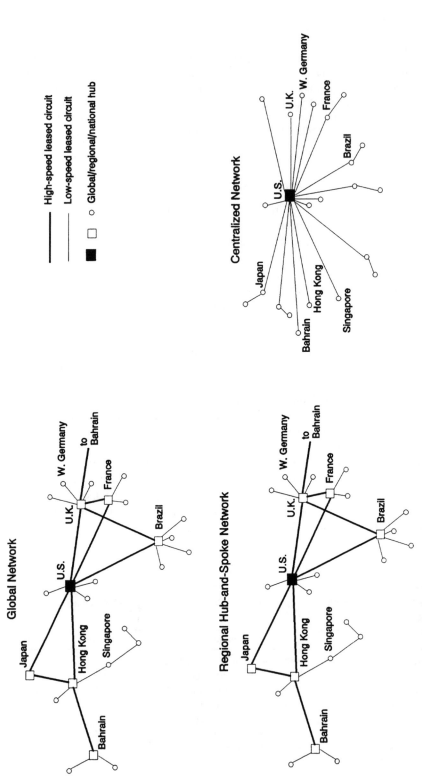

Figure 3.8. Major types of international leased telecommunications networks for United States TNCs. *Source:* Langdale, 1989: Fig. 1.

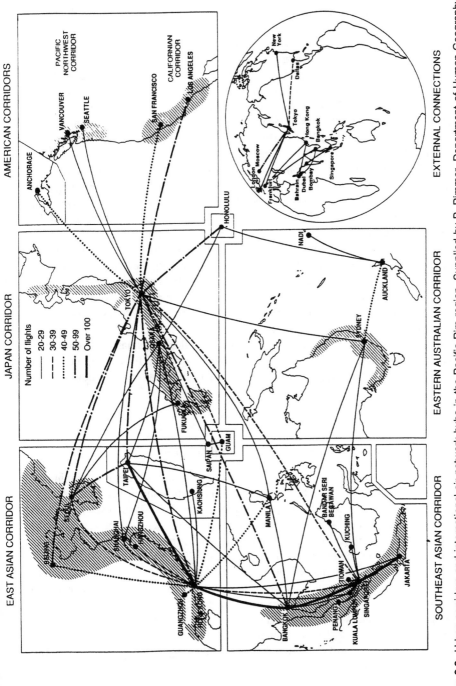

Figure 3.9. Urban corridors and international air network hubs in the Pacific Rim region. Supplied by P. Rimmer, Department of Human Geography, Research School of Pacific Studies, Australian National University.

89

3.4.4 World Cities

Only a relatively few metropolitan areas have emerged as *world cities*, though the number has increased over the last decade or so, especially in the Asia–Pacific regions. They are the centers of concentration of headquarters of large corporations. These cities have become the "geographical control points" (Dicken, 1992:197) of the global economy. Three global cities stand out—New York, Tokyo, and London—and below them are a significant and increasing number of key cities within the North American, Western European, and Asian *triad* of regions that are the focus of trade and capital flows in the world economy, with outliers in South America and Australia. This network of world cities of major concentration of corporate and regional headquarters is illustrated in Figure 3.10.

There is a *time–space convergence*, particularly between the key nodes in the global communications and production networks, which link these increasingly important places. As a result, there is a reinforcing effect that enhances and increases the importance of the world cities that are nodes in those global networks, and they become more and more places where activities agglomerate.

Hall (1966) was one of the first researchers to write about world cities, and he produced an interesting list of attributes that they tend to exhibit. This is shown in Table 3.2. The need for global transactions is central and requires the provision of advanced infrastructure in communications and transportation systems, including international airports. In addition, infrastructure crucial to the success of a world city includes educational and cultural facilities and services of a world-class standard. Emphasis is placed on the provision of flexible, smart, and elite infrastructure and services generating new dimensions of place formation, incorporating attributes of amenity, and providing access to information and knowledge resources.

3.4.5 A Model of Internationalized Operation for a Firm

Dicken (1992) has proposed a general model, shown in Figure 3.11, outlining the modes of foreign investment and organizational relationships that typify flexible production systems and flexible specialization in the TNC. It involves "extremely complex internal and external transactions and relationships . . . and . . . where the major modes of foreign involvement are related to the more important contributory influences, both economic and political" (pp. 221–223). These factors are the main influences on organizational arrangements and locational decisions. The continual state of flux of network relationships makes these relationships extremely complex and dynamic. It is difficult, therefore, to separate out the organizational from the geographic in this, and it is best to treat the location of activities in space as the interaction of strategic alliances, subcontracting relationships, and organizational forms with geographic networks structured around linked agglomerations or concentrations of activities (Walker, 1988; Dicken, 1992).

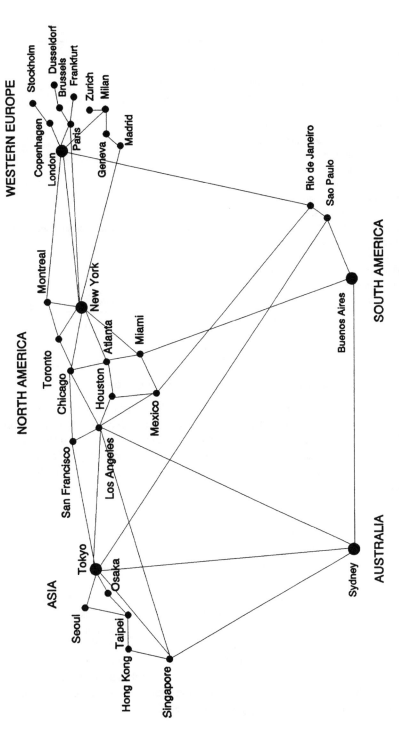

Figure 3.10. Major global nodes of concentrations of corporate and regional headquarters of transnational corporations. *Source:* Dicken, 1992:198.

TABLE 3.2.
The Typical Attributes of World Cities

Major centers of political power

Seats of national and international governmental and nongovernmental agencies

Major headquarters of transnational corporations

Headquarters of major professional organizations and trade unions

Major centers of industrial activity

Major railroad, highway, port, and airline centers

Headquarters of major banks and insurance companies

Major medical and legal centers

Headquarters of national courts of justice

Major universities and research institutes

Major libraries, museums, and churches

Major theaters, opera halls, restaurants, etc.

Headquarters of information dissemination—publishers, advertisers, radio, television, and satellite data

Large populations and an international labor force

Increasing share of employees in services as opposed to industry

Centers of very specialized goods and services

International markets

Sites of global conferences: governments, industries, and volunteer organizations

---

Source: Hall, 1966:7–9.

3.5   Structural Economic Change within Economies

Since the 1970s, the economies of nations throughout the world have been transformed by economic restructuring that has generated massive shifts in the levels and the distribution of employment across industry sectors as a result of the impacts of the processes of globalization, internationalization, and technological change discussed above. Basically, what we have seen is the shift in employment from the manufacturing to the services sectors of the economy, plus the accompanying social phenomenon of increasing participation of women in the labor force. In addition, in most advanced western nations, the phenomenon of "full" employment has disappeared, and it would seem that between 5 and 10% permanent unemployment now is regarded as being the norm. This is an outcome both of structural unemployment and of high levels of youth unemployment with the highest incidence often found among specific racial and ethnic groups.

The impacts of structural economic change related to the processes discussed above have resulted in a dramatic change over the past two decades in the structure and distribution of employment across industry sectors and between public and private sectors of the economy. There are commonalities between advanced postindustrial economies of the developed world, and within the developing nations, and especially in the NICs, there are similar trends occurring as employment shifts increasingly to the services sectors of the economy. Some case studies from a number of nations will illustrate the magnitude of these changes.

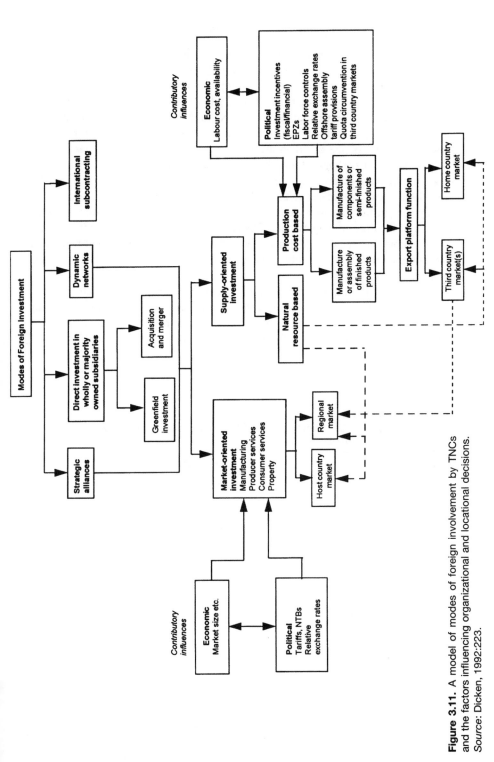

Figure 3.11. A model of modes of foreign involvement by TNCs and the factors influencing organizational and locational decisions. *Source:* Dicken, 1992:223.

3.5.1 Changes in Employment in the United States Since 1970

In the United States during the 1980s, over 19 million new jobs were added to the nation's employment base, and the vast majority were located within the larger urban system. Between 1970 and 1990, the proportion of the nation's employment located in urban areas in fact increased from 79 to 81%. There have been significant impacts of economic restructuring with respect to shifting shares of employment by industry sectors, public and private sector shares, and the pattern of employment by sector according to size of urban areas. These data are shown in Tables 3.3–3.5. Hicks and Rees (1993) have drawn the generalizations that are summarized below.

Employment change by sector and industry group shows that, in the 1980–1990 decade compared to the 1970–1980 decade, major reversals occurred in employment in the goods-producing manufacturing sector and the mining sector, with major declines in employment. The areas of rapid growth in jobs were in the services-providing sector, and especially in the finance, insurance, and real estate (FIRE) sector and in community services, while growth in retail and construction remained strong.

There has been a major shift in jobs from the public to the private sector in the 1980s, and although much of this reflected increased levels of federal spending in the defense sector, employment in the private sector reflects the natural location of jobs rather than the source of demand responsible for creating and sustaining jobs. But public sector employment dropped from 18.2% in 1970 to 15.2% in 1990,

TABLE 3.3.
U.S. Total Employment Change Patterns by Sector and Industry
Group: 1970–80 and 1980–90

	Employment Change (%)	
	1970–80	1980–90
U.S. Total	25.1	22.2
Private Sector	27.3	24.1
Public Sector	15.2	12.5
Goods-Producing	11.4	1.1
Services -Providing	32.1	31.1
Manufacturing	5.5	-4.9
Construction	28.3	28.2
Mining	71.8	-22.6
FIRE	53.8	34.8
Trans/Public Util.	16.4	16.3
Retail	30.7	27.5
Wholesale	37.7	15.5
Services	46.8	54.0
Government	15.2	12.5

Source: Hicks and Rees, 1993:2.65.

TABLE 3.4.
Metro Regional Employment Shares by Sector 1970, 1980, and 1990

	U.S. Total Employment (%)		
	1970	1980	1990
U.S. Total	100.0	100.0	100.0
Private Sector	81.8	83.3	84.6
Public Sector	18.2	16.7	15.4
Goods-Producing	27.6	24.6	20.3
Services-Providing	67.4	71.2	76.4
Manufacturing	21.9	18.5	14.4
Construction	4.9	5.0	5.3
Mining	0.8	1.1	0.7
FIRE	5.5	6.7	7.4
Trans/Public Util.	5.4	5.0	4.8
Retail	15.2	15.9	16.6
Wholesale	4.6	5.1	4.8
Services	18.5	21.7	27.4
Government	18.2	16.7	15.4
Nonagricultural	95.0	95.7	96.8
Agricultural	5.0	4.3	3.2

Source: Hicks and Ross, 1993:2.66.

while private sector employment increased from 81.8% to 84.6% in metropolitan regions.

The services-providing sector accounted for 97% of total growth in the U.S. urban system during the 1980s, and service industries expanded their employment base by over 31%. Goods-producing employment growth fell from 11.4% during the 1970s to only 1.1% in the 1980s. In the metropolitan regions, the share of manufacturing fell from 21.9% in 1970 to 14.4% in 1990, while services grew from 18.5 to 27.4% and the FIRE sector from 5.5 to 7.4%. The broad set of services sectors raised their shares of employment from 67.4% in 1970 to 76.4% in 1990.

During the 1980s, the rates of employment growth in construction, the FIRE group, retail, and general business and consumer services surpassed that of the U.S. economy as a whole, and most of this growth was distributed broadly across the metropolitan regions, with only 13 of the 282 metropolitan regions experiencing net employment loss in the FIRE group during the 1980s, and only seven regions lost employment in retail jobs.

At various levels of city size in the urban hierarchy, employment changes varied. The two largest cities, New York and Los Angeles, had 16% growth in employment in the 1970s and 22.4% in the 1980s. The 1- to 10-million population size cities had higher levels of growth in employment: 31% in the 1970s and 31.7% in the 1980s. The patterns of growth in employment showed slowing rates with de-

TABLE 3.5.
Metro Regional Employment Change Patterns by Sector 1970–80 and 1980–90

	Total Employment (%)		Private Sector Employment (%)		Public Sector Employment (%)		Goods-Producing Employment		Services Producing Employment	
	1970–80	1980–90	1970–80	1980–90	1970–80	1980–90	1970–80	1980–90	1970–80	1980–90
U.S. Total	25.1	22.2	27.3	24.1	15.2	12.5	11.4	1.1	32.1	31.1
>10 M	15.9	22.4	17.5	23.8	7.0	13.8	-0.8	-2.0	21.8	29.8
1–10 M	31.0	31.7	33.2	34.5	21.0	17.1	10.6	5.5	38.9	40.0
500 K–1 M	31.7	21.1	35.0	22.2	17.4	15.6	13.3	-6.7	39.9	30.8
250 K–500 K	29.5	17.0	30.5	19.5	25.4	6.0	12.7	-3.1	37.5	23.3
<250 K	25.7	24.4	15.6	12.4	-1.9	0.6	5.0	-4.4	15.6	16.8

Source: Hicks and Rees, 1993:2.67.

creasing size of the metropolitan region. There was a widespread slowing of growth rates throughout the urban system outside the largest metropolitan regions over these two decades, and by 1990, it was only the size categories including the 38 urban areas with populations over 1 million where employment growth exceeded that of the nation as a whole. These patterns were reflected in private sector employment, with rates of growth showing decline with decreasing size of metropolitan regions. Patterns of employment change in the public sector differed slightly, with the decline in employment growth being evident in all size categories except for the largest, where the rate of public sector employment expansion doubled from 7% in the 1970s to almost 14% in the 1980s.

During the 1980s, goods-producing employment growth had filtered incrasingly to nonmetropolitan rural and exurban locations outside the U.S. urban system, reflecting a variety of factors, including shifting tax treatment of construction investments in the major central and satellite business districts of metropolitan areas and the increasing shift of manufacturing employment growth in the technology-intensive sectors and an increasing emphasis on productivity improvements.

Growth in the services-providing sectors actually slowed substantially outside metropolitan regions with populations less than 1 million, with the larger metropolitan regions displaying the strongest growth in employment.

Hicks and Rees (1993:2.16–2.17) identified three broad trends in total employment change across the United States:

1. In both the 1970s and 1980s, employment growth was fastest in the South and West, with the Northeast and North Central regions lagging, reflecting the much touted sun belt/snow belt dichotomy.
2. There were some limits to this convergence process in the 1980s with the South and West experiencing expansion at rates greater than the Northeast and North Central regions, but at rates in the farm regions markedly lower than was the case in the 1970s; whereas rates in the latter regions were higher in the 1980s than they were in the 1970s.
3. During the 1970s, the relatively rapid growth in employment that characterized the middle-size metropolitan areas in the United States slowed, with growth shifting to the largest metropolitan regions, indicating the increasing importance of urban size and scale as magnets for employment growth.

3.5.2 Services within the Economies of the Newly Industrialized Countries

The trend toward increasing dominance of the services sectors in the economies of the developed nations is reflected also in the NICs of Asia. In Singapore, for example, the indexed growth of GDP by industry categories between 1960 and 1986 clearly demonstrated the emergence to dominance of the financial, business, and property services and the transport and communications sectors, as shown in Table 3.6.

In Taipei, the major city in Taiwan (another of the "little tigers"), over the period 1971 to 1988, the commerce sector share of employment rose from 16% to over

TABLE 3.6.
Growth in Singapore's GDP by Industry Sector, 1960 to
1986 (at 1986 Market Prices)

Industry	Change (%)
Agriculture and Fishing	43.8
Quarrying	800.0
Manufacturing	1172.4
Utilities	956.8
Construction	1160.4
Commerce	445.9
Transport and Communication	1255.6
Financial and Business Services	1673.4
Other Services	412.2
Less: Imputed Bank Service Charges	6398.6
Add: Import Duties	63.5
Total Gross Domestic Product	692.2

Source: Singapore Chief Statistician.

32% while the share of manufacturing fell from 42% to 25% and the community and personal services sector grew from 18 to 23% (Wu, 1992).

These trends are the rule, not the exception, and it is evident that urbanization in the developing world, and especially in the NICs, is being driven by *tertiarization* much as it is by industrialization.

3.5.3 Toward Small Business Growth and Female Employment in Part-Time Jobs

A significant phenomenon that has emerged since the 1970s, and one that has proceeded rapidly in the economic recession of the late 1980s and early 1990s, has been the impact on employment of downsizing and outsourcing that has characterized big business and the growth of small and medium-size businesses, particularly in the producer services sector activities.

A dramatic illustration of the magnitude of the loss of jobs in large companies and the growth of jobs in small businesses (and especially in self-employment) is provided in Figure 3.12 showing data for Australia in the early 1990s.

In addition, it is evident that much of the growth of new jobs has been for part-time employment. For example, in Australia between 1986 and 1993, about two-thirds of the 737,000 new jobs created were part-time, and many of these were female jobs. Table 3.7 shows that job growth has been dominated by part-time employment, especially in the older and larger cities of Sydney and Melbourne and also in the industrial city of Adelaide; while in contrast, the consumer services dominated "sun belt" growth metropolitan regions of Brisbane and Perth had a more balanced mix of full-time and part-time job growth that was fed by in-migration-generated demands.

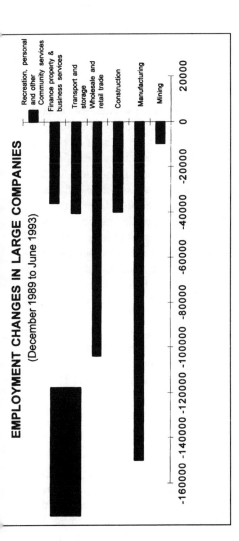

EMPLOYMENT CHANGES IN LARGE COMPANIES
(December 1989 to June 1993)

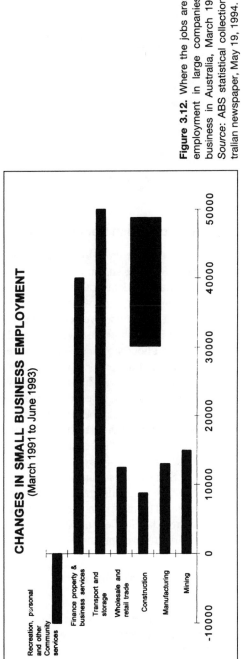

CHANGES IN SMALL BUSINESS EMPLOYMENT
(March 1991 to June 1993)

Figure 3.12. Where the jobs are: changes in employment in large companies and small business in Australia, March 1991 to 1993. *Source:* ABS statistical collections; The Australian newspaper, May 19, 1994.

TABLE 3.7.
Employment Change in Austrailian Cities by Full- and Part-Time Jobs, 1986 to 1993

Place	Jobs 1993 (000s)	Job Change 1986–1993 (000s)		
		Total	Full Time	Part Time
Sydney	1651	85	-3	87
Melbourne	1405	59	-5	64
Brisbane	644	123	71	52
Adelaide	469	27	-1	28
Perth	561	85	43	43
Hobart	77	3	1	2
Canberra	160	31	19	12
Australia	7716	737	258	478

Source: O'Connor, 1993.

3.6   Social and Demographic Change

During the last three to four decades in advanced western societies—such as the United States, Canada, the United Kingdom, many western European nations, and Australia—fundamental structural demographic changes have been occurring that have been the causes of, as well as being caused by, significant changes of a social nature. The results have been dramatic at the national, regional, and local levels of scale. As a result, there have been fundamental changes in the structure and characteristics of households, the nature of gender relations, the racial and ethnic diversity of cities, and the patterns of movement and mobility of populations across spatial systems. When these changes are coupled with the processes of economic and technology change discussed above, we witness a dynamic situation in which the structure of populations, the activities of households, and the roles of individuals are quite different in the late 20th century compared to earlier times when many of the theories developed in economics, geography, and sociology about cities and regions and their functioning were conceptualized.

Some of these fundamental *demographic* and *social* changes are summarized below. They are illustrated by reference to two recent studies—one in Australia, conducted by the Australian Housing and Urban Research Institute (AHURI, 1994), and the other in the United States, conducted by the U.S. Department of Housing and Urban Development (Sommer & Hicks, 1993).

3.6.1   A Mobile Society

In the United States, the rate of mobility of people declined in the late 1970s and 1980s from 20 to 17%, but still just under one-fifth the population changes residence every year, and more than half the population moves in a 5-year period. Renters move more often than owners, and younger people move more often than older people. As a result, about half of the population does not actually move in a given period. However, there is a high degree of age selectivity in the residential mobility of populations. Figure 3.13 shows that in Australia, for example, mobility rates increase dramatically for both men and women between the years 15 and 20, and more so for men than women. Thereafter mobility rates drop with increasing age to level off at about 50 years before they rise slightly with older age, where mobility is often associated with retirement.

However, as Clark (1986) has shown, the United States is a restless society, and people move frequently for reasons related to jobs, to improve their housing, to escape the crime in inner-city neighborhoods, as well as for a myriad of other reasons.

3.6.2 The Growth and Decline of Regions: Components of Population Change

The dynamics of aggregate population change over time in nations, regions, and local areas is the outcome of the interplay among three demographic processes:

1. *Natural increase or decrease* in population, which is the balance between births and deaths and is affected by changes in the levels of *fertility* (which have generally declined markedly in the post-World-War-II years) and age-specific *mortality* rates (with life expectancy having risen dramatically to around 80 for women to the mid-70s for men).

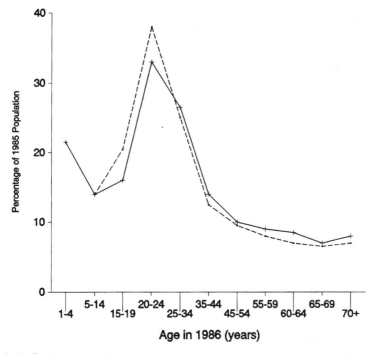

Figure 3.13. Residential mobility rates by age and gender in Australia, 1985–1986. *Source:* Hugo, 1994.

2. *Internal migration*, which is the balance between gains and losses of population in regions as a result of movements into and out of a region from and to other regions. Usually such movement is age selective, as well as gender differentiated. The outcome is either a net gain or a net loss in population due to internal migration.

3. *Immigration*, which is the balance between people migrating into a nation from overseas countries and those emigrating from a nation to overseas countries.

Table 3.8 shows the components of growth in the population for the United States since 1960. What are significant are the shifts in the components of growth, with the most pronounced being the greater role played by international migration, due both to undocumented aliens and refugees as well as normal immigration programs. In the 1980s alone, over 8 million immigrants entered the United States. In addition, fertility levels in the 1980s were far below those of the 1950s and 1960s. But it is interesting to note that the fertility rate did not dip as had been expected when the large baby-boom cohort entered the younger child-bearing ages, a result in part of the unexpected increase in fertility among women in their late 30s. In addition, the Hispanic, Asian, and black populations experienced higher fertility than the white population, thus the "previously forecast "zero population growth" scenario is no longer anticipated" (Frey, 1993:3.4).

There are considerable regional differences at various levels of scale in the patterns of population growth and decline. A major factor within nations is the pattern of net flows in internal migration, which typically displays considerable spatial biases. For example, as shown in Figure 3.14, the patterns of internal migration among the states in Australia show a pattern of net gains to the "sun belt" states of Queensland and Western Australia, with the old core states of New South Wales and Victoria experiencing the greatest net losses through internal migration.

TABLE 3.8.
Components of Change in the United States Population, 1970–1990

			Annual Average Rate per 1000			
			Natural Increase			Net Civilian
Period	Population at Start of Period (1000s)*	Net Growth	Rate	Birth Rate	Death Rate	Immigration Rate
1960–64	179,386	14.8	13.2	22.6	9.4	1.9
1965–69	193,223	10.7	8.8	18.3	9.5	2.1
1970–74	203,849	10.6	6.8	16.1	9.3	1.7
1975–79	214,931	10.5	6.3	14.9	8.7	2.0
1980–84	226,451	10.1	7.1	15.7	8.6	2.9
1985–89	238,207	9.8	7.0	15.8	8.7	2.7

*Includes overseas admissions into, less discharges from, Armed Forces and includes for 1960 to 1980 errors of closure (the amount necessary to make components of change add to the net change between census).
Source: Frey, 1993:3–52.

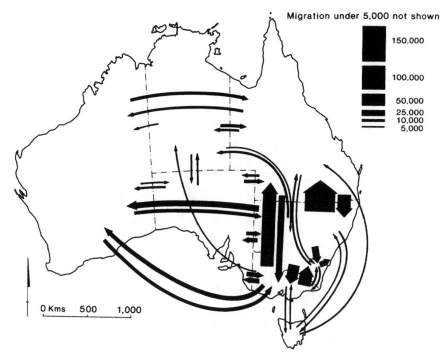

Figure 3.14. Patterns of internal migration between states in Australia 1986–1991. *Source*: ABS Census; Hugo, 1994.

3.6.3 Changing Age Structures: The "Graying" of Society

For a given geographic entity, an impact of the interplay between these dynamic demographic processes is that the age–sex distribution will change dramatically over time. Figure 3.15 graphically illustrates this for Australia, showing that between 1989 and the year 2021 the age structure will shift markedly toward the middle and older age groups, representing a trend that is typical of most advanced nations.

The changing age distribution of the population will produce a marked shift in the relationship between both the young and the old components of the dependent population. Such changes have significant public policy implications. An aging population is one of the most significant demographic trends and social issues facing advanced nations. While the proportion of the population that is in the older age categories is increasing significantly, and is likely to reach about 20% of the population in Australia early next century, this still will be proportionately less than the levels of aging that have been reached in many European countries for some time. Nonetheless, the rapidity of the increase and the actual number of additional aged persons represent major issues that have significant policy implications for governments, private sector marketing, the provision of human services, the demand for goods, and the requirements of the aged for retirement housing that is oriented specifically to the needs of the elderly.

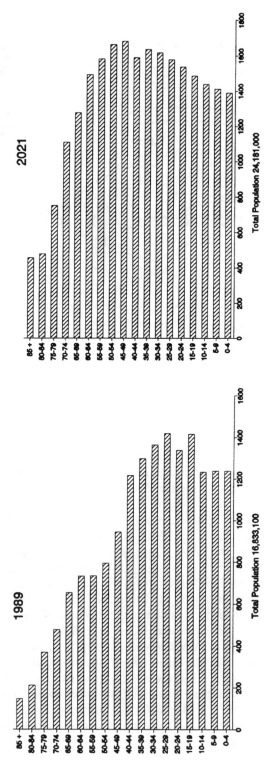

Figure 3.15. The changing age distribution of the Australian population; actual 1989 and projected 2021. *Source:* ABS; Hugo, 1994.

3.6.4 Changing Household Structures

The structure of households has changed dramatically since the 1950s. For example, in Australia over the period 1954 to 1991, there has been a dramatic decline in the proportion of households with five or more members from 25.8 to 13.3%. The number of one- or two-person households has increased from 22.2 to 51.2%. Similarly, there has been a major change, particularly since the 1970s, in the family structure of households. There have been a marked growth in couple-only households, a particularly strong rate of growth in one-parent households, and in particular in single-female-parent households, and a lesser level of growth of the more traditional nuclear-family household, with a marked decline in their proportionate shares. Non-family households have increased substantially.

The trend away from the traditional husband-and-wife-with-child households has been strong also in the United States. In the 1980s and 1990s, couple households represented only 56% of all households, whereas couple households with children represented only 26%. Table 3.9 shows that it was the female-headed households and non-family households that had the greatest numerical gains in the 1980s. About the future, Frey (1993) suggested:

> While these trends will probably continue, the 1980s shifts were not as dramatic as those of the 1970s. This was due to the fact that the large baby boom cohorts have largely passed through their household formation stages. Moreover, some boomers have begun to settle into more 'traditional' households as they aged into their 30s and 40s. It is anticipated that succeeding cohorts of young adults will follow the boom patterns of delayed marriage, independent living, and single parenthood. However, the smaller size of those cohorts will continue to slow household growth. Nonetheless, the growth of households outpaced the growth in population by 50% during the 1980s. Mean household size became reduced from 2.75 to 2.63 persons per household. (p. 3.6)

Frey (1993) has pointed to an important implication of this changing household composition, with the shift away from traditional households resulting in lower

TABLE 3.9.

Changing Component of Households by Family Type in the United States, 1970 to 1990

Type of Household	Total Households 1990 (1000s)	Percent of U.S. Households			Percent Change	
		1970	1980	1990	1970–1980	1980–1990
Family Households	66,090	81.1	73.8	70.8	15.7	11.0
Married Couples	52,317	70.5	60.8	56.0	9.8	6.5
Male Householders	2,884	1.9	2.2	3.1	41.1	66.4
Female Householders	10,890	8.7	10.8	11.7	58.3	25.1
Nonfamily Households	27,257	18.8	26.3	29.2	77.7	28.4
Male Householders	11,606	6.4	10.9	12.4	116.8	31.8
Female Householders	15,651	12.4	15.3	16.8	57.6	26.0
Total	93,347	100.0	100.0	100.0	27.4	15.6

Source: U.S. Bureau of the Census; Frey 1993:3.55.

incomes and higher rates of poverty. In this transition, it has been the female-headed families that have fared worst, and in the United States in 1991, this group had a poverty rate of 35.6%, compared to 6% for married-couple households and 11.5% for all families. The rate of poverty went up markedly with race, reaching 51.2% for black female-headed households and 49.7% for Hispanic female-headed families.

An interesting recent phenomenon, as illustrated by Australian data, has been the rapid increase between 1981 and 1991 in the proportion of 20- to 29-year-old persons living with parents, a phenomenon that might reflect deteriorating economic circumstances. Interestingly, the proportion of the youth population aged 15 to 19 years living with parents actually declined slightly, particularly for men, reflecting perhaps the increasing incidence of homeless youth.

3.6.5 Increasing Participation of Women in the Labor Force

A particularly important phenomenon of the years since World War II has been the increasing participation of women in the labor force. This is a result of major changes in the nature of gender relations in society, and it reflects also the increasing dependence of households on more than one income, particularly in the 1980s era of high interest rates and increased levels of unemployment.

Figure 3.16 illustrates this changing pattern of female participation in the la-

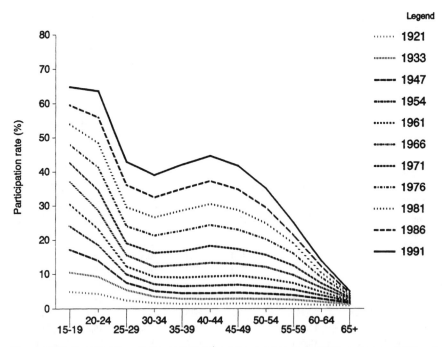

Figure 3.16. Female labor force participation rates by age in Australia between 1947 and 1991. *Source*: ABS Census data. Figure supplied by G. Hugo, Department of Geography, University of Adelaide.

bor force in Australia over the period 1947 to 1991, demonstrating not only an overall increased rate of participation but also the reentry to the labor force of women of
child-bearing and post-child-bearing age. And there was a tendency for women to
stay in the labor force once they entered it.

3.6.6 Racial and Ethnic Diversity

Over the years, large-scale flows of immigrants, particularly into many large cities
that are the major recipients of immigrant settlers, have created increasing ethnic diversity in cities. Within the United States, this has been reinforced by the large-scale
movement of black populations into the cities over many decades. More recently,
the large-scale flow of new settlers from Mexico and the increasingly important if
smaller flow of people from a range of Asian nations (including Chinese, Vietnamese, Korean, and Filipinos) have resulted in yet further ethnic diversification.

   These immigration flows, both legal and illegal, have caused a strong nation-
wide growth in the Hispanics–Asians in particular vis-à-vis the black population
(Figure 3.17). Across the cities and regions of the United States, however, there
were considerable disparities in these rates of growth. In general, Hispanics, blacks,
Asians, Native Americans, and other minorities tended to disperse to a greater degree in the 1980s than they had in previous decades, but the bulk of the growth in

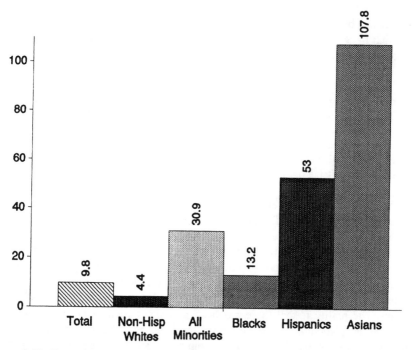

Figure 3.17. Percent change for total population, blacks, and other minorities in the United
States, 1980 to 1990. *Source*: Frey, 1993:3–50.

the minority population has remained concentrated mainly in the southern and western regions and in selected metropolitan areas; only nine metropolitan areas accounted for 54% of the growth in minority populations in the 1990s, with the Los Angeles Consolidated Metropolitan Statistical Area (CSMA) alone accounting for 20%. In contrast, over two-thirds of the 284 metropolitan areas have lower minority percentages than the United States as a whole (24.4%), and 96 metropolitan areas have more than 90% of non-Hispanic white minorities. Frey (1993:3.3) concluded that: "while each minority group exhibits different distribution tendencies, the sharp majority–minority distinction across broad regions and metropolitan areas will affect the social and political character of these areas. It also gives rise to new patterns of racial change at the community and neighborhood levels within metropolitan areas that experience sharp minority gains."

3.7   The Victims of Globalization: An Emerging Underclass

The impacts of globalization, internationalization, and economic restructuring on society have been profound in social and in equity terms. The magnitude of these impacts is difficult to specify, but they have been universal in cities, and they have been occurring since the 1970s. For example, many major cities, such as New York, went through fiscal crises in the 1970s that have resulted in an increasing managerialist approach to governance, with a major withdrawal of funding for social support programs. The inner cities in particular, and especially in the United States, are characterized by ghettos and drugs, and an urban underclass has emerged. It is estimated that in central New York City about 70,000 people have constituted an urban underclass of permanent outcasts for almost two decades.

Richardson (1989) has referred to:

> the bifurcation of the labor market, the development of "citadels" and "ghettos"; the widening fluctuations in metropolitan economies as exposure to the hypermobility of capital, exchange rates and resource price changes; erosion of the metropolitan tax base as a result of competitive subsidies to attract and retain internationally mobile economic activities; a shift from investment in social overhead capital to economic overhead capital; a rise in CBD land and property values.

Peter Marcuse, a lawyer, planner and social analyst at Columbia University, has suggested that the socially differential outcomes of these processes of change have exhibited different dimensions in the North American and the Western European urban contexts (1996). He has proposed the model represented in Figure 3.18. This conceptualization is interesting in that it differentiates between the "potential" and the "reality" of outcomes that may be derived from the processes of internationalization and technological advances in the post-Fordist economy of the era of globalization. The reality is that there has been *a concentration of private economic power*, and an *abdication of public power*, the latter being to evident through the withdrawal in the 1980s of governments from social and welfare programs and the moves toward privatization that typified political regimes under the governments of

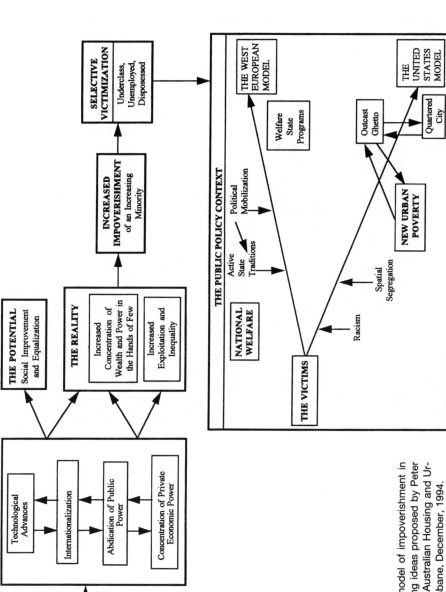

Figure 3.18. The Marcuse model of impoverishment in cities. *Source:* Compiled using ideas proposed by Peter Marcuse in a seminar to the Australian Housing and Urban Research Institute in Brisbane, December, 1994.

Reagan in the United States and Thatcher in the United Kingdom. Although the "potential" is for processes of internationalization and technological advances to generate widespread social improvement and provide equalization, the "reality" may well be increased concentration of wealth and power in the hands of few and an increased exploitation of many resulting in increased inequality. The outcome is seen by some as being increased impoverishment of some segments of society and by others as selective victimization, manifesting itself in an urban underclass consisting of long-term or permanently unemployed persons and people who are dispossessed of material well-being.

The model proposed by Marcuse suggests a dichotomous public policy outcome or response, reflecting in general terms the broad difference between the West European model of welfare state programs and the North American Model that has seen a withdrawal of government from welfare programs, resulting in the victims becoming outcast in ghettos and the new urban poor being spatially segregated within what has been termed a *quartered* city. This quartered city comprises the differentiated winners and losers in the post-Fordist city through the processes including globalization. The city is now characterized by four social groups: the *professional, technical, managerial winners*; the *middle class survivors*, but those doing less well relatively; the *working class survivors* who are doing much less well relatively, and many of whom are the new urban poor; and the *outcasts*, who constitute the urban underclass.

Important elements in this model are the role of the public sector, through the presence or lack of interventionalist policies and programs, the different spatial outcomes, and the different social outcomes that result from the Western European and the United States approaches. This raises public policy considerations and choices in how to address the social impacts of the processes of globalization, internationalization, and economic restructuring.

4

Urban Patterns and Trends

4.1 Focusing on Cities

The processes of change discussed in Chapter 3 are not affecting all cities and regions within nations in the same manner. In this chapter we turn attention to considering the way in which the processes of economic restructuring and patterns of population migration have been affecting both the structure of national settlement systems and the internal structure of cities over the last couple of decades. We focus attention mainly on patterns and trends in the United States. Particular attention is given to the rise of polycentric metropolitan forms as cities become dominated by knowledge-based and information-intensive services activities. These are reshaping the internal structure of city regions, giving rise to new factors that affect the location decisions of economic activities, changing the internal structure of urban labor markets, and impacting on spatial behavior.

The chapter also provides an overview of the social characteristics and patterns of residential areas within cities with respect to socioeconomic factors, household structure, and ethnic segregation dimensions. The issue of locational disadvantage is discussed in the context of the way it affects the access opportunity of certain groups of city residents to services and employment.

The metropolitan region and its internal structure and functioning need to be reconceptualized in the era of globalization and internationalization. A new paradigm is outlined in which the focus is on the needs of the information and services urban economy, quality of life issues, and smart infrastructure. The chapter concludes with a consideration of institutional and planning issues, and the demands generated in an urban services-based economy. It suggests also that traditional land use zoning and regulation approaches to planning be replaced by strategic planning that provides an integrated approach to urban development, planning, and management.

4.2 Changes in National Urban Systems

The processes that drive metropolitan urban development and growth and that shape the form and structure of cities and regions change over time. A new hierarchy of cities is emerging in many nations that relates a set of metropolitan areas to one another across national borders as well as changing the relationships between levels of the hierarchy within countries. Within the hierarchy of cities, a dynamic set of processes influence how functions compete and determine how space is used. These processes are creating new agglomerations of activities in urban space; they allow new patterns of decentralized activities; and they generate adaptive uses of existing spaces.

4.2.1 An Overview

Early work in the United States by Borchert (1972) clearly demonstrated how the pattern of urban settlement in the United States changed as *transportation systems* evolved, particularly as a result of the post-1950s national highways network and later with the development of the hub and spoke air networks. Work by Vining, Dobronyi, Otness, and Schwinn (1982) suggested that there had been a shift in the basic axis of economic activity, with the pattern changing as the national urban system became integrated, resulting in all its parts being linked as a single operating unit rather than as a series of regional or local units. This has been reinforced in the past two decades with the development of new telecommunications networks. Inevitably it will become reinforced with the advent of the *information superhighway*.

In Europe, too, work by Cheshire (1989) and Kunzmann and Wegner (1991) has quantified the pattern of change and pointed to a new urban structure that has emerged from the growth and change in recent decades. The new structure reflects the pattern of continental-scale transportation systems and especially those associated with high-speed rail networks.

Three broad themes appear to dominate the research literature on broad-scale changes that are occurring in urban systems:

1. *The shift away from the industrial heartland to new locations.* In the United States, this was documented by Abbott (1981) and in Europe by Sabel (1989). An interesting theme relates to the changing forces in favor of particular regions, with special local institutional contexts, or special clusters of high-tech industries.

2. *The demographic shifts to "sun-belt" locations.* New patterns of migration, often associated with the economic restructuring process, but also linked to retirement migration, have created new patterns of population growth in some areas. In the United States this occurred predominantly in the South and the West; in Australia it has occurred in Queensland and the coastal areas of New South Wales, and the Southwest of Western Australia; and in Europe the shift has been to the "higher-amenity" areas such as in the southeast of England and the London–Bristol corridor, and southern France. The extent to which this shift is leading or following economic restructuring is not clear. Popular interpretations see the move to the South in the United States, for example, as a response to warmer climates, with more detailed analyses having highlighted the location of job-rich industries, and especially those linked to the defense industries (Hall & Markusen, 1990). Interestingly, in the early 1990s, former growth regions such as southern California and Florida, now

face out-migration and are suffering severe economic recession as a result of military base closures and defense spending cuts.

3. *New housing market conditions.* Some researchers have focused on housing market changes, which have been seen as involving a mix of suburban spill-over and some intensification in the center of old city regions. On balance, in the United States the spill-over seems to be the strongest numeric change, generating major new patterns of intraurban traffic and locally focused labor markets (Frey, 1988; Hartshorn & Muller, 1992). In Europe inner-area redevelopment has been more prominent (Van Den Berg, Drewett, Klassen, Rossi, & Vijverberg, 1982). The spill-over of activity toward and beyond the metropolitan fringe also has provided a strong boost to nonmetropolitan locations within commuting times of major cities.

From these three approaches it is evident that, at the national level, processes of economic and demographic change can produce a range of outcomes. Understanding them is important for planning longer-term financing and provision of infrastructure and services, and it provides a context in which the plans for individual cities and regions can be formulated.

It has been suggested by O'Connor and Edgington (1991) that the processes of globalization and internationalization and the associated linkages, flows, and interactions that are generated between cities have resulted in the emergence of three orders of cities:

1. *First-order "world cities,"* for example, New York, Los Angeles, Tokyo, Hong Kong, London, and Sydney. They are concerned with global market systems and are international financial centers, air gateways, and global ports, with strong CBDs and highly driven communication and financial systems spaced across their regions. They are the foci of the emerging information superhighways.

2. *Second-order cities,* for example San Francisco, Boston, Chicago, Dallas, and Miami in the United States, Osaka in Japan, Melbourne in Australia, and Manchester in the United Kingdom. They are linked to global markets via major world trade centers and control their own special service production links; their space economy is oriented to their international functions, such as air, sea, and ground cargo, tourism, and technology.

3. *Domestic and regional centers,* for example, Pittsburgh, Seattle, Houston, San Diego, and Orlando in the United States, and Brisbane in Australia. They are part of international production systems and/or flows of tourists, but they do not exercise significant control over producer services within the system. Sometimes they do not have a strong center, and they tend also to be multinucleated.

What is clear is that *producer services* and *information* are now crucial for the economic vitality of the modern metropolitan economy. The lesser degree of locational constraint of services, and the concentrations that emerge through the networks of individuals and firms that comprise design–production–marketing–distribution chains based on information transmission and interpretation, are major forces creating this new form of urban hierarchy, in which the *world city* and its region dominates.

4.2.2 Suburbanization and Inner-City Decline

The growth of the metropolis in the 20th century resulted in increasing concentrations of population in urban areas. In the rapid metropolitan expansion of the post-World-War-II years, the internal structure of cities changed dramatically. As a result, so too did the pattern of population distribution. *Suburbanization* has become the dominant process of growth. Figure 4.1 shows the increasing dominance of the suburbs and the decline of the inner city in the United States in the period from 1960 to 1990. The proportion of population living in the suburbs increased from 30% to about one-half of the nation's population, while the proportion living in the central cities remained at a little under one-third.

The railways initially permitted outward growth of the city and the development of suburbs. However, the diffusion of ownership of the automobile was the crucial factor, and, along with the assembly-line production system, it generated suburban growth and deconcentration of population from the inner city to the suburbs. The assembly-line system of manufacturing consolidated all facets of production in one physical location through processes of vertical integration of activities and functions at suburban locations. Car ownership enabled the labor force to disperse to spaces separated from work locations. Increasingly, both people's residences and their places of work were suburbanized. Inner-city populations declined, and manufacturing activities relocated to greenfield sites. In American cities in particular, the development of freeway transport networks enhanced the suburbanization process (Frey, 1990; Knox, 1987).

Decline in the central cities was particularly strong in the 1960s and 1970s, and some, such as St. Louis, Buffalo, Cleveland, and Detroit, lost more than one-fifth of their populations in the 1970s.

But as Frey (1993:3.30) has reported, "the impact of these patterns on central cities has been mixed. The strong racial and social status selectivity that characterized the massive immediate post-war suburbanization started to dissipate as black suburbanization began in earnest and as pockets of white gentrification evolved in some of [the] more cosmopolitan cities. The 'black city–white city' image showed some signs of fading, though not enough to curtail the emergence of other kinds of pocket-ghetto poverty."

The link between suburbanization and rising affluence has been strong (Chinitz, 1991; Frey & Speare, 1988; Frey, 1992). The suburbs grew in the post-World-War-II period within an urban context of population growth and rising incomes. Governments aided the development of suburbia by financing new transportation and communications infrastructure (Baldassare, 1992; Harvey, 1989) and by making housing more affordable through mortgage aid, tax incentives, and veteran home-loan schemes (Baldassare, 1992; Kemeny, 1983; Maher, 1993). In the United States these were, however, highly selective suburbanization rates, based on race as well as wealth (Chinitz, 1991).

Suburbanization has presented a challenge to the central city that is perhaps insurmountable (Stanback, 1991). Many inner-city areas are now the domain of an urban *underclass*. In Australian, Canadian, and European cities, this phenomenon does not exist to the same degree, and though population decline is endemic in the

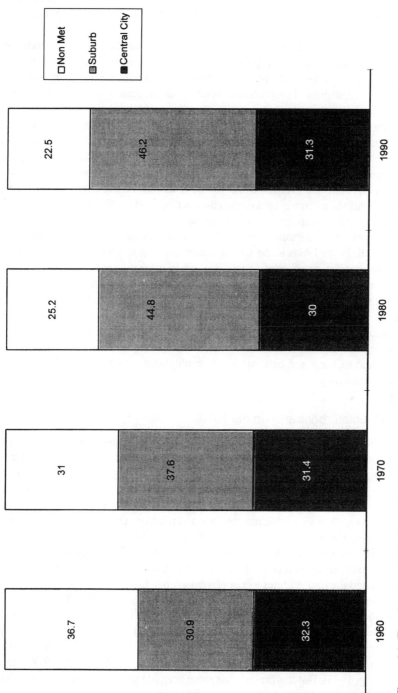

Figure 4.1. The changing distribution of the United States population by central city, suburb, and nonmetropolitan areas, 1960 to 1990. *Source:* U.S. decennial censuses.

inner cities, there are areas of rejuvenation through processes of gentrification and renewal. Rarely do these activities turn around the net loss of population to the suburbs.

The growth of the suburbs continues, seemingly unabated, and extends well into the hinterlands of the metropolitan cities. In the last two decades or so, this has taken the form of rural-residential living, which has spread widely into the commuting zone of large cities. These and the outer suburban areas often have the highest rates of population growth.

The new flexible production systems have enhanced the continuing dispersal of economic activity to the suburbs (Chinitz, 1991). Within that process there is also agglomeration of like industries, which enables firms to take advantage of agglomeration economies, such as a rising critical mass of demand for inputs, services, and skilled labor (Malecki, 1985). Central to this is the increasing tendency for corporate organizations to move backroom and ancillary office functions out of CBD headquarters into cheaper suburban locations and also to second- and third-tier cities. Storper and Christopherson (1987) saw the resurgence of metropolitan regional growth in the 1980s in the United States as being an outcome of this process.

Suburbanization of employment has led to dramatic increases in suburb-to-suburb commuting by car that is significantly greater than that of city-to-suburb travel. Chinitz (1991) has argued that suburban congestion has the potential to affect future land use patterns in the suburbs. An outcome of this could be higher-density suburban living as residents attempt to reduce the time spent commuting by car, but this will not necessarily be in the inner suburbs alone. Most likely it will be focused around existing and new suburban employment centers and agglomerations of economic activity.

4.2.3   Counterurbanization in the 1970s

In the 1970s, research on the geography of population in advanced industrial nations began to question the dominance of metropolitan concentration on the basis of evidence of a *population turnaround*. This was identified first by Beale (1975) who noticed a trend of movement of people from the big cities to nonmetropolitan cities and towns. This was hailed as evidence of decentralization, and it was characterized in the United States as "the simultaneous absolute and relative decline of many large metropolitan areas and by a resurgence of growth in many rural small-town areas, which had experienced long-term decline" (White et al., 1989:267). Figure 4.2 shows how it was the other (smaller) metropolitan and the nonmetropolitan areas in the United States that had the highest rates of population growth in the 1970s. This ran counter to the prevailing expectations of continued metropolitan growth, suburban expansion of large cities, and concentration of national populations. The phenomenon was identified by researchers in other nations as well, including Australia (Burnley & Murphy, 1993; Serow & O'Cain, 1992), Canada (Joseph, Keddie, & Smit, 1988), the United Kingdom (Champion, 1989), and West Germany (Kontuly & Bierens, 1990; Kontuly & Vogelsang, 1988). But this process has since been identified also as being more an outcome of *urban spillover* (Joseph et al., 1988; Maher, 1993; Sant & Simons, 1993) than a reorientation toward country living.

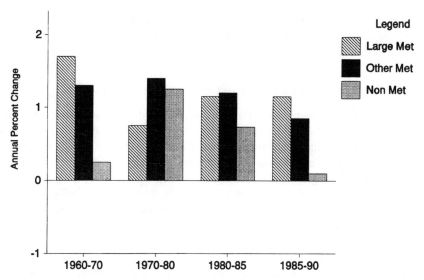

Figure 4.2. Annual percent change in population in large metropolitan, other metropolitan, and nonmetropolitan areas of the United States, 1960 to 1990. *Source:* Frey, 1993:3.58.

Both economic and noneconomic explanations were advanced to explain this *counterurbanization* phenomenon. Research in the United States showed that "both employment growth in non-metropolitan America as well as preferences for smaller places in attractive environmental settings underlie the turnaround" (Fuguitt, 1985). It was common also for rising housing costs to push some new metropolitan home buyers out to the urban fringe and beyond into *exurban* estates and *satellites*. For example, in Australia in the late 1980s, large numbers of immigrants, particularly to Sydney, increased the demand for housing beyond housing-stock supply. The outcome was escalating housing costs that effectively pushed low-income, would-be home buyers, already faced with high interest rates, out of the metropolitan housing market (Vipond, 1992). Without knowing the circumstances surrounding individual moves, it is not appropriate to identify this type of population movement as being counterurbanization.

But White et al. (1989:267) noted in the United States that "just as counterurbanization became accepted and population deconcentration trends were being extrapolated into the future, the patterns shifted back in the early 1980s, with metropolitan areas once again growing faster than non-metropolitan areas." In the Australian context, the term *counterurbanization* is somewhat inappropriate, as "metropolitan areas continue to grow while rural areas on the whole continue to decline" (Maher, 1993). While noting that the major rates of increase were occurring in a limited number of relatively specialized urban and urbanizing regions (Flood, 1992; Maher, 1993), Maher saw their focus in the "narrow coastal strip beginning north of Sydney and continuing to the Sunshine Coast north of Brisbane as another form of concentration, close to existing metropolitan regions" (Maher, 1993). In the United Kingdom, Lewis, McDermott, and Sherwood (1991) found that the highest growth rates were often in nonmetropolitan zones. They were also regionally focused, primarily within a 60-km semicircular belt in the southeast of England and located within 90 to 150 km of London.

4.2.4 Resurgence of Metropolitan Concentration in the 1980s

The 1980s saw a resurgence of metropolitan growth in the United States, Canada, and Australia, as well as in many European countries. In the early 1990s, amidst the recession, metropolitan concentration continued (Maher, 1993, Frey & Speare, 1992). There is now less emphasis placed on counterurbanization in the literature on regional population growth (Frey & Speare, 1992; Maher, 1993; Sant & Simons, 1993). However, the German literature (Kontuly & Vogelsang, 1988) identifies counterurbanization as a continuing feature of regional population growth in that country. What has happened in the last 15 years or so has been a decline in this phenomenon and a return to higher rates of growth in the large metropolitan areas.

From the time of the first OPEC oil crisis in 1973 as a benchmark, the mode of global production and capital accumulation began to switch from *Fordist* to *post-Fordist* flexible production systems. This was signified by: the vertical disintegration of the processes of production; decreased employment opportunities in industry as many manufacturers moved offshore; the outsourcing of many functions of the industrial conglomerate; and the increased importance of the services sectors, and particularly of the producer services, in employment growth. The processes of globalization and technological change discussed in Chapter 3 have tended to increase the importance of *centrality* for many economic activities in linked network production systems, and the *mega metropolitan regions* and *urban corridors* have assumed increasing importance as the engines of economic growth.

In the 1980s in the United States, the largest metropolitan areas (with populations over 1 million) grew at a faster rate than metropolitan areas as a whole. Their growth was significantly faster than that of the nonmetropolitan areas whose growth rate was much reduced compared to the previous decade. It is significant that this faster rate for the metropolitan areas became even more accentuated in the second half of the 1980s (Figure 4.3).

Frey (1993:38) has shown that the fast-growing metropolitan areas in the United States (defined as growing at 2.5 times the national growth rate) numbered 59 in the 1970s, 47 in the early 1980s, and 49 in the late 1980s. The number of large metropolitan statistical areas (MSAs) classed as fast-growing areas increased from 7 in the 1970s to 12 by the late 1980s, and the number of small metropolitan areas that were "fast growing" were reduced in number from 38 in the 1970s, to 22 in the early 1980s, and to 14 in the late 1980s. The large central cities that rebounded from their 1970s population losses—cities such as New York, Boston, Minneapolis–St. Paul, Kansas City, San Francisco, Oakland, and Seattle—did so for two reasons: (1) their economic functions dovetailed with secular patterns of corporate growth and the related advanced producer services industries that boomed in the 1980s, as in New York; and (2) they grew through accelerated immigration to the major port-of-entry cities, such as Los Angeles, New York, San Francisco, and Miami.

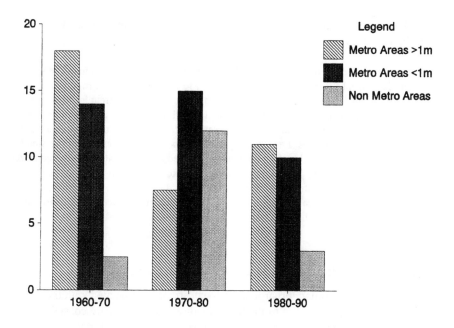

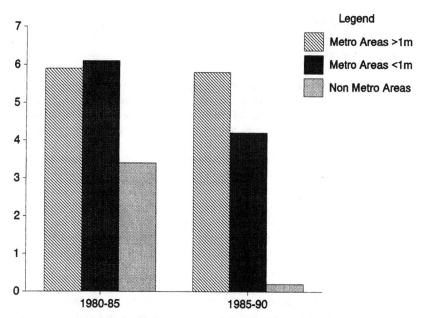

Figure 4.3. United States metropolitan growth trends, (a) 1960 to 1990 and (b) 1980–1990. *Source:* Frey, 1993:3.49.

Frey (1993) provides a succinct rationale for this reemergence of the large cities in the 1980s:

> Within the United States traditional 'heavy industries' became less labor-intensive as production jobs were phased out or exported to other nations . . . Proponents of this view anticipated the re-emergence of suburbanization but in different locations than in the past. Growing metropolitan areas would be those that successfully transformed their economies towards advanced services; high tech research and development; and growing new industries. Less stable growth prospects were foreseen for the smaller 'subordinate cities' and metropolitan areas engaged in peripheral, routine production activity that could be phased out by decision makers located in corporate or (in the case of defense activities) Government centers . . . In short, this 'regional restructuring' perspective forecast a return to population clustering in large areas constituting specific economic niches. (p. 3.9)

4.2.5 Concentration and Dispersal: A Matter of Scale

Part of the nonmetropolitan turnaround has been strongly associated with the directional dispersal of populations to the high-amenity—and especially coastal—locations that typify "sun belt" regional growth, of which retirement migration is an important component. But the reality often is that the absolute levels of population growth in regions associated with the *counterurbanization* processes and *nonmetropolitan turnaround* are substantially less than the growth of population numbers in the traditional concentrations of the large metropolitan cities.

White et al. (1989) provide us with a timely reminder:

> The economic geographic theory of concentration via returns to scale and agglomeration always implied the far end of the curve—that of diseconomies from costs of congestion, excess travel, generation of unavoidable externalities, and the like. Remember that in all probability all societies, and certainly in capitalist ones, firms make the basic location decisions. Thus, under certain social, economic, and political conditions, it simply became more profitable to shift some kinds of activities to regions and countries having lower costs and offering higher returns. Under other conditions, core areas became more profitable. (p. 268)

Whereas for some firms and people, nonmetropolitan locations have enabled them to maximize their utility and well-being, for many, shifts in industrial and occupational structures and widespread deregulation have reestablished the attractions of the big city.

It would be dangerous to extrapolate into the long-term future based on short-term trends, because it would seem to be the case that there will continue to be relatively frequent changes in the nature and the force of processes that generate concentration and agglomeration as against deconcentration and dispersal. When the issue of the level of scale aggregation or disaggregation at which we analyze trends and patterns in population growth and decline is added, the complexity and uncertainty are exacerbated.

What we need to realize is that differential growth of population is a fundamental dynamic in any society. However, the forces of change, and some of the outcomes of that dynamic, have altered substantially over recent decades:

1. On the one hand, there has been *a continuation of metropolitan concentration* and an *intensification in the process of suburbanization*. There has been also *a growing diversity in nonmetropolitan areas*, which has been given the term *counterurbanization*. These are processes that are well documented if not well understood.

2. On the other hand, *new forms of spatial patterns have emerged* in response to some of the dramatic economic, technological, and social changes that have been experienced in recent decades. Some of these are related only peripherally to population growth per se, although they have a great deal to do with population composition and the evolving character of social and economic processes and their spatial outcomes. These include some of the economic trends of investment and innovation in new plants and equipment, the concentration of *knowledge-based* rather than production-based, activities, and the development of high-level producer services. Although these are not related directly to population changes, they are important in underpinning some of the observed trends because it is many of these that provide the conditions freeing population from traditional constraints. Thus, advanced communications and information technologies have made spatial proximity less significant. The wealth generated from some of these activities allows some people greater freedom of choice of location. At the same time, there has developed a growing *polarization in the labor force* between types and locations of employment, and even access to employment itself. Thus changing population distribution is the outcome of a very complex set of changes.

O'Connor and Stimson (1994) have conceptualized the trends and processes that generate metropolitan regional and nonmetropolitan regional growth and that are shaping spatial outcomes in Australia. Figure 4.4 provides a general model of these processes and urban forms.

4.3   The Polycentric Urban Form and Structure of the Information and Services City

Within cities the transformation of urban economies from a *manufacturing* production base to a *services* and *information* base, along with the continuing process of suburbanization, are generating a new urban geography that adapts old spaces to the process of change in addition to creating significant new urban spaces (Blakely, 1992).

Daniels, O'Connor, and Hutton (1992) have shown how the processes of change have "unleashed locational forces that have given rise to new urban structural arrangements." In particular, *consumer services* have become *dispersed* throughout metropolitan space, reflecting the dispersal of residential populations. But *producer services*, in particular, have shown strong tendencies toward locational *centrality*, with a preference for the CBD and the emerging regional business dis-

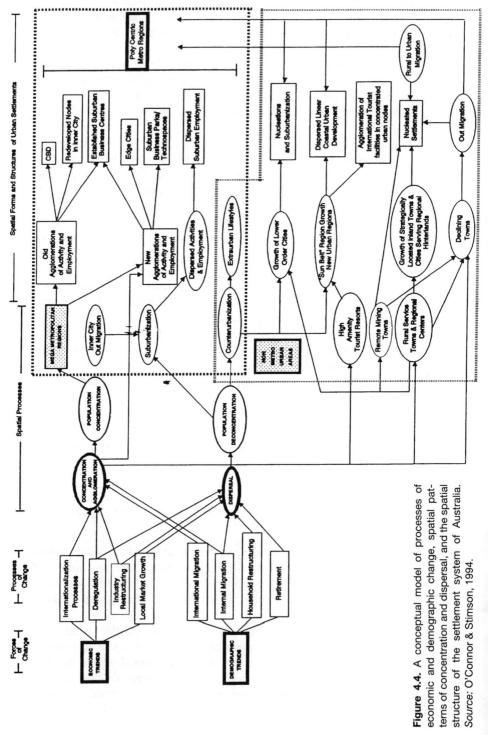

Figure 4.4. A conceptual model of processes of economic and demographic change, spatial patterns of concentration and dispersal, and the spatial structure of the settlement system of Australia. *Source:* O'Connor & Stimson, 1994.

tricts that display many of the characteristics of what Garreau (1991) has termed *edge cities*. And there also has been an accelerated growth in the trend for them to disperse to small nucleated sites. In many ways, the edge cities with their middle class, professional, and white-collar, suburban labor sheds, are now outperforming the central cities (Fishman, 1990). In addition, *technospaces*, which are specialized technology-based nodes that are located both within existing high-activity locations as well as at new nodes outside core urban areas on greenfield sites, have developed (Blakely, 1992). Often they develop in association with universities and government research laboratories, and they rapidly form new clusters of R&D activities in primarily suburban locations. Also, new manufacturing and commercial complexes are evolving associated with international airports (Kasarda, 1991).

The processes of change that are impacting cities are rendering redundant many of the traditional theories of urban form and structure. In addition, they are challenging the appropriateness and the efficiency of many public policies, management practices, and planning frameworks, and in particular those that are based on single-use zonings and development control practices that may have been appropriate to the age of the mass-production and high-energy-consumption city with its antecedents in the industrial revolution.

4.3.1   New Locational Forces in the Information and Services City

Hall (1991:5) has stressed how the emerging new metropolitan economy is governed increasingly by access to *information* that occurs in two ways:

1. By *direct face-to-face communication*, encouraging *agglomeration* of activities in locations such as the CBD.
2. By *electronic transfer*, which enables *dispersal* of activities but which also leads to the development of strong *concentrations* of specialized information-generating, information-exchange activities at key nodes for national and international transportation and communications.

Theoretically and practically, technological innovations, especially in telecommunications, have made it possible to allow a substantial proportion of office work to be performed at spatially scattered nodes. Castells (1989:43) claimed that the physical consequences for urban systems, including impact on the social life of people, will be phenomenal. Trends such as those toward the *information superhighway* "would signal the end of the basic material from which the industrial culture emerged in the 19th Century." And Garreau (1991:134) has painted a picture of a new metropolis as one whose prime purpose is "to make distance irrelevant," so that one "wonders why we need to build cities at all."

But, the evidence seems to be that although there has been a considerable *dispersal* of economic activities to the suburbs (following as well as leading the suburbanization of the population as discussed earlier in this chapter), nonetheless, these economic activities remain powerful forces of *agglomeration* which both perpetuate many old concentrations of activities, such as the CBDs of cities, and spawn the development of new suburban centers of concentration of business, high-tech activities,. This gives metropolitan regions their polycentric structures.

The importance of the producer services is such that they are now critical for the economic vitality of the modern metropolitan economy. The high degree of locational constraint for many of those services, and the concentrations that emerge through the networks of individuals and firms that comprise a production chain based on information transmission and interpretation, have both been identified by Friedmann (1986) as major factors in the emergence of the so-called *world cities*. Increasingly, this is the case for all metropolitan cities of any significant size.

What is interesting is that the trend toward decentralization of services activities does not necessarily involve large-scale relocation of firms and activity from the CBD to suburban centers. O'Connor and Blakely (1990) claim that the nature of the services-sector activities in the CBD and in suburban centers is fundamentally different. The CBD producer services are dominated by functions that are high-contact intensive and export oriented with a tendency to focus on global activities. The suburban centers are characterized by more standard services activities, including local business services, data processing, and public and nonprofit services. Many of the major employment activities in them are "back office" functions.

4.3.2 The "Edge City" Phenomenon

The term *edge cities* has been coined by Garreau (1991) to describe the evolution of rapidly expanding, multifunctional, urban conurbations on the periphery of traditional "downtown" areas and beyond in the new suburbs of North American metropolitan cities. Cities examined included Washington, Los Angeles, Atlanta, New York, Phoenix, Detroit, San Francisco, Boston, and Dallas. Edge cities are new major business districts that have developed as low-density agglomerations based on office parks, but they include also the more standard higher-density, high-rise office districts. There are several hundred such edge cities that have evolved in metropolitan regions in the United States. The edge city is defined as having 5 million square feet of office space; it is the workplace of the information age; it has 500,000 square feet of retail space; and it has more jobs than bedrooms. It is perceived by the population as one place—a regional end-destination with mixed uses: jobs, home, shopping, and recreation.

The essence of edge-city growth stems from locational decision-making processes of firms in the post-1970s era in which activities sought relatively low-cost locations, sites giving proximity to good-quality affordable housing, with good transport infrastructure, and proximity to international airports. This reflects a change in the nature of the raw materials of commerce. The new locational processes have resulted in the creation of spaces that have become self-sustainable agglomerations as activities relocate to follow key leaders. Their new product is *cleverness*. The workplace has become overwhelmingly centered on offices rather than factories, and cities have been transformed in responding to the needs of this new product. The product does not have to be near rail or water, and the edge city typically has risen at the interchange of freeways, as the car is the preferred transportation mode of the modern age.

Garreau attempted to throw light on the evolution and function of edge cities by examining their social, spatial, and economic morphology, their relationship with traditional downtown centers, and their interaction with other established emerging "urban nodes." He sees the emergence of edge cities as the most significant change in urban spatial development since the industrial revolution:

> These new hearths of our civilization—in which the majority of metropolitan Americans now work and around which they live—look not at all like our old downtowns. Buildings rarely rise shoulder to shoulder. . . . Instead their broad, low outlines dot the landscape like mushrooms, separated by greensward and parking lots. Their office towers, frequently guarded by trees, gaze at one another from respectful distances through bands of glass that mirror the sun in blue or silver or green or gold like antique drawings of "the city of the future." (Garreau, 1991:3)

And Garreau (1991) has noted a distinct change in emphasis from the traditional "downtown":

> The hallmark of these new urban centers are [sic]not the sidewalks of New York of song and fables, for there are usually few sidewalks. These are jogging trails around the hills and ponds of their characteristic corporate campuses.
>
> These new centers are no longer tied together by locomotives and subways but by jetways, freeways and rooftop satellite dishes.
>
> Their characteristic monument is not a horse mounted hero, but [they are] the atria reaching for the sun and shielding trees perpetually in leaf at the cores of corporate headquarters, fitness centers and shopping plazas.
>
> These new urban centers are marked not by the penthouses of the old urban rich or the tenements of the old urban poor. Instead their landmark structure is the celebrated single-family detached dwelling, the suburban home with grass all around. . . . (pp. 3–4)

4.3.3 Technospaces

While technology has rendered distance less relevant, some locations within cities are becoming increasingly important as specialized technology nodes. They have been dubbed *technospaces* by Blakely (1992). These are places that have the greatest concentrations of high-technology firms agglomerated in subregions of large cities outside the core city area. Globally they are relatively few. Technospaces are the "Silicon Valley" phenomenon. They have long, symbiotic relationships with the established research bases of universities and government and private research laboratories, and they have spawned new high-tech activities as incubator areas. They are predominantly suburban. But it is significant that the modern, information-based metropolitan economy has not eliminated the functions of larger and larger central cities and indeed, the rise of information-intensive service activities has reinforced the role of the CBDs of the global cities. In the 1980s, there was a major boom in "smart" office buildings and a major expansion of floor space as the downtowns were redeveloped and expanded.

There has been a considerable amount of research conducted into the special land use characteristics of high-tech activities in metropolitan city regions (Saxenian, 1985; A. Downs, 1985; Moss, 1988). In the industrial era, it used to be common for governments and communities to use land use tools and land provision to attract manufacturing industries. But these land use approaches have not captured the diverse rationales associated with the specialized, technology-based developments (Blakely 1992:7).

High-tech industries have tended to display considerable spatial bias and concentrations in their geographic distribution. In the United States, this has been related closely to the geographic distribution of federal R&D spending. Typically, the 20 leading high-technology states received 80 to 90% of total R&D funds, with the dominant states being California, Maryland, and Virginia (linked to the national capital Washington, DC), Massachusetts, New York, Florida, Texas, Pennsylvania, and Ohio dominating. National Science Foundation research has shown how federal agencies tend to allocate funding to locations that have concentrations of aircraft, aerospace, and electronics industries, concentrations of university research talent, and missile, aircraft, spacecraft, and explosive testing facilities. Early studies by Malecki (1980, 1981) showed how R&D spending and high-technology activities are even more concentrated geographically than both population and industrial activity. This results from a *breeder reaction* (incubator effect) where agglomeration of R&D personnel fosters further local spin-offs of innovation activity that attract more research-oriented companies and funding to an area. Thus, many of the old declining industrial regions, such as the Northeast, still remain important as areas of concentration of high-technology industries. Melbourne in Australia is another example of this phenomenon. Thus not all growth regions, however, are high-technology growth regions.

In the United States, it has been shown that there are a set of factors that are related to the pattern of regional growth and the diffusion of high-technology industries. These include (not in any specific order of importance):

- Headquarter–branch plant relationships and locations.
- A region's relative share of growth industries.
- A region's relative share of high-technology industries.The pool of potential entrepreneurs.
- Costs of doing business (wage rates, utility costs, local taxes).
- Characteristics of the labor force.
- General attractiveness of a city.
- Rates of urban business formation.
- Relative levels of local economic performance.
- Low levels of unionization of labor force.
- Business climate (presence of specialized business services, research facilities).
- Infrastructure factors (transportation networks and utility services).
- Community factors (cultural facilities, housing prices, pollution levels, climatic characteristics, good schools).
- Socioeconomic factors (minority populations, conservative voting behavior).

Certainly the "high-tech" sector is providing for new and evolving forms of spaces in cities, combining the quality of office space found previously only in the CBD locations with a more attractive working environment and the ability to undertake a range of efficient activities in one location. Figure 4.5 illustrates the range of property developments that typify such environments.

4.3.4 Airports and Seaports

One of the most significant strategic spaces and locations in the new globally linked international metropolitan city is the international airport. As a result of the juxtaposition of the changing production system and innovations in transportation technology, we are witnessing a new era in which JIT delivery cycles, customized goods on short notice, uniform batch production, and geographic separation of individual stages in the production process have resulted in a growing requirement for air and surface–air lift transportation.

Kasarda (1991) has described the emergence of international airports as integral components of this linked global production system. Increased volumes of air freight, plus the phenomenal rate of growth in international business and tourist air travel, have created powerful forces requiring the expansion and redevelopment of

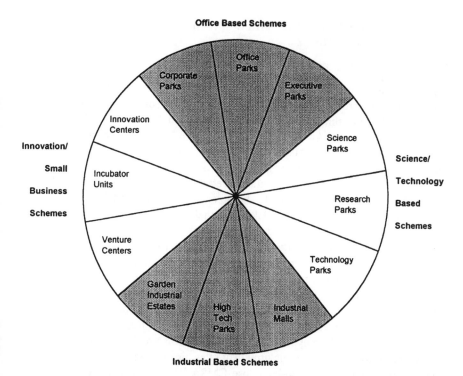

Figure 4.5. The variety of high-tech developments in metropolitan regions. *Source:* Jones Lang Wootton, 1990.

existing, and the dedication of large, new spaces, for the development of airports and their related functions in metropolitan regions. Inevitably, regional proximity to an international airport is cited as a locational factor in executive decision making for both the relocation of existing and the location of new enterprises. Airports have become the loci of major industrial, technology, and commercial complexes, incorporating large spaces for transshipment activities.

Similarly, technological change and globalization processes have resulted in the rapid growth of container shipping. Major metropolitan regions have become the focal locations in globally linked networks of trade. This has generated the need for large spaces to be dedicated to new container shipping ports and has created the need for major new *transportation corridors* and *communications networks* linking air, sea, and road systems.

Increasingly, airports and seaports of the highest international quality are the essential components of the land use and economic structures of metropolitan regions to provide the required strategic global linkages that underpin much of the economic development of, and employment in, cities. Significant conflicts inevitably arise between the need for new or expanded airports and seaports (and their associated activities) and their sometimes deleterious environmental and social impacts.

4.4 Economic Performance and Employment in Cities: Trends and Patterns

4.4.1 The United States Experience in the 1980s

It is common for considerable differences to exist between the economic performances and the nature of labor markets in large cities. For example, in the United States in 1980, the three largest metropolitan regional economies (Consolidated Metropolitan Statistical Areas, or CMSAs) were the New York–Northern New Jersey–Long Island, the Los Angeles–Anaheim–Riverside, and the Chicago–Gary–Lake County. This reflected their global city roles as well as their prominent roles within the national economy and national settlement system. But by 1991 the nation's urban economy had adjusted further to changing global and domestic circumstances, with the San Francisco– Oakland–San Jose region displacing the Greater Chicago region as the nation's third-largest metropolitan regional economy. This reflected the relative growth and development of the information and services sectors. Tables 4.1 through 4.3 show the fastest-growing and -declining centers in terms of gross economic output and regional employment. The changes portrayed in these tables are "symbolic of the relative gains made by producer services at the expense of consumer services within the rapidly transforming services producing sectors of metroregional economies throughout the nation" (Hicks & Rees, 1993:2.27).

During the 1980s, only the 10 largest metropolitan regional economies in the United States displayed economic growth at rates higher than that of the nation as a whole. San Francisco–Oakland–San Jose CMSA led with nearly a two-thirds increase in economic performance, much of it due to its high-tech Silicon Valley ac-

TABLE 4.1.
Ten Largest U.S. Metro Regional Economies (Ranked by Gross
Regional Output)

1980	1991
New York–Northern New Jersey	New York–Northern Jersey
Los Angeles–Anaheim–Riverside	Los Angeles–Anaheim–Riverside
Chicago–Gary–Lake County	San Francisco–Oakland–San Jose
San Francisco–Oakland–San Jose	Chicago–Gary–Lake County Philadel-
phia–Wilmington–Trenton	Philadelphia–Wilmington–Trenton
Detroit–Ann Arbor	Detroit–Ann Arbor
Houston–Galveston–Brazoria	Boston–Lawrence–Salem
Boston–Lawrence–Salem	Dallas–Forth Worth
Dallas–Forth Worth	Houston–Galveston–Brazoria
Cleveland–Akron–Lorain	Washington, DC

Source: Hicks & Rees, 1993: 2.98.

tivities, but also because of a diversified economic base. It was followed by the Boston– Lawrence–Salem (55%) CMSA in New England, again because of the role of the famous Route 128 high-tech complex and its industry and university research clusters. Third was Dallas–Fort Worth (48%), followed by Los Angeles–Anaheim–Riverside (43%). Overall, metropolitan regions displayed both a broadening of their manufacturing base, and rapid growth in high-level producer services (especially in financial, management, consulting, medical, legal, and educational services) and high-tech driven commercial and defense activities. The slowest rates of economic expansion among the top 10 metropolitan regional economies were in greater New York, Chicago, Houston, and Cleveland CMSA.

Over the period 1980–1991, employment growth was registered in all but 13 of the 282 metropolitan regions of the United States, but the patterns of perfor-

TABLE 4.2.
Ten Fastest Growing Metro Regional Employment Centers
(Ranked by 1980–1991 Metro Regional Output Change)

Region	Rate of Expansion 1980–1991 (%)
Orlando, FL	85.7
Las Vegas, NV	75.3
West Palm Beach, FL	74.7
Austin, TX	59.8
Phoenix, AZ	56.8
Raleigh–Durham, NC	53.7
Tampa, FL	53.7
San Diego, CA	53.6
Sacramento, CA	50.1
Augusta, GA	49.8
United States	20.0

Source: Hicks & Rees, 1993:2.98.

TABLE 4.3.

Ten Fastest Contracting Metro Regional Employment Centers
(Ranked by 1980–1991 Metro Regional Employment Loss)

Region	Rate of Decline 1980–1991 (%)
Davenport, IA	–7.3
Worcester, MA	–6.1
Peoria, IL	–3.5
Youngstown, OH	–1.8
Beaumont–Port Arthur, TX	–1.4
Huntington, WV	–1.3
Pittsburgh–Beaver Valley, PA	–0.2
Shreveport, LA	0.5
New Orleans, LA	0.9
Flint, MI	1.7
United States	20.0

Source: Hicks & Rees, 1993:2.98.

mance varied greatly around the national figure of 20%. Among the 10 largest metropolitan regions, seven had lower rates of employment growth. Thus, growth in employment was highly concentrated with greater Los Angeles alone adding 1 in 10 of the nation's new jobs. Its labor force gained 1.9 million jobs. The five fastest-growing metropolitan regions accounted for almost one-third of the job growth. Florida had 8 of the 10 fastest-growing metropolitan regions in the nation, and California, Nevada, Arizona, Texas, Georgia, and North Carolina all had metropolitan regions in the top 20 for employment growth. Hicks and Rees (1993) concluded that:

> This employment growth in large urban areas may be especially remarkable to the extent that it is likely that the Nation's largest metropolitan regions nurtured economic growth as much through gradually redeploying their assets into mixes of higher value-added and more productive industries as through simply putting more people to work. In short, their local economies tended to evolve in ways that raised new labor requirements. Eventually, the redistribution of labor among regions and the more subtle transformations of what constitutes "work" itself within them has the power to rework slowly the Nation's overall system of urban and rural economies and even blur the boundaries between them. (p. 2.29)

There was considerable diversity among the employment growth experiences of cities between 1980 and 1990 with all of the top 10 being in the South and West, reflecting high population gains through internal migration and reflecting long-term, multistate regional growth that has shifted from the Southeast and Midwest to the South and West. But, there was not a direct relationship between a high rate of employment growth and regional productivity. The latter actually declined in some cases, as with Las Vegas. Thus, "in urban circumstances employment growth in and of itself may not necessarily constitute much of an economic elixir for a region or much of a foundation for healthy regional development" (Hicks & Rees,

1993:2.37). For metropolitan regions with the lowest employment change, the greater losses were in cities that began the 1980s at productivity levels above the national average, indicating that job loss might have been a vehicle for regions to rise in productivity.

Hicks and Rees (1993) also analyzed the degree to which metropolitan regions in the United States had diversified or specialized their economic base during the 1980s. They used an eight-sector model to measure restructuring effects. Their results showed that for 227 (70.3%) of the 323 metropolitan regions their basic economic structures grew or became more diversified during the 1984 to 1988 period and that only 96 (29.7%) grew more specialized by experiencing rapid growth in already dominant sectors. In the Northeast, the general trend was increased diversification. In the North Central sector, there was considerable economic restructuring and a shift toward diversification in all major cities except St. Louis. In the South, the largest metropolitan regions of Washington and Dallas began the period more specialized than other metropolitan regions, but diversification was apparent across the board. In the West, there was evidence of considerable heterogeneity across metropolitan regions but relatively little evidence of restructuring shifts, with some cities becoming a little more diversified. Overall, it was evident that, as metropolitan regions went through transitions from a heavy goods manufacturing orientation to a services orientation, there was a shift toward diversification. A significant issue is the degree to which such shifts will be permanent or cyclical (Hicks & Rees, 1993:2.22).

4.4.2 Employment Patterns within a World City: Sydney

Within cities there are marked differences in the patterns of employment growth and decline that reflect the processes of structural adjustments discussed above. An example is Sydney, Australia's largest and only "world city."

As the Australian economy becomes more internationalized, it is to be expected that Sydney will become even more dominant as the location for internationally linked functions for both overseas and domestic firms. It is the dominant international gateway airport, the leading tourist destination, an international finance and trade center, a center of R&D, and a center of international cultural and entertainment activities. Sydney's winning bid for the 2000 Olympics will focus even more international attention on that city as the gateway to the nation. It is a highly cosmopolitan city, being the main destination for new immigrants to Australia (a role once held by Melbourne).

But within the metropolis of Sydney, relatively few activities, people, and places are part of this world city phenomenon. As Sydney has become part of the network of cities linked through processes of globalization and internationalization, a new form of duality has emerged within its regional economy and across its social space. Sydney's labor market and economic structure are bifurcating into high-flying industries such as finance, property and business services, and low-paying, often part-time occupations in retailing, entertainment, and community services. Tens of thousands of manufacturing jobs have gone, and large segments of the region have very high levels of unemployment. These differences express themselves strongly in space, and this differentiation is compounded by high levels of diversified immigration. Not only is there broad differentiation among zones according to socioeconomic states, life-cycle characteristics, and ethnic mix, but also the scale of the developments has produced identifiable cities-within-cities that exhibit microeconomic and social characteristics distinctive from the rest of the metropolis (Daly & Stimson, 1992).

For the Sydney metropolitan region, while about 80,000 new jobs were generated over the decade 1981 to 1991, there were large losses of jobs in the manufacturing/transformative sectors of the metropolitan economy with a loss of over 70,000 jobs. This is illustrated in Figure 4.6. Within the city, this pattern of loss was particularly marked in the inner suburbs and the suburban areas that based their growth after World War II on energy-intensive, assembly-line manufacturing activities and employment in the distributive activities (including retail and transportation). These declined in the inner-city zones and in some outer, suburban areas. There was strong job growth in social services, especially in the lower socioeconomic areas in the outer suburbs. Growth in jobs in producer services was widespread throughout the metropolitan region; this was especially so in the higher socioeconomic status locations in both inner and outer suburbs.

This pattern of suburbanization of employment growth in the consumer and community services as well as producer services has been noted by other researchers, including Gordon in Los Angeles. Figure 4.7 shows how, between 1982 and 1987, employment growth in the outer counties of the greater Los Angeles area was overwhelmingly dominant across all industry sectors compared to the City of Los Angeles (inner city) and the five county regions.

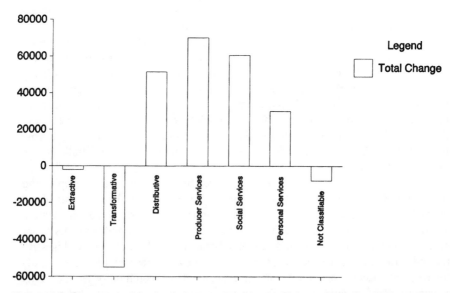

Figure 4.6. Changes in jobs by industry categories in Sydney, Australia, 1981 to 1991. *Source:* AHURI, 1994.

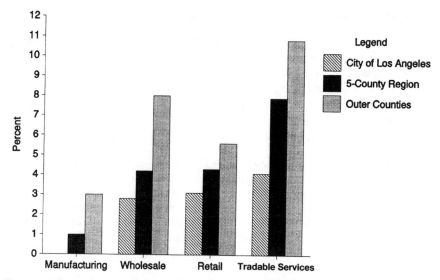

Figure 4.7. Average annual growth rate in employment in the Greater Los Angeles region, 1982 to 1987. *Source:* Peter Gordon, University of Southern California.

4.4.3   Implications for Commuting

Within metropolitan regions, the pattern of work–residence commuting relationships is far from simple. They are becoming more complex as a result of changes in the processes of economic and demographic evolution discussed in Chapter 3. In the polycentric-structured metropolitan region, we see the emergence of subregional labor markets, the increasing incidence of multiple-purpose trips, increased participation of women in the labor force, and growth in the incidence of part-time and multiple-job employment. These trends, along with the increase in jobs that involve considerable travel and the conduct of tasks at many locations and the increasing incidence of home-based work (telecommuting), all combine to produce increasingly complex patterns of intraurban travel.

Current work on commuting by Peter Gordon, a planner at the University of Southern California, shows that by 1990, for the first time, over half of all households were generating two- or more person daily work trips and that the time distance of the commute was becoming longer for women than it was for men. Gordon's work and that of Cervero (1989) show that cross-metropolitan region and suburb-to-suburb trips are by far the dominant commute patterns in cities, making it difficult, if not impossible (and certainly making it economically impossible), to provide public transport for the vast majority of trips. Gordon and Richardson (1993) show that for the Los Angeles region, which is the prototype mega metroregion, between 1980 and 1990 the actual amount of commuting travel times decreased as a result of the evolution of subregional labor markets and the increasing importance of the many edge cities.

4.5   Urban Social Space and Residential Patterns

Within cities the nature and patterns of residential areas have for long been a major focus of research by geographers and sociologists. The characteristics of the population of residential areas vary considerably across the city, and they change over time within suburbs because of the demographic transition of the population.

An important question confronting geographers as social scientists is "How do people sort themselves out in [urban] space?" Jakle, Brunn, and Roseman (1976) noted the following characteristics of groups of individuals in space:

> The individual gains knowledge about places from his own perception and by his interaction with others. When this information is processed into behavior, it will affect his views about particular groups of the population, preferred places to live, and acceptable modes of personal communication. The groupings of like-minded individuals in a specific area, whether a distinct social area of a city or a social region of a nation, will lead to various patterns of clusterings. These are clusters of people who have decided to occupy similar spaces for economic, social, or political reasons. (p. 185)

The consequences of these processes may be regarded as the aggregate location decisions of people. At the *intra urban* level, one may differentiate residential areas according to the social and economic characteristics of their inhabitants and the characteristics of the dwelling stock within suburbs. In other words, the *social space* characteristics of cities and their *territorial* expression provide a mosaic of residential areas in any city.

4.5.1   The Notion of Social Space

It was the French sociologist Chombart de Lauwe (1952) who referred to *objective social space*, which he defined as "the spatial framework within which groups live; groups whose social structure and organization have been conditioned by ecological and cultural factors." *Territorial space* may be stratified by the social dimensions of the population of the city. Buttimer (1969, 1972) discussed the notion of social space and its applications in urban geography, and countless geographers have mapped the spatial distribution of various social groups in cities.

However, there has been considerable semantic confusion over the notion of social space. The term *social space* identifies relationships between groups or other social phenomena chosen as points of reference. This was defined by sociologists in the 1920s, in terms of a system of coordinates whose horizontal axes referred to group participation and whose vertical axes referred to status and roles within groups. In this way the social position of any person in society could be defined. The sociologist Park (1924) stressed the subjective dimensions of social space. Somewhat later social space became defined in terms of psychological variables. Later Theodorson and Theodorson (1969:394) contended that "social space is determined by the individual's perception of his world and not by the objective description of his social relationship by the observer." This interpretation had its roots in Durkheim's (1893) concept of sociological space and said nothing of physical or territorial space. Sorre (1955, 1957, 1958) attempted to link these, as did Chombart

de Lauwe (1952, 1965). Basically, their propositions were that the patterns of organization were techniques of social life and that political, economic, and kinship
spaces were dimensions of social space. Sorre (1957) envisaged social space as:

> a mosaic of areas, each homogeneous in terms of the space perceptions of its inhabi-
> tants. Within each of these areas a network of points and lines radiating from certain
> points could be identified. Each group tended to have its own specific social space,
> which reflected its particular values, preferences, and aspirations. The density of social
> space reflected the complementarity, and consequently the degree of interaction be-
> tween groups. (p. 114)

Subsequently, techniques such as *social area analysis* and *factorial ecology*
have allowed the identification, description, and objective mapping of social space.
While theoretical and philosophical considerations of social space were being pursued largely by French geographers and sociologists, it was not until the late 1960s
and the early 1970s that Anglo–American geographers began to understand the
complexity of the concept and the processes underlying social space in cities. At the
same time, many urban and social geographers in North America were beginning to
disaggregate the normative economic models developed by location analysts in an
attempt to investigate, and hopefully explain, human spatial behavior.

During the 1950s and 1960s, the pioneers of social area analysis and factorial
ecology had looked at the isomorphism of social participation patterns and social
spaces in cities, and they had matched activity patterns with the spatial morphology
of social characteristics. However, they made little attempt to examine the isomorphism of place identification with so-called social spaces. *Factorial ecology* developed to identify the dimensions of urban social space and to map their spatial distributions, but no consideration was given to whether or not the populations resident
in those social spaces were satisfied with their lot or how they came to be there.
Later, behavioral researchers in geography became concerned with studying, among
other things, the relationships between and among social space and individual
and household action spaces and the time–space budgets and activity patterns of
people.

It is interesting to see the relationship between the study of human spatial behavior in urban areas and Chombart de Lauwe's concept of *urban social space*, in
which a hierarchy of spaces had been proposed within which groups and individuals
live, move, and interact. Chombart de Lauwe, (1960:403–425) identified four specific spaces:

1. *Familial space*, or the network of relationships characteristic of the domestic level of social interaction.
2. *Neighborhood space*, or the network that encompasses daily and local
   movement.
3. *Economic space*, which embraces certain employment centers.
4. Urban sector, or *regional social space*.

These are progressively larger and overlapping dimensions of space that constitute the nominal spatial framework within which an individual or a group acts. They may be related to the four spaces developed by other researchers, namely:

1. *Personal space* (Sommer, 1969).
2. *Neighborhood space* (Lee, 1964, 1968).
3. *Activity space* (Horton & Reynolds, 1969, 1970).
4. *Action space* (Horton & Reynolds, 1969, 1970).

Obviously, migration and residential location decision processes of individuals and households will have an effect on shaping the territorial social space characteristics of urban areas. As indicated by the work of Buttimer (1969:423), geographers have approached the investigation of residential mobility and the residential location decision process of households in cities and the study of social space using three levels identified by Chombart de Lauwe, namely:

1. The *behavioral level*, which relates to where and how people live and move.
2. The *level of knowledge*, which relates to where people know what alternative opportunities are available.
3. The *aspirational level*, which relates to where people would like to go if they had the opportunity.

Until the 1970s, geographers tended to focus their attention on the territorial expression of social space, using census data for small areas to plot the distributional characteristics of the *dimensions of social space*, such as socioeconomic status. They rarely examined explicitly the influences of life-style, social stratification, stage in the family life cycle, status, role, tradition, symbolism, and values in studying the relationships between the behavior of individuals and groups and the elements of the structures that comprise the spatial environment. Buttimer (1972:285) identified five distinct levels of analysis that attempt to integrate *objective* and *subjective* social space in urban analysis. These were:

1. A *social psychological* level investigating a person's position within society—that is, *sociological space*.
2. A *behavioral* level investigating activity and circulation patterns—that is, *interaction space*.
3. A *symbolic* level investigating images, cognitions, and *mental maps*.
4. An *affective* level investigating patterns of identification of *territory*.
5. A purely *morphological* level in which population characteristics are factor-analyzed to yield *homogeneous social areas*.

This provided a useful framework for analysis, and it was possible to take a number of the aspects of Chombart de Lauwe's social spaces modified by the terminology developed by Buttimer, as an interpretive base. These included:

1. *Social space* could be taken to refer to Chombart de Lauwe's objective social space or to Buttimer's social areas. It is the territorial expression of descriptive

dimensions extracted from a factor analysis of a set of population census variables for small areas. Such variables are surrogate measures or constructs such as socioeconomic status, family/life-cycle status, and ethnic status. These were used to classify or describe the position of people within society. The approach incorporated Buttimer's sociological space in a territorial framework.

2. *Preference space* is taken to be comprised of Buttimer's *symbolic level* of investigating images, cognitions, and mental maps. It relates to the residential aspirations of subgroups of the population and is influenced by constraints such as economic factors and institutional barriers.

3. *Affective level* and *interaction space* were taken to be related to that form of spatial behavior called *migration and the residential location decision process.* This incorporates the identification and description of things such as spatial search and the accumulation of knowledge, and levels of satisfaction and dissatisfaction. The process involved both spatial and nonspatial considerations, which relate to those elements that constitute and shape urban social space and space preferences of people in urban areas.

4.5.2 Ecological and Psychological Fallacy

It is in this type of context where geographers sought to bring together the normative type aggregate analysis of spatial patterns and the behavioral-based disaggregate analysis of individual decision making which, when aggregated, produce spaces of differentiation, that gave rise to concerns over what are known as the *ecological fallacy* and the *psychological fallacy.*

These two fallacies were produced by the felt need to work at aggregate or disaggregate scales. Building theories and models to help understand individual behavior was seen to be less powerful and desirable than building theories and models that explained the behavior of aggregate populations. While progress was made (and continues to be made) on each of these fronts, particular concerns arose to predict individual behavior on the basis of aggregates (ecological fallacy) or where models of individual behavior were extrapolated in an attempt to account for the behavior of aggregate populations (psychological fallacy). These fallacies generally emerged where intransigent positions were taken as to the value of macro or micro scale approaches.

The more general approach, which we endorse in this book, is to work toward the merging of the *micro* and *macro*, the *individual* and the *society*, the *single unit* and the *aggregate*. This has proven to be an extremely difficult task and is not solved to date. However, we feel that by combining both of these approaches in a single volume, we can bring forth the problem and potentially highlight some of the difficulties and barriers that are likely to be encountered in such a merging process.

4.5.3 The Constructs of Territorial Social Space: Social Area Analysis

The analysis of the social structure of urban society, and the spatial ramifications of the interaction of socially differentiated population subgroups within it, have both been enhanced by: (1) the use of statistical techniques, such as *ecological correlation* and *residential segregation indices*; (2) by the recognition of processes of *ecological competition and succession*; and (3) by the formulation of theories of urban structure such as Burgess' *concentric zone model* (1925), Hoyt's *sector model* (1939), and Harris and Ullman's *multiple nuclei model* (1945). These models related to the structure and form of cities and the way they evolved over time (see Figure 2.8 for a representation of such structures and forms of urban areas).

A major advance in the study of the social structure of the city came in the 1950s with the development of social area analysis. Attributed to two West Coast sociologists in the United States, Shevky and Bell (1955), this provided a theoretical and methodological framework by which urban social space could be described and analyzed.

Social area analysis, as proposed originally by Shevky and his various associates, provided a systematic classification of residential areas in cities. A social area schema attempted to specify the nature and effects of structural change in urban industrial society as it had evolved up to the 1950s, in which the metropolis emerged as the major living space of people. It was hypothesized that social differentiation in residential areas could be viewed from the basis of three broad trends in urban society that produced three broad components of social differentiation in cities, namely:

1. *Changes in the distributions of skills*, and subsequently the arrangement of occupations based on function, which give a *socioeconomic status (social rank) construct*.
2. *Changes in the way of life*, involving the movement of women into urban occupations and the diversification of family patterns, which give a *family status (urbanization/familism/household structure) construct*.
3. *Changing composition of the population* through the redistribution of space of ethnic and religious groups, due to both internal and international migration, giving an *ethnic status (ethnicity/segregation) construct*.

Shevky and Bell maintained that these three constructs could be identified and described by means of selected indices taken from the census tract data on the characteristics of populations and dwellings. The use of the census tract as an areal unit of analysis meant that the *social area typology* for inquiring into the changing structure of urban industrial society was made synonymous with the study of social differentiation in residential areas of cities.

Despite criticisms of the conceptual basis and methodology of *social area analysis*, and in particular the lack of explicit recognition of gender roles that are discussed in Chapter 15, Section 15.3, empirical studies of the schema, particularly when modified to permit the identification of social dimensions derived from multivariate analysis of a wide battery of census tract population and dwelling data, have led to general validation of the hypothesis that the three constructs, *socioeconomic status*, *family status*, and *ethnic status*, are necessary to describe the social differentiation that occurs in urban ecological systems.

4.5.4 Factorial Ecology Studies

In the 1960s the increasing availability of computer facilities led researchers to develop a *factorial ecology analysis* to identify and describe the dimensions of urban social structure. Typically they analyzed a list of census date variables relating to: gender, occupation, income, and education of the labor force; gender and age distribution and marital status of the population; ethnic structure of the population; and nature of tenancy and dwelling characteristics of the residential areas of a city. The multivariate statistical tool called *principal components* or *factor analysis* was used to reduce this data matrix and to identify a set of descriptive dimensions relating to the social structure of a city. Usually between 4 and 10 components/factors explained up to about 80% of the total variance in the data matrix. These factors were described in terms identical to, or derived from, the three social area constructs proposed by Shevky and Bell (1955). It was usual to compute component or factor scores for residential areas on these dimensions, and to analyze the spatial distribution of the scores, often with the intention of testing for the presence of zonal and sectoral variations in the spatial patterns of dimensions. Thus, it was possible to identify dimensions that describe the nature of urban social space and to map the territorial patterns of these spaces using small-area census data.

4.5.5 The Relationship between Social Space Dimensions and Territorial Space in Cities

Figure 4.8 conceptualizes the relationships between social space and geographic space in cities. In the light of hundreds of empirical studies conducted in the 1950s, 1960s, and 1970s, this relationship may be described in the following manner:

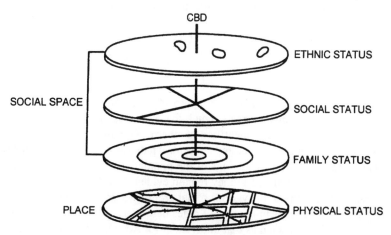

Figure 4.8. Urban social space constructs and their territorial relationships to physical space. *Source:* Murdie, 1969:9.

1. *Territorial (or geographic) space* provides a living space within the city in which, or over which, its inhabitants reside and behave. The space contains elements that are point located and area located and that form paths along which movements take place. For example, there is a network of transportation routes (often sectoral in arrangement) that converge on locations at which urban functions are concentrated, such as the CBD, regional shopping centers, and industrial nodes. These movement paths criss-cross the city. The spaces between are functional, and they contain varying mixtures of commercial, retail, residential, and recreational land uses. At some focal or nodal points, and along some movement paths, there are areas of concentration of specialized functions, but the major parts of these spaces, particularly at locations away from those focal points and paths, have a residential function.

2. Overlying this geographic space is the *social space* of the city, which is multidimensional and comprises three basic constructs, social status, family status, and ethnic status.

Typically, the territorial space expression of these components of social space varies. For example:

1. *Family status space* is *zonal* in its geographic space occurrence, with the older, inner-city areas being characterized by older populations and many nonprimary family unit households. With movement out of the inner-city areas, there is a concentric or zonal progression through increasingly more youthful populations and areas characterized by higher fertility and mostly primary family units.

2. *Social status space* is seen to vary in a *sectoral* pattern, the hypothesis being that once established, high- and low-status residential areas tend to extend along sectors defined by movement paths (transport routes) as the city grows.

3. *Ethnic status space* is theoretically ghettoed or *enclaved* geographically, with the population subgroups that have specific ethnic status tending to be residentially segregated in these enclaves, which have a tendency to move through time. Thus, social status spaces are sector-cutting across zonally arranged family status spaces, while ethnic status spaces are enclaves usually within low-social-status space.

A well-known application of this approach was Murdie's (1969) study of Toronto, Canada. Here, major urban social space dimensions and their physical space occurrences were used to identify a number of communities or social areas, which were arranged in both sector and concentric patterns and which had enclaves and areas of dominant ethnic groupings of the population. Such patterns are common in North American cities, and they remain evident over time, despite changes in preference and mobility of the various social groups within metropolitan areas. In this way, *typologies* of residential areas of cities based on their social space characteristics could be formed, such as that undertaken for Adelaide, South Australia, by Stimson (1978, 1982), as illustrated in Figure 4.9.

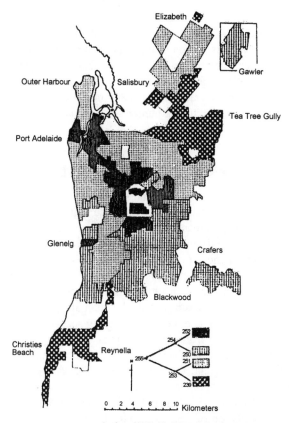

I. Group 239 (N=16): Outer suburban areas in the southern, northeastern and northern sectors. Areas of rapid growth; very high levels of familism; low to medium socio-economic status; low levels of urbanization; and high concentrations of British migrants.

II. Group 251 (N=50): Industrial suburbs of the northern northeastern, western and southern sectors, plus the new town of Elizabeth. Generally low socio-economic status; high incidence of Housing Trust housing; variable levels of familism, but generally above average; above average levels of ethnicity, both British and southern European; below average growth.

III. Group 250 (N=32): The eastern, southern hills, and southwestern coastal sectors. Areas of high socio-economic status; variable levels of of familism; low ethnicity; variable growth.

IV. Group 252 (N=30): The old inner-city suburbs and the northwestern sector. Generally low to medium socio-economic status, with some pockets of above average levels; high ethnicity; high urbanization; low familism; low growth.

Figure 4.9. A typology of residential areas in Adelaide, South Australia: social space characteristics of territorial spaces. *Source:* Stimson & Cleland, 1975.

4.6   Socioeconomic Disadvantage in Urban Space

A major thrust in the 1960s and 1970s among human geographers was to investigate the nature, incidence, and distribution of social well-being, particularly within
cities (D. M. Smith 1973, 1977; Stimson, 1982). In the above discussion on urban
social space and the earlier discussion of the internal structure of cities, and in particular on the hierarchical distribution of urban services, it was noted that there is
usually (1) a marked spatial differentiation within a city in the distribution of constraints such as socioeconomic status, and (2) an uneven pattern of distribution of
the location of urban services, including human services such as health services and
educational facilities.

4.6.1   Differences in Access Opportunity and Social Disadvantage

As a result of these spatial variations in social spaces and in the incidence of provision of services and facilities in cities, there are wide spatial inequalities in the access opportunity of people to urban services. This is so because services are not
ubiquitous in their location, and some subgroups of the population are spatially segregated, whereas others are more evenly distributed. Thus, a person's or a household's socioeconomic status, for example, is likely to constrain the range of residential locational choices that he or she may exercise because of spatial variations in
the nature and cost of housing markets; and the location of a person's residence will
influence both the time-distance and the cost of gaining access to specific services.

   For example, it is evident from Figure 4.10 that in Sydney, Australia, there is
a marked pattern of concentration of socioeconomic disadvantage within the inner
suburbs, the western suburbs, and the outer urban fringe areas to the southwest,
west, and far north of the metropolitan region.

   Data in Table 4.4 for both Sydney and Melbourne, the two largest Australian
cities, show that the population subgroups with the highest *index of access* difficulties are older single persons, couples with children under 10, and sole parents. Within various zones of those cities there were major differences for the indices in particular subgroups across the inner, middle, and outer suburbs and the fringe areas.
In general, it was the fringe areas of the metropolitan region where indices of access
difficulties were the greatest.

4.6.2   Duality within Cities

In Chapter 3, Section 3.2 we noted that the processes of globalization and economic
restructuring have produced inequitable social impacts. This is particularly so within cities. There have been substantial differential impacts on land markets through
competition for prime office space and quality residential locations, and there have
been major impacts on labor markets, including competition for quality staff. As a
result, highly differentiated housing markets have developed, and nonprofessional
and technically lesser-skilled workers increasingly are segregated and pushed into
lower land-cost areas on the fringe of cities. As a result often they face long commutes to work. And in many of the older industrial cities of the United States, capi-

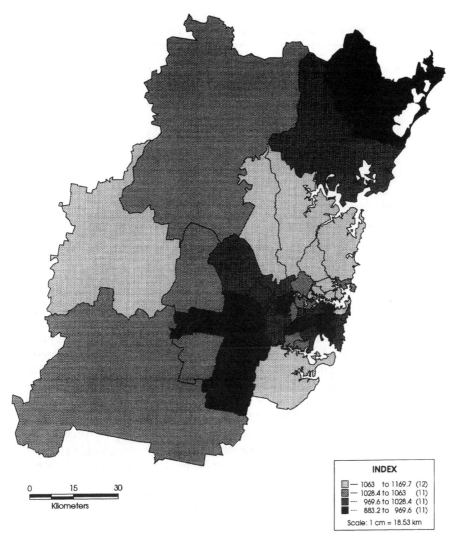

Figure 4.10. The pattern of socioeconomic disadvantage in metropolitan Sydney, Australia, 1991. *Source:* ABS; Fincher, 1994.

tal and production have fled the decaying inner-city areas, as have the middle-class white populations, leaving these areas as the domain of a disadvantaged black population, other disenfranchised minority groups, and the homeless. In addition, deregulation of labor arrangements and differentials in wage and working conditions have occurred, exacerbating further the social differentiation within cities.

Although social and economic segregation always have existed in cities, a new form of differentiation appears to have emerged. Now a duality is common in cities (Mollonkopf & Castells, 1991), and there is considerable debate over whether planners, developers, or the market have created this spatial segregation.

TABLE 4.4.
Index of Access Difficulties by Location within Sydney and Melbourne, 1991

Location within Cities	Core	Inner	Middle	Outer	Fringe	Metropolitan Region
Prefamily singles, couples <35	0.5	0.7	0.7	0.6	1.3	0.7
Singles and couples aged 35–64	0.6	0.7	0.8	0.8	1.2	0.8
Couples and young children <10	0.5	0.8	0.8	1.1	1.5	1.1
Couples and older children	0.4	0.5	1.0	1.0	1.0	0.9
Sole parents	0.5	0.7	1.3	1.2	1.9	1.1
Older singles 65+	0.9	1.7	1.9	1.4	1.9	1.6
Older couples 65+	0.5	1.2	0.9	0.9	1.4	1.0
Other households	0.7	0.8	0.7	1.2	1.0	0.9
All households	0.6	0.9	1.0	1.0	1.3	1.0

Source: National Housing Strategy, 1992:38.

Daly and Stimson (1992) show the magnitude of this duality in a case study of Sydney, Australia. As a result of the processes of internationalization, the growth of the services sector, and a loss of employment through restructured manufacturing, the city has been bifurcated into high-paying industries, such as finance, property, and business services, and low-paying, often part-time and female-dominated, jobs in retailing, tourism, and community services. These differences express themselves strongly in space. Their impacts are compounded by high levels of immigration from diverse sources. For example, the local government areas of Fairfield and Liverpool, have a combined population of over 300,000 making them larger than any nonmetropolitan city in Australia. They have 125 different ethnic groups. In some localities most people do not speak English. Their populations are among the youngest and fastest growing in Sydney, but unemployment is double the average, and youth unemployment stands at around 50%. Investment has been directed at special local projects, including cultural, sporting, retail, and office developments of considerable scale. The area has a plan to attract tourists, and the Vietnamese shopping center attracts hundreds of thousands of visitors a week. These local authorities have developed an economic and social profile that is distinctive within Australia, but they represent vast areas of economic and social disadvantage in the suburbs.

Similar processes have been identified in major cities such as Los Angeles, which has evolved as a system of cities, with local labor markets and with distinctive areas of ethnic domination encompassing blacks, Hispanics, and various Asian minorities, such as Koreans and Vietnamese.

4.7   New Paradigms for the City

4.7.1   A "Metroplex" of Spaces and Places

It is necessary to rethink the traditional models of the city in the new era of the global economy and society in the information and services age. We have seen how the economic base of the city system has become decentralized because it is now increasingly knowledge-based and information-intensive. As a result, the basic product now is new *information* rather than goods.

This gives rise to a new city paradigm as proposed in Figure 4.11. It starts with *human resources,* rather than the needs of industry for raw material locations, which in turn leads to a pattern of *preferred* rather than required locational considerations. Human capital choices for high-amenity, often low-density life-styles, and more opportunities to exercise social and institutional choices drive the configuration of the city. More and more jobs have moved to the suburbs, to the new edge cities, and to more widely dispersed locations, including home offices. Amenity becomes a basic component of infrastructure. This new form of city does not have the traditional hierarchical spatial forms of the older city structures; new synergies evolve through the overlap among environment–technology–humanity components of a complex set of socioeconomic systems. These are defining the evolution of new urban functions in the spatial context of a polycentric, dispersed urban form and structure. Blakely and Stimson (1992) have proposed the notion of a *metroplex* of places and spaces of functions and people, operating in an international as well as a regional or local environmental context. This is conceptualized in Figure 4.12. It shows, in a two-dimensional way, the multidimensional interrelationships that drive the modern city. It is not solely a spatial model; rather it shows how work, home, recreation and life-style diversity are all transformed into a single living system. The built form of the city evolves to accommodate the socioeconomic–technologi-

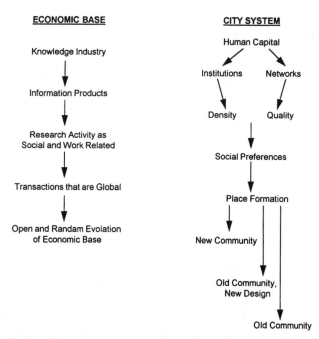

Figure 4.11. The basis for the new city. *Source:* Blakely & Stimson, 1992.

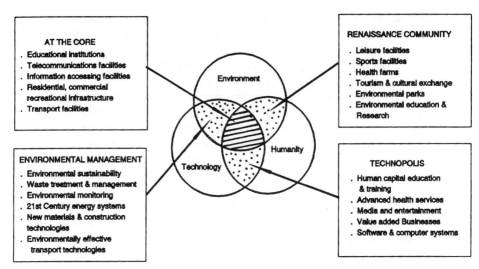

Figure 4.12. Synergies for the new information city. *Source:* Blakely & Stimson, 1992.

cal–environmental–humanity factors that are driving the changing processes of production, social interactions, roles, and life-styles.

4.7.2 Institutional Rigidities and the New Demands of the Services and Information Age

Important issues for the citizens of any community in modern urban society, whether it be a neighborhood or a city or a region, are the appropriateness and suitability of traditional institutional arrangements of government for managing the provisional services. How to manage change and how to fund and provide the services and facilities that people and business need are major challenges.

Knight (1989:332) has claimed that traditionally cities have "tended to be handicapped by their limited investment in institutionalized learning and by the short time-frame within which they plan them by their corporate charters." In general there is an inability of tradition-bound bureaucracies to cope with the continuously shifting economic, social, and demographic contexts. There is an inability to augment public expenditures, and in particular capital expenditures, at rates commensurate with the rising needs for public investment in urban and regional infrastructure and services. This problem has been exacerbated by the success in the 1970s and 1980s in the United States of voter-initiated restrictions on fund-raising and expenditure programs of governments and city administrations. In addition, right-wing rationalist economic ideologies have held considerable sway in governments, resulting in curtailment of the fiscal base and the capacities of cities and communities to provide infrastructure and services. Generally there exist inadequate capacities and capabilities (in particular human resource skills) in government at the city and local levels to deal with the increasingly technical and complex job of managing large and rapidly growing city regions. Bureaucracies are suffering under the strain of cumbersome, vertically arranged functional departmental structures, which represent constraints to embracing effectively new communications and information technologies, undertaking joint planning, and developing intergovernmental collaboration. And typically there are inadequate time frames (usually annual) for budgeting and decision making in government. This is exacerbated by an ad hoc approach to planning and program delivery.

What we have in most if not all nations are inherited 18th- or 19th-century systems and structures of government that were developed in an era of totally different economic, social, demographic, and technological forces and processes to those operating at the end of the 20th century. Governments are struggling to cope with the quite different and complex contemporary exigencies.

Daniels et al. (1992:10–11) have pointed out how the growth of the services sector has had important ramifications for the financing, upgrading, extension, maintenance, and management of urban infrastructure. The infrastructure of many cities is old and often inadequate to provide the capacity and efficiency of the "smart infrastructure" that is essential for the sustained development of the services sector activities of the modern metropolitan city. Langdale (1991) and Moss (1988) have shown how the services sector has replaced manufacturing as the most significant component of the nonresidential revenue and tax base in many metropolitan cities in Australia and the United States. But they are also generating unprecedented and often unexpected demands for new forms of infrastructure that require innovative approaches for funding, provision, and operation.

The following have been identified by Daniels *et al.* (1992) as reasons why the producer services are of crucial importance in this policy context:

1. *Services generate demand for physical infrastructure*, including office floorspace, advanced telecommunications facilities, international air services, public transport services, and efficient highway/freeway networks within metropolitan regions; much of this infrastructure is for "intelligent" office buildings and state-of-the-art facilities, including large-volume voice and data transmission systems utilizing intermittent networks.

2. *Services have profound impacts on local labor markets*, generating increased requirements for highly educated/skilled/professional specialized labor, plus less-skilled, predominantly female, and part-time labor for data processing.

3. *Services have high multiplier effects on the demand for specialized services*, especially in computing, personnel recruitment, and high-quality printing, plus multiplier effects on retail services (quality and quantity), housing (location and accessibility), education (especially private schooling and quality universities), social services, health services (especially private), and cultural facilities and activities.

4. *Services impact on physical planning*, raising questions about the most appropriate spatial patterns of different types of tertiary industry relative to changing attitudes of consumers with their increased mobility, and raise expectations about the quantity, quality, and range of services available, both producer and consumer.

5. *Services have cost impacts* through competition for prime office space leading to price escalation, competition for quality staff requiring deregulated and competitive wage and salary structures, and competition for quality residential environments that are very different to those where the workers in other sectors live.

The policy and planning implications of the emergence of the services-dominated city are profound.

4.7.3 Smart Infrastructure

The information-driven city requires new ways to think about infrastructure. "Some cities require *smart infrastructure* to link talent, technology, capital and know how" (Smilor & Wakelin, 1990:53). Figure 4.13 shows how these four factors may be mobilized for industrial and technological development. It illustrates how this will depend on the existence of important environmental conditions that facilitate innovations and new production systems. This requires enlightened policy initiatives and actions to be taken by entrepreneurs and governments that include fiscal measures,

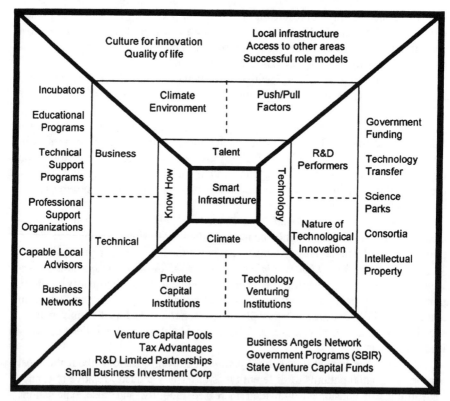

Figure 4.13. Smart infrastructure for cities: key factors, environmental conditions, and policy implications. *Source:* Smilor & Wakelin, 1990:67.

planning- and development-control mechanisms, legal arrangements, public–private partnership initiatives, and the provision of facilitatory and enabling infrastructure.

What is implicit in this model is the notion that policies influence conditions affecting the availability of capital, the development of networks, and the availability of business and technical know–how. Institutional infrastructure is crucial: This includes universities, advanced producer services, and social and civic groups, in addition to the physical infrastructure needed to provide efficient and timely access to places within and outside the city (through transportation links including freeways, fast trains, and airports) and also access to knowledge (through communications networks).

4.7.4 The NIMBY Syndrome

Often it is the case that local desires and preferences are such that a community or an interest group will strive to exercise influence or even control over political representatives in order to stop growth and development so as to exclude outsiders from gaining access to live in a community, or to prevent a particular development, or even to oppose any change to the built environment. These local interests are exclusionary. Actions are aimed at protecting what the community or group enjoys and to ensuring that its members' current values are maintained. Often this is to the detriment of the wider interests of a region or the wider community.

These attitudes and their rigorous pursuit are known as NIMBYism (not-in-my-backyard), and they are widespread throughout cities and regions. An example at a city level is the no-growth policy of Santa Barbara in California. At the local neighborhood level there are countless examples of citizens banding together through neighborhood associations, and the like, to fight and block rezonings, to oppose commercial and tourist developments, and to protect the incursion of development into local bushland. Molotch and Warner (1992) have noted the "extremely exclusionary but increasingly popular" no-growth and stop development movements and groups across the United States. Increasingly citizen groups and even local authorities are mobilizing to oppose change and to protect local environments. Dear (1991:33) has suggested that NIMBY attitudes are becoming more extreme, to the stage where these sentiments are so pervasive that many people now talk about CAVE groups (i.e., Citizens Against Virtually Everything).

A major challenge for governments and agencies responsible for the planning and development of the built environment of cities and regions is to ensure that the self-interest of local groups is not so pervasive as to divert attention from the wider and longer-term interest of a city or region and that state or national requirements are not subverted by local interests, such as the provision of strategic infrastructure such as an international airport. It is ironic that usually it tends to be the better-off communities that have the potential for growth and development where NIMBYism is frequently encountered in the most extreme forms. In some cities, such as Seattle in the Pacific Northwest of the United States, NIMBYism has become a major political and management issue.

4.7.5   Strategic Approaches to Regional Development and Planning

There is mounting evidence that the traditional focus of planners and planning agencies on the physical consequences of urban growth and development, through the use of blunt processes and mechanisms, such as single-use zoning schemes and nested hierarchies of service centers, represent response to urbanization processes that belong to a bygone era. These approaches to planning have little relevance for the contemporary world with its processes of change that are restructuring and reshaping the nature of cities and regions. Regional plans often have identified and specified the locations of hierarchies of business centers; green belts have been designated; and population targets have been proclaimed. Rarely have these prescriptive approaches materialized because of changing market circumstances and changing economic and social processes. Today one would be skeptical of the desirability and feasibility of developing and implementing rigid physical regional land use plans that continued to reflect trends and conditions of past years. We are unlikely to be able to predict accurately the nature, magnitude, and implications of changes of a technological, social, and economic nature that cities and regions will experience in the future.

Fundamental to the development, planning, and management of urban environments are the processes used, particularly by public agencies, in undertaking planning. In the last 15 to 20 years there has been an adoption by communities of all sizes (from small towns to major metropolitan regions) of strategic planning processes, which initially were developed for applications in the corporate sector.

Many policy issues and planning concerns need to be addressed coherently, comprehensively, and sensibly in the context of the processes of change that are generating today's polycentric metropolitan regions. These are the real operational environments of people, households, businesses, and public agencies in the current world. Environmental impact, environmental quality, public health, social equity, alienation, infrastructure provision and management, security and safety, housing affordability, structural economic change, jobs and employment, gender relations, and so on, collectively represent the complex range of issues that must be considered in strategic planning for the future of cities. A continuing emphasis on physical planning that seeks to control and direct urban form and structure is simplistic. Furthermore, it neglects the roles of the multitude of actors involved in the urban and regional development process. Still further it fails to appreciate the basis of decisions in space and time that collectively create the built form of the environment and generate the complexity of economic and social activities and the trips, movements, and flows of people and goods within spatial environments.

Blakely (1989:60) suggested that "communities must use their current human, social, monetary, and physical resources to build a self-sustaining economic system." He conceptualizes development in the following functional terms:

Regional Development $= f$(natural resources, labor, capital investment, entrepreneurship, transportation and communication, industry competition, technology, size, export market, international and national situations, local government capacity, national and state government policies, development support)

Strategic planning provides a useful methodology for the development, formulation, and implementation of strategies to plan and manage the development of a city or region. Figure 4.14 sets out the logical flow of processes involved in this task. Stimson (1992) suggests that it requires the following:

1. An understanding of the dynamic processes of structural economic and demographic change that underpin patterns of growth and an awareness of the impacts of global and national processes of change as they impact at the regional and local level.
2. An understanding of the nature of external forces and of local attributes that have created a region's competitive advantage and model likely future trends and potential outcomes to maximize that advantage.
3. Identification of the interrelationships, performance, and potential conflicts in the process of development, maintaining environmental quality, achieving social equity, and the effective and timely provision of infrastructure and services.
4. Development of institutional capacity, capability, and preparedness to manage change and implement strategies to achieve successful and sustainable development of a balanced form.
5. Provision of strategic, enlightened, and sustained leadership utilizing public–private–community partnerships for formulating and implementing strategies to manage growth.
6. Monitoring, evaluation, and adjustment of strategies to ensure the region is flexible and is fast in responding to changing environmental circumstances.

Strategies for the development and planning of a city or region essentially are about "coordinating activities and facilitating the achievement of desirable outcomes in the manner in which land is used to serve the economic and social requirements of a city or region; ensuring that the development process is achieved effectively and efficiently and to minimize the deleterious impacts on the environment; and [providing] an overall outcome of sustainable development for the metropolitan city or region" (Stimson, 1992:22). A conceptual framework for this integrated approach is proposed in Figure 4.15. Basically, the components include:

1. *Actors* involved in decision making and consultative processes to develop a vision for the city or region are identified.
2. *Principles* for guiding development and managing growth are developed.
3. Agreed *goals* and *objectives* with respect to economic, social, and environmental (and other) issues are specified and are used to develop *desired spatial outcomes* in broad terms with respect to existing and new urban areas.
4. *Institutional arrangements* are identified for the *management, allocation, funding, and pricing* of infrastructure and services, land, and transport and traffic.
5. The nature of *government intervention* and role of *market forces* are considered.

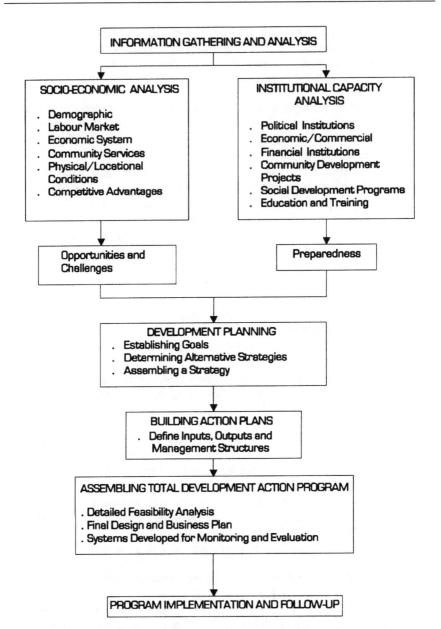

Figure 4.14. The local and regional development process flow chart. *Source:* Blakely & Bowman, 1986:23.

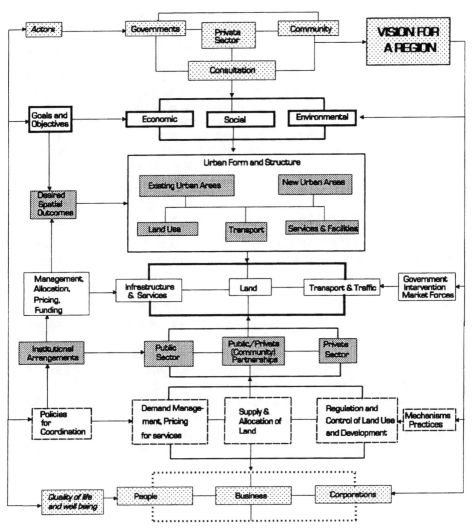

Figure 4.15. Conceptualizing the urban development and planning tasks for regional management. *Source:* Stimson, 1992.

6. *Policies, mechanisms, and practices* are proposed to *coordinate development and planning.*

The process involves developing a vision for the city or region and the outcomes of the implementation of policies and plans may be evaluated against agreed *quality of life* and *well-being* criteria and measures to determine progress towards achieving the goals and objectives.

Knight (1989) provides us with the timely reminder that:

City building is becoming increasingly challenging as more and more cities position themselves to take advantage of opportunities created by the global society. In order to sustain development, a city will not only have to expand the global economy but it will also have to improve the quality of life it offers. In the long run, the only way that a city can maintain its advantage in the global society is by offering a quality of life that is, in the eyes of its citizens, better than that offered elsewhere. The power of the city will then depend less on powers from above and more on powers from within, that is on the effectiveness of its civic processes. In short, in order to remain viable in the global society, the city has to be self-government and self-invigorating. (p. 333)

4.8 The Interface of Macro and Micro Milieux

It is in the contexts of the changing macro and micro operational milieu that the decisions of institutions, firms, firms, households, and individuals are made. What is evident is that we live in a rapidly changing and competitive world in which the processes of change—economic, demographic, social, and technological—are complex. When coupled with the political and institutional environments that operate at the national, regional, and local levels, the dynamics and the uncertainty of change are compounded.

But the processes of change have important spatial impacts in terms of the form and structure of cities and regions, on how spaces and places are used and for what purposes, and on the decisions of people and institutions in time and space. In reading the chapters that follow, in which we focus on specific types of spatial behavior and decision and choice processes, it is important to keep in mind that these behaviors are taking place in a changing urban space. The macro processes of change, the changing patterns and characteristics of cities and of their functions and internal structures outlined in these two chapters, thus become essential background on which to paint the mosaic of individual, group, and societal behaviors.

5

Acquiring Spatial Knowledge

5.1 Shifting to a Micro Behavioral Perspective

In this chapter we shift from the *macro* level approaches presented in the previous chapter, to a *micro* scale exploration of the behavior of individuals or small groups. This emphasis on disaggregate behavior will extend through the next few chapters as we explore concepts involved in human spatial decision making and choice behavior.

Human spatial behavior can be defined as any sequence of consciously or subconsciously directed life processes that result in changes of location through time. Although broad, this definition focuses on sensate behaviors, accepts the need for direction, eliminates behaviors that are fluctuations of inanimate systems elements (such as rocks rolling or rivers flowing), and covers all possible behaviors from the weakly motivated to the strongly stereotyped. *Learning* is the process by which an activity originates or is changed through responding to a situation—provided the changes cannot be attributed wholly to maturation or to a temporary state of an organism. Learning is thus defined as a process that is involved in a wide range of acts from innovation to repetitive imitation, while discounting effects such as aging, or the influence of temporary bodily system states.

5.2 Types of Behavior

It makes little sense to try to categorize types of human behaviors except in the grossest sense. We can for convenience however, recognize three broad behavior types:

1. *Weakly motivated and random behaviors.* These often are associated with the exploratory phase of a learning process. It is often accepted that apart from the first sensory motor experiments of the newborn child, there is probably no such thing as random behavior of humans. Within this context, however, we shall include

weakly motivated behavior for which it may be impossible to determine the critical derivative forces or which cannot meaningfully be differentiated from random acts, such that the behavior might reasonably be called irrational, arbitrary, or relatively unpredictable.

 2. *Problem-solving behaviors.* These behaviors occur when a sensate being is confronted with a problem that requires deliberate thinking in a specified direction prior to making a choice among sets of feasible alternative solutions to the problem. This type of behavior may involve vicarious trial and error behavior and uncontrolled or patterned search activities. A component of problem solving is the spatial manifestation of the process or the overt response.

 3. *Repetitive learned behaviors.* These are habitual behaviors that are relatively invariant, difficult to extinguish, frequently follow the paths of minimum effort, eliminate conscious consideration of alternatives, and are designed to reduce uncertainty and effort in the decision-making process. This is behavior assumed in virtually every geographic model involving human activity.

5.3 Spatial Abilities

5.3.1 Definitions

Definitions of spatial abilities usually are tied to performance on spatial aptitude tests and the dimensions of visualization and orientation contained within those tests (Eliot & McFarlane-Smith, 1983). More expanded definitions, including a dimension of spatial relations, have been offered by Lohman (1979), and Self and Golledge (1994) who have suggested that spatial ability includes the following:

1. The ability to think geometrically.
2. The ability to image complex spatial relations such as three-dimensional molecular structures or complex helices.
3. The ability to recognize spatial patterns of phenomena at a variety of different scales.
4. The ability to perceive three-dimensional structures in two dimensions and the related ability to expand two-dimensional representations into three-dimensional structures.
5. The ability to interpret macro spatial relations such as star patterns or world distributions of climates or vegetation and soils.
6. The ability to give and comprehend directional and distance estimates as required in navigation and path integration activities used in wayfinding.
7. The ability to understand network structures.
8. The ability to perform transformations of space and time.
9. The ability to uncover spatial associations within and between regions or cultures.
10. The ability to image spatial arrangements from verbal reports or writing.
11. The ability to image and organize spatial material hierarchically.
12. The ability to orient oneself with respect to local, relational, or global frames of reference.

13. The ability to perform rotation or other transformational tasks.
14. The ability to recreate accurately a representation of scenes viewed from different perspectives or points of view.
15. The ability to compose, overlay, or decompose distributions, patterns, and arrangements of phenomena at different scales, densities, and dispersions.

Spatial abilities are developed by geographers so that they can answer a wide range of geographic questions. As in other sciences, geographers continually ask questions of: *identification* ("What?" and "Where?"), *definition* and *meaning* ("What?"), *description* ("What?" and "Where?"), *classification* ("What group?"), *analysis* ("Why?"), *process* ("How?"), *temporal existence* ("When?"), and *theoretical or practical relevance* ("Why?").

In general, the *what* and *where* questions are those of identity and produce the declarative knowledge base of the discipline. *How* questions emphasize the processes that result in the facts of existence including the values and beliefs and feelings of people. *When* questions add a temporal context to the occurrence of and relations among phenomena. *Why* questions require analysis and interpretation in the search for geographic understanding.

Throughout this book, we use a variety of question types and tap a wide range of spatial abilities in our discussion of the interactions between and among society, space, and behavior.

5.3.2 The Psychological Definition of Spatial Abilities

Psychologists have long investigated questions of spatial aptitude or spatial ability, and entire books listing hundreds of spatial aptitude tests have been published (e.g., Eliot & McFarlane-Smith, 1983). There is some discussion over the critical number of dimensions of spatial ability. For the most part, this discussion is split between whether there are two or three dominant dimensions.

1. The first dimension, *spatial visualization*, is perhaps the most widely acknowledged and accepted. This is the ability to mentally manipulate, rotate, twist, or invert two- or three-dimensional visual stimuli. This ability is important for solving many mathematical problems and is a factor in comprehending geometric structures. Since many argue that viewing the environment is just another facet of viewing three-dimensional geometric representation of phenomena, this dimension does appear to be of critical importance for geography; its map and image base of data now being used with great frequency in the discipline. Perhaps more than anything else, differences in performance by male and female subjects on aptitude tests emphasizing this factor have given rise to the argument that male spatial ability is superior (Masters & Sanders, 1993; Stumpf, 1993). The argument put forward by Self and Golledge (1994), however, is that no single dimension can justify the claim of overall superiority of one sex over the other. Other researchers have shown that even on visualization tests, female subjects often score better than male subjects, particularly if accuracy and not response time is the key variable.

2. The second most commonly accepted dimension of spatial ability is *spatial orientation*. This involves the ability to imagine how configurations of elements would appear from different perspectives. It is important in many geographic tasks including those of map reading or image processing. It is also important in wayfinding and navigation. Often male subjects score higher on abstract tests of orientation problems, but female subjects often score equally as high, if not higher, on real-world orientation and direction tasks. In many other circumstances, no significant differences between the sexes can be found. Recently geographers Gale, Golledge, Pellegrino, and Doherty (1990) and Golledge, Ruggles, Pellegrino, and Gale (1993) have shown that preteenage girls performed better on distancing, sequencing, and location tasks than did preteenage boys or male or female adults. They found also that female geographers performed better than male geographers and male and female nongeographers on tasks such as angle estimation and orientation (Golledge, Dougherty, & Bell, 1995).

3. The dimension that is least clearly defined and that is most contentious is the catchall dimension called *spatial relations* (Lohman, 1979). Self and Golledge (1994) suggest that this dimension includes: abilities that recognize spatial distribution and spatial patterns; identifying shapes; recalling and representing layouts; connecting locations; associating and correlating spatially distributed phenomena; comprehending and using spatial hierarchies; regionalizing; comprehending distance decay and nearest-neighbor effects in distributions; wayfinding in real-world environments; landmark recognition; shortcutting; orientation to real-world frames of reference; imagining maps from verbal description; sketch mapping; comparing maps; overlaying and dissolving maps; and many other of the activities commonly found in the geographic classroom. Few if any of these abilities have been tested in the spatial aptitude test developed in psychology. Self and Golledge (1994) argue also that these relational and associational components of spatial knowledge are among the most significant for everyday spatial behavior. The fact that few detailed examinations of whether male and female performance differences exist on these components leads to the conclusion at this stage that there is no clear dominance of one sex over another in the geographic domain.

5.4 Place Learning

Whereas some of the initial interest in learning theory in geography was provided by models that focused on learning *movement sequences* or *movement habits*, perhaps the most progressive stream of research using learning concepts has been firmly based in cognitive theory.

Place-learning theories hold that organisms learn the location of paths or places rather than movement sequences. This theory is inherently appealing to the geographer. This place-learning theory suggests that learning is a cognitive process guided by spatial relationships rather than by reinforced movement sequences. Given that an origin and destination is known, variable movement may be undertaken between the origin and destination with the same results in terms of goal gratification. Thus, there are clear implications that places are learned, that possible connections between them are built up over time, and that individuals develop a capacity for linking previously unknown destinations to given origins by referring to a general spatial schema that incorporates concepts of proximity, contiguity, clustering or separateness, sequence, and configuration or pattern. Tolman (1948), in fact, argued that sensate organisms develop *mental maps* that allow them to navigate in any given environment. We rely heavily on place-learning concepts later in this chapter (as well as in subsequent chapters).

5.4.1   Search and Learning

Behavior changes over time from motivated search to fully learned activities. Obviously, such a transition can be spread over greater or smaller lengths of time depending on the type of behavior, the purpose of the behavior, and the system constraints.

   If we were examining the responses of an individual in terms of achieving a goal, it may appear that alternative responses appear successively in an unorderly fashion until a certain level of knowledge is achieved and the beginnings of a satisfactory response sequence emerge. Thus, commonly it is assumed that an individual placed in an unfamiliar environment and stimulated to seek a particular goal will exhibit a tendency to vary responses that are made under conditions of uncertainty in an attempt to achieve a specified goal object. Once a correct or most satisfying response has been achieved, then experimentation diminishes and incremental learning proceeds.

   An essential component of this theory is the notion of *search*. Examination of aspects of search, from provisional try to stereotyped procedures, has produced a spate of geographically relevant models, including those relating to job search (Rogerson & MacKinnon, 1981), the search for new housing (Clark, 1993), locational search (McDermott & Taylor, 1976), consumer behavior (Rogers, 1970; Timmermans, 1980), and wayfinding behavior (MacEachren, 1992; Golledge, 1993). All have developed from this useful and powerful learning concept. Notions of *directional bias* and *distance bias* (Adams, 1969; Humphreys & Whitelaw, 1979; Brown & Holmes, 1971; Costanzo, Hubert, & Golledge, 1984) have focused on spatial regularities in human search processes. Almost inevitably, the traditional distance decay effect enters into discussions of search activity regardless of the search context.

5.4.2   Developmental Theory

The developmental theories of Piaget and Inhelder (1967) have become a critical part of the literature of environmental learning and spatial knowledge acquisition. The stages of development and appropriate sequences in the development of spatial knowledge according to Piaget's theories, are shown in Figure 5.1.

   In brief, Piaget argues that development progresses through four stages from infancy to adolescence and beyond:

   1. The first of these is called the *sensory–motor stage*, and covers the period

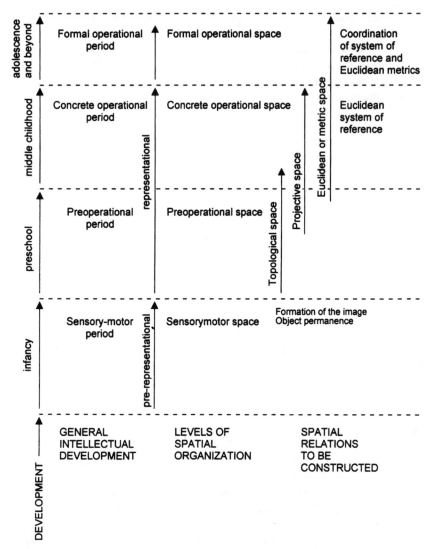

Figure 5.1. Piaget's theory of development of spatial schemata. *Source:* Hart & Moore, 1973:265.

from birth to 2 years old. Development proceeds from reflex activity to hand–mouth coordination to representation and sensory–motor solutions to problems.

2. A second stage, *preoperational*, is said to exist covering the period from about 2 to 7 years. At this stage, development proceeds from sensory–motor representation to prelogical thought for solutions to problems. Problems are solved through representation, which is in part dependent on language development. Thought and language are both dominated by egocentric controls, and children at this stage are not believed able to solve conservation problems.

3. The third stage, *concrete operational*, lasts from about 7 to approximately 11 years, and at this time development proceeds from prelogical thought to logical solution to concrete problems. Concepts such as reversibility occur, and the child is able to solve conservation problems using logical observations but cannot solve complex verbal problems.

4. A fourth stage, *formal operational thinking*, appears around 11 years and beyond where development proceeds from local solutions to concrete problems to logical solutions of all classes of problems. The individual appears capable of scientific thinking, can solve complex verbal problems, and all the cognitive structures mature.

In a spatial context, Piaget argues that the representation of space arises from the *coordination* and *internalization* of actions. In this view, representations of space result from manipulating and acting in an external environment rather than from a perceptual copying of it. Thus, interaction in space, not perception of space, is a fundamental building block for the acquisition of spatial knowledge. It is suggested further that the initial actions of the sensory–motor child include copying or imitation of other people's actions. As imitation schemas are internalized, responses become more symbolic. In the specific relational context, three classes are seen to conform to the general notion of spatial cognition, and, along with the progression through development stages, one proceeds from an egocentric prerepresentational frame of reference, through topological, projective, to a Euclidean or general metric relational structure.

Although some geographers have turned to special populations to examine the completeness of inferences about the acquisition of spatial knowledge, it is agreed generally that most of the geographer's subjects come from the group located in the formal operational reasoning stage. This means that individuals are capable of removing abstract spatial operations from real action, real objects, or real space and that any specific act can be placed in the context of the universe of spatial possibilities. However, it is obvious also that spatial understanding and knowledge are not only cumulative but comprehensive. Thus, newcomers first exposed to an area, in their efforts to learn about it, may revert to a preoperational level of spatial comprehension. At this level, the fundamental topological properties of space (such as *proximity* and *separation*, *openness* or *closure*, *dispersion* or *clustering*, and so on) will dominate in their cognitive representations. As interaction with the environment increases, progression from pure topological to projective relations may occur and be readily expressed in terms of perspectives or points of view. These are relatively invariant under projective or perspective transformations. According to Shemyakin (1962), there is a combination of *preoperational* and *route-type* knowledge schemas that develop in the early stages of understanding or comprehending an environment. As more knowledge accumulates about the environment, a transition to knowledge of abstract relational spaces, which may or may not have coordinate frames of references, occurs. Again in Shemyakin's terminology, this represents the development of *survey-type schema* where abstract configurational properties of space can be understood by the subject.

Perhaps the final relevant comment from the geographer's viewpoint concerns the changing geometries used in spatial representation as development occurs. In the prerepresentational levels, set theoretic concepts of *clustering, inclusion, exclusion,* and *intercept* may be sufficient for establishing the relevant spatial characteristics of an environment cue or object. As development occurs and an egocentric frame of reference develops, the concepts of *linearity* and *monotonic sequence* or *order* appear, along with some first notions of *distance, direction,* and their various transforms. Beyond projected relations at the next (concrete operational) stage, coordinate structures are comprehended and may be developed and used. At the final (formal operational) level and beyond, general and mixed metrics may be more representative of an individual's understanding and use of spatial concepts than any single formal metric space (such as Euclidean space).

Liben (1981) provides a modified view of the role of development in environmental cognition by examining it across the life-span. She stresses that there are three components of life-span change: (1) *physical change,* such as increase in size, musculature, and changes in body proportions; (2) *socioemotional changes,* which include recognition of body–body, body–object, and object–object relationships as well as the developing role of attachment; and (3) *cognitive change.*

With respect to the last, Liben (1981) raises the question of whether or not changes in underlying cognitive skills in general affect changes in the ability to *encode, remember,* and *manipulate* information about environments. She argues that the Piagetan theory provides the most complete model of qualitative cognitive change from infancy through adolescence. Liben focuses on the period from sensorimotor to preoperational thinking in her analysis, particularly emphasizing the change in ability to recognize simultaneous relationships among logical operations such as *addition* and *subtraction* and *categorization* along more than one dimension. Other problems involve an inability to understand *proportionality* and a difficulty in understanding *transitive relationships* (i.e., A > B, B > C, therefore A > C). She points also to the problems of comprehending *seriation* and *conservation*—such as the fact that quantities can remain unchanged despite changes in location or position in a spatial arrangement. Further, Liben (1991) points out that changes in linguistic skills are yet another cognitive change that is tied into life-span development. And finally, she suggests changes occur in metacognitive skills (i.e., changes in the ability to attribute success on tasks to their own increasing skill). This is also an essential part of *life-span development* and the cognitive change across the life-span. Liben concludes by saying there is little work in which basic cognitive skills are measured independently from environmental cognition so that specific differences between the two are not obvious. Liben argues in conclusion: "However, environmental cognition is much more than identifying elements of an environment and/or of its representation. It extends beyond to understanding relationships implied by, or latent in, the representation" (Liben, 1991, p. 267). Despite some reservations, when we consider *cognitive change across the life-span,* some fundamental conclusions can be stated with confidence:

1. Older children do better than younger children on tasks involving environmental cognition.

2. Elderly adults often show some deterioration of functioning with respect to environmental cognition.
3. Experience with representation of environments plays a powerful role in the development of spatial understanding.
4. There are some functions served by environmental representations that cannot be served by direct environmental experience.
5. While the ability to experience, code, store, recall, and use environmental information generally increases across the life-span until old age begins a deterioration of ability, observed differences are sometimes more qualitative than quantitative.

5.5 The Nature of Spatial Knowledge

5.5.1 Basic Components

At a very general level, we can conceptualize spatial knowledge as having the following components:

1. First is a *declarative* component, which includes knowledge of objects and/or places together with meanings and significances attached to them. This is sometimes referred to as *landmark* or *cue* knowledge, and it requires knowledge of *existence* (i.e., ability to recognize an element when it lies within a sensory field). This knowledge is the minimum required for object and pattern recognition and discrimination.

2. A second component of spatial knowledge may be called *relational* or *configurational*. Here information about *spatial relationships* among objects or places develops. Concepts such as *proximity* and *sequence* allow one to develop various knowledge structures that may include hierarchical networks, knowledge chunking, and an understanding of the notion of configuration in a multidimensional sense.

3. A third component is called *procedural knowledge*. This is required for the development of *object* and *procedure association* for processes such as the development of locomotive ability, and it consists of sets of procedural rules (often expressed in "If . . . then . . ." statements) that link bits of the declarative structure. This type of procedure is necessary for the development of wayfinding behavior and route learning.

5.5.2   Acquiring Spatial Knowledge

In the process of acquiring knowledge, a number of relevant questions come to mind. Perhaps the most obvious and important of these relate to finding features of objects and/or locations that cause them to be selected as recognizable elements of environment. Examples are given in Figure 5.2.

These characteristics provide the differential attributes of objects. At the *procedural level*, features such as the nonlinear relationship between subjective and ob-

Size

Visual Form

Clarity

Dominance

Color

Contour

Architectural Design

Location

Proximity to Other Cues

Functional Class

Shape

Figure 5.2. Environmental characteristics perceived in environmental learning.

jective distances, asymmetries in distance cognition and path segmentation, directional distortions, knowledge gaps, and fuzziness (or variance in locational estimates) all contribute to the extent to which accurate procedures develop in one context or another. For example, route knowledge by itself consists of a series of procedural descriptions involving a sequential record of starting point or anchorpoint, intermediate choice point recognition, directional selection, path segment identification and sequencing, travel mode selection, and destination choice and endpoint recognition.

Thus, *procedural knowledge* involves identification of places on a path or landmarks on or near a chosen route segment where a decision is to be made relating to continued navigation of the next segment of the path. Obviously the more detailed knowledge one has of a route, the more intermediate landmarks or junction points can be identified and hierarchically organized into secondary, tertiary, and lower-order nodes and segments (see Figure 5.3). Similarly, the more one knows of a route, the more accurately one can evaluate path lengths regardless of the criteria used in this evaluation process. Route segments of equivalent lengths, for example, may require different procedures to navigate in either direction because of environmental changes such as moving uphill or downhill. Again, route segments of equivalent length can vary in terms of degree of difficulty of navigation depending on whether they are contextually simple or complex (i.e., whether or not their end points are within a visual field) and whether directional change is absent, gradual, or abrupt. In a sense, therefore, *route segmentation* characteristics may be related to

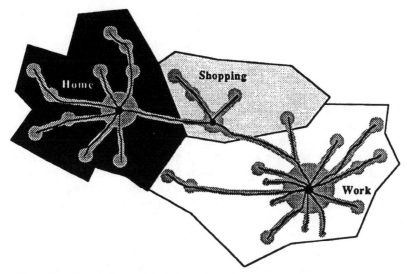

Figure 5.3. Descriptive model of the anchorpoint theory. *Source:* Golledge, 1976.

the mental effort involved in recalling and/or recognizing cues to identify where one
is in the segment or along the route.

5.6   Theories of the Development of Spatial Knowledge

5.6.1   From Landmark to Route to Survey Knowledge

An embryonic theory relating to the transition from *landmark* to *route* to *survey
knowledge* can be found in the works of Shemyakin (1962), who argues that the acquisition of spatial knowledge progresses consecutively from landmark recognition
to path definition and the understanding of general relational characteristics of
areas. As knowledge accumulates and as interpoint information becomes more precise, thus reducing notions of variance or fuzziness, precision also accrues to concepts of angularity, direction, proximity, and separateness, as well as to other fundamental spatial properties. The similarity between this conceptualization and Piaget's
notions of movement through the various developmental stages is obvious.

5.6.2   The Siegel and White Hypothesis

In Siegel and White's model (1975), the first stage in the acquisition of spatial
knowledge involves some *landmark recognition*. *Paths* or *routes* then develop between the landmarks, with route knowledge progressing from *topological* to *metric*
in terms of properties. Subsequently, sets of landmarks and paths are organized into
clusters based on metric relationships within a cluster and the maintenance of general topological relationships between clusters. In the final stages, an overall *coordi-*

nated frame of reference develops with metric properties being available within and across clusters producing survey knowledge.

This theory, along with others (e.g., Golledge 1978b; Kuipers, 1978), assumes that a memory representation of a complex external environment contains hierarchically organized knowledge including landmarks or salient locations, routes, and configurations representing the integration of information about routes into coherent structures organized about the relative location of landmarks.

There is an alternative to this dominant developmental framework for understanding the process of acquiring spatial knowledge that generally suggests that there is a qualitative transition among three distinct types of knowledge: *landmark*, *route*, and *survey*. Montello (in press) argues that the first two types contain nonmetric forms of spatial knowledge (e.g., order or *monotonicity* and *topological relationships*); the third (*survey knowledge*) encodes metric information about the layout of places.

The common basis of previous theories about environmental knowledge acquisition suggests that early knowledge structures lack metric information, particularly concerning distances and relative directions. Montello disputes this claim and offers a conceptual and empirical formulation that argues for a quantitative rather than a qualitative model of spatial microgenesis. He argues that metric knowledge is accumulated and refined from the beginning and that nonmetric understanding is not a necessary precursor to mature layout knowledge. Instead, nonmetric information is often encoded in linguistic or prelinguistic sets of information and may exist concurrently with (or in addition to) metric knowledge. Rather than expand his alternative model in the life-span context, Montello focuses on the process of acquiring knowledge about new environments by mature adults.

Montello's framework consists of five major tenets:

1. There is no stage in which only pure landmark or pure route knowledge exists: Metric configurational knowledge begins to be acquired on first exposure.
2. With increased familiarity and exposure, there is a continuous increase in the quantity, accuracy, and completeness of spatial knowledge.
3. Integration of separately experienced bits of knowledge into a meaningful whole is an important and sophisticated step in the microgenesis of spatial knowledge.
4. Even when individuals have equal levels of exposure to a place, their internal representations of the place and its features will differ.
5. A relatively pure topological knowledge can be found in linguistic systems used for storing and communicating spatial knowledge about places.

Thus, Montello argues that the process of acquiring spatial knowledge in large-scale environments is one of quantitative accumulation of metric information and its consequent refinement rather than a qualitative shift from nonmetric to metric forms.

Montello's formulation has a temporal basis, as does traditional qualitative developmental theory. But the comprehensive structure of Montello's argument raises significant questions such as:

1. Are adults able to integrate knowledge of places learned separately; and, if so, how well can they do this?
2. Does knowledge integration occur spontaneously with the passage of time and thought, or is it a function of increased travel experience?

The question concerning the degree to which spatial knowledge is qualitative or quantitative is an important one in today's GIS literature (e.g., Egenhofer, 1991; Frank, 1992; Freksa, 1992). The question of knowledge integration is an equally perplexing question, with some evidence showing that integration of separately learned routes in familiar environments does take place with reasonable accuracy (Siegel, 1981), whereas the earlier process of integrating separately learned bits of information from an unknown environment is but poorly achieved (Golledge, Dougherty, and Bell, 1995; Golledge et al., 1993). The latter argue somewhat against the spontaneous viewpoint and support the more experiential viewpoint.

Regardless of which viewpoint ultimately emerges, a critical problem for the geographer is the question of how the final knowledge structure can be known (i.e., represented externally). We examine this question in more detail in Chapter 7.

5.6.3 Anchorpoint Theory and Knowledge Hierarchies

This characterization closely resembles the anchorpoint theory suggested by Golledge (1975, 1978b), in which a *hierarchical ordering* of *locations*, *paths*, and *areas* within the general spatial environment is based on the relative significance of each to the individual. Initially locations that are critical in the interaction process—such as home, work, and shopping places—anchor the set of spatial information developed by an individual and condition the search for paths through segments of space capable of connecting the *primary nodes* or *anchorpoints*. Both node and path knowledge is organized hierarchically with primary, secondary, tertiary, and lower-order nodes and paths forming a *skeletal structure* upon which additional node, path, and areal information is grafted. Home, work, and shopping tend to serve as the initial primary nodes and are among the major anchorpoints from which the rest of the hierarchy develops. Other anchorpoints may include commonly recognized, known, and often-used places in the environment.

A total set of anchorpoints then is a combination of those selected in common with others as part of a general pattern of recognition of critical things in an environment (i.e., the common cues that identify cities as being distinct one from another), together with the principal idiosyncratic sets of points that are relevant to any single individual's activity patterns. As interactions occur along the paths between the primary nodes, there is a *spillover* or *spread effect* and the development of the areal concepts of neighborhood, community, region, and so on. Neighborhoods surrounding the primary node sets become known first, and continued interactions along developing node-path networks strengthen the image of segments of the environment for each individual at the same time as they formalize the content and order of the basic common knowledge structure.

When anchorpoints are misspecified (i.e., their locations are inaccurate), the segments of space they dominate and the lower-order nodes contained therein are

similarly distorted (Figure 5.4). Couclelis, Golledge, Gale, and Tobler (1987) determined three types of distortions, which they called *tectonic plates*, the *magnifying
glass*, and the *magnet*. These three models are commonly observed in cognitive
maps.

   1. In the *tectonic plate* model, a dislocation of the principal anchorpoint produces a common directional shift in the location of other cues, primarily in the same
direction as the anchorpoint.
   2. In the *magnifying glass* model, there is an overestimation of interpoint distances from the anchorpoint to nearby locations and an underestimation of the interpoint distances to more distant places. This results in a distortion sometimes described as a "fisheye lens" distortion.
   3. The *magnet* model simply recognizes that attractive force will decay over
distance so that nearby cues will be strongly pulled toward the anchorpoint regardless of their position in space, whereas further cues will be less affected by the mag-
netic pull. In the latter case, the amount of displacement will be smaller, and the direction of displacement will depend more on the relative attractive strengths of
adjacent anchorpoints.

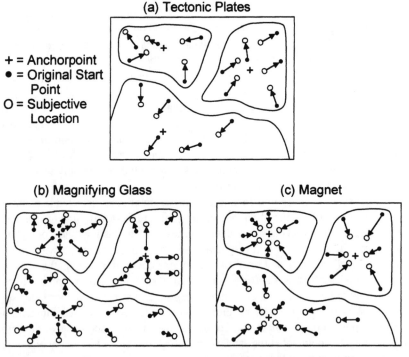

(a) Tectonic Plates

+ = Anchorpoint
● = Original Start
   Point
O = Subjective
   Location

(b) Magnifying Glass   **(c) Magnet**

Figure 5.4. Types of anchorpoint models. (a) Tectonic plates. (b) Magnifying glass. (c) Magnet.

5.7 Children's Wayfinding Behavior: Case Studies

5.7.1 Learning about One's Neighborhood

Empirical literature on the acquisition of spatial knowledge in children typically has examined what is known about a familiar environment such as the child's neighborhood. Such studies focus on tasks such as the accuracy of distance judgments for pictorially presented routes and tasks with significant spatial components such as wayfinding and route reversal. For the most part, they have been restricted to small-scale spaces and short routes, often within buildings. More recently, extensive empirical work in variously sealed environments by Matthews (1992), Blades and Spencer (1994), Unger, Blades, Spencer, and Morsley (1994), Cunningham, Jones, and Taylor (1994), and Golledge, Smith, Pellegrino, Doherty, and Marshall (1985) has focused on the processes of place recognition, wayfinding, and layout comprehension in children acting in known and unknown urban and rural neighborhoods.

There are three critical assumptions underlining much of this empirical work. These have been stated by Golledge *et al.* (1985) as:

1. Individuals tend to code more information at spatial decision points than nondecision points in a wayfinding task.
2. During place recognition, wayfinding, and layout comprehension, automatic processes of perception, concatenation, pattern matching, and generalization respectively extract, store, access, and reorganize information received from the environment.
3. Any given environment consists of sets of objects or features, their properties, and values or saliences attached to each one. These are seen to vary from individual to individual.

Given the above set of assumptions, it is possible to derive sets of hypotheses such as:

1. Environmental cues or other features of the environment have the highest probability of being perceived and recognized if they are in the immediate vicinity of choice points.
2. Any given route tends to become segmented with respect to choice points.
3. The amount, variety, and accuracy of information extracted at choice points vary according to the nature of the action required at the choice point (e.g., simple vs. complex responses).
4. Information recalled about features of the environment is related to the significance of the choice points anchoring any particular segment in or near which the feature is located.
5. Route segments and routes themselves become concatenated in long-term memory and can be used in other tasks.

6. Generalization and spread effects inevitably lead to hierarchical representations of environment.

7. Areas of recognition, action, and expectation are likely to occur with decreasing frequency on successive representational trials, or on successive route-following trials, as behavior becomes proceduralized.

8. Young children find it difficult to integrate knowledge obtained from specific routes into a configurational whole; rather they tend to retain routes in memory as isolated events and thus have difficulty performing abstract reasoning tasks such as shortcutting or configuration reproduction.

9. Some preteenage and more teenage children can point accurately to occluded places (particularly in familiar environments such as the home neighborhood or schoolyard), indicating that an abstract representation of the *task* environment has been encoded and stored in memory.

10. There is increasing evidence that Piaget's stages offer guidelines rather than present milestones of strict adherence (Liben, 1981, 1991).

5.7.2   Wayfinding in a Residential Neighborhood in Goleta, California

An empirical study based on the above assumptions, and designed to test the above-stated hypotheses, was undertaken in a moderate-density residential neighborhood in Goleta, California (Golledge et al., 1985). The area was completely residential in character, thus simplifying the tasks of recognizing, categorizing, and using environmental cue information. Most of the empirical features related to housing, street, or cross-street characteristics. There were no visibly dominant cues in the neighborhood to aid in orientation, and spatial knowledge had to be acquired in a sequential manner as a specific route was followed on successive trials through the neighborhood.

The route chosen was one where the origin and destination points were not simultaneously visible but, rather, were identifiable only from the first and last segments of the route. Along the route were a number of choice points requiring decisions concerning left- or right-hand 90° turns, street crossings, bypassing of inconsequential side streets and cul-de-sacs, navigation of long straight-aways with good visibility, and navigation along gently curving sections (Figure 5.5). Although the route length was 0.8 mile, the origin and destination were only 0.4 mile apart, indicating that the route was approximately "U" shaped. Some general references provided background and possible orientation information—mountains to the north and a noisy freeway to the south.

The basic cue structure of the neighborhood was obtained from cue lists and sketch maps compiled from a large sample of neighborhood children. These were contacted through local school and scouting groups. Frequency of feature or cue mention was considered the main criterion for including cues in a commonly recognized neighborhood cue set. Pilot testing of route difficulty was undertaken with male and female children from ages 9 to 11, and navigation in both a forward and reversed mode was undertaken. The pilot data indicated that information about the route was still partial and/or fragmentary even after several exposures.

Multiple-trial testing was then carried out with an 11-year-old male subject over five successive days. The boy undertook wayfinding in a forward and reversed

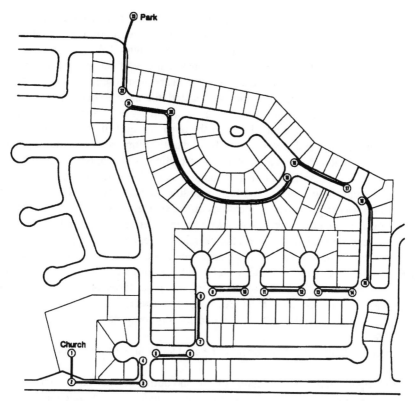

Figure 5.5. Street maps of neighborhood showing route from church to park.

mode, and after each wayfinding task was completed, extensive debriefing occurred in a mobile laboratory at the beginning of the route. As part of this debriefing, the child was asked to describe the strategy for navigation and to list salient features recalled about the route. He was also asked to generate a set of verbal directions that would allow another child to navigate the route successfully; this was followed by a map-drawing task on a blank sheet of paper with the origin and destination nodes indicated in their appropriate relative and absolute positions. Finally, a complete video tape of the route traversed in the forward direction was shown, and the child was asked to identify any particular cues, locations, features, and so on that were remembered and what actions had to be taken at specified loci. Expectations about what would appear next in sequence and what behavior would be required were also probed. This multiple trial testing provided a wealth of data.

By the 10th trial, the route appeared to be known well, was segmented in the same way consistently, was traveled with confidence, and was relatively error free. Little new information was produced either during the navigation task or in the debriefing period, and the experiment was consequently terminated. Sketch maps and verbal descriptions matched very well in terms of the features and characteristics of the route represented.

Referring back to the first assumption concerning information and choice points, the authors constructed a composite representation of cue and feature knowledge for loci along the entire route, as shown in Figure 5.6.

Examination of this cumulative frequency diagram of cue and feature recognition for the entire route shows that, whether data are plotted separately for forward and reverse trials or whether they are aggregated, the most frequently referenced points represent loci at or near choice points—particularly near intersections or corners. There was some difference between the absolute level of cue or feature frequency in the forward and reverse mode, but the critical importance of choice points is made quite obvious in such a graphic representation. Also it appeared that choice points with complex operations (e.g., change in direction and crossing the street) had the potential for greater cue and feature coding than similar points where only a single action was required. Other locations or cue segments, where no major action or choice was mandated, were less well known, less frequently mentioned, and less accurately represented in either verbal or sketch-map representation mode.

An examination of data across a sequence of days reveals a progressive differentiation of cues along the route, with the initial information primarily tied to the origin and destination, and with second, third, and other lower-order nodes emerging on subsequent exposures. Cues devoid of any specific action other than the maintenance of the current state remain consistently low with respect to degree of knowledge.

In general, the following appear to be evident from these experiments:

1. Origin and destination nodes anchor and define the task environment.
2. Interstitial second, third, and lower-order nodes identifying choice points emerge on successive trials, with those points mandating multiple actions providing the critical intermediate anchorpoints.
3. Lower-order nodes in the vicinity of higher ones help to define expectations with respect to behavior or to potential behavior at choice points (priming).
4. Some cues were trial- or episode-specific and were used as peripheral or supplementary reinforcing information only, and their absence or presence

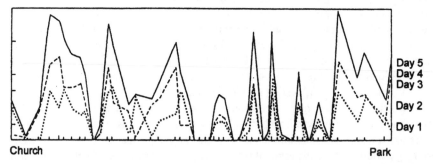

Figure 5.6. Knowledge of cues and features as a function of route location and day of testing. *Source:* Golledge et al., 1985.

(e.g., a barking dog) did not influence substantially the knowledge level in their vicinity.

Examination of the verbal reports, videotape reactions, and sketch maps throws some light on the segmentation and sequencing of components of route knowledge acquisition. For the route to be segmented appropriately and concatenated in correct sequence, the individual must have sufficient knowledge of origin, destination, and major choice points along the route. Knowledge must develop too of the way places and route segments are linked or proceduralized. This knowledge emerges over sequential trials, and the lack of such knowledge in the early trials results in an inability to generate appropriate sequences of route segments. Thus segment order is violated, directional changes at intersections or street crossings may be reversed, and sections may be curved in an inappropriate manner on these early trials (see Figure 5.5). The sketch map sequences shown in Figure 5.7 indicate the degree of completeness of sketches produced on a selection of both forward and reverse trials. They show some well-identified route segments in the vicinity of the primary nodes on the initial trials with incomplete or missing segments typifying the remainder of the early structures. A gradual elaboration occurs over successive trials with some distortions, reversals, and lack of segment differentiation. But the final maps, after five trials in each direction, are relatively complete with respect to segments and in terms of detail about cues and features at choice points and in their vicinity.

An interesting phenomenon occurred related to the appropriateness of the sketch-mapping technique. There was a marked difference in the trial sequence at which information occurred (Table 5.1). For many second- and third-order nodes, the subject was well into the sequence of trials before he was able to add such detailed information to his sketch maps. However, this cue and segment information regularly appeared much earlier in the verbal descriptions and discussions based on video tapes of places along the routes. Statistical tests confirm these impressions. Differences between the video and field experiences on Trial 5 (last navigation trial) were found to be significant with respect to all of the variables: Recorded Behaviors, $t = 4.39$, $p < 0.01$; Total Mistakes, $t = 3.52$, $p < 0.05$; Unrealized Mistakes, $t = 2.73$, $p < 0.05$; Time, $t = 2.50$, $p < 0.05$. No significant differences were found when the video data were compared with the Trial 2 (first navigation trial) field results. Seeing the route five times in the laboratory had essentially the same effect on navigation as one presentation of the route in the field. This observed difference in navigation performance is an important finding, especially in light of the other experimental results, for it suggests that different types of knowledge were acquired in the two learning experiences. Similarly, slides of cues taken at various places along the route could be identified earlier than they were found on the sketch maps. Thus, while at the end of the trials the type and quantity of information on sketches, verbal reports, and video tape or slide recognition were equivalent, in the early states, there were significant differences among results obtained from each of these methods.

Obviously one must beware of using sketch-mapping techniques as the only means of summarizing a person's existing knowledge structure in situations where subjects have little prior knowledge of an environment or where exposure or prac-

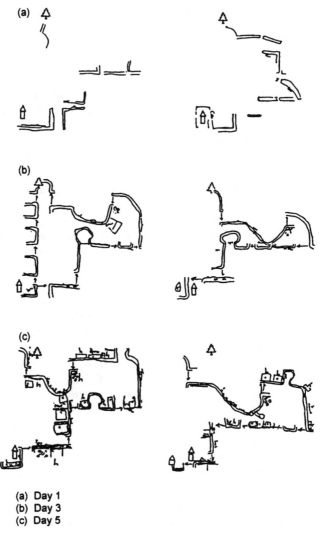

(a) Day 1
(b) Day 3
(c) Day 5

Figure 5.7. Sketch maps by 11-year-old boy of route tasks: forward and reverse. *Source:* Golledge et al., 1985.

tice with this representational form is minimal. Errors occurring during the space-learning process can be classified into two types—action and expectation errors.

1. *Errors of action* consisted of navigational or procedural mistakes that caused the child to deviate from the established route.
2. *Expectation errors*, or *errors of confusion*, may not directly result in a mistake in action but indicate incomplete knowledge at a particular choice point. Confusion errors included allocating particular nonexistent cues to specific locations, substituting cues from one route segment to another,

TABLE 5.1.

Errors by Trial and Media Type*

Performance Measure	Field Learning ($n = 16$)		Video Learning Only ($n = 8$)
	First	Last	
All behaviors	8.25	0.75	7.25
All mistakes	3.50	0.13	3.63
Unrealized mistakes	2.94	0.06	2.71
Time (minutes)	22.22	16.09	21.07

*Navigation performance during field learning (first and last navigation trials) and after video learning (only navigation trial).

Source: Gale et al., 1990.

nonrecognition of scene sequences, and an inability to couple cues and actions undertaken during the wayfinding task.

Figure 5.8 illustrates the decreasing number of errors in action or expectation, as well as total errors, as the trials proceeded. Many errors had disappeared completely after the third day. If we examine where the errors occurred by plotting the frequency of error at potential choice points, we see that the peak of error making occurred at a location in the middle of the route where multiple actions were mandated. Relatively few errors occurred during the beginning or end of the route and in the vicinity of other major anchorpoints (nodes 9 and 12). In contrast, errors at locus #10 persisted until trial 5 and are the last recorded errors. Obviously, the probability of error occurrence is related to the relative significance of a cue or route segment in the total task situation.

5.8 Experiencing Environments

5.8.1 Learning from Maps versus Learning from Travel

Presson and Hazelrigg (1984) compared learning from a physical map (secondary or symbolic learning) with learning by primary or direct experience. They argued that *map learning* resulted in a *figural representation*, which is characterized by precision but also by specific orientation. Direct learning by traveling through an environment produced a more flexible internal representation or cognitive map that was not orientation specific.

Warren and Scott (1993) used simple path maps and "You-Are-Here" maps to investigate hypotheses concerning map alignment and to test whether or not wayfinding performances improved with aligned as opposed to nonaligned maps. In these cases, alignment was made with the local environment not with an abstract frame of reference such as global coordinates or cardinal compass directions. Warren and Scott used two laboratory experiments in a structured environment traversed by four-segment paths; they then conducted a final experiment in a larger, wooded hilly environment in which subjects carried a map and consulted it at will in

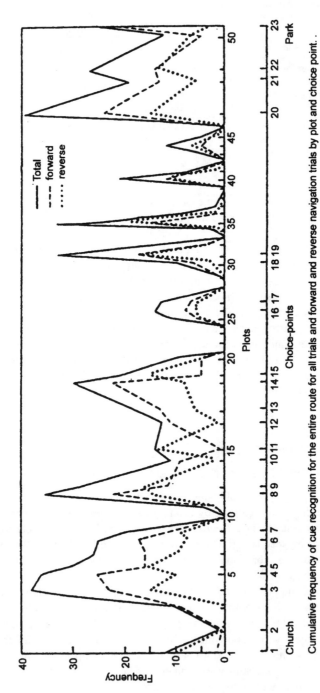

Figure 5.8. Correct behavior at plots and choicepoints: forward, reverse and total trials.

Cumulative frequency of cue recognition for the entire route for all trials and forward and reverse navigation trials by plot and choice point. .

order to find a route from a given origin to a destination. They found in the final task that the subjects consulted the map constantly, particularly at each turn. When they did so, subjects turned the map to accord with a map/environment alignment 66% of the time. When the map was not aligned, an increased quantity of errors occurred. The researchers also discovered that even though the task was route following, use of the map encouraged the acquisition of incidental knowledge about the spatial structure of the total environment.

In another spatial alignment problem, Warren, Rossano, and Wear (1990) asked subjects to view projected photographic slides of a building and then to indicate the location from which each photograph was taken. They were required to do this by marking a location on a perimeter floor plan map of the building. As with other research, subjects showed a tendency to perform best when *alignment* of map and environment was undertaken. In addition, orientation of the aligned map to conform with the subject's orientation or viewpoint of the building also increased success. In this study, Warren et al. (1990) found that the nature of environmental features and the environment itself played a role in task performance. They found that buildings have specific features that vary in salience, and when subjects make judgments with respect to the buildings, they narrow a range of locations by making a sequence of decisions using first the most salient features and then moving to those less salient. In this study, similar results to those achieved in earlier studies were obtained with respect to the importance of map alignment. Here, however, given the choice of aligning the map with the environment or the map with the observer, error conditions were significantly smaller with the map/observer alignment. But the majority of people preferred overwhelmingly to align the map with the environment than with themselves.

In yet another study, Tversky (1981) examined *orientation and location errors* at a continental scale. She asked subjects questions such as: "Which is further west—Santiago, Chile or New York?" "Which is further east—the Caribbean or the Pacific entrance to the Panama Canal?" "Which is further north—Washington, DC or Rome, Italy?" At a more local scale, she has asked: "Which is further east—UC Berkeley or Stanford University?" Where the east–west questions were considered, Tversky found that many errors (such as choosing Santiago or the Caribbean entrance to the Panama Canal) probably resulted from alignment errors (Figure 5.9). In this case, there was a tendency to line up the N–S axes of North and South America, thus producing longitudinal judgment errors. However, latitudinal errors could not be similarly explained. These seemed to relate more to popular views of factors such as climate and cognized distance from well-known areas of warm climates (e.g., Rome/Mediterranean Sea vs. Washington/Caribbean Sea). In both cases, however, map memory and memory from travel (or a travel substitute such as movies or television) produced incorrect answers.

5.8.2 Pointing

Part of the general process of interpersonal communication involves relaying information about the environment. This is true equally as well for blind people as for sighted individuals. In the sighted domain, pointing focuses attention in a specific

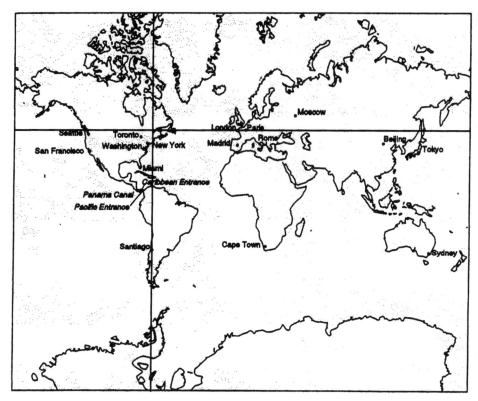

Figure 5.9. Mercator projection showing common specification errors. *Source:* Tversky, 1992.

segment of space and on specific landmarks or objects being referenced. Within the *perceptual* domain, one can project a line along the given pointing vector and discover an object or feature that is being targeted. In the *nonvision* domain, pointing relies on the internal representation of one's environment, and pointing accuracy depends on the ability with which a target is recalled and placed in the context of a frame of reference.

The relationship between a person and a target can be indicated by pointing, by globally describing where the target is, by sketching a map showing the relative locations of self and target, or by locomoting to the target site.

Pointing is one of the simplest, direct, and most effective ways for humans to indicate direction. It appears to be a natural act found in many different cultures in different stages of development and in different regions of the world. It can be argued that this is an activity eminently suited to recovering spatial information from blind people, since it forces them to access directly their cognitive map for characteristics such as the location of self in an environmental layout and the location of the target object, so that they can estimate the angular relationship between self and target. Estimating such an angle can draw on egocentric or polar coordinate systems, or it can depend on mental geometry if a relational or global coordinate system or frame of reference is not used.

Examples of the use of the most accurate methods for recovering direction include research by Da Silva (1985) and Loomis, Da Silva, Fujita, and Fukusima (1992), who used arm extensions and finger pointing to indicate target location. Klatzky et al. (1990) required subjects to rotate their body inside a hula hoop to reproduce angles and found that individuals could reproduce angles of varying magnitude with a uniformly low error averaging less than 5° (Table 5.2). Rossano and Warren (1989) had people indicate directions by orienting their body toward a target. Klatzky et al. (1990) and Barber and Lederman (1988) used external instruments such as rotating a pointer dial or arm on a simulated compass to help people recall angular directions from origins to targets. The former found that this produced varying errors as the size of the angle varied (Table 5.3). Bryant (1982) and Kozlowski and Bryant (1977) had subjects mark the locations of remembered objects on paper, whereas Dodds, Howarth, and Carter (1982) and Millar (1988) used the process of sketching a path between an origin and destination on a flat sheet of paper. Lindberg and Gärling (1981) simply requested people to name an angle.

Of these various methods, *pointing*, *sketch mapping*, and *walking to targets* parallel research methodologies that have been used extensively in *cognitive mapping*. It is significant to note that perhaps the most prominent of these methods—*sketching*—was found by Haber, Haber, Penningroth, Novak, and Radgowski (1993) to be among the least reliable! After evaluating nine different ways of indicating direction, Haber et al. showed that simply using a finger or a device extended like a finger (e.g., a stick) was the most accurate technique.

Byrne and Salter (1983) asked subjects to point toward the direction of target buildings and to estimate their distances from selected origins. They found little difference between performance accuracy of sighted and visually impaired subjects when the task was executed in a familiar residential setting. Herman, Chatman, and Roth (1983a) also have shown that vision-impaired and blindfolded individuals can point to destinations and that they can perform triangulations via pointing to locate features in familiar environments.

TABLE 5.2.
Average Errors (Degrees) for Estimates and Reproductions as a Function of Turn Angle

	Errors	
Angle (degrees)	Estimate	Reproduction
60	-5.76	1.16
90	-1.60	3.50
135	1.02	1.57
180	6.78	3.57
225	10.57	5.67
270	21.31	2.86
300	20.42	-3.02

Source: Data compiled by Loomis et al., 1989.

TABLE 5.3.
Angular Error as Size of Angle Increases

Angle	Approximate Average Unsigned Error
60°	11°
90°	12°
135°	18°
180°	15°
225°	23°
270°	28°
300°	29°

Source: Data compiled by Loomis et al., 1989.

5.8.3 Learning Shortcuts

Passini, Proulx, and Rainville (1990) report on a task that involved pointing to distinctive locations visited during a journey through a labyrinth. A *shortcutting* process was tested by taking subjects around a labyrinth and requiring them to find a shortcut back to the point of departure through the labyrinth.

Passini et al. (1990) examined route-learning tasks when three and five decision points were included on the routes. Successful completion of the exercise occurred when no errors were evident in wayfinding from origin to destination. A second task involved reversing a route (route retrace). A third task consisted of combining previously learned routes into new combinations. In this case, two independent routes were used; the routes intersected at the second decision point. The task consisted of requiring subjects to combine the first half of route one with the second half of route two. Transfer of scale involved learning a route on a small-scale model and undertaking behavior in the real environment represented by the model.

One of the most widely accepted tasks for indicating the possible existence of complex spatial understanding in blind individuals is the *triangle completion problem* (Worchel, 1951). In this problem, an individual walks two legs of a triangle, turning the included angle, and then is required to travel back directly to the origin while not retracing the already experienced legs. This can be achieved only by making a correct final turn angle and correctly estimating and walking the distance back to the origin. This task has been accomplished by leading people over the first two triangle legs (Worchel, 1951). It has been undertaken also by having subjects haptically experience in sequence the first two legs of the triangle (in both forward and reverse directions) then, after removing the symbol representing the origin, requiring the subject to go directly from the end of the second leg to the origin (Klatzky, Golledge, Loomis, Cicinelli, & Pellegrino, 1995).

Klatzky et al. (1995) and Loomis, Klatzky, Golledge et al. (1993) have examined congenitally blind, adventitiously blind, and blindfolded sighted subjects in three different types of wayfinding tasks. These included: simple reproduction and estimation of turns and distances; table-top tasks involving spatial skills that included tasks such as mental rotation; and more complex locomotion tasks requiring spatial inference (e.g., undertaking shortcutting activities to complete a trip). Group

differences were not found for some of the simple locomotion tasks, such as maintenance of heading, turn reproduction, and distance reproduction. One might expect
to see differences in performance between those with and without vision when tasks
require an understanding of the abstract properties of space that require spatial inference, or that appear to require the solution of mental geometric or trigonometric
problems. One such task is *pathway completion*.

Following procedures first used by Worchel in 1951, Klatzky et al. (1995) had
subjects perform a triangle completion task by walking the third leg in a path consisting of isosceles right triangles. Worchel previously had concluded that blindfolded sighted observers performed significantly better than blind observers, though he
found no significant difference between congenitally and adventitiously blinded
persons. In contrast, the Klatzky et al. (1995) study found no consistent differences
among blindfolded sighted, adventitiously blind, or congenitally blind subgroups.

To have been performed successfully, a pathway completion task requires that
a subject maintain a representation of the location of the origin in memory. The failure to find a group difference led to the conduct of further experiments using a
more difficult task. This task previously had been used by Rieser et al. (1985) and
required a subject to develop a representation of multiple locations. Rieser et al.
(1985) had found that congenitally blind subjects were significantly worse, both in
accuracy and response latency, than were blindfolded sighted or adventitiously blind
subjects, when they were required to walk to an unseen but known location and then
to point to other objects from that location. Although Klatzky et al. (1995) found a
trend in that direction, they did not find significant differences when comparing
blindfolded sighted and congenitally or adventitiously blind on such tasks. Their
conclusion was that at least some of the congenitally blind were able to update locations of known objects during locomotion equally as well as blindfolded sighted
persons could. A similar conclusion had been proposed by Landau and colleagues
(Landau, 1988; Landau, Spelke, & Gleitman, 1984). She used a similar task to
study the ability of a young congenitally blind child (Kelli) to take shortcuts between familiar locations within a room. Although one might suspect immediately
that such an ability was developmentally beyond the capabilities of the subject, the
experiment showed that Kelli was quite good at the task and, thus, that vision was
not necessary or was not required in order to develop an ability to make spatial inferences. But, Liben and Downs (1989) among others have expressed skepticism
about this work and its conclusion.

Klatzky et al. (1995) reported also on the results of congenitally, adventitiously, and blindfolded sighted subjects performing a variety of manipulative and
ambulatory tasks (using both tabletop manipulatory and locomotion tasks). They
analyzed the entire data set relating to those tasks in an attempt to investigate commonalities in task performance. In this study, congenital, adventitious, and blindfolded sighted subjects recruited from the Los Angeles and Santa Barbara branches
of The Braille Institute of America (all of whom were experienced independent travelers in their own environment either by walking or using public transportation),
were matched in terms of age and educational level. For each subject, the testing
(including rest periods) took about 4 hours. The session began as an interview for
personal data collection; this was followed by the tabletop tasks that took approxi-

mately 30 to 45 minutes and were administered in a random order, counterbalanced across subjects. The tasks included: (1) the same/different judgments of rotated stimuli; (2) assembly of whole shapes from parts; and (3) estimation of distances. Feedback was not given on any task. After a break, locomotion tasks were undertaken. These included: (1) maintenance of a heading and the walking of a constant distance; (2) replicating a walked distance; (3) replicating and estimating a turn; (4) complex locomotion involving completing a triangle after walking two legs; and (5) consequently completing or retracing a two- or three-sided figure. Thus there were both simple and complex locomotion tasks. The simple tasks involved a linear segment or turn and could be performed with direct proprioceptive memory. The complex task combined two or more segments with turns between them and required some degree of spatial inference. The results of the experiment show that the average signed error went from overestimation for the triangle with the shortest leg length to underestimation for triangles with longer leg lengths. There was an indication also that the congenitally blind showed an overall overestimation error while the blindfolded sighted tended to have an overall underestimation error. But there was no significant difference among groups when absolute error was investigated. Klatzky et al. (1995) reported that generally the blind and blindfolded sighted subjects performed the tasks quite well.

In tasks requiring assembly of parts into a particular shape, errors and non-completions were very similar across groups averaging 28%, 25%, and 27% for the congenital, adventitious, and blindfolded sighted groups respectively. Response times were slightly lower for the adventitious group and somewhat higher for the blindfolded sighted group. However, Klatzky et al. (1995) found no main effect of response time, but there was an effect of shape. The times for assembling the triangular, trapezoidal, chevron, and house shapes were 70.0, 64.1, 47.4, and 42.3 seconds respectively, an order suggesting that the degree to which the components were discernible in the whole was a factor in the performance.

Whereas Worchel (1951) showed that performance on similar types of assembly tasks produced significantly more error by the blind population (47% vs. 26% respectively), the Klatzky et al. (1995) tasks showed comparable error rates but slightly higher response times by the congenitally blind subjects. In a rotation task, correct responses were obtained by 62%, 70%, and 46% for congenitally, adventitiously, and blindfolded sighted groups respectively. Klatzky et al. found no evidence that response latency increased with the orientation difference between the two stimuli being compared. This was markedly different from the results obtained from other studies. Results generally, however, slightly favored the blind groups over the blindfolded sighted group.

There were likewise few significant differences in task performance among the groups on locomotion tasks. Using absolute magnitude of veer or simply a measure of the ability to perform simple replications and estimations in turns and linear segments, Klatzky et al. (1995) found no group differences. In the triangle completion task, and using either subjects turning toward the origin after the first two legs or from analyzing the subsequently walked distance to complete the third leg after the turn, there was a pattern of systematic regression to the mean. Subjects tended to overrespond when the required distance was small and to underrespond when it

was large. These tendencies did not vary significantly by groups. Retrace performance was similar across groups, and all groups found that a three-segment pathway, involving a crossover of two legs, produced the greatest problems in terms of
returning directly to the home base.

After combining all these results into a multiattribute table and subjecting it
to a principal components factor analysis with orthogonal rotation, Klatzky et al.
(1995) isolated three factors with eigenvalues greater than 1 which accounted for
67.3% of the explained variance. Factor 1 reflected competency in the locomotion
task and involved spatial inference rather than replication. Factor 2 correlated negatively with performance on the tabletop tasks. Factor 3 included the locomotive reproduction of simple linear segments and turns and had a positive loading. Discrim-
inant analysis and jackknifed classification produced similar results in that
approximately 7 of the 12 congenitally blind, 8 of the 13 adventitiously blind, and 8
of the 12 blindfolded sighted were correctly classified. Klatzky et al. (1995)
claimed that classification on the basis of performance on these tasks was almost
twice as accurate (62%) as chance expectations, and what errors were made tended
to place individuals into adjacent rather than remote groups. Overall, the three factors seemed to indicate that past visual experience was not necessary for successful
completion of a variety of spatial tasks that drew on different abilities and different
spatial processes. The first factor encompassed complex navigation tasks (triangle
completion and pathway retrace), which emphasized spatial inference and offered a
hypothesis that errors on these tasks derived from the stage of encoding a representation of the pathway configuration. This has been examined further in Fujita,
Klatzky, Loomis, and Golledge (1993). The second factor loaded highest on the
tabletop tasks, suggesting a common component for the processes needed to perform small-scale spatial tasks. The third factor related to simple locomotion.
Klatzky et al. (1995) found that perhaps the most significant outcomes overall were
the lack of difference among groups across the tasks and the failure to isolate specific spatial deficits among the blind. Results of the factor analysis did suggest that
the blind and blindfolded sighted groups are likely to do best on tasks involving spatial inference rather than proprioceptive memory. However, the findings of the degree of individual variability and the high level of confidence shown by some blind
subjects on specific tasks present an argument against making generalizations about
the effects of prior visual status on task performance.

5.8.4 Learning Layouts

Herman, Herman, and Chatman (1983) used an interesting mix of methodologies to
evaluate the ability of subjects to understand *layouts*. They used maps, table models,
and free movement as tasks in the layout learning experiments. The layout consists
of four objects arranged in a diamond shape with one object designated as a home
base. A tabletop model was used to show the configuration so that distance and directional relationships among the other three objects could be comprehended. The
subject was then taken to a gymnasium where objects were placed in the same layout as on the tabletop, and the task was to walk to the supposed location for each of
the other three objects. Thus, this included scale effects. Each point in turn was used

as a home base. The learning performance of subjects was found to be superior to what would have been expected by chance alone.

Learning layouts by integrating information obtained when traveling separate but overlapping routes also has been shown to be a viable procedure by Moar and Carlton (1982). In this case a very familiar environment was used. However, when they examined environmental learning by route integration in unfamiliar environments, Golledge et al. (1993) found that subjects had great difficulty integrating such information. When they were asked to produce a single sketch map of the overlapping routes, usually the subjects represented these as separate entities. This was true even after multiple trials.

Hollyfield and Foulke (1983) examined spatial learning of novel routes using adults over 20 years of age. Two sighted groups were used, one performing with vision and the other blindfolded. Another group of adventitiously or later-blind persons, with at least 12 years of sighted experience, and a group of early blind, with less than 1 year of sighted experience, also were used. The route covered five city blocks, and it was about half a mile in length. Initial exposure was via an experimenter-guided tour, followed by five subject trials. Subjects' verbal comments along the route were tape recorded during each trial, and after each trial, a modeling kit was used to map the route. On the final trial, a subject was stopped at a midway point, was told that an obstacle blocked the route, and was requested to find an alternate route to the destination. The results showed that the early and late blind performances were comparable although not as good as the sighted groups. Initially the blindfolded sighted subjects were very poor, but they improved to equal approximately the adventitious blind by later trials. The detour task significantly interfered with performance in all the nonvisual groups.

5.9 An Artificial Intelligence Model of Wayfinding

Given the quantity of evidence from the empirical experiments discussed in the previous section, it is possible to highlight critical features of the wayfinding decision process and to model it in an artificial intelligence context.

A model representation of this decision-making task can be constructed (Gopal, Klatzky, & Smith, 1989; Gopal, 1988) with two major components:

1. The first is a representation of a *static spatial task environment* (SSTE).
2. The second is a representation of the *cognitive structures* and *processing* of the *individual performing tasks* in such an SSTE.

The essence of the model is to develop the possibility for a decision maker to move around this representation of the environment, obtaining information for making navigational decisions. In modeling the SSTE, a representation of a given real environment is produced in which subjects can be tested. The task area is modeled in such a way as to incorporate the salient cues that subjects employ during spatial navigation.

In constructing the representation of the SSTE, two basic units are required.

The first (*PLOT*) is essentially a parcel of land on which a given dwelling and appurtenant structures are located, together with the pavement and roadside adjacent to the plot. A *decision point* represents a place at which a navigation decision must be taken. Each plot can be represented as two data structures—the *PLOTFRAME* contains information about the visual cues on a plot including structures, vegetation, parked vehicles, and so on. The *VIEWFRAME* contains information characterizing other plots as they are seen by a subject standing in front of the given plot; it may include context information such as curvature and slope of the street, visible intersections, and unique features such as hills, trees, mailboxes, or so on that may serve as cues during navigation.

A simple logical language is developed focusing on elements known as feature–property–values (FPV triples). Each FPV consists of: one member of a class of objects (e.g., houses, mailboxes, trees, cars); one member of a class of properties characterizing the object (e.g., shape, size, color, orientation); and a member of the class of values that can be taken on by the property (e.g., large, blue, sculptured, untidy). Relational properties (such as the logical connectives "and," "or," and "not") are used to link object, properties, and values in order to produce complex forms (house/object-next to/property-big tree/value).

Along with the representation of the SSTE is required a *computational process model* (CPM) of the individual who interacts with the SSTE during navigation tasks. The individual modeled by the CPM obtains information from direct experience of the immediate environment. This is then employed in accessing other information stored in the individual's knowledge structure to allow the making of decisions regarding what actions must be taken to complete a given task. Actions include information-processing tasks such as perception, storage, accessing, and transformation of environmental information. Obviously the modeling process at this stage gets complicated, and the structures and processes that are invoked are only models of the cognitive structures and processes in human decision makers. A test of the validity of any given CPM is the ability of the model to predict behavior and error making in given experimental situations.

Figure 5.10 summarizes the major components of the CPM.

1. The first is a *conceptual component* involving three *cognitive structures*—a *perceptual buffer*, a *lexicon*, and a *short-term* or *working memory*. The lexicon contains each individual's lists of feature, properties, and values, together with saliencies for each F, P, or V, reflecting idiosyncratic and individual differences in the relative ordering or significance of environmental information. Short-term memory is assumed to have two substructures, one containing FPV triples that one expects to find in the SSTE and the second containing FPV triples that are actually found in the SSTE.

2. The second is a *storage component*. This component focuses on actions and states, with A_i representing a particular action, and S_j representing a sequence of states of the system. An appropriate sequence to represent a state–action–state sequence is SAS. The storage component involves a short-term memory, a long-term memory, and a buffer containing an activated portion of long-term memory. This

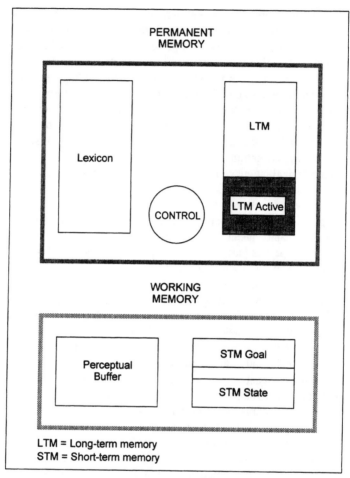

Figure 5.10. Diagrammatic representation of knowledge structures in the cognitive model. *Source:* Smith, Pellegrino, & Golledge, 1982.

buffer is a temporary store for those elements of LTM that have been accessed recently in connection with other processing. The transfer of information from STM to LTM involves a process of spreading activation in an attempt to accomplish pattern matches between elements in STM and LTM. A second process takes the most recent sequence of state–action–state from STM and places it in a buffer such that it can be concatenated with other actual or expected SAS sequences. Transfer from STM to LTM occurs when action is taken, thus admitting possible change of saliences associated with features or strings of plots represented in the states, and hence to possible errors of knowledge acquisition.

3. The third component is an *action–choice component* (ACC). This accesses LTM and selects the most appropriate action to take in any given point in a wayfinding task. Here for any SAS sequence in LTM, the current state and goal stored in

STM are matched up, and a specific action is chosen for execution. Part of this total procedure involves action, choice, and goal-setting behavior, and a goal comparison process that compares the state achieved after taking the action with the goal state expected.

4. A fourth component of the computational process model is a *set of processes* restructuring LTM by allowing long concatenations and by generalizing over selected SAS combinations.

The above model can be formalized as a set of interacting computer programs that allow for any given spatial task environment (e.g., selecting origin–destination pairs) to be given to a decision-making unit that is expected to find a route between them. Thus, the model is capable of predicting a given navigational task for a specific individual whose saliences have been obtained from prior experimentation. The test of the validity of the model consists of a comparison of the model selected route and its appropriate segments and error components with the actual route selected and errors made by a given subject. Visible intersections and unique features, such as hills, trees, mailboxes, or so on, may serve as cues during navigation.

6

Perception, Attitudes, and Risk

6.1 The Experience and Conceptualization of Space

Within the general objective geographic environment is a subset known as the *operational* environment. This consists of that portion of the world that impinges on any given human. It influences behavior either directly or indirectly. That part of the operational environment of which a person is aware is called the *perceptual environment*. Awareness of it may be derived from:

1. Learning and experience of a segment of the operational environment.
2. Sensitivity to messages from environmental stimuli.
3. Secondary information sources not necessarily related to direct experience.
4. Evaluation or assessment of environmental qualities.
5. Individual or social beliefs and values about specific environments, environmental conditions, or environmental behavior.

Part of the perceptual environment may be regarded as being symbolic rather than a mapping of an objective evaluation. Thus, when a person uses the word *space* it is usually a reference to what is being conceptualized. Space as *conceptualized* will be treated in the next chapter in the context of environmental cognition. Space as *experienced* is often reflected in examinations of affective responses to environment, literary perspectives on environment, and in terms of the roles of attitudes, emotions, and personality factors of the person operating in an environment.

Whereas in the past geographers tended to emphasize the spatial structure and spatial relationships in human activities, with the evolution of the behavioral or *process-oriented* approach, it has been recognized that these structures and relationships do not appear the same to all people. Amedeo and Golledge (1975) have indicated that:

The extent of our information about a system, the perception of barriers to movement in the system, the specific needs and values of the individuals and groups located therein will influence our cognitions of the spatial properties of such a system, distorting it in some places through illusion or incomplete information, and perceiving the structure very close to a scale model of physical reality in other places. . . . Recognition that different needs sets spawn different action spaces and that these action spaces are related to different elements of the objective world, has led to a search for an understanding of the processes that form the images of spatial reality that are retained in the human mind. (p. 381)

It is the search for an understanding of the perceptual and cognitive processes that create this *image* of operational environments that has received much attention from behavioral geographers.

6.2 A Problem of Defining Terms: What Are Perception and Cognition?

The terms *perception* and *cognition* have been employed in a confusing variety of contexts. In this and the following chapter we attempt to clarify these terms and examine their use in geography.

6.2.1 Perception

Among geographers, the term *perception* often has been used differently to the way it is used in psychology. Geographers have tended to use the term in the sense of how things are remembered or recalled by people—as with respect to "perception" of resources or hazards. Planners or architects have used the term to describe the mutuality of interests among various groups of actors in the design process (Rapoport, 1977: 30). However, psychologists have tended to treat perception as a sub-set or function of *cognition*. For example, Werner and Kaplan (1963) see *perception* as an *inferential process* in which a person plays a maximal and idiosyncratic role in interpreting, categorizing, and transforming stimulus input. Ittelson (1960) also has emphasized the central role of simultaneity in perception of a stimulus situation. As the real world is complex and is sending out millions of information signals about all aspects of life and environment, we can only be aware of a small portion of them. An individual receives these signals through his or her *senses* —by sight, hearing, smell, taste, and touch.

The senses are viewed as functional systems designed to provide feedback to the body's motor system and to seek out information about the surrounding environment (Gibson, 1966). As active information seekers, the senses are thought to record only those stimuli that have a bearing on the needs of the individual and to ignore the rest. Occasionally, a strong stimulus may impose itself on the system, but normally the system decides what it will look for. Information is experienced and recorded as part of the array of potential stimuli that present themselves to the body. It is contained in differentials of color, heat, motion, sound, pressure, direction, or whatever else may be present in the environment and is within range of the sense organs. The perceptions of two individuals vary as a function of differences in the content of the information presented and the differences in the ability of individuals to pick up the information messages.

These senses do not, however, play an equal part in spatial perception, as suggested in Figure 6.1, with only sight, hearing, and smell being able to receive stimuli from parts of the environment beyond the so-called tactile zone. However, much of our environmental information is secondary, culled from the media and through hearsay via communication with fellow human beings. This is also viewed as *perception*. Here its meaning can be expanded to include the *immediate apprehension of information* about the environment by one or more of the senses. It occurs because of the presence of an object. It is closely connected with events in the immediate surroundings and is, in general, linked with immediate behavior.

6.2.2 Cognition

Cognition refers to the way information, upon reception, is coded, stored, and organized in the brain so that it fits in with a person's accumulated knowledge and values. Thus, cognition is developmental. Wapner and Werner (1957) argue that as development proceeds, perception becomes subordinate to higher mental processes. Further, an organism's available cognitive structures influence perceptual selectivity, which leads to a reconstruction of the world through selected fields of attention. To give a simple example, we may perceive the street where we live by physically being there, but knowing the route to work depends on cognitive organization of a set of perceptions experienced through frequent travel.

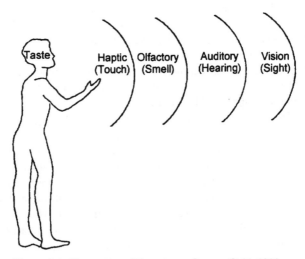

Figure 6.1. The ranges of the senses. *Source:* Gold, 1980.

6.2.3   Immediacy and Scale Dependence

This distinction by psychologists between perception and cognition is one where *perception* is linked to *immediacy* and is *stimulus dependent*, whereas *cognition* need not be linked with immediate behavior, nor need it be directly related to anything occurring in the proximate environment. *Cognition* concerns how we *link* the *present* with the *past* and how we may *project* into the *future*. Cognition is, then, the more general term, subsuming the more specific concepts and surrogates of sensation, perception, imagery, retention and recall, reasoning and problem solving, and the making of judgments and evaluations (i.e., decisions and choice).

However, this distinction is not an absolutely clear dichotomy, the difference being one of degree and focus (Downs & Stea, 1973:14). Stea (1969) has offered a more useful distinction from a spatial point of view, suggesting that cognition occurs in a spatial context when the spaces of interest either are occluded or are so extensive that they cannot be perceived or apprehended at once. Thus these large-scale spaces have to be committed to memory and cognitively organized to contain events and objects that are outside the immediate sensory field of the person. This is a scale-dependent difference and is intuitively acceptable to the geographer.

The end product of perception and cognition is a *mental representation* of the objective environment. Information signals are filtered through perception, then further filtered through cognitive processes and existing cognitive structures in the brain in the manner suggested in Figure 6.2. Thus, people respond not directly to their real environment, but to their mental representation or *image* of it, and as a result, the location of human activities and the spatial pattern of their movements will be the outcomes of perceptual and cognitive structuring of their environment.

It is important to realize that all experiences with elements of environments external to the individual take place within a framework of space and time. Amedeo and Golledge (1975) stress that the *cognitive processes* (such as perceiving, learning, formation of attitudes, etc.) operate to produce in people an individual *spatiotemporal awareness* or *knowing* about environments. Learning affects the completeness with which some things will be perceived and understood by individuals, and their attitude to them helps determine their clarity and relevance to a person. Thus, "The image of say, an urban environment, or of a shopping center, or of a hazard, is the product of both the immediate sensation experienced when [one is] con-

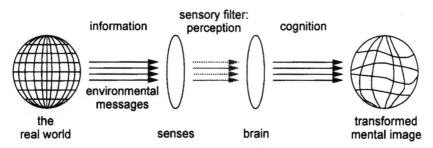

Figure 6.2. The formation of images. *Source:* Hayes, 1980:2.

fronted by an object and the memory of past experiences with the same or similar objects" (Amedeo and Golledge, 1975: 382).

Different people may give different interpretations to the same spatial structures and phenomena, which then take on individual meanings. We may postulate that individuals impose a mental ordering on environmental information to provide *identity*, *location*, and *orientation* for the *elements* perceived in the objective model. But it is the variations in the accuracy of these cognitive orderings that will furnish some explanation of the variations evident in the behavior of different people in the same environment.

6.3 Factors Influencing the Nature and Structure of the Perceived Environment

The roots of the concept of perception lie firmly in the realm of psychology, and within that discipline a range of interpretations are given to this sensory process.

6.3.1 The Functionalist Approach

One school regards perception in *functional* terms arguing that perception serves a function by which individuals convert information about what is "out there" into meaningful terms. It thus serves as a guide to action. This functional view includes the idea that perception is a matter of identifying something in the environment and anticipating its properties. The identification places phenomena into some category of equivalent objects and, by implication, distinguishes it from other categories. The criteria used in identification may be single or multiple in nature.

6.3.2 Perception as an Encoding Process

An alternate viewpoint regards perception as an *encoding process*. It has been argued that the environment may be regarded as a mass of "to whom it may concern" messages (Figure 6.3). In this case, perception is a process by which we select those messages that are of concern to us. In other words, we use perception to encode input in the environment—we see a building and encode it as a house; or perhaps we see people acting boisterously and encode the event as a riot. In this situation, it is argued that through a vast amount of recurrent experience we build up sets of concepts that are like *templates*. Perceptual input is then compared to the set of templates stored in the mind. If the input fails to match a conceptual template or pattern, then it is compared with another and another until finally one is found that matches it sufficiently well to aid in categorization. Input is assimilated in terms of this particular template, and a bit of information is thus encoded in the brain. In selecting the appropriate templates for the pattern-matching process, there is a preliminary sorting activity that eliminates many of the potentially irrelevant ones and concentrates on the few that are thought to be most suitable in helping with the current task. In artificial intelligence terms, this process is termed *activation spread*, and the search for a potential pattern matching is undertaken via a complex hierarchically organized network of encoded information.

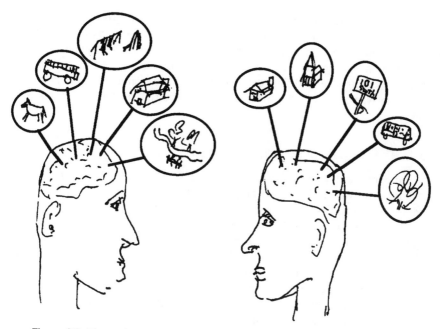

Figure 6.3. The environment: "A mass of 'To Whom it May Concern' messages."

Regarding perception as an encoding process involves arguing that the amount of information we are capable of encoding depends on the extent to which the concepts or categories with which we are familiar are elaborated and differentiated sufficiently one from another. Whereas we may have difficulty distinguishing among a series of shapes of approximately the same size or approximately the same shade of color, particularly if some are rotated, we would not have the same degree of difficulty in differentiating among separate parts of our general external environment (e.g., houses vs. shops vs. offices vs. factories). Critical components of the environment such as the distinction between large and small buildings, between commercial and noncommercial areas, between built-up and non-built-up areas, and so on, appear to be perceived readily. Of course the templates used under these circumstances for the pattern-matching process are relatively gross. It would be much more difficult to distinguish between every house on the basis of its size, age, value, color, location, and other perceptual characteristics, although it is quite possible that, given a particular experimental context such distinctions may be achieved. For example, it has been suggested that people develop stereotypes of housing types and can easily distinguish between say, a typical Australian house and a typical American house, despite the fact that there are so many regional differences in house styles in Australia and America that would make it difficult to construct "typical" examples. Some combination of housing attributes seems to provide a clue that allows such a distinction to occur. It is obviously easier, depending on cultural circumstances, to distinguish houses from factories, homes from parks, factories from schools, or houses from churches.

As we increase the *scale* at which information is collected for encoding purposes, there appears to be an increase in the tendency for *selectivity* to take place. *Abstraction from reality* becomes another common process, and incompleteness and generalization processes also occur. In many cases, the accumulation of information consists of adding finer and finer detail to an initial gross template such that a more comprehensive picture of the attributes and dimensions of an external area is compiled.

6.3.3 The Concept of Scale in Perception

It is sometimes useful to study environmental perception according to the *scale* at which it is undertaken. This scale ranges from the most elementary level, the personal bubble, to the less tangible perception of characteristics and features of units such as national groupings. Much recent work, for example, has been concerned with perception in unusual environments such as in space capsules. These artificial environments are precisely designed to fit the human physical form and its needs, with the aim of keeping stress at a minimum. Some of the most widely publicized and significant research in space perception today focuses on the perceptions of and reactions to environmental extremes. Thus, conditions of stress, conflict, crowding, or abnormal cramping are becoming more and more the norm in terms of research into the perception and use of space.

6.3.4 Perceptual Thresholds

When dealing with space perception, one most frequently focuses on the visual senses. Regardless of which of the senses are the focus of attention, it is recognized that each has a minimum of sense-organ stimulation that is necessary before sensory experiences are evoked—that is, a *perceptual threshold* exists. Factors influencing the threshold level in any particular sensory situation include not only the physiological receptor limits but also the constraints imposed by social, economic, cultural, nationalistic, or other sets of factors. For example, when do we perceive something to be dilapidated? When is something too far away? When do we perceive an area as a ghetto? When do we consider an environmental disturbance to be a hazard?

To answer the *when*, *why*, and *how* of these questions it is necessary to incorporate the notion of a perceptual threshold. It is accepted that thresholds may vary among individuals or may vary consistently across a broad class of individuals (e.g., as in the ability of rural and urban residents to perceive subtle changes in land uses). Perceptual thresholds may, however, be consistent across individuals at the same location.

A significant part of the discussion that follows, therefore, focuses on the effect that our position in both the physical and human environments has on the degree to which environmental stimuli impinge on our senses and are classified in various event categories. For example, the occurrence of snow at certain times of the year in Fargo (North Dakota) is expected. Residents fill their cars with antifreeze, carry snow chains, and put snow tires on their vehicles. Consequently, a 6-

inch snowfall there might be tolerated with no major catastrophic consequences. However, a 6-inch snowfall in Atlanta (Georgia) is another matter. Neither individuals nor the city administration would be prepared for such an event, and many physical, social, economic, or personal disasters might occur. In the Fargo case, the event would not reach a perceptual threshold that would warrant extraordinary action. In Atlanta, individuals, institutions, firms, and governments might well all agree that a threshold had been reached and emergency action was required.

6.3.5   Perception and the World of Identifiable Things

Perception assumes considerable importance to the geographer if we accept the argument that individuals are concerned with constructing a *perceived world* while maintaining, as far as possible, *stability, endurance*, and *consistency* of the images they develop. This argument suggests that there is an attempt to overlook some of the variability of objective reality and to systematize knowledge along the line of greatest invariability.

In another context, it is argued that the built environment is the spatial manifestation of human decision making and that many of these decisions are related to the way in which we perceive space, evaluate the elements of space, and image the potential use of it.

For example, considerable furor arose in the United States in the 1980s over the substantial differences in the way environmental resources were perceived by Secretary of the Interior James Watt on the one hand, and by the many groups of environmentally concerned individuals on the other hand. Obviously, the conflict over resource development and natural environmental preservation existed well before this particular controversy, but it did exemplify the degree to which conflict resolution depends on differing perceptions of the worth of various segments of objective reality.

In general it appears that any order retained in the perception of spatial reality is the result of perceiving enduring objects or the result of enduring sociocultural interpretations or constraints. Perception, then, can be regarded as helping to build a world of identifiable things. Were the individual not sensitive and responsive to environment, he or she would be unable to satisfy needs, communicate with others, or derive pleasure from his or her surroundings.

6.3.6   Perceptual Constancy

When one considers the very complex problem of individual differences, and also the complexity of objective reality, one cannot help but wonder about how we can obtain *perceptual constancy* across large groups of people. It is a fact, however, that the complex real world is handled by people with limited capacity for information storage, manipulation, and retrieval. It is also obvious that each individual must make certain simplifications and adjustments in his or her perceptions of the real world in accordance with needs and experiences. These simplifications and adjustments are a direct result of the way that reality is conceptualized.

There is no simple, unique, one-to-one correspondence between a stimulus when presented to an individual and a stimulus perceived by an individual. But the same stimulus presented to a large number of individuals will result in a similar type of perception across that group. For example, in many cases, adults of similar social class and status perceive much the same things because their perceptions are constrained in similar ways by social norms. Thus, a Marxist viewing the housing market in a capitalist system may perceive the built-in inequalities favoring those with capital and power and disadvantaging those without such resources. In contrast, a capitalist may see only the normal working of an economic system in which stratification is based on differentially earned resources.

Perception is thus a creation of the observer. The observer gives characteristics to what is seen, such as size, shape, and position, in accordance with information supplied to the senses. Perception can be seen as a critical component of the development of cognitive structure. We also see that perceptual characteristics are allocated to stimuli based in part on past experiences with the stimuli that help determine their known attribute dimensions.

6.3.7 Perceptual Focusing or Attention

Perhaps the most obvious factor influencing perception is *attention*. This is sometimes called *perceptual focusing.*

At any given time there are numerous potential stimuli impinging on our senses. All parts of the environment external to mind can potentially act in this way. We do not react equally to all stimuli; rather we tend to focus on a few stimuli that are found relevant in a specific purposeful context. At any point in time, and at any given place for any individual, there is competition among stimuli for attention. Selecting one or more stimuli from the mass that are available represents the resolution of a form of conflict. The form of the conflict is to discriminate among the stimuli being presented and, hopefully, to select those that are the most relevant and most significant in any given situation.

Some of the critical factors involved in the resolution of this conflict include the *size, intensity, repetition*, and *clarity* of any given stimulus. Sometimes the attention-getting qualities of stimuli are dependent on the context or the situation in which they are presented. For example, a green lawn would stand out to a much greater degree in a semiarid or desert area than in parts of Sussex or Iowa. Stimuli that stand out from the general contextual background appear to be perceived more clearly than those that merge with the background. Stimuli that are unique or rare in an environment appear to have more attention-getting qualities than those that are common in the environment. The maxim "One can't see the forest for the trees" is a good representation of this problem.

6.3.8 Preparatory Sets

Preparatory sets comprise a further influence on perception. Here anticipatory changes take place prior to the presentation of a stimulus and influence the ways things are perceived. In other words, the preparatory sets influence preconceptions of a given stimulus.

For example, we all have ideas of what a ghetto looks like even if we have never actually experienced one. Our *preconceptions*, then, influence the type of characteristics that we expect of a stimulus such as this, and in turn, they may influence the degree to which we characterize a given situation as ghetto-like. Thus, many people would reject the possibility of living in a certain place because they associate with it the undesirable characteristics of a ghetto-like existence—even if they have never had the opportunity to confirm these preconceptions. Similarly, shopping behavior is often a result of the preparatory information sets that people have about places. These sets are a major reason why advertising is such a significant factor in selling goods, locations, and store images. The same factor appears to be of some significance in the process of residential site selection. In this context it is frequently incorporated in the notion of *levels of expectation*. These levels of expectation about a given house type or potential living environment appear to have a major influence on search patterns and the eventual choice of sites. Of course search activity itself may be undertaken if guided by an interested third party (such as an estate agent) in order to overcome sets of preconceptions and to assist in the site-selection process.

6.3.9 Individual Needs and Values

Perception is regarded also as a function of the *needs* and *values* of individuals. Social values may accrue to objects or places in an objective environment and, as such, may influence the probability of a stimulus passing a perceptual threshold. For example, it has been shown that poorer children overestimate the size of coins that were presented as stimuli, whereas children from richer families were able to estimate coin sizes much more accurately. Deprivation appears to have influenced the nature of a perceptual image. Often it is argued that it is better to own the poorest house in a rich area than to own the richest house in a poor area. The attributes of the area are imputed to the house in question and may inflate or deflate its value.

6.3.10 Cultural Values

When individual differences are manifested in human behavior and are not assignable to structural causes, the question arises as to whether and to what extent the differences are assignable to *cultural values* and to the nature of *social interaction*. This action is important in determining perceptual thresholds. It is readily accepted that physiological needs, and conditioning, along with other motivational forces, influence what may be perceived. Personal beliefs, social values, and other social connotations also appear to influence the attributes of what is perceived. Working in combination, these factors influence perceptual thresholds, or at times they determine the extent to which parts of an existing knowledge structure are retained, modified, or deleted.

Nasar (1984) examined whether there were significant cultural differences between subjects in Japan and in the United States in terms of selection of "essential" features that typify urban streetside environments. He found that the following features were most frequently recognized as being essential or typical features of cities in both countries. New buildings were mentioned with the highest frequency in the United States, but with only moderate frequency in Japan. Vehicles were mentioned with moderate frequency in the United States, but with very high frequency in Japan. Vegetation was mentioned with low frequency in the United States but high frequency in Japan, whereas pedestrians were mentioned with relatively low frequency in both environments. Although Nasar's study was small in size, it did provide some empirical evidence of perceptual differences that might be found in the Japanese and U.S. urban environment, and it supported suggestions by Canter and Canter (1971) and Rapoport (1977) that even though the physical structures of cities in different countries today are becoming more similar, different cultures will perceptually focus on different features as characteristic and differentiating attributes.

6.3.11   Ecological and Anthropocentric Constraints

It is sometimes argued that one can classify the relevant attributes of a given stimulus into sets that are *ecologically constrained* and those that are *anthropocentrically constrained.*

Ecological factors include things such as the dominance of the perceptual form (e.g., visible dominance), clarity or degree of ambiguity, and simplicity of structure. Each of these factors is inherent in stimulus objects and their general environmental context.

Perhaps equally important, however, is the *anthropocentric dimension* that includes factors such as frequency of exposure to a given stimulus, the source of information about the stimulus, its relevance to the activity space of the perceiver, its position in the general knowledge structure of the perceiver, and its sociocultural importance.

6.3.12   Location and Orientation of Individuals

The *location* and *orientation* of individuals, as well as the clarity of environmental phenomena, will affect the level of awareness we have of *objects*, *places*, and *events*.

We have seen that people are selective in constructing their images of the environment. It is important that these images have *stability, endurance,* and *consistency,* because humans respond to environments to satisfy their needs, communicate with their fellows, and utilize their surroundings.

Two alternative approaches have been suggested concerning how humans store the information perceived from an environment and the accuracy with which environmental elements are positioned in the cognitive representations of a large-scale environment such as a city.

1. The first approach suggests that humans give elements in space *coordinate locations* on a large *mental map*, and additions to the cognitive structure get allocated their position, size, shape, and so on with respect to an existing structure.

Thus, the accuracy of the positioning of new elements will depend on the accuracy of the positioning of *existing* elements in this structure.

2. The second approach gives to these new structural elements *attributes* that are specified with respect to *surrounding* characteristics within the structure; that is, they will have relational properties such as larger than, smaller than, closer to, more distant from adjacent elements, rather than be judged purely on their own characteristics.

Inherent in the above are a number of very specific spatial properties, such as *locational constancy, proximity, similarity, position, continuity,* and *closure.* According to Amedeo and Golledge (1970),

> The degree to which each of the above elements is recognized and utilized by individuals will determine the accuracy and consistency with which an image is formed on successive trials. If information of any sort is given to individuals (or is obtained by them) which affects perception of proximity, position, similarity, and continuity, or destroys or assists a desire for closure, then the resulting perceptual image changes and consequent spatial behavior might change. (p. 384)

If one were to stop and look around and ask the question "What do I see?" the answer is likely to be very complex and consist of a collection of people, events, and things. With perceptual focusing, one might pick out specific things instead of making a general statement such as the one above. In addition to selecting specific things, one might cite some of their individual characteristics or sensory qualities. We might see the green trees, the snowy road, or the ugly building rather than the greenness, the snowiness, or the ugliness of the entire scene. In making such an analysis, we record a perceptual experience of segments of the environment. Each experience focuses on parts (or sometimes whole patterns) of things, and recognition of the interrelationships of parts and wholes helps to build the world of identifiable things.

After perceiving something, one may remove oneself from the perceptual field but still remember or think of the things perceived in their original places and context. The extent to which our recall of the initial stimulus situation reflects its true structure will, in turn, depend on our ability to extract critical elements of the perceived scene, and on our ability to encode, store, and recall them in an appropriate and efficient manner.

On a superficial basis we may be perceiving the physical structure and be unable to determine the hidden structural dimensions of a scene—for example, the underlying socioeconomic structure, ethnic, or other relations, and the interactions and activities that take place within that scene and give prominence to and recognition of individual scene elements. These hidden dimensions may also be responsible for allocating to a scene a characterization such as mystery or beauty.

Of course the way in which we perceive a situation may influence the way we react to it. Assume that we are standing on a busy street waiting for a gap in the line of traffic to occur so we can cross the street. The size of the gap judged to be sufficient to allow us to cross the street can vary with the intensity of traffic. On a very busy street, a small gap may be perceived as adequate enough for a crossing to take place. On a street not so busy the perception of the frequency of cars passing and their rate of progress may be such as to substantially change the size of a traffic gap imaged as being necessary before a crossing is attempted.

In general if we perceive a structure to be complex, it is possible that our behavior, with respect to that structure, may also be complex. Similarly, the perception of simplicity in a structure may influence a simple type of reaction toward it. An individual who daily experiences the complexity of life in the downtown area of a large metropolis may become so familiar with those activities that they appear to be quite uncomplicated. In contrast, the newcomer to the same place may be faced with the same spatial structure but may regard it as so complex an environment as to be almost impossible to negotiate.

Thus, if we accept the fact that perception is an important factor likely to influence behavior, then we must pay attention to it, just as we would pay attention to any of the structural properties of a situation in which an explainable behavior is undertaken.

6.4   Attitudes

Regardless of the exact definition chosen, it appears that *perception* involves an interaction or *transaction* between an *individual* and an *environment*. The individual receives *visual*, *auditory*, *tactile*, or other *information* from an external environment, and he or she *codes*, *records*, and *uses* this information to *constrain* or *modify* potential *behaviors*. Whereas in the classic sense perceptions are regarded as flexible and transitory phenomena that occur only in the presence of the stimulus, Downs (1981a,1981b) has pointed out that in the discipline of geography perception has been interpreted in a broader context with a distinct evaluative component. In many cases the term *perception* has been confused with the concept of *attitude*, which is seen to be a relatively permanent structure that may hold in the absence of any particular stimulus. Attitude, therefore, can be regarded as a *learned predisposition to respond* to a situation in a consistent way. In the case of some of the geographic literature on "environmental perception," the concept of perception is interpreted in terms of this learned predisposition to respond.

6.4.1   Attitude and Uncertainty

We suggested in the first two chapters of this book that it is generally accepted that humans make decisions in relatively uncertain environments. This uncertainty occurs because humans as decision makers have incomplete knowledge about the various environments in which they exist. Thus decisions are made in an atmosphere of conjecture. Humans are faced with choosing between the alternatives of which one is aware, about which one has imperfect knowledge, and in the context of uncertainty as to what decisions other people are making and which may have an effect on decision making. However, choice or decision is made with an expectation as to what the outcome is likely to be. The concept of *attitude* of the decision maker is a key variable influencing this decision process. The concept of attitude is important because it *brings together the internal mental life* of a person (i.e., cognitions, motivations, and emotions) and *overt behavioral responses* within one framework (Gold, 1980: 23).

The concept of attitude and the nature of attitude formation have received relatively little attention by geographers concerned with investigating human spatial behavior. There have been some references to *attitude formation* as a causal factor in the analysis of locational decision making. For example, Isard and Dacey (1962) argued that as an individual's attitude toward the role played in an economic system changed from one of optimizing to satisficing, then that individual would adopt different strategies in coping with an environment. Attitudinal information was thereby incorporated by Isard and Dacey into models using the theory of games in the context of investigating new approaches to central place theory and location theory. It was incorporated also by Gould (1963) in his analysis of decision making by subsistence farmers in uncertain physical environments, and by Hägerstrand (1952) in the development of simulation studies of the diffusion of innovations.

An alternative approach emerged in the work of Golledge (1970b) and Kotler (1965), who examined attitudes from a psychological and marketing view. If, for example, a producer regards consumers as *Marshallian* (economically rational) beings who are price conscious, then locations will be sought that allow favorable competition in a competitive market, and the producer will manipulate price to penetrate adjacent market areas. However, if the producer regards consumers as being *creatures of habit*, then a less than economically optimal location might be accepted. At this point the producer will attempt to attract spatially rational customers, and aggressive marketing policies need not be pursued. Thus, the attitude producers adopt toward consumers will influence their choice of business locations and will affect any equilibrium pattern of producers that may emerge in a spatial system. For example, in Hotelling's classic example of the effects of duopolistic competition in a linear market, the optimal locational strategy under conditions of active competition was for the competitors to locate back to back at the center of the market (Figure 6.4). But, if the producers acted in collusion, optimal locations (i.e., profit maximization) would be achieved by locating at the quartiles. In the central location case, consumers had to travel up to half the length of the market to reach a producer. In the context of collusion, no consumer had to travel more than one-quarter of the length of the market to reach a seller. Thus, depending on the seller's view of the criteria customers used to make decisions, central or dispersed locations would be negotiated.

In contrast, consumers who act according to *Pavlovian* principles will respond immediately to the introduction of a stimulus set, such as an exciting advertisement or a pressure salesman, and the behaviors of such consumers in space may become distorted from the patterns predicted by the economically rational models. Yet different behavior could be expected from a *Veblenian* consumer, who would be

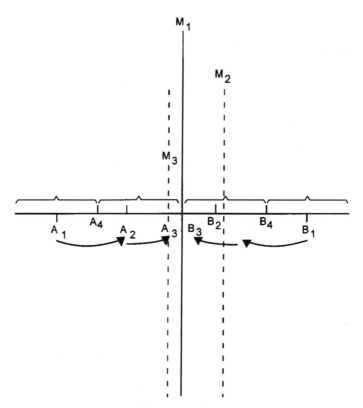

Optimal Location under Competition : A3B3
Optimal Location under Collusion :   A4B4

M1 = initial market area boundary
M2 = boundary between A2B1
M3 = boundary between A2B2

Figure 6.4. Hotelling: duopolistic competition.

much influenced by a peer group and would tend to patronize those outlets used by peers even though this strategy may produce some unusual spatial patterns of behavior.

These types of behaviors may be seen as reflecting attitude toward a given situation. Kemeny and Thompson (1957:284) argued that human attitudes toward *risk taking* under uncertain conditions can explain, in large part, different behavioral outcomes. Diagrammatic representations of various individual utility functions according to the attitude of the individual toward taking risks are shown in Figure 6.5. These are in the form of a functional relationship between perceived value and monetary payoff. The differences among the individuals are clearly evident. For example, the *reckless person* (A) exaggerates gains and discounts losses, whereas the *cautious person* (B) tries to avoid large losses and tends to disparage and discount large gains. The socioeconomic status of a person is seen to influence attitude to risk, with the poor person (C) exaggerating both wins and losses, while the rich per-

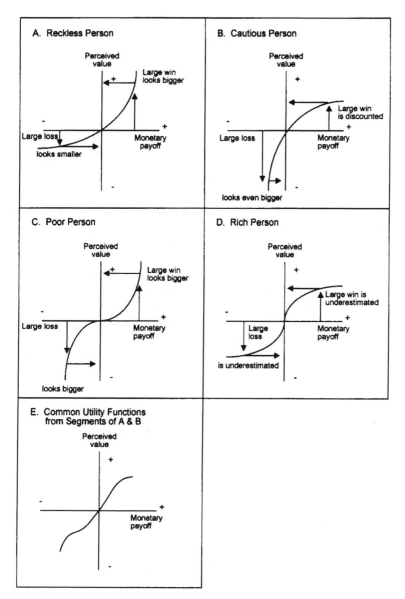

Figure 6.5. Individual utility functions and attitudes to risk taking. *Source:* Kemeny & Thompson, 1957:284.

son (D) underestimates large gains and losses. The actual *utility function* of people is thus seen to reflect their attitudes to risk taking and their perception of value or utility. In reality people are likely to be a complex mixture of these simple extreme types of decision makers, and a common utility function is likely to incorporate elements of all the above, as demonstrated for person (E). The important point is that the attitude of individuals to specific spatial stimuli in choice behavior situations is an important element to consider in explaining the behavioral response and strategies thus adopted in coping with uncertainty. Attitudes are important in furnishing explanations of difficult-to-explain spatial behaviors in the more deterministic models used by geographers.

6.4.2   The Nature of Attitudes and Attitude Formation

The concept of attitude is a complex one, and there is considerable dispute over its definition. One widely accepted definition is: "Attitude is a learned predisposition to respond in a consistently favorable or unfavorable manner with respect to a given object, person or spatial environment" (after Fishbein & Ajzen, 1975).

As such, *attitude* contains the notion that it is *learned*, that it *predisposes* a person to perform an *action*, and that it is *relatively invariant* over time. Fishbein formalized this traditional view of attitudes specifically as having three components.

1. *Cognitive*—involving perceiving, knowing and thinking—processes by which the individual gets to know an environment.
2. *Affective*—involving feelings and emotions about an environment, motivated by desires and values that are embodied in environmental images.
3. *Conative*—involving acting, doing, striving, and thus having an effect on the environment in response to (1) and (2). (This may also be termed a *behavioral component* of attitude.)

Figure 6.6 demonstrates this view of attitude components as intervening variables between stimuli and behavior.

It is important to realize the limitations of this traditional view of attitude. The three components should be seen to interact. Thus, a person's feelings toward

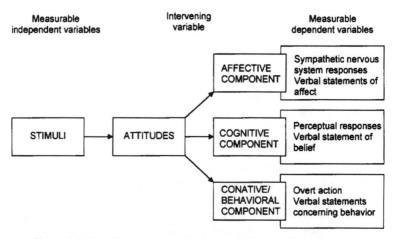

Figure 6.6. The three components of attitude. *Source:* Triandis, 1971.

an environment will be determined, at least in part, by what is known about it. *Learning* also plays a role, as all the components of attitude may change with experience. However, the view is useful, as it allows us to see what aspects of some total attitude we may be measuring at a given time in a particular situation. Furthermore, it points to possible sources of error if we attempt to predict subsequent behavior of persons exposed to certain situations. The view is generic, as it takes attitude to include a range of overlapping concepts—beliefs, biases, doctrines, faiths, ideologies, judgments, opinions, stereotypes, values, and so on.

Attitudes are evolutionary through the learning process, but it has been suggested that they are learned mostly in childhood and that, once formed, they are pervasive. Attitudes can withstand extensive and concerted attempts to change them, as has been discovered in military propaganda and mass communication studies. O'Riordan (1973) suggests that even if attitudes are changed, a change in behavior is not necessarily going to follow because there is a gap between stated attitudes and actual behavior. This is an interesting proposition as it has important implications for the media, advertisers, the military, and governments, all of which tend to believe that attitude can be modified. It is, then, not surprising that attitudes and attitude formation comprise an area of considerable concentration of research in psychology.

6.4.3 Attitudes, Values, and Stereotypes

Of the concepts that relate to the above generic view of attitudes, *values* and *stereotypes* are particularly important.

Values may be regarded as enduring beliefs about specific modes of conduct or specific outcomes that are either personally or socially preferable to alternatives.

Stereotypes are sets of beliefs about something (people, places, events, objects, etc.) that contain no more than a "grain of truth" but are the basis of opinions and are used to justify or rationalize conduct toward that thing, place, or person (Belcher, 1973).

Thus, values and stereotypes, as types of attitudes, may be seen to act as mechanisms whereby people erroneously come to terms with their complex environments. Through values and stereotypes, we arrange information into categories, and the individual items are assumed to have the same attributes as the whole category. For example, we ascribe certain values to areas that look basically the same, such as run-down housing areas being thought of as slums and rather unsavory places to visit. This is important in understanding people's attitudes to environmental phenomena, for, as Gold (1980:24) states, "once a person can identify the category to which an entity belongs, he then has the basis for rapidly assimilating new information and rapidly reaching decisions about behavior."

6.4.4 Attitudes, Motivation, and Emotions

It is useful also to consider *attitudes*, and particularly their *conative* component, to be linked to *motivation* and *emotions* as these are central in human–environment interaction because our behavior is said to be goal-oriented.

Motivation can be thought of as the force leading one to seek specific goals in order to satisfy needs, and in so doing to pursue a specific *strategy* and to take certain *actions*. The strength of motivation behind an act will depend on the level of need attached to that act. Thus, the notion of need is an important one, and it has attracted considerable attention among social scientists. Evans (1976) suggested that people are motivated by a desire to overcome the problems and stresses imposed by an environment in order to reduce uncertainty and tension. Wohlwill (1968) suggested that people desire novelty, excitement, and stimulation and that they seek to increase the level of tension in the environment.

People have different types of needs, ranging from *survival-type needs* (for shelter, food, clothing) that must be satisfied to sustain life, to *social* and *personal needs* to do with acceptance and rejection and self-actualization. These are not necessary to sustain life in a physical sense, but they can create mental anguish if they are unfulfilled. Maslow (1954) proposed a *hierarchy of needs*, in which basic (survival-type) needs had to be satisfied before higher-order needs. In this context, the link between needs and motivation is self-evident. But, there is considerable debate over whether human needs are genetically determined or environmentally acquired through learning.

A version of this schema is incorporated in Chapin's (1968) activity scheduling model, which suggests only two dominant categories—*obligatory* and *discretionary* activities. In developing daily or other episodic schedules of potential behavior, obligatory needs take precedence in a person's schedule.

Linked to motivation is *emotion*, which encompasses a wide range of both physiological and mental conditions. Emotions are a state of excitement that is characterized by strong feeling and definitive behavior. Thus, emotion relates strongly to the affective component of attitude. Mehrabian and Russell (1974:8) have postulated the link between environment and behavior through emotion, in which personality variables, combined with perceptions of environmental stimuli, arouse primary emotional responses, such as *pleasure, arousal,* and *dominance,* leading to a behavioral response that may include *physical* action, *affiliation,* and expression of *preferences*.

These *affective responses* by people to their environment have been widely investigated to help the design of the built environment and to minimize impacts on human behavior. For example, continuing earlier work on the role of interior (room) design in hospitals to encourage rapid recovery after illness, Ulrich and Simons (1986) examined the effect on human physiology of everyday settings in the home, office, or in special environments where recovery is important (such as hospitals). They looked at how to evaluate the human costs and benefits of contacts with different environments. Using selected physiological measures, they examined the extent to which exposure to everyday settings can facilitate or hamper recovery from stress. This was undertaken by having subjects view a stressful movie then exposing them to color and sound video tapes of different outdoor environments. Recovery during the environmental presentations was assessed by recording physiological activity such as reduction in muscle tension, skin conductance, pulse transit time, and blood pressure. Ulrich and Simons (1986) found that individuals recovered significantly faster and more completely from stress when exposed to natural settings as opposed to color or sound presentations of urban environments that include pedestrians or traffic. The results show that exposure to peaceful, natural (outdoor) scenes helps to produce physiological well-being and reduce stress. Ulrich and Simons (1986) concluded that exposure to harmonious and peaceful environmental settings may be a useful ancillary design tool for recovery areas in both private and public health facilities.

6.5 Uncertainty and Risk

Uncertainty is a description of the precision and accuracy with which something is known or predicted from knowledge. It is not an abstract hypothetical construct used to help define formal models, but rather a natural fact of living in an environment that is complex enough and large enough to prevent the instantaneous dissemination of all knowledge. Uncertainty is, therefore, characterized by incomplete knowledge, suboptimal decision logic, and the selection of idiosyncratic criteria as a basis for decision making, and it is an inevitable product of social and economic systems where individual, institutional, and systems constraints exist.

Risk is a concept associated with the evaluation of uncertain events. Risk can be evaluated on a multidimensional continuum from *tolerable* to *intolerable*, from *acceptable* to *unacceptable*, and from *significant* to *insignificant*. Risk is perceived to exist in the following circumstances:

1. When in the presence of physical agents that have definable and known effects (e.g., radiation).
2. When society elects a course of action that is said to contain a known risk (e.g., the selection of energy production by nuclear power plants).
3. When the specific disadvantages of particular geographic areas become known (e.g., the environs of inner-city New York or the environs of the wilds of Montana).
4. When economic and social systems allow the presence of hazardous industry (e.g., asbestos making or production of life-threatening chemicals).
5. When social and economic systems allow the frequent use of product classes that can have harmful effects on humans (e.g., pesticides).
6. When uncertainty occurs with respect to the relationship between accepted social activities and specified illnesses (for example, tobacco smoking and cancer).
7. Where events in both the natural and built environment and hazards resulting from people's use of those environments produce the threat of mortality (e.g., traffic accidents, damage from earthquakes, fire, volcanoes, or other hazards).

Otway and Thomas (1982) identify five issues relating to risk perception and the hidden agendas behind those issues. These are:

1. What are peoples' perceptions of the frequency or probability of societal risk? Are they accurate? Biased? Are their errors predictable? Can we understand the cognitive processes underlying such judgments?

2. Should risk perceptions be changed? (This is probably the most significant of hidden agendas in risk research.)
3. What does risk mean to people?
4. What underlies risk judgments? How many dimensions are needed to describe risk?
5. How can risk perception studies be used and by whom?

People do not always make probabilistic judgments that agree with statistical estimates of risk; nor do they always make choices that can be predicted by normative models based on rational behavior. This implies that people are inferior processors of probabilistic information and that they make errors and irrational choices. The psychologist Clyde Coombs argued that people have personal risk preferences that determine their choices among certain outcomes. These preferences are resolutions (sometimes perceived as optimal) of approach–avoidance conflicts. An example of this is a choice situation where one is stimulated by greed and fear.

Tversky and Kahneman (1974) have offered a series of heuristics that people use to help make sense of an uncertain world. These include:

1. The *availability heuristic*, which suggests that people perceive an event as more probable if they find it easy to remember instances of the event or if instances of an event are easy to imagine.
2. The *representedness heuristic* in which reliance is placed on judgment of the degree of similarity between possible outcomes of an event and earlier experiences.
3. The *anchoring heuristic*, which suggests that people make probabilistic estimates by a process of revision from an initial subjective reference point (i.e., an iterative procedure somewhat similar to that involved in Bayesian decision theory).

These issues about risk are elaborated below.

6.5.1   Perception of Risk Frequency and Probability

There is some suggestion that people can be classified as *risk averse* or *risk seeking*. On the one hand, some people estimate relative frequencies of say, causes of death, in a way resembling that which would be observed in the real world. On the other hand, they may overestimate rare causes of death and underestimate common causes. This emphasizes the vital role of experience in risk assessment in that if one's experiences are biased, one's perceptions are likely to be inaccurate. Unfortunately, much of the general information to which people are exposed provides a distorted picture, particularly of the world of hazards, risky situations, and situations with uncertain outcomes.

In general, people's perceptions of risky events are rarely close to the objective probabilities of occurrence of those events.

6.5.2 Changing Risk Perception

Risk perceptions can be changed, but generally they are changed slowly and often not in a predictable way. This is often the case where public information campaigns and advertising become heavily involved. A specific example of this is the public's slow acceptance of nuclear energy programs.

Risk is obviously a multidimensional concept. As such, it results in different perceptions of risk by different population groups. For example, one stream of risk research emphasizes mathematical or numerical measures of safety, hoping to lead to general laws by which risk and safety criteria will become acceptable to the public. The question remains as to whether these essential risk models match the internal representations that people have of risk. Since populations invariably are nonuniform, the likelihood is small that successful adoption of a model of risk (or an attitude toward risk) will be achieved. Similarly, economies and societies differ markedly in what they accept as risky situations (e.g., from capitalist to noncapitalist societies, from developed to less-developed economies, from the rich to the poor, and among different nationalistic or ethnic groups).

6.5.3 Judgment of Risk

Research into risk has examined the question, what underlies the judgment of risk? Nine such factors can be elaborated. The following factors are critical:

1. The degree to which the risk source or risky activities are seen to be *voluntary*.
2. The degree to which risk is perceived to be *controllable*.
3. The degree to which risk is *known* to the scientific, engineering, or professional communities.
4. The degree to which risk is *known to those exposed to the risky event*.
5. The degree to which risk is *familiar*.
6. The degree to which risk is *feared or dreaded*.
7. The degree to which risk appears to be certainly *fatal*.
8. The degree to which risk appears to be *catastrophic*.
9. The degree to which risk is *immediately manifest* or to which it is perceived to be a future problem.

Slovic (1993) has examined research concerning these factors and suggests collapsing them to two primary dimensions—*unknown risk* and *dread risk* (see Figure 6.7).

6.5.4 Using Risk Perception Studies

An important issue is how risk research can be used. One way is to use knowledge of risk assessment to develop an awareness of the risk component in various decision-making processes. This is particularly important when decision processes involve technical decisions at a scale likely to impact significant segments of society.

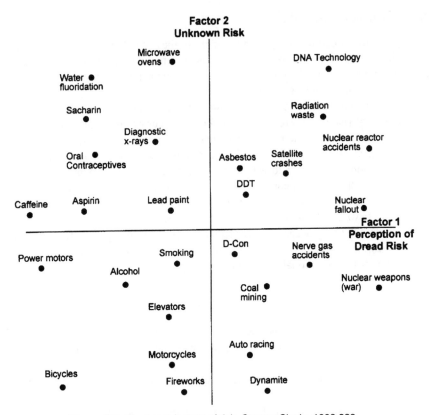

Figure 6.7. Dominant factors of risk. *Source:* Slovic, 1993:229.

Needless to say, there are plenty of opportunities to attempt to influence perception via public attitude polls, biased information dispensed through mass media, downplaying of costs as opposed to benefits associated with risky activities, or other manipulative devices capable of being used by information dispensers and policy makers.

The bottom line of risk perception appears to be that it depends upon:

- The information that people have *received.*
- The information people have *chosen to believe.*
- The *social experience and values* to which people have been exposed.
- People's *cognition or image* of the world.

However, in addition to these more individual values, there appear to be social or group values in which risk perception is related to the dynamics of interest groups, the legitimacy of institutions, the vagaries of the political process, and the historical moment at which the risky event occurs.

6.5.5   Risk Assessment

Beliefs about uncertain events often are expressed in a personal or numerical evaluative format such as: "I think that . . ."; "Chances are . . ."; "It is unlikely that . . ."; "The odds in favor of. . . ." Thus, risk assessment is concerned with how people assess these probabilities of uncertain events or the value of an uncertain outcome. Kahneman, Slovic, and Tversky (1982) argue that there are a limited number of principles that reduce complex tasks of assessing probabilities and predicting values to a set of simple judgmental operations. They also point out that when people rely on a limited number of heuristic principles this leads to severe and systematic errors. This is often called *common sense knowledge*. Its degree of reliability, validity, and accuracy is usually suspect.

Kahneman et al. (1982) point to the similarity between the problem of assessing risk and the problem of judging distance. A common heuristic rule for judging distance is that the apparent distance from an object to an observer is determined by its clarity; that is, the more sharply the object is seen, the closer it appears to be. However, this rule breaks down and leads to systematic errors in the estimation of distance when visibility is poor and the contours or outlines of an object become blurred. Under these circumstances distances often are overestimated under poor visibility and are underestimated when visibility is good. This clarity heuristic therefore leads to a bias in distance *estimation*. Apparently similar types of bias exist when we attempt to estimate subjective risk.

6.5.6   Heuristics for Evaluating Risk

Heuristics have been proposed to evaluate risk. The *representativeness heuristic* typically asks the question "What is the probability that object A belongs to class B?" or "What is the probability that *event* A will occur with *frequency* B?" In other words, what is the estimated judgment of how well *A represents* members of the *set* B. This is also called the *similarity heuristic*.

Using such a rule one would attempt to get an estimate of the degree to which two events, objects, situations, or processes are perceived to be similar. From such a judgment, one may be able to estimate the odds that would be allocated to the probability that *object* A belongs to *class* B. When judgments like this are made with some prior knowledge of event probability, a formal mathematical model known as the Bayes Theorem can often be used to estimate current and future probabilities as event frequency knowledge is fed back to the person making the estimates.

6.5.7   Errors in Risk Assessment

Risk assessment is not an exact science. It is fraught with errors. Some sources of error include:

1. *Insensitivity to sample size*. This is particularly so since smaller samples often have wider variance and are less likely to produce outcomes with graded deviance from an average response.

2. *Misconceptions of chance.* Many people expect the global characteristics of a process to be present in a small sample. Thus they would give a greater probability to the outcomes in a random coin toss of HTHTTH than they would to the sequence of HHHTTT. This is because the latter does not "appear" to be random. In general, common sense knowledge incorporates very little understanding of the notion of random events.

3. *Insensitivity to predictability.* For example, given information on two stock companies, one of which is more favorably represented in a brochure than the other, but neither of which provide any indication of profit-making potential, often the company that is more favorably described in advertising material will be assessed as having a higher probability of being profit making.

4. *The illusion of validity.* This is defined as the unwarranted confidence produced by a good fit between the predicted outcome and the event occurrence. The illusion of validity sometimes comes about because of redundancy in the input variables. For example, if a person were asked to predict the probability of rain in two environments, one with a fairly uniform rainfall and one with a wildly erratic one, the person would tend to have more confidence in the prediction of uniform rain because of the *redundancy* in the role of information about the amount and frequency of rain.

5. *Availability.* At times, high probabilities are allocated to events because of the availability of data on instances of event occurrence. For example, one's subjective probability of the occurrence of heart attacks in middle-aged individuals may be inflated because of the experience of one or more of one's close associates or acquaintances having had such an event.

In summary, there are significant problems associated with risk assessment. These include problems of relating sample data to population estimates or generalizations, problems of appropriate anchoring of responses, difficulty in obtaining information relating to prior distributions, and the tendency to assume that a priori estimates are more reliable and valid than they really are. Remember, when asked to estimate the risk associated with a given situation, few people have readily available statistical evidence on hand to help them with their judgment processes. Inferences must be made on what they remember or recall or what they have observed about events.

6.5.8 Examples of the Use of Heuristic Rules in Risk Assessment

Previously we suggested that frequently occurring events generally are easier to imagine and recall than are rare events (i.e., the *availability heuristic*). Often the recency of information introduces biases about the judgment process. For example, the recency of the Northridge earthquake (1994) may influence estimates of earthquake severity and frequency in southern California. Recent occurrences tend to set an awkward bound to the size of losses imaged as being associated with an event, and particularly with hazardous events. People seem to have some idea at a global scale as to what type of events produce the greatest quantity of deaths. But many estimates often are distorted by a person's most frequent exposure to causes of death.

Thus, because of their frequency accidents are judged as causes of as many deaths as diseases, whereas in fact diseases cause about 16 times as many deaths as accidents. Because of the media exposure, homicides are usually judged to produce more deaths than diabetes or stomach cancer, whereas the latter produce about 11 times more death. Homicides appear frequently in television and daily newspaper accounts, whereas death from diabetes, cancer, or stroke rarely produce headlines. Table 6.1 represents those events most frequently overestimated in terms of producing deaths.

People tend to have a predilection to view themselves as personally immune to risky events—the "it won't happen to me" syndrome. Most people imagine themselves to be better than average drivers, to be more likely than average to live past 80, to be less likely than average to be harmed by products they use, and to be less likely than average to be assaulted, raped, murdered, or exposed to a major environmental hazard. The risk associated with these events looks very small from an individual's perspective.

Experience of risky events, especially when associated with fear, surprisingly is often tucked away in relatively inaccessible parts of long-term memory. The maxim "Out of sight, out of mind" is a popular one. Thus planners and policy makers put nuclear power plants and hazardous waste-producing industries in areas less accessible to the general public. The simple rule is that "if you can't see it, it won't bother you." Although this frequently works as a planning and policy tool, it is a tool that considers only minimally the full impacts of hazardous activities on human populations.

Often we are convinced by massive propaganda that we know with certainty the probability of specific events occurring. This form of over confidence is sometimes produced by a failure to consider the ways in which human errors can affect technological systems. An example would be the experience at the Three Mile Island nuclear plant in the United States or the Chernobyl nuclear disaster in the former U.S.S.R. Overconfidence is often produced because of faith in the current level of scientific knowledge. Often, however, that scientific knowledge is incomplete and is not fully inclusive of the secondary, tertiary, and other indirect effects in-

TABLE 6.1.
Bias in Judged Frequency of Death

Most Overestimated	Most Underestimated
All accidents	Smallpox vaccination
Motor vehicle accidents	Diabetes
Pregnancy, childbirth, and abortion	Stomach cancer
Tornadoes	Lightning
Flood	Stroke
Botulism	Tuberculosis
All cancer	Asthma
Fire and flames	Emphysema
Venomous bite or sting	
Homicide	

Source: Slovic, Fischoff, & Lichtenstein, 1979.

volved in making a decision. The classic example of this is the problems produced by the widespread and uncontrolled use of DDT before the full side effects of that pesticide were known.

Problems also arise when there is a failure to appreciate how technological systems function as a whole. An example of this was the failure of a DC-10 aircraft in flight because designers had not realized that decompression of the cargo compartment would destroy vital control systems. Other problems concern slowness in detecting chronic cumulative effects such as occurred with the very gradual realization that black lung disease was a direct result of confinement in coal mines. Yet other risky situations continue because there is a failure to anticipate human responses to safety measures. Year after year we see evidence of people being flooded out of house and home because of the inflated perception of safety afforded by dams and levees that may give the surrounding populace a false sense of security and safety.

6.5.9 Informing People about Risk

The most common source of information about risky situations is public information dispensed by the mass media. Other legislated information includes labeling controls of food and drugs and early warning systems for hazardous events such as hurricanes, floods, or tornadoes.

One problem associated with informing people about risk is that those not interested in participating in such warnings often provide counter information. For example it has been suggested that the risk of death associated with one hour of riding a motorcycle is about the same as the risk of death associated with living an hour as a seventy-five-year-old person! Averages associated with different events are usually statistically independent and cannot be used in this manner. Some other statements are simply misuses of statistical information. An example would be to state that every takeoff or landing in a commercial airline reduces one's life expectancy by an average of 15 minutes. Comparisons such as this usually convey little or no meaning.

We turn now to consider some case studies of attitudes towards risk and hazardous events.

6.6   Attitudes to Technologically Produced Hazardous Events

6.6.1   Nuclear Waste

Many people judge the benefits of nuclear power as being quite low at the same time they judge its risks to be unacceptably great. On the benefit side, such individuals do not see nuclear power as a vital link in the total national network designed to serve energy needs. Rather, they may view it as a supplement to other energy sources that are themselves adequate and less risky. On the risk side, nuclear technology evokes a greater feeling of risk than almost any other technological advance. Part of this is due to the fact that people *image* the effect of a nuclear accident over an extremely large geographic area as opposed to the more localized effect of accidents associated with other types of power sources. They also *perceive* the magnitude and severity of such an accident to be greater because of the possibility of radioactive fallout. From such a base of inadequate information, scare stories, and irrationality comes judgment of risk. Events that have a high potential dread index associated with them or are perceived to have severe psychological impacts are often judged extremely risky—regardless of the technological claims of safety associated with them.

In such cases as these, the *availability heuristic* is a dominant one. Any information dissemination about nuclear accidents in the near temporal environment increases the imagability of such an accident occurring in one's own environment and the intensity of the perceived risk associated with the occurrence of a nuclear power plant. Thus, the spread effect of Chernoble was global.

It appears, therefore, that *risk* and *uncertainty* are two different things. Faulty judgments can be made about risk associated with a given environment or the occurrence of a particular event because of overconfidence in one's knowledge of the event or because of the general availability of information about the phenomenon in question. This produces major problems for management. Management's reactions to hazards, whether caused by humans or nature, require cooperation by large numbers of people. In many cases these people are quite ill-informed about the hazard and have their own subjective probabilities of frequency or severity and occurrence based on personal exposure. Obviously communication of correct and accurate information to as many of the people at risk as possible is an important part of the management process.

6.6.2   The Case of Yucca Mountain

In December, 1987, the United States Congress amended the Nuclear Waste Policy Act to allow Yucca Mountain, Nevada, to be designated a nuclear waste depository. It was argued that this was a geologically sound and technically feasible site for the disposal of high-level nuclear waste. The site had to pass a set of proscribed technical criteria before it officially could be designated such a depository.

There appear to be both physical and biological risks potentially associated with the development of this type of facility. The depository must safely contain radioactive material for a period that is twice as long as the earth's recorded human history.

Slovic et al. (1991) investigated the social risks of this set of actions. They focused on the probability that this decision would produce adverse economic effects on the city of Las Vegas and the state of Nevada. Such impacts were said to include: reduction in tourist visits to the city and state by both vacationers and conventioneers; reduced immigration to and increased outmigration from the state of Nevada; and reduced ability to attract businesses. All were discussed as part of the attempt to evaluate societal risk for this particular decision. Slovic et al. attempted to overcome the problem of asking people to make judgments about things they knew little of by using *indirect strategies* based on the notion of environmental imagery. Their work (1991) tested three propositions:

1. Images associated with environments with known diverse positive and

negative factors effect meanings that influence the preferences people develop for places at which to vacation, retire, or find employment.

2. The nuclear waste depository conjures up extreme perceptions of risk and strongly negative images.

3. The fact that a hazardous waste repository exists at Yucca Mountain will (over time) increase the negative imagery of Nevada and Las Vegas.

Evaluation of these propositions was undertaken using survey research. For example, questions were asked so as to provide images of the cities of San Diego, Las Vegas, Denver, and Los Angeles. Similar imagery was solicited for the states of Nevada, Colorado, New Mexico, and California. The survey was conducted via telephone.

An initial question was based on word association. Respondents were asked to listen to the name of a place (e.g., city or state) and then provide six thoughts or images about that place. The word associations they gave were then repeated to them and were scaled from "very positive" to "very negative" on a five-point scale. Another task required respondents to rank order the places with respect to their preference for a vacation site. Rankings were also elicited for other characteristics such as retirement sites or migration destinations. Informational questions included queries as to whether the respondent knew in which state the federal government had proposed locating a nuclear waste facility and in which state a nuclear test site was located. Of critical importance was the imagery elicited by the phrase "underground nuclear waste storage facility," which was invariably negative, eliciting images of dangerousness and death, pollution and radiation.

From the preference scores, images for cities and states were then developed. For example, San Diego was described as having "good beaches," "zoo," and "pretty town"; whereas Las Vegas was more negatively described by "rowdy town," "casinos," "too much gambling," and "out of the way." The negative imagery associated with Las Vegas was paralleled by imagery associated with the state of Nevada. In both cases respondents made negative associations with either nuclear depository or nuclear test site and gave correspondingly lower preference ratings for cities or states they associated with these problems. However, Slovic et al. (1991) also indicated that the idea of a nuclear waste depository at Yucca Mountain had not yet penetrated the awareness of the greatest number of their respondents and consequently had not yet had much effect on actual behavior.

Overall, the three initial propositions were supported. The results of the Slovic et al. (1991) study indicate that the actual development of the Yucca Mountain repository would force Nevadans to gamble with their future economy. They must be prepared to face substantial losses from tourist and leisure-time activities as well as from the business economy. Yet, because of the uncertainty inherent in projecting future behavior from these images, there remained a difficult decision on the part of policy makers as to whether or not to go ahead with the decision to locate. There appears little doubt that as more information about the event filters through the population at large, the images and perceptions of the desirability of the state as a place to recreate or to conduct business will go through further reevaluations.

6.7 Natural Hazards and Perceived Risks

Extreme events are designated as *hazards* because their magnitude and frequency of occurrence are less known than normal events. Thus if people are unprepared for them, they cause greater than normal havoc. Typical hazards that are associated with the human occupancy of the environment include air, water, and soil pollution, hazardous waste spills, nuclear accidents, and, at a larger geographic scale, denudation and famine. In evaluating the extent to which an occurrence might be deemed a hazard, one must determine the *event probability* and the *probability of severity* associated with such an event. In addition, one must define the *potential population* that is at *risk* by the event and be aware of the possible *adjustments* and *adaptations* that people can make to *mitigate* the occurrence of the event. In the built environment in particular this involves knowledge of existing controls or policies that may have a direct effect on event severity.

6.7.1 Risk Perception in Natural Environments

Extreme environmental events are usually designated as hazards because of the lack of temporal information about their magnitude and frequency of occurrence. They include wildfire, flood, drought, windstorms such as tornadoes or hurricanes, volcanoes, and earthquakes.

For example, a tropical *cyclone* hazard arises when violent action of wind, rain, sea, and waves make an impact on human environments. The cyclone is an atmospheric depression with a clearly defined minimum value at the center, a steep gradient of pressure, and winds blowing in a spiral toward the center (clockwise in the southern hemisphere and counterclockwise in the northern hemisphere). The intense form of the depression is called a cyclone, but it is also known by the name of hurricane or typhoon.

Floods are defined as hazards when the stage of river flow involves considerable channel discharge of surplus water over the banks into surrounding areas. Floods are probably the most universally experienced natural hazard. They tend to have a larger spatial impact than most other hazards, and they frequently involve larger loss of life than do other hazards.

An *earthquake* is a sudden release of accumulated strain energy by the earth. This release takes the form of shock waves and elastic vibrations called seismic waves. These are transmitted through the earth resulting in oscillatory and sometimes violent movement. They may also involve deformation of the earth's crust and may produce offsets along faults, compaction of unconsolidated soil materials, landslides and mudflows, liquefaction of soils, avalanches, and fracturing of rock masses. Earthquakes can be triggered by slippage along fault segments, change of pore pressure of soils as a result of fluid injection, or they can be triggered by nuclear devices and releases of volcanic energy.

6.7.2 An Example: Reaction to Cyclones

With respect to cyclonic activity, risks are often associated with expected damages. Damages are usually a function of the size and intensity of the storm, the density

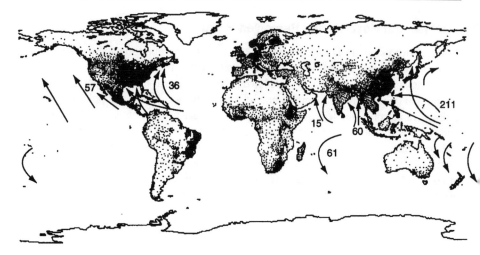

Figure 6.8. Population distribution of the world (1954) and approximate tracks of regular cyclonic activity.

and distribution of the population at risk, and the type of local economy. Since some of the higher population densities in the world are located in coastal areas (Figure 6.8), potential damage of a cyclonic event is great. Between 1960 and 1970, for example, 17 cyclones worldwide took the lives of approximately 350,000 people and caused immense damage to local livelihoods. Within the United States, the dollar damage of cyclones has escalated throughout this century despite better warning systems that have substantially lowered death rates.

The effect of cyclonic activity can be summarized in three categories: the source of damage; the type of damage; and the effect on humans and their environments. Let us discuss each of these in depth.

1. *Source of damage.* With cyclones, damage is perceived to be most associated with wind and water. High winds strip trees, blow down signs, and power and communication lines, facades, and roofs of buildings, rock skyscrapers, smash windows, and do other visible damage. Water in the form of driving rain, coastal tidal surges, wave action, and local flooding also produces much visible damage. Both wind and water are *direct agents* of destruction. *Indirect agents* include fires (e.g., from downed power lines) and explosions (e.g., from ruptured gas lines). Physical agents of damage are supplemented by psychological and social agents of despair, depression, sickness, loss of jobs, loss of employment hours, loss of economic base, and personal and financial ruin.

2. *The type of damage.* Damage can occur in the natural, built, social, or cognitive environments. In the natural environment, vegetation may be leveled or stripped; soils may be eroded or salted; dunes may be destroyed; tidal or river channels filled or scoured; crops and livestock may be destroyed. Buildings, recreation areas, lines of transportation, bridges, airports, shops, and financial, and education-

al institutions can be damaged and closed for varying periods for repair or reconstruction.

   3. *Effect on humans*. As humans continue to occupy portions of the earth's surface that are on the cyclonic tracks, they have attempted to adapt and adjust to the events in a number of ways. Active adaptation involves the search for ways to alter the nature of the hazard or protect one's life and possessions. Passive activity includes adjusting one's thinking to facilitate living with the psychological and real threats of hazard. With respect to cyclones, humans have tried to modify them by cloud seeding, have attempted to modify potential damage by constructing levees or shore works, dune control, reforestation to protect shorelines and low-lying coastal areas, and early warning systems and evacuation procedures designed to allow people to escape the onset of a hazard. The revision of building codes to exclude certain uses from hazard-prone areas, the building of shelters, and the development of more aggressive insurance programs have also been encouraged to help reduce damage loss. Federal and local subsidies to cover damage and reconstruction costs, usually called Emergency Relief Funds, are also ways perceived to mitigate the effect of this natural hazard.

6.7.3   The Case of Bushfires in Australia

Cunningham, Kelly, Greenway, and Jones (1993) examined the level of understanding a community has of a bushfire hazard that threatens it. They also examined the extent to which this type of knowledge changes over time.

   A hundred and seventeen adults were selected from households that were located in the most fire-vulnerable streets of towns in the Blue Mountains area of New South Wales in Australia. Respondants were first questioned in 1984 and then followed up in 1993, with responses from 96 of the original 117 households who were still able to participate. This provided Cunningham et al. (1993) with temporal data for comparative purposes. Households were divided between those who were located in the area prior to a 1977 fire disaster and those that entered the area in a time consequent to that. Cunningham et al. (1993) showed significant differences between the two groups in terms of the level of understanding they had of fire as a potential hazard. Previous experience provided residents with a much better understanding of the hazard and its probability of occurrence. Results also showed, however, that even in this group, knowledge diminished considerably with time. Understanding of and compliance with fire mitigation strategies were low in both groups. Most householders saw themselves as powerless in the face of the fire hazard. Rather than attempt to influence the potential hazardous situation by their own actions, they tended to delegate responsibility for their protection and for the protection of their property to official bodies.

   The basic history of fire incidents that destroyed property is shown in Table 6.2. The purpose of the study was to find how much knowledge members of a hazard-prone population had of the threatening event.

   The survey method used to collect information involved distributing a brochure in the study area 2 weeks before the survey took place. If a household could not be interviewed, a kit of information was left together with a questionnaire

TABLE 6.2.
Fire Incidents That Destroyed Property: Blue Mountains, Australia, 1910–1991

Area	Number of Incidents since 1910	Mean Interval of Incidents (Years)	Maximum Interval of Incidents (Years)	Minimum Interval of Incidents (Years)
All urban areas	11	4.5	16	1
Upper mountains (A)	4	16.5	36	1
Central mountains (B)	6	13.6	20	6
Lower mountains (C)	7	9.4	16	6

Source: Cunningham et al., 1993.

to be returned to researchers by prepaid mail. Two surveys were conducted: one in 1984 and one in 1993. In 1984, only one-third of the survey households had experienced a bushfire in their home area. By 1993, the proportion of people experiencing a fire hazard in the area had dropped to 26%. Forty percent of the population had moved into the area since the 1984 study had been conducted, even though the area was well known as one of the principal fire hazard areas in the entire country. Only 27% of the households in the 1993 survey accepted the fact that they faced a very high risk of fire, compared to 70% in 1984. In 1984, 100% of householders had nominated their area as one of fire risk, but in 1993, this proportion was only 58%. This clearly reflected a change in the perception of risk by the resident population. In addition to this, many fewer of the 1993 sample had information about the fire history of their area and could nominate the year of the previous wildfire. Among the fire-experienced residents, 77% in the 1984 survey were able to predict the correct year of the previous fire hazard, whereas only 46% in the 1993 survey could make a correct prediction.

In general, knowledge of critical weather and climate factors that often play a major part in the spread of a bushfire was but poorly known. Very few people were able to interpret the daily televised weather maps that incorporated key environmental indicators of potential hazard. Similarly, only few residents realized that the peak for the fire hazard was in late spring and not in high summer. Similarly there was only little knowledge about the information distributed by local authorities explaining how to behave when fire hazard was high or when a bushfire eventually occurred. In summary, the authors revealed a classic pattern of lack of knowledge about the potential hazard by local residents. This was so even in the face of quantities of knowledge being freely available. Even those with prior experience of a fire event took relatively few precautions and did not anticipate a new occurrence of a hazardous event in the near future. As each year passes, the risk of another disaster increases; in this case, as each year passed, the propensity of the community to prepare itself for a disaster decreased. Even individuals who had experienced the previous hazard tended to take fewer precautions and rely more and more on local authorities or state and federal assistance if an event did occur.

Cunningham et al. (1993) suggested that, if a household was well informed, then there was propensity to take individual precautions that would allow residents to better adapt to the occurrence of a fire hazard. Characteristics of such a household might be:

- One that kept itself informed.
- One that would make attempts to become better informed about weather processes and their contribution to hazard buildup.
- One that would involve able-bodied members in bushfire prevention training activities.
- One that would undertake prevention actions around its own property such as cleaning off brush and dead vegetation.
- One that helped involve neighbors in learning and practicing fire prevention tactics.

Cunningham et al. (1993) argued that community or societally backed strategies frequently were ignored despite the fact that this information was clearly the most widely distributed and readily available of all strategies. Given the lack of response to community- and society-generated warning information, the authors predicted that future hazard would again be disastrous to property and perhaps life. They argued that unless the individual households were more active in an individual or neighborhood group committed to maintaining preparedness and monitoring the state of the local environment, future effects of fire hazard occurrence were unlikely to be mitigated.

6.8 Perception of the Built Environment

6.8.1 Fear of Crime: A Case Study

In research examining characteristics of *perceived environments*, Nasar (1982) used simple regression models to predict crime as a function of judged attributes of residential exteriors. He did this using photographic slides of thirty residential sites as the environmental cues. Using an expert panel of planners and architects he obtained ratings of the visual attributes in each setting on 18 bipolar descriptive scales. He then used subgroups of elderly and other adults to evaluate each setting on a bipolar scale. This was used to produce an expected rate of vandalism, robbery, burglary, and assault. Both population subgroups found significant correlations between visual attributes in the represented scenes and fear of crime.

Examples of the different bipolar attributes used include: uniform–diverse; ornate–plain; ambiguous–clear; commonplace–unusual; disorganized–organized; closed–open; colorful–dull; well-kept–dilapidated; fitting–unfitting; much mystery–little mystery.

Nasar also solicited evaluations of the degree of prominence or occlusion of features such as natural elements, buildings, shape, surface texture, verticals, cars, and poles with wires and signs. For the elderly he obtained an $r^2 = 0.77$ for a simple model of fear and crime expressed as follows:

low crime = 5.61 − 0.52 dilapidation − 0.26 diverse.

For the general population the model was:

low crime = 0.60 + 0.38 nature prominent + 0.27 shape prominent

+ 0.30 well-kept.

It yielded an r^2 of 0.78.

Nasar interpreted these regression results as supporting his hypothesis that evaluations of neighborhood settings and perceived fear of crime can be related. In particular, they can be related to the occurrence of sets of visual attributes present in the residential exteriors. From these results he offered the following recommendations:

- Improvements in the upkeep and decreases of the diversity of building will produce a reduction in the fear of crime.
- Increases in the prominence of nature and improvements in the upkeep of buildings will produce a reduction in the fear of crime.

Overall this research suggests that definition of the environmental attributes that are associated with emotions or feelings of different population subgroups can be woven into planning and policy-making activities. In this case, changes in the environment are hypothesized to reduce levels of anxiety and fear of criminal activity.

6.8.2 Visual Evaluations by Residents and Visitors

In another comparative study, Nasar (1979) previously had shown a significant difference between residents and visitors to Knoxville, Tennessee, when they were asked to identify and specify areas that they liked or disliked visually. He also elicited descriptions of the environmental reasons for the pro and con evaluations. Examples of elements specified in need of improvement by residents and visitors respectively included the following: (1) signs and billboards (emphasized more by visitors); (2) buildings (emphasized slightly more by residents); (3) industry (heavily emphasized by residents and only lightly by visitors); (4) highways (heavily mentioned by residents and much less so by visitors); (5) utility poles and wires, parking, river front, and other characteristics all of which were referenced only slightly by residents and visitors alike.

Nasar concluded that familiarity with the environment produced a focus on somewhat different attributes for improvement then did nonfamiliarity. He also pointed out substantially different emphasis on elements associated with potential use of the environment, with residents emphasizing more those things that they experienced on a regular basis and visitors emphasizing environmental elements that assisted them in wayfinding rather than helping them adapt to or live in such an environment.

The importance of *affect* and *emotion* in dealing with space and place perhaps lies in the contention that affect can direct the course of cognition, influence the retrieval of stored information (memories), enter into the learning process, and guide the nature of experience. In other words, affect and emotion are seen to intervene in many of the subprocesses involved in the general condition of information processing. This is particularly important for approaches such as the transactional and the interactional–constructivist, which stress the importance of the subprocesses in the reasoning and inference used to give meaning to person–environment–behavior episodes. The bottom line is that there exists strong sentiment that it makes more sense to explore emotional occurrences as they unfold in environmental settings than to examine them as isolated occurrences in controlled environments such as in laboratories. The relationship among affect, emotion, and actual experience in different settings allows prevailing sociocultural effects to exert their full influence in subprocesses such as perception, conception, learning, and cognitive mapping.

7

Spatial Cognition, Cognitive Mapping, and Cognitive Maps

7.1 Background

Cognitive mapping is defined as "a process composed of a series of psychological transformations by which an individual acquires, stores, recalls, and decodes information about the relative locations and attributes of the phenomena in his everyday spatial environment" (Downs & Stea, 1973: 7).

Cognitive mapping is usually considered to be a subset of *spatial cognition* which can be defined as "the knowledge and internal or cognitive representation of the structure, entities and relations of space; in other words, the internalized reflection and reconstruction of space and thought" (Hart & Moore, 1973: 248). In turn, spatial cognition is sometimes seen as a subset of *environmental cognition*, which refers to "the awareness, impressions, information, images, and beliefs that people have about environments. . . . It implies not only that individuals and groups have information and images about the existence of these environments and of their constituent elements, but also that they have impressions about their character, function, dynamics, and structural interrelatedness, and that they imbue them with meanings, significance, and mythical–symbolic properties" (Moore & Golledge, 1976: xii). Thus, environmental cognition adds a set of *affective* components to the *cognitive* components emphasized in spatial cognition. These affective components include feeling, attitude, belief, value, and other emotional characteristics (Hart & Conn, 1991). Other definitions that include an emphasis on the term *place* instead of *space* are given in Kitchin (1994), but in this chapter we emphasize space, in all its multidimensionality, as the key concept.

The end product of a *cognitive mapping process* is called a *cognitive map*. Usually accepted to be a device that helps to simplify and order the complexities of human–environment interactions (Walmsley, Saarinen, & MacCabe, 1990), the cognitive map is essentially our individual model of the world in which we live.

The term cognitive map is now widely accepted in a number of disciplines. It had its origins in the work by Tolman (1948), who described why a rat running a maze in search for food might transgress the boundaries of the maze and go directly (as the crow flies) to the food source. This shortcutting procedure attributed to Tolman's rat has become a standard indication that animals—and by analogy humans—can proceed directly from an origin to a destination without being forced to retrace a previously learned path. It was suggested later by Shemyakin (1962) that Tolman's rat chose a solution to the food search problem that would be quite logical if one viewed the task environment from a birdseye or *survey* view. This overview would provide simultaneous information about the starting location, the target destination, and the sets of potential paths linking origin with target destination.

By the mid 1960s, geographers had become exposed to this inherently appealing concept of the representation of spatial and other environmental information in the mind in some maplike form. Behavioral geographers adopted the idea, with Gould (1963) suggesting that people formed *mental maps* of their environment. He illustrated this concept by producing sets of place-preference surfaces that were constructed from rankings of the residential desirability of places and regions in different countries (Figure 7.1). By the end of the 1960s, other behavioral geographers, such as Downs (1970a,1970b) and Golledge, Briggs, and Demko (1969), had extended the idea of the mental map/preference surface into the broader domain of spatial cognition. Downs (1970b) identified cognitive dimensions of shopping centers, while Golledge and colleagues examined point, line, and areal components of cognitive maps. For example, Zannaras (1968) discussed the idea of perceived neighborhoods; Golledge et al. (1969) and Briggs (1972) examined the nature of cognitive distance and the role that cognitive distance played in wayfinding; Briggs (1969) used preferences and cognitive attributes to define place utilities for shopping centers; and Golledge and Rushton (1972) introduced the idea of using metric and nonmetric multidimensional scaling (MDS) to recover the latent spatial structure underlying people's preference and evaluations of proximity or similarity, constructing from these the first examples of cognitive configurations—that is the maplike externalization of implied or latent spatial knowledge derived from MDS output. Other behavioral geographers, working together or with psychologists (Blaut & Stea, 1969; Beck & Wood, 1976a, 1976b; Hart & Moore, 1973; Downs & Stea 1973; Cox & Golledge, 1969), introduced and investigated the use of the concept of cognitive mapping in such disparate research as: determining whether young children could cognize and understand maps and aerial photographs; examining children's behavior in play environments; analyzing spatial concepts such as proximity, direction, and orientation; and examining the rationality of observed spatial behavior when it was compared to behavior mapped into configurational representations of different environments.

To the geographer the concept of a cognitive map was a logical one. Those interested in spatial decision-making behavior had already argued that spatial behavior was most frequently *boundedly rational* (Wolpert, 1964) or *satisficing*, rather than being utility maximizing and optimal. The argument was advanced that what was perceived to exist and what was already experienced or known was often more important in the decision-making process than the objective reality of a problem situation. Since behavior in space was seen to be the outcome of decision-making processes that rely on combining stored information with ongoing experience, the significance of the cognitive map as the mechanism for storing, recalling, and using such information was an appealing one and began to spread widely.

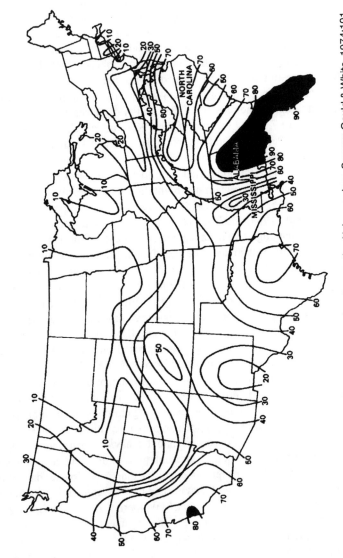

Figure 7.1. "Mental maps" of preference for places—the Alabama view. *Source:* Gould & White, 1974:101.

7.2 Characteristics of Spatial Cognition

7.2.1 A Developmental Base

Much of the initial research in environmental cognition and later spatial cognition (particularly among geographers and environmental design professionals) was heavily influenced by the development sequence theories expressed in Piaget and Inhelder (1967) (Figure 7.2). In particular, their work on the child's conception of space was extensively used. In this developmental context, they identified several stages of perceiving:

1. The first stage is one of *vague awareness*, when the perceiver is aware that something is impinging on the senses and that the sensory input appears to be coming from specific segments of the environment. At this stage, the geographer might develop expectations that the initial sensory input may relate to physical structure, location, or potential use of perceived objects.

2. At the second stage of this developmental process, *spatial characteristics* are added to the *perceptual set*. Thus, the object is given an *existence or location* in space and time and is fitted into a *class or category* of objects that may have similar spatial characteristics. At this stage, the beginning of differentiation among objects on the basis of spatial characteristics commences.

3. A third stage involves *recognizing* the *relevant parts* of the *perceptual objects* and being able to *specify* these components. Such components, or attributes, become differentiating characteristics used to distinguish a given object from others in its class. Thus, identity, size, condition, color, function, and so on, are allocated to a stimulus object.

4. A fourth stage is one of *identification*, or more realistically, the *attachment of meaning* to the stimulus object. The meaning and significance of an object apparently influence its durability and usefulness to an observer. Once an object becomes an identifiable entity, it may thereafter occupy a regular niche in the cognitive structure of an individual. It, along with other members of the set, can be used as a reference point against which new stimuli are matched.

To elaborate on the process of spatial cognition, we now turn to a brief dis-

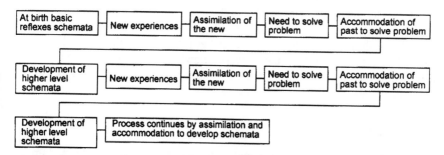

Figure 7.2. Sequence of spatial understanding. *Source*: Kaluger & Kaluger, 1974:75.

cussion of how information is coded, stored, reconstructed, and processed in memory. Lloyd (1982) reviewed a number of approaches associated with spatial cognition. These included radical image theory, conceptual propositional theory, and dual-coding theory. His findings are summarized in the following sections.

7.2.2   Radical Image Theory

Radical image theory postulates that the perceptual processes reduce information to a simpler and more organized form. The form suggested is called an *image*. Images are generated from underlying abstract representations. Images are seen to be pictorial in nature. This theory suggests that images, once formed, are wholes that may be compared to percepts in a template-like manner.

Evidence for the existence of radical image theory comes largely from studies of reaction times. Such studies suggest that images are scanned by the "mind's eye" in the same way one would use vision to scan an object. Response times for determining which of two named objects is larger were found to be inversely related to the actual size differences between the objects. For example, the response time to differentiate between a cat and a donkey was faster than that taken to differentiate between a goat and a horse. Similarly, response times for comparing which of two places is further north are found to be proportional to the difference in latitude between the places. In addition to response-time studies, however, other supporting evidence comes from rotation studies. Shepard and Metzler (1971), for example, undertook an experiment in which the results showed that response time to the question of whether or not two objects were the same was proportional to the rotational angle with which members of the pairs of objects were viewed.

7.2.3   The Conceptual Propositional Theory

This theory suggests that verbal and visual knowledge are stored as abstract conceptual propositions. Unlike imagery theory, which seems on the surface to be more tied to visualization, this theory has no specific tie to any particular sensory modality. It argues that part of the meaning of a concept can be represented in terms of its connections to other concepts. In other words, the concept is embedded in a *configurational meaning structure*. The closer together concepts are in a propositional network, the better cues they are for each other's recall.

Evidence for the existence of this theory came from experiments showing that, when subjects were given prior labels or meanings for squiggley designs ("droodles"), they could better reconstruct the pictures that the "droodles" were supposed to represent (e.g., Label: "Worm crawling over a razorblade"; Image: slices of worm piling up on one side of the blade and the balance of the whole worm on the other side).

7.2.4   Dual-Coding Theory

This suggests that there are two interconnected memory systems—verbal and imaginal. These interconnected systems operate in parallel (Paivio, 1969). For example, this theory suggests that pictures and concrete words that are easily imageable may be represented in both *verbal* and *visual* memory, but abstract material is represented in the verbal system. From this point of view, imageability makes things easier to remember.

A long debate has taken place in cognitive science between the image theorists and the conceptual–propositional theorists. This controversy impacted research on cognitive mapping in many disciplines. In geography, image theorists were prone to accept the idea that environmental knowledge was represented in maplike form and that it could be recalled and or externally represented by cartographic-like presentations (e.g., sketch maps or distorted Euclidean configurations). Conceptual–propositional theory had a significant impact on studies concerning the development of spatial knowledge. Anderson (1982) suggested that spatial knowledge consisted of a declarative base and a set of procedural rules that specified how components of the declarative base could be linked together. This was seen by some geographers to be in opposition to many of the spatial cognition theories that were developmentally based, and they argued that spatial knowledge evolved through developmental stages as well as through different metric structures. Today geographers interested in spatial cognition still evince no clear preference for one or the other of these theories.

In general, interest in spatial cognition has permeated a number of areas of human-geographic research. Specific problems that have generated significant geographic input include: the examination of cognitive distance and the asymmetric distances that appear to typify this concept; the warping of space because of asymmetric distance effects; the examination of hierarchical structures at different scales and at different levels of generalization; the concept of regionalizing at the macro level and chunking at the micro level; and the concept that the critical part of environmental structures includes the designation and definition of anchorpoints. In the sections that follow, as we discuss the nature of cognitive mapping and cognitive maps, we explore some of these concepts.

7.3 Cognitive Maps and Cognitive Mapping

7.3.1 Cognitive Mapping

We now turn to consider more structured approaches using a variety of experimental methods aimed at uncovering and representing the cognitive structure of both small-and large-scale environments.

The *process* of *cognitive mapping* is a means of structuring, interpreting, and coping with complex sets of information that exist in different environments. These environments include not only the observable physical environment, but also memories of environments experienced in the past, and the many and varied social, cultural, political, economic, and other environments that have impinged both on those past memories and on our current experiences. The nature, structure, and content of these many environments also influence our expectations whether these be in terms of immediate shortrun expectations that are essential to the more frequent episodes of our daily activity patterns (e.g., route selection in a daily trip to work or shopping), or those less-frequent episodes with greater temporal intervals between their occurrences (e.g., annual holidays).

This mapping process requires each individual to undertake a cognitive taxonomic process that is culturally constrained and that results in the filtering of the varied "to whom it may concern" messages emanating from the many environments in which we live.

Unlike much of the literature on perception, which emphasizes the acquisition of information through the senses in the presence of a stimulus, it is generally accepted that the process of cognitive mapping has no tie to a particular sensory modality but, instead, spans all of them.

7.3.2 The Mapping Process

The first assumption is that information extracted from large-scale external environments exists in an undetermined *psychological space*. This is the space in which characteristics, meanings, and configurational relationships about elements in the world are held as mental constructs. A second assumption is simply that individuals must have *internal knowledge* about external environments in order to exist in them. A third assumption is that internal representations will be part *idiosyncratic* and part *common* in structure (i.e., people in general will know the difference between a street and a stream, the significance of a traffic light, and the difference between privately and publicly owned transportation systems).

As indicated in Table 7.1, Golledge (1976c) has categorized methods for extracting and representing imagery into the following classes:

- Experimenter observation in naturalistic or controlled situations.
- Historical reconstructions.
- Analysis of external representation.
- Indirect judgmental tasks.

The above conceptualization includes a range of external representational formats such as verbal reports, sketch maps, tables, profiles, word lists, analog models, slides, novels, poems, paintings, diaries, interviews, protocols, toy play, proximity judgments, scalings, and creative stories or writings. Thus, at various times individuals have been asked and have provided an impressive array of written and oral descriptions, pictures, sketches, cartographic representations, and grouped or clustered, scaled, and otherwise modified or transformed, bits of environmental information. These have come from experiments using recall of interpoint distances and analysis by multidimensional scaling, from tasks requiring haptic (touch) exploration of layouts, from solving problems requiring auditory localization, and from simple wayfinding tasks.

One of the most widely referenced comprehensive works using a variety of the above methods was Lynch's (1960) pioneering work *The Image of the City*. In a later section we examine his concepts in detail and expand on them at different spatial scales.

TABLE 7.1.
Methods for Extracting Environmental Cognition Information

Method	Procedure	Subject Skill	External Representational Form	Example
Experimenter observation in naturalistic or controlled situations	Experimenter observes or tracks movements through actual environments (e.g., crawl patterns, search behavior overt spatial activity, wayfinding)	Cognitive Concrete Psychomotoric	Observations Reports Maps Tables	Lynch (1960) Marble (1967) Ladd (1970) Jones (1972) Devlin (1973) Zannaras (1973)
	Experimenter infers degrees of cognitive knowledge from behavior in unstructured "clinical" situations	Cognitive Concrete Motoric	Charts Profiles	Werner (1948) Piaget and Inhelder (1956) Hart (1974)
	Subjects reveal environmental knowledge in the process of sorting or grouping elements of actual or simulated environments	Cognitive Abstract Relational	Lists Tables Composite maps	Downs (1970a) Wish (1972) Zannaras (1973) Golledge et al. (1975)
	Subjects adopt roles or perform acts in simulated and/or real environments	Cognitive Abstract	Photographs Tables	Ittelson (1951) Milgram (1970) Acredolo (1976)
	Subjects arrange toys or objects representing environmental elements or model environments, and experimenter observes the sequence of acts in positioning elements and/or using the environment	Cognitive Concrete Motoric	Analog models	Piaget et al. (1960) Blaut and Stea (1969) Laurendeau and Pinard (970) Mark (1972) Hart (1974)

(continued)

TABLE 7.1 (continued)

Method	Procedure	Subject Skill	External Representational Form	Example
	Subjects draw sketches or sketch maps representing environments	Affective Graphic Relational	Pictoral sketches Sketch maps Quantitative and structural analyses	Lynch (1960) Shemyakin (1962) Stea (1969) Appleyard (1970) Ladd (1970) Moore (1973) Wood (1973)
	Subjects arrange toys or make models representing environments	Affective Cognitive Concrete Motoric	Models Arrangements of toys	Piaget and Inhelder (1967) Blaut and Stea (1969) Mark and Silverman (1971) Stea (1973) Hart (1974) Stea (1976)
	Subjects show existence, location, proximity, or other spatial relations of environmental elements; use of symbols to represent such elements	Cognitive Graphic Abstract Relational	Base maps with overlays Notation systems	Lynch (1960) Thiel (1961) Appleyard (1969) Wood and Beck (1990)
	Subjects are asked to identify photographs, models, etc.	Affective	Verbal	Piaget and Inhelder (1956)

Category	Description	Type	Format	References
Indirect judgmental tasks		Motoric Abstract Relational	Protocols	Laurendeau and Pinard (1970) Stea and Blaut (1973) Zannaras (1973)
	Selection of constructs that reveal environmental information; adjective checklists, semantic differentials, repertory grid test, etc.	Cognitive Abstract Relational	Word lists Tables Graphs Grids	Kelly (1955) Downs (1970a) Honikman (1976) Harrison and Sarre (1976) Golant and Burton (1969)
	Paired proximity judgments and other scaling devices that allow specification of latent structure in environmental information	Cognitive Abstract Relational	Maps Tables	Briggs (1973) Lowrey (1973) Golledge et al. (1975) Cadwallader (1973a) Golant and Burton (1969)
	Projective tests (e.g., T.A.T.)	Affective Abstract Relational	Verbal stories	Burton et al. (1969) Saarinen (1973b)

233

7.3.3 Cognitive Maps

A *cognitive map* may be defined as "long-term stored information about the relative location of objects and phenomena in the everyday physical environment" (Gärling, Böök, & Lindberg, 1979:200). Cognitive maps thus represent information about environments that are either known to exist or are imagined but not necessarily present. Any given map, therefore, may be a mixture of information received at quite disparate time periods, and at any particular point in time may be incomplete, more or less schematized, or distorted, and may contain fictional or hypothetical information, or relics of the past which no longer exist.

This model of reality is a complex one and should not be interpreted as a simple one-to-one internal mapping of discrete things that exist in an external environment. Thus, it is not assumed that a cognitive map is an equivalent of a cartographic map, nor is it assumed that there is a simple Euclidian one-to-one mapping between a piece of objective reality and a person's cognitive map of that reality. Cognitive maps are generally assumed to be incomplete, distorted, mixed-metric representations of real-world environments, but they can also be maps of the imaginary environments represented in literature, folk tales, legends, song, paintings, or film.

Even simple tasks, like going from home to work, to school, or to a store, or directing newcomers to places they have never been, require information to be stored, accessed, and used in a convenient and easy way. To perform such tasks it is necessary to use one's memory representations of spatial information—that is, one's cognitive map. Over the past three decades, scientists in a variety of fields have been examining questions relating to the nature of these cognitive maps, the process of acquiring and forming such mappings, and their role in everyday spatial activity. In what follows, we draw on concepts, theories, and empirical evidence from a variety of disciplines as a means of examining the process of developing cognitive maps and illustrating the product of such knowledge acquisition.

7.4 Cognitive Map Metaphors

7.4.1 Basic Metaphors

With more widespread acceptance, the concept of a cognitive map has become more clearly and more accurately defined. In the course of such clarification, the metaphor most often used in describing cognitive maps has changed.

Gould's (1963) early use of the preference surface metaphor for mental maps was next replaced by the rubber sheeting metaphor favored by Golledge et al. (1969), Golledge, Rayner, and Rivizzigno (1975), and Tobler (1976) when using nonmetric multidimensional scaling or trilateration procedures to construct cognitive configurations (Figure 7.3). S. Kaplan's (1973, 1976) development of an information-processing metaphor led to further elaborations of the cognitive map as a computerized data base and processing system. Concurrently with these metaphors was the less favored, but appealing, notion of a cognitive map as an internal cartographic-like representation or indeed as an atlas (Lieblich & Arbib, 1982). Cognitive scientists then suggested that cognitive maps were list processors or images,

Long Term
Resident

Short Term
Resident

Newcomer

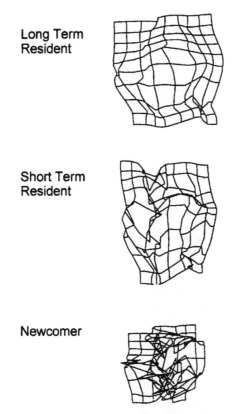

Figure 7.3. Distorted grids as examples of rubbersheeting.

metaphors that were closely examined for geographic relevance by Lloyd (1982). The rubber sheeting and cartographic representation metaphors were linked by Tobler's (1976) elaboration of the geometry of mental maps and by his development of bidimensional regression procedures for correlating objective and subjective configurations (Table 7.2).

Cognitive maps that include not just spatial information but also attributive values and meanings (Wood & Beck, 1990) provide yet another metaphor (i.e., the cognitive map as a belief system). The notion that cognitive maps involve the integration of images, information, attitudes, and values was also offered by Spencer and Blades (1986). In other words cognitive maps are not to be thought of as isolated entities but as being contextual, dynamic, and providing the interface mechanism between sensed information and behavior.

Golledge and Timmermans (1990) suggested that cognitive maps are a series of knowledge structures consisting of different levels of detail and integration. These knowledge structures develop with age and education and, in a sense, may be described more by the metaphor of a cognitive atlas than a cognitive map (see also Lieblich & Arbib, 1982: 640). From this atlas, individual cognitive maps can be resurrected and used for specific tasks (Golledge et al., 1985)

TABLE 7.2.
Bidimensional Correlations between Objective and Subjective
Configurations (Control Group of Long-Term Residents)

Control Group of Long-Term Residents	Bidimensional Correlation between Objective and Subjective Maps
C1	0.891
C2	0.952
C3	0.855
C4	0.944
C5	0.938
C6	0.882
C7	0.889
C8	0.901
C9	0.709
C10	0.882

Source: Golledge, Rayner et al., 1982.

The evolution of understanding of the concept of cognitive maps is now at the stage where it is assumed that they represent the store of knowledge an individual has about an environment. Implicitly and explicitly, they contain spatial relational data along with environmental attributes and individualized and socioculturally conditioned beliefs, values, and attitudes. Context-specific and problem-specific cognitive maps can be constructed at will, each having a particular space–time context and containing some common information (e.g., common anchorpoints) and some idiosyncratic or personalized elements. The common elements facilitate communication with others about the characteristics of an environment; the idiosyncratic elements provide the basis of the personalized responses to such situations.

7.4.2   Cognitive Maps as Internal GIS

The idea that sets of environmental information might be represented as layers with differential attribute weights attached to hierarchically organized components within and between layers promotes yet another metaphor of the cognitive map. Recently Golledge and Bell (1995) suggested that the cognitive map should be viewed as an *internalized geographic information system* (GIS). In both systems, data are symbolized and coded. Decoding and recall are often focused on selective components of the total knowledge set. The recalled information can be represented in a variety of verbal, graphic, acoustic, or haptic forms. A range of manipulative processes can be accessed to help solve a task and produce a behavior (Figure 7.4).

It is difficult to think of a single functionality embedded in the GIS that does not have a parallel in human-information processing capability. The one difference, however, is that in the GIS, once activated, a manipulative procedure should be carried out quickly and accurately, whereas in humans, although the ability to perform a manipulative activity may in principle be within their intelligence realm, there are many personal and societal barriers that may induce error and inhibit the use of such a skill. Thus while the GIS is a practical tool for processing, analyzing, and

Constructing Gradients and Surfaces
Layering
Regionalizing
Decomposing
Aggregating
Correlating
Evaluating Regularity or Randomness
Associating
Assessing Similarity
Forming Hierarchies
Assessing Proximity
Measuring Distance
Measuring Directions
Defining Shapes
Defining Patterns
Deterimining Cluster
Determining Dispersion

Figure 7.4. Samples of manipulative processes used in cognitive mapping and GIS.

displaying spatial information, the comparable human processes may be error prone and poorly developed. And given the extensive literature on development and life-span theories of intellectual growth, one would expect these abilities to be more or less present depending on age and education.

7.4.3   Debatable Alternatives

In geography, different views on the cartographic nature of cognitive maps are elaborated in a debate between Graham (1976, 1982) and Downs (1981a,1981b), with the latter clearly identifying the metaphoric significance of the use of the term "map." With respect to this and other related debates, Kitchin (1994) summarizes the four critical debatable questions as follows:

1. "Is it the case that the cognitive map *is* a cartographic map (*explicit* statement)?
2. Is it the case that the cognitive map is like a cartographic map (*analogy*)?
3. Is it the case that the cognitive map is used *as if it* were a cartographic map (*metaphor*)?
4. Is it the case that the cognitive map has no real connections with what we understand to be a map (i.e., a cartographic map) and is neither an explicit statement, analogy, or a metaphor but, rather, an unfortunate choice of phrase: *'a convenient fiction'* (Siegel, 1981), or in effect, just an hypothetical construct." (Kitchin, 1994:3)

The argument for the cognitive map *being* a map (explicit case) derives from the work of O'Keefe and Nadel (1978), who argued that the hippocampus, a part of the brain associated with long-term memory, *is* a cognitive map. With today's emphasis on neural mapping, there is renewed support for this idea. The cognitive map interpreted as analogous to a cartographic map generally relies on the argument that both have Euclidean spatial properties. Although there has been substantial empirical evidence that cognitive configurations constructed in two-dimensional Euclidean spaces correlate highly with Euclidean representations of objective reality (Golledge, 1975; Golledge et al., 1975; Golledge & Spector, 1978; Gärling, Böök, & Lindberg, 1985), there is also much evidence that, for the most part, cognitive maps can violate some if not all of the basic Euclidean axioms (Gale, Doherty, Pellegrino, & Golledge, 1985). The elaboration of the fact that cognitive maps exist in multimetric spaces (Baird, Wagner, & Noma, 1982), along with some evidence that curvilinear spaces are perhaps more suited to represent the spatial relationships contained in cognitive maps (Golledge & Hubert, 1982), have caused this debate to lose force.

The use of a cognitive map as a *hypothetical construct* or *convenient fiction* implies that the use of the term map provides no literal meaning. Moore and Golledge (1976: 8) suggest that "as a hypothetical construct the term cognitive map and its approximate synonyms refer to covert, nonobserved processes and organizations of elements of knowledge." As principles of symmetry and transitivity, for example, are violated (Tversky, 1981; Baird et al., 1982), then the term "map" becomes a convenient fiction that provides a hook on which to hang interpretive structures but that is tied to no particular representational form.

But, for better or for worse, from medical science to engineering sciences, from the humanities to the social sciences, the term cognitive map is now widespread.

7.5   The Use of Cognitive Maps

7.5.1   Cognitive Maps and Spatial Behavior

If we accept the broadest definition of a cognitive map (i.e., as our internal model of the world in which we live), then cognitive maps play a role in all behaviors, both spatial and nonspatial.

Traditionally, geographers have focused on the role that cognitive maps play in deciding what overt spatial behaviors are to be performed in any problem-solving situation. This means that the cognitive map becomes a critical component of general spatial problem-solving activity. It is embedded in the process of spatial choice and decision making. This role was made explicit by Briggs (1973) after being implicitly recognized by geographic researchers in the 1960s. The cognitive map plays a role in deciding what choice to make and whether one has to travel or not to achieve a goal; it helps decide where to go, which route to take, and what travel mode to take to get there. These questions have been addressed in a variety of contexts such as consumer behavior (Coshall, 1985a, 1985b; Timmermans, 1979; Pacione, 1978, 1982). Many other examples can be found in a review by Timmermans and Golledge (1990).

Cognitive maps have also played a role in research on movement patterns both in a migration and a mobility context (Johnston, 1972; Briggs, 1973), as well as for movement associated with recreational and leisure choice (Pigram, 1993; Golledge & Timmermans, 1990). The latter have exhaustively reviewed the use of cognitive maps in many different environmental situations from establishing preferences for shops, shopping centers, apartments, and modes of transportation, to the planning of specific residential environments, and the use of cognitive maps as devices to aid wayfinding and navigation. They are in common use when one is examining the learning of unfamiliar layouts, at scales varying from the neighborhood to the international arena, and they are essential for the development of any examination of the nature and degree of fundamental spatial concepts, such as location, distribution, pattern, shape, direction, orientation, region, hierarchy, network, and surface. In the sections that follow we explore specific examples of some of these uses.

7.5.2 Cognitive Maps as Planning Aids

Often it has been suggested that cognitive maps are an essential part of the making of policy and the development of plans. This belief is held alike by geographers and planners. The underlying hypothesis here is that the quality of human decision making (whether individually or in congress) will be improved as the quality of knowledge about the task environment is improved. And if we are aware of people's preferences for, perceptions of, and attitudes toward different environments, then better matches between planning and policy making and the felt needs of the populations, for whom plans are being made, can be achieved (Gärling & Golledge, 1989).

This sentiment had been expressed much earlier by Lynch (1976) whose written work often stressed that better planning, design, and management of environments could be undertaken *for* people if they were done *with* people (i.e., participatory planning). Improved modeling, Lynch argued, would also occur if we understood the images of the world in which people perceive they live and the world in which they would like to live. Aitken, Cutter, Foote, and Sell (1989) more recently reiterated this sentiment, suggesting that the most adequate planning in the context of the built environment will reflect the behavioral propensity of residents and other users. Knowing something about people's perceptions, preferences, and images provides information that complements the designers' and planners' intuition, guidelines, and legal restrictions.

7.5.3 Cognitive Maps and Disability

One area where this advice is being taken into consideration is the design of environments to meet the needs of populations with disabilities (Passini, 1984). Investigation of the policy guidelines needed to encourage the development and improve the quality of life for the deinstitutionalized mentally retarded constitutes yet another area of successful integration of policy design and cognitive mapping (Golledge et al., 1979a and 1979b; Golledge, Richardson, Rayner, & Parnicky, 1983), as is the policy area represented by the work of Carpmen, Grant, and Simmons (1985) who explored the effects of hospital design upon the wayfinding activities of the hospital staff, patients, and visitors. In a similar environmental context, Ulrich (1984) discovered that rearranging hospital beds in a room could increase the ambiance of the room and reduce convalescent time for hospital patients.

In addition to the mentally retarded and the hospitalized, other special populations that have benefited from research focused on cognitive maps include children's environments (Hart, 1981, 1984) and elderly environments (Ohta, 1983; Kirasic 1991; Kirasic, Allen, & Haggerty, 1992; Golant, 1982). Some of this research is developed at length in Chapters 14 and 15 of this book.

7.5.4   Cognitive Maps and Crime

Cognitive mapping also has become a useful device in combating criminal activity. Rengert (1980) and Rengert and Waselchick (1985) have examined the cognitive maps of residential burglars. More recently, Rengert (1994) has developed representations of what people think they know about the location of illegal drug sales and different forms of violence in urban areas. Canter and Larkin (1993) have used cognitive maps to track down criminals for the police. They showed the limited range of cognitive maps and commented on their circularity. Extrapolating to a real-world setting, they successfully predicted the probable location of a serial rapist, thereby allowing the police to trap the perpetrator. It has been suggested also that cognitive maps produced by friends and acquaintances may be useful in the process of tracking down missing persons.

Even though we have presented a variety of alternative conceptualizations of the process of cognitive mapping and the nature and use of cognitive maps, it should be apparent that no definitive answer exists at this stage to the question of how best to describe cognitive maps. It appears that the selection, coding, storage, decoding, reconstruction, and use of cognitive information potentially are fertile topics for further geographic investigation.

7.6   Methods for Externally Representing Cognitive Maps

7.6.1   Externalizing Information: Cognitive Configurations or Spatial Products

The use of many different metaphors for the term cognitive map has produced continuing confusion in the various disciplines that use the concept. Perhaps the most enduring of these is the notion that a cognitive map must have the same properties as a cartographic map.

The early 1970s saw much cross-disciplinary discussion about whether or not such a configural matching existed, with the overwhelming evidence pointing to the fact that cognitive maps were *not* simply internalized cartographic maps. In an early effort to differentiate between the internalized cognitive map and the external representation, Golledge (1975) differentiated between the cognitive map as the internal representation and a "cognitive configuration" as an externalization of information gleaned from a cognitive map. The cognitive configuration could be task specific and could have more or less properties of conventional cartographic maps. Thus, sketch maps were considered primarily nonmetric cognitive configurations, whereas multidimensional scaling outputs were considered to be more metric cognitive configurations. Although this differentiation continues to be used in some geography literature, in other disciplines it is more common to differentiate between the internal representation (cognitive map) and an externalization called a "spatial product"—a term used by Liben (1981) to describe an externalized representation.

Methods developed to recover cognitive configurations are as varied as the purposes behind such research. One of the earliest methods suggested by Lynch (1960) was the use of sketch map techniques (Figure 7.5). Other procedures

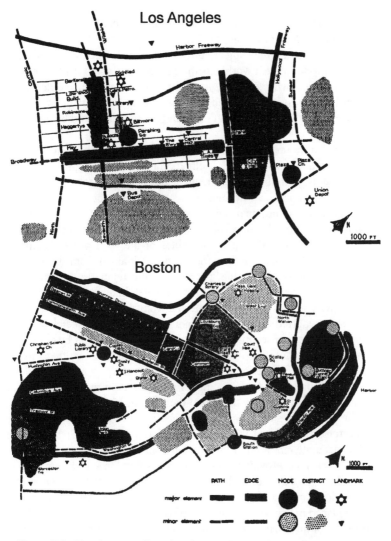

Figure 7.5. Sketch maps of Los Angeles and Boston. *Source:* Lynch, 1960.

include: requesting subjects to image scenes from different perspectives; to list the best recognized or most frequently visited places; to reconstruct images of unseen objects; to estimate lengths of streets and angles of intersections; to use various unidimensional scaling procedures to obtain interpoint distance judgments; and, by using proximity judgments in a paired comparison context, to develop cognitive distance estimates from multidimensional scaling configurations. More recently, table top modeling, interactive computer experiments, and even the use of aerial photos, all appear to be capable of producing reasonable configurational representations.

7.6.2   Sketch Maps

Sketch mapping has long appeared to be a useful instrument for recovering information about environments, if the maps are properly interpreted. This technique suffers from the assumptions that the subject understands the abstract notion of the model and its relationship to the real world has the sufficient motor skills to portray accurately in sketch format what she or he is attempting to complete, and that a uniform metric is applied across the sketched information. Where violations of these assumptions occur, the best information that can be obtained would be topological in nature.

Although rarely undertaken as a solo exercise anymore, sketch mapping is used often in conjunction with other methods for extracting cognitive information about environments. Although sketch maps usually cannot provide reliable metric information, they can provide information, such as the number of features included on the map, the mix of point, line, and area features used in the sketch, an indication of dominant functions in a locale as perceived by the sketcher, and ordinal information such as the sequence of cues along routes or the sequence of segments and turns along routes. Additional information can be obtained from the system constructed as the basis for the sketch, particularly the regularity or irregularity of frameworks such as street systems and environmental features. Frames of reference used in local environments can also be inferred from sketches.

It has been reported often that the amount of material on sketch maps and its accuracy increase over consecutive trials. An example of this can be seen in Figure 7.6, which shows sketches of a route in an unfamiliar environment that was learned in a forward and reverse direction on five consecutive days by a 12-year-old boy. It is clear that the sketch went from a minimal representation to a quite accurate and comprehensive representation by the end of the 10 trials (5 forward, 5 reversals). Although in the early stages there were characteristics such as segment reversals and directional errors, by the fifth trial these were generally eliminated, and the sketches converged on accurate rendition of the task environment.

Blades (1990) has provided evidence to show that the sketch-mapping procedure is quite consistent over time. This does not imply that metricity accrues to the sketch, but it does imply that when used in a multiple trial–task situation, confidence can be placed in the sketch-mapping procedure. Sketches can, therefore, indicate the quantity of information that comes readily to mind when one is given a task

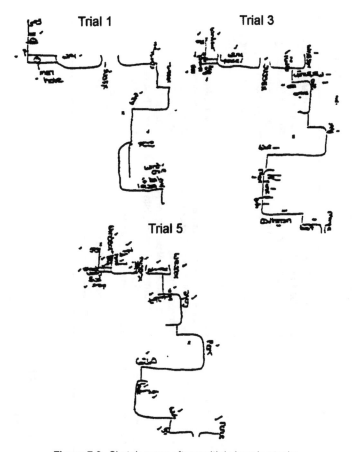

Figure 7.6. Sketch maps after multiple learning tasks.

of representing local knowledge within specified time constraints, and if the environment is known, the sketches produced are reliable—that is, they are consistent over time.

As with cognitive maps generally, sketches are incomplete, distorted, mixed-metric, or nonmetric modes of representation; they are schematized and are often full of blank spaces and nonconnected networks. In examination of selections of sketch maps for a given environment, frequency counts of the appearance of different features allow the development of composite maps on which are located those places known best by the largest number of people. The classic example still remains the map constructed by Milgram and Jodolet (1976) representing multiple levels of a common hierarchical knowledge structure of Paris, France. A listing of the most commonly cited landmark cues in selected urban areas is given in Figure 7.7.

	Paris [a] (1970)	Columbus [b] (1974)	Santa Barbara [b] (1978)	San Francisco [c] (1989)
1.	Etoile	Graceland Shopping Center	La Cumbre Plaza	Fisherman's Wharf
2.	Notre Dame	Eastland Shopping Center	Santa Barbara Airport	Chinatown
3.	Tour Eiffel	Westland Shopping Center	County Court House	Golden Gate Bridge
4.	Seine	Lazarus, downtown	Mission Santa Barbara	Union Square
5.	Bois de Boulogne	Port Columbus	UCEN Building at UCSB	Cable Car Ride
6.	Champs Elysees	I-71 N and I-270 intersection	Goleta Beach	Golden Gate Park
7.	Concorde	Veterans Memorial Auditorium	FEDMART, Goleta	Museums/Galleries
8.	Louvre	Ohio State Fairgrounds Coliseum	Santa Barbara Harbor	Alcatraz
9.	Chaillot	Western Electric, East Industrial Plant	Botanical Gardens	Union Street
10.	Cite	I-71 S. and I-270 intersection	Robinson's Department Store	Broadway/North Beach
11.	Luxembourg	Northland Shpping Center	Dos Pueblos High School, Goleta	Japantown
12.	Montmartre	N. High Street and I-161 intersection	Arlington Theatre	Mission District
13.	Montparnasse	Columbus State Hospital	Santa Barbara Museum of Art	Haight-Ashbury
14.	St. Germain	Anheiser Busch Brewery	Magic Lantern Theatre, Isla Vista	Live Theatre
15.	St. Louis	Morse Road and N. High Street intersection	Picadilly Square	Castro District
16.	St. Michel	Lane Avenue and N. High Street intersection	Bank of America, Isla Vista	The Zoo
17.	Tuilleries	Capital University	YMCA	Sports Event
18.	Bastille	Ohio Historical Society	Rob Gym at UCSB	Symphony/Concert
19.	Buttes Chaumont	Riverside Hospital	Greyhound Bus Depot	Opera

[a] Ranked by importance. Data from Milgram and Jodelet (1976).
[b] Not ranked by importance. Unpublished data from surveys conducted by Golledge.
[c] Ranked by visitation. Data from Economics Research Associates' 1989 Survey of San Francisco Visitors.

Figure 7.7. Lists of major landmarks in cities.

7.6.3   Multidimensional Methods

Increased sophistication in specifying the design of experiments for recovering cognitive information, and the use of powerful *multidimensional methods*, have given researchers a great deal of confidence in their ability to recover useful spatial information from what appear to be nonspatial knowledge structures. Many people have proven to be amazingly accurate in remembering the essential details of the spatial layouts of large-scale environments with which they are familiar (Golledge & Spector, 1978; Golledge, Rayner, & Rivizzigno, 1982). To test this contention, emphasis was placed on the development of methods for comparing recovered configurations with some representation of objective reality (e.g., cartographic or other such representations).

As an example of the use of multidimensional methods for recovering and interpreting cognitive spatial information, we now examine a set of experiments designed to recover cognitive spatial information and examine it for accuracy of its spatial properties. This set of experiments was undertaken in the city of Columbus, Ohio (Golledge, Rivizzigno, & Spector, 1974). Over a period of 9 months, sets of subjects (subdivided into newcomers, intermediate residents, and long-term residents) participated in an experiment involving the grouping or clustering of pairs of locational cues and the allocation of scale scores based on their subjective estimates of the relative proximity of each pair. The scale scores obtained from each subject were analyzed using a nonmetric multidimensional scaling algorithm KYST (Kruskal, Young, & Seery, 1976) in which interpoint distance estimates proportional to the scale scores were developed. These were then used in an iterative procedure based on a gradient method for achieving convergence, and a two-dimensional point configuration of the locations of the target cues was developed. The interpoint distance information contained in this configuration was monotonically related to the original scale scores, and the fit between the data and the developed configuration was assessed through a monotonic regression procedure, which minimized the error variances between recovered points and the original scale scores. To produce interpretable configurations, an initial configuration representing the actual locational arrangement of the cues in the city was used as the starting (initial) configuration. Thus, the final configuration outputted from the scaling program represented a distortion of the actual pattern based on the proximity scales developed by each individual. Samples of such configurations showing different types of axial biases are shown in Figure 7.8. The stress (badness-of-fit) statistic in this way becomes an interpretable goodness-of-fit statistic proportional to the error obtained in the outputted configuration. Thus, by using a measure of spatial correlation, it illustrates the extent to which (after rotation, translation, and uniform stretching or shrinking of axes) the cognitive and objective configuration coincided.

Each two-dimensional output configuration obtained for each subject was matched with a two dimensional Euclidean configuration of the cues in their urban space via a process of standardizing, centralizing, and rotating the two configurations until the closest possible match had been achieved. The degree of coincidence associated with this procedure was recorded in the form of a bidimensional correlation coefficient (Tobler, 1978). Thus, individual configura-

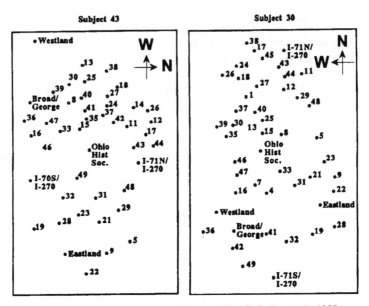

Figure 7.8. Typical axial biases. *Source*: After Golledge et al., 1969.

tions that closely match the two-dimensional Euclidean map would have high bidimensional coefficients.

A simpler visual interpretation of the degree of correspondence between subjective and objective configurations can be obtained by warping a standard grid to fit the subjective configuration (Figure 7.9). The grids are obtained in a manner similar to the way one would compile a contour map—that is, by interpolating grid lines between points in each configuration. Obviously, some of the finer displacements and distortions cannot be shown using this grid distortion technique, but the use of this simple generalization helps interpretation. Although each simplified grid has its own distinctive pattern of distortion, there are similarities repeated among many of the subjects, permitting some subjects to be grouped on the basis of similarity of distortions. Many grids, for example, indicated a pronounced exaggeration of the shorter distances. Overall, there was a pronounced localization effect with distortion being least in the daily activity space of sets of individuals. The use of grid representations of the configurations provide further evidence that people's knowledge structures of urban information is incomplete, schematized, and filled with holes or folds or warps. Obviously there are pronounced local effects, with the general knowledge surface declining exponentially away from places that could be designated as primary nodes or anchorpoints.

7.6.4   Anchors and Errors in Cognitive Maps

In discussing the concept of correspondence, we have used the notion of bidimensional regression and grid pattern matching as indicators of structural agreement. In

Columbus, Ohio

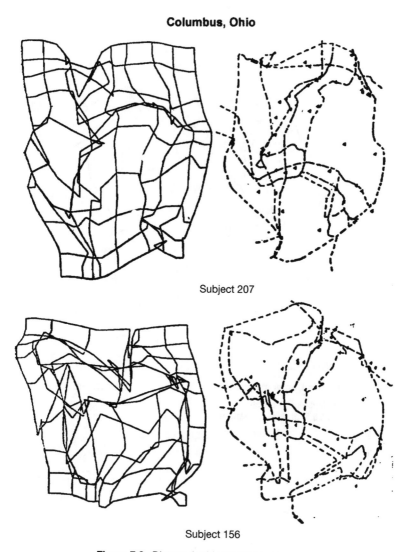

Subject 207

Subject 156

Figure 7.9. Distorted grid and street maps.

a more precise context, it is possible to estimate the degree of distortion (or displacement of subjective cues from their objective locations) and fuzziness (or the degree of variability of subjective estimates of cue locations).

One way of examining information on cognitive maps and/or assessing degrees of coincidence has been developed by Tobler (1977). For example, for the population as a whole one may locate all subjective estimates of each location and summarize the resulting locational set in terms of error ellipses (Figure 7.10).

Such a construction indicates the degree to which the location of individual cues in the total set is more or less known by the subject population. Obviously, the

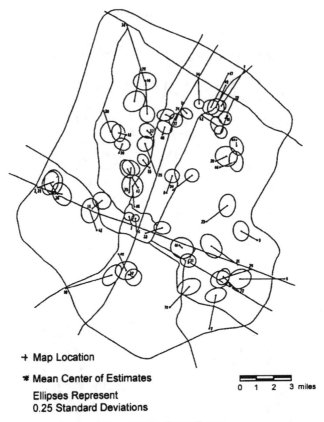

+ Map Location

✳ Mean Center of Estimates

Ellipses Represent
0.25 Standard Deviations

0　1　2　3 miles

Figure 7.10. Error ellipses.

larger the ellipse and the more circular the ellipse, the less accurate is the general knowledge structure of the cue's location. Similarly, by examining the magnitude of the distance displacement between a cue in a matched subjective and objective configuration pair, and estimating its directional distortion, the fuzziness component previously calculated can be complemented with an individual's distortional component (Figure 7.11). From examining distortions of the cues used in the Columbus experiment, it appears that distortion is least in the cue set contained in an individual's activity space or associated with primary nodes and greatest associated with places visited less frequently. Generally, a strong relationship exists between the amount of fuzziness associated with each location cue and the amount of distortion. Using such procedures, one can identify the least distorted and fuzzy cues as *anchorpoints* in the environment. Increasing error properties then indicate less concise knowledge about other cases and imply the cues are lower in a knowledge hierarchy. Also, by examining the directional bias in the ellipses and relating these to characteristics of the anchorpoints, we can develop some idea of hierarchical order or dominance. This provides some ordering principles for understanding the knowledge structure revealed in the configuration.

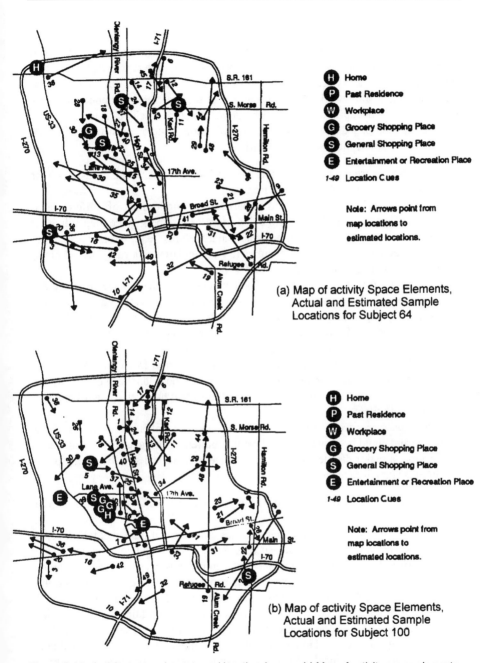

(a) Map of activity Space Elements, Actual and Estimated Sample Locations for Subject 64

(b) Map of activity Space Elements, Actual and Estimated Sample Locations for Subject 100

Figure 7.11. Activity space elements and locational errors. (a) Map of activity space elements and actual and estimated sample locations for subject 64. (b) Map of activity space elements and actual and estimated sample locations for subject 100.

7.7   Images of Cities

A major advance in work investigating *city image* was made by Kevin Lynch (1960). His book, *The Image of the City*. This book not only served to focus attention on perceptual and cognitive qualities of urban environments, but it also provided a conceptual framework for the discussion of the structural components of city images that still occupies a primary place in the literature on city structure.

Lynch proposed a five-element classification system consisting of paths, boundaries, districts, nodes, and landmarks:

1. *Paths* were the basic channels along which persons would occasionally or customarily move around a given environment.

2. *Boundaries* were conceived as barriers tending to mark off differentiated segments of space, with the barriers being more or less permeable, depending on whether they consisted of natural environmental features, (e.g., rivers, parks, freeways), or social, political, cultural features (e.g., boundaries of a gang turf, administrative boundaries, ghetto boundaries, and so on).

3. *Districts* were the main areal component of the urban image and were defined as recognizable areas contained within sets of well-defined edges. Districts were generally larger than neighborhoods and often were allocated an official or unofficial descriptive title such that they were readily recognizable by the local population and by many others living outside their boundaries.

4. *Nodes* were defined as strategic spots that acted as foci for human behavioral activity. They may be things such as bus stops, outdoor luncheon or recreational areas, critical street corners, or specific buildings that had well-known functions associated with them.

5. *Landmarks* were visibly dominant and easily identifiable elements of the landscape and could be drawn from the natural, built, or cultural environments (e.g., an overlook, a major distinctive building, a place of historical significance, a place of religious significance, and so on).

These five elements were illustrated earlier for Los Angeles and Boston in Figure 7.5.

While the initial research on city images incorporated fundamental concepts from the psychology of perception—such as figure-background composition, clarity, singularity of form, visual dominance, perceptual closure, and so on—many of Lynch's imitators moved beyond these fundamental perceptual qualities to add the values, meaning, and significance of city elements in both a societal and a cultural sense. For example, Appleyard's (1969) research on why buildings are known included a discussion of basic characteristics such as singularity, level of visibility, community significance, and intensity; but he also emphasized societal and cultural context, historicity, and local significance as critical features of recognition. Other research focused on a search for reasons behind the departure from objective "accuracy" in city images and included things such as residential history of the respondents, levels of familiarity with places, basic elements of form, ethnic or cultural heritage, or mode of travel (e.g., Maurer & Baxter, 1972; Orleans & Schmidt, 1972; Ladd, 1970; Ley, 1972). It must be remembered, however, that although the emphasis on affect and form dominated a considerable part of the research in this area, other research focused on attempts to refine modes of representing cognitive information other than via sketch maps and verbal descriptions (e.g., Golledge & Zannaras, 1973; Golledge, Rivizzigno, & Spector, 1976).

Although Lynch's model is a valuable conceptual tool, it falls foul of the set of problems involved in aggregating and disaggregating data. Even though any city can be decomposed into its component parts, detailed analysis of the parts raises questions of how they can possibly be reassembled. Gale et al. (1985) show that cognized distances violate Euclidean axioms, that nodes and landmarks are often poorly located in memory, that paths often cannot be integrated, and that perceived districts overlap. The separate decomposed elements cannot be reassembled into a whole using conventional geometric or cartographic methods.

7.8 The City as a Trip

A transactionally based hypothesis concerning our knowledge of urban environments would be that one obtains knowledge about the city according to the type of interactions that one has with it. Thus, urban knowledge accumulates as a result of the various trips undertaken as part of the everyday process of living. Whereas other conceptualizations focus more on the node and landmark structure or areal pattern of urban knowledge, this conceptualization is path based.

As an example of a *path-based* view of city images, we present a discussion of the city as a trip by Carr and Schissler (1969). They interviewed a number of commuters, both car drivers and passengers, in the city of Boston in an attempt to identify the types of environmental cues noticed and used by individuals when they are undertaking trips within an urban environment. To obtain information on the recognized and used cue structure, subjects were examined prior to the day of the field trip and asked to state their preconceptions of the trips they were to make, including regular trips undertaken by the individuals and specific tasks set them by the experimenters. In addition, subjects were laboratory tested to determine such things as the accuracy of recall and depth of memory and the angle of eye movements. The next day, subjects were taken on a trip with a camera mounted on their head to record where they were looking, and optical devices focused on the corneas of the subjects' eyes enabled the researchers to identify the directions in which the subjects were looking. This helped to identify possible landmarks seen at various places along the route. Thus, by establishing points of fixation and also requesting a verbal summary of what was noticed and used as navigational aids along the trip, the researchers attempted to define those elements of the city that were observed, recognized, and used in a trip-making context. The subjects were exposed to a variety of verbal, graphic, and descriptive materials associated with the trips and films of the trip in order to provide a data bank. Data banks were then searched using content analysis to find those places, cues, or areas that provided anchors for an individual's cognitive representation of the environment. The features recovered using these processes were then ranked according to the frequency with which they appeared in the subject's responses (Table 7.3).

TABLE 7.3.
Individual Items Memory List for Drivers, Passengers, and Commuters in Order of
Best-Remembered Item

Rank	Drivers	Passengers	Commuters
1	Mystic River Bridge	Mystic River Bridge	Mystic River Bridge
2	Toll Booth	Toll Booth	Overpass (early)
3	Overpass (late)	Prudential Building	Prudential Building
4	Sign for Haymarket	Three-Deckers: Chelsea	Bunker Hill Monument
5	Overpass (early)	Overpass (early)	State Street Bank
6	Second Bridge	Bunker Hill Monument	Three-Deckers: Chelsea
7	Three Deckers: Chelsea	Overpass (late)	Government Center
8	Billboard: John Hancock	Government Center	Custom House Tower
9	Sign for Downtown Boston	John Hancock Building	Soldier's Home
10	Sign for Charlestown	Charlestown residences	Toll Booth
11		Custom House Tower	U.S.S. Constitution
12		Charles River Park Apts.	Naval Hospital
13		State Street Bank	Signs for Storrow Drive
14		Sign: Chelsea	William's School
15		North End residences	American Optical Co.
16		U.S.S. Constitution	John Hancock
17		Colored oil drums	Charles River Park Apts.
18		Billboard: John Hancock	Colored oil drums
19		Twin Tower Church	Sign: Fitzgerald Expressway
20		Billboard: Seagrams	Bradlee's Shopping Center
21		Residences on Soldier's Home Hill	North Station
22			Sign for High Street
23			Sign for Dock Square
24			Sign for Chelsea
25			Grain Elevators
26			Wallpaper Factor
27			Barrel Factory
28			Cemetery

Source: Carr and Schissler, 1969:21.

Some of the results from this study focused on the difference between the number of items recorded by regular commuters, as opposed to newcomers in the area, and the number of features recorded by passengers, as opposed to drivers. For example, the average driver remembered only 10 objects over a test section of 4 miles of road, but passengers remembered an average of 21 objects. Commuters who regularly traveled the route remembered an average of 28 objects. It was also found that individuals performing different functions tended to remember similar things in the same order of importance. Thus, the features remembered by drivers were included in the set of features remembered by the passengers, which in turn were contained within the set of features recalled by regular commuters. It appeared that familiarity with the route did change the range of things that were recorded in terms of adding more detail (and less significant features) rather than adding other major anchorpoints. This is indeed a critical feature because it suggests that even newcomers to an area will focus on a fundamental set of primary cues or anchorpoints. These will then commonly be used to anchor cognitive maps and to provide detail in the vicinity of paths, such that expectations with respect to major decision points along the path can be defined with a great deal of reliability.

The above findings, in turn, suggest that some structural properties of a city impress themselves more on memory than do others. This appears to be the case either for casual or regular observers, and it appears likely that these points of fixation become *orientation nodes* (focal points) in the structuring of a path-dominated image of an area. Given that regular commuters observed and recalled the greatest range and variety of elements in the city, it appears that individuals learn about the environment over time and that they constantly modify their knowledge set as they travel throughout the local environment and accumulate further information. The connection of major nodes or landmarks by paths adds a critical spatial dimension, for it gives an individual an idea of spatial separation and connectivity that helps define the concept of relative location or layout among all points in the environment.

As represented in memory, and as evidenced by behavior, the city as a trip appears to be a highly meaningful concept to all individuals. Whether it is the young child first exploring a neighborhood, or the bored long-time commuter, the structure, sequencing, and dominance of features of the environment related to trip-making are critical in terms of specifying the spatial extent of knowledge at any given point in time. The acquisition of information as one selects and travels a subset of all possible paths in the city helps define the concept of a node path structure that provides a basic framework for cognitive maps. In short, for people to be able to relate to any given environment, they must be able to "read" it, or comprehend what is there. Apparently this is what all individuals must do in their day-by-day interactions with the external environment. This helps them build coherent knowledge structures that, in turn, allow them to quickly store and access the information needed to operate on a day-by-day basis with a minimum amount of confusion and a minimum amount of effort. The path or network structure used in episodic activities thus becomes a critical feature of an image of a spatial environment.

7.9 The City as a Hierarchical Structure

Lynch (1960) identified landmarks and nodes as two of the major components of city images. Much of the early work on urban cognitive structures also focused on landmarks and other features that could be represented as points.

7.9.1 Anchorpoints

Typical of this research was Golledge's (1978a) *anchorpoint theory* of urban cognition. In this theory the acquisition of simple declarative knowledge (understanding individual nodes and landmarks in relation to self) is followed by procedural knowledge in which primary nodes are linked by primary paths in a hierarchical linking of places. As information about place accumulates, spread effects in the vicinity of primary nodes and paths allow the development of simple areal concepts such as neighborhoods. Geometrically, the primary node–path network (or *skeletal structure*) of the emerging cognitive map is filled out by the addition of intermittent areas surrounding the primary nodes and paths. *Distortions* in the relative location of primary nodes and their associated areas produce foldings or warpings in the fabric of understanding of space. A lack of knowledge of segments of the environment may cause folding in which known places are drawn closer together in the cognitive map structure or may result in the emergence of holes or blank areas that can contribute to locational and relational errors in the spatial information acquired. The final set of spatial information encoded, therefore, may include simple Euclidean properties among places that are well known, and hierarchically conditioned topological relations for those places less well known in a general spatial context. Given this conceptualization, it is easy to see how a definition of a cognitive map as being incomplete, schematized, distorted, and without continuous regular metric or geometric structure may emerge and be found acceptable.

7.9.2   Regionalized Hierarchies

Empirical evidence for the hierarchical organization of spatial knowledge generally, and the arrangement of knowledge about cities in particular, has been offered also by Stevens and Coupe (1978) and Hirtle and Jonides (1985).

The former used a classic geographic problem to illustrate how spatial knowledge was embedded in a dominant–subordinate structure. This was tested by questions asking for clarification about location within and between states. A basic question was: "Which is further east—Reno, Nevada or San Diego, California?" Stevens and Coupe (1978) found that more people chose Reno as the answer to this question. They explained it by suggesting that Reno was embedded in the state concept of Nevada, whereas San Diego was embedded in the state concept of California. Since for the most part Nevada is perceived to be east of California, inference would lead one to suspect that cities in Nevada are further east than cities in California. Thus, a landmark embedded in a region, or dominated directly by another landmark, may take on the characteristics of the dominant place. The result of course is distortion of spatial information.

Hirtle and Jonides (1985) examined how people organized information about specific places in Ann Arbor, Michigan. Lists of places were given to subjects with the task of reproducing them from memory. Using a variation of clustering techniques called ordered tree analysis, Hirtle and Jonides showed that recall proceeded by a focusing on critical anchorpoints or landmarks and then a listing of places subordinate to them (Figure 7.12). Thus a set of key anchorpoints were defined throughout the city, each directly controlling sets of secondary places, which in turn dominated subsets of lower-order places. The city was seen as a multinodal, hierarchically organized system. Such results provide some empirical support for the anchorpoint hypothesis and provide insights into how individuals structure information about the cities in which they live.

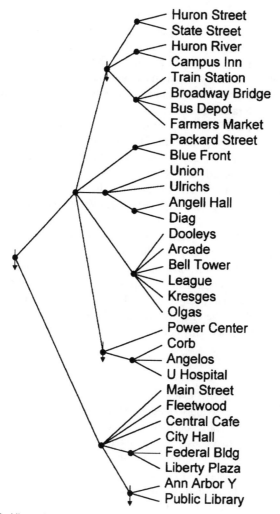

Figure 7.12. Hierarchy of cues in Ann Arbor. *Source*: Hirtle & Jonides, 1985:213.

7.9.3 Hierarchies of Paths

In a manner similar to that used in landmark organization, Péruch, Giraudo, and Gärling (1989) suggested that the road and highway network of most cities was also organized hierarchically. They identified at least three levels that were commonly used when one is driving between origins and destinations. The three levels consisted of major highway and freeway segments, which dominated regionalized segments of arterial and main roads, and which in turn dominated local community and neighborhood street systems. Pétruch et al. (1989) found that from any origin the usual process would be to proceed as quickly as possible through the lower orders of the hierarchy to get to the nearest primary level. This would then be used as the path to an exit as close as possible to the potential destination. A final step would involve going down through the hierarchy again until the specific target had been reached. This cognitive structure parallels the traditional Department of Transportation classification of routes into categories such as interstate, state highway, limited access, multilane arterials, arterials with unimpeded bidirectional flow, and local streets.

Much of the literature on urban, economic, and social structure also identifies hierarchical organization. Within the business domain, a hierarchy of business centers can be identified ranging from the CBD to major and minor regional shopping centers, suburban shopping areas or business districts, community centers, neighborhood centers, and isolated store clusters. Socially, cities are often divided into regions, districts, communities, and neighborhoods. Both of the latter hierarchical structures involve the embedding of lower orders within the sphere of influence of higher orders.

Thus, it can be shown objectively that hierarchical systems exist within urban areas and that people either recognize them or create their own hierarchies to help make sense of the complex information system that typifies cities. It is interesting to note that hierarchical organizations appear to be a fundamental process in the cognitive domain also. Here objective and subjective categorization procedures complement each other. The end product is a simplified and more easily comprehended cognitive map of a highly complex and multidimensional environment.

7.9.4 Priming

Holdings (1994) offers further evidence of the existence of hierarchical ordering of spatial information by referring to studies using *priming*. Priming occurs when the presentation of one cue leads to quicker recall or recognition of another cue. For priming to happen, the two cues must be linked somehow in memory. It is suggested that if there is a hierarchical organization embedded in the memory structure, then the distances between pairs of cues in the same cluster will be less than the distances between pairs of cues in different clusters. Thus, *spreading activation* will result in quicker recall of the paired intracluster cues than for the intercluster cues.

In her analysis, Holdings created two maps with similar structure but with one map being a rotated version of the first. Labels on the map were also switched. On her maps, specific cue names were placed with a constant distance between selected pairs of cues located at different places across the map but with unequal distances between members of different pair groups. Given a hierarchical representation of the map, the pairs should fall progressively further apart within the hierarchy. Holdings concluded that the physical structure of the maps produced the anticipated clustering of cues such that buildings that lay together on the same street were stored closer together in memory than were buildings located on separate streets. Similarly, cues within a given township were stored closer together than were cues from separate townships. She concluded that the maps were mentally represented in a hierarchical form and that these hierarchies mirrored the latent spatial hierarchical structure embedded in the map. Also it was found that it was possible to make predictions as to the level at which features such as townships, streets, and houses enter in a person's hierarchical ordering.

The assumption underlying this latter conclusion was that cues that were removed from one another by more than two links in the hierarchical network are unlikely to follow one another on any simple recall list. For example, if one recalled a house, it might call to mind the street on which it lies or the township in which it is located, or it could call to mind one of the other houses on the same street. Thus, a house might call to mind a wide variety of items. The street also could call to mind the houses on the street, the other streets in the township, or the township itself. Likewise, a township could call to mind any of the houses or streets within it or any of the other townships. Thus, the level of entry of any cue into the hierarchy was expected to mirror the relationship among the numbers of cues that it was likely to activate.

As an example of the basic intercluster distances, Holdings suggested that houses on the same street would be joined by two links (house–street–house). Houses on separate streets in the same township would be joined by four links (house–street–township–street–house). Houses in separate townships would be joined by seven links (house–street–township–map–township–street–house). Thus, the more links that separate any two cues in the hierarchy, the greater should be the subject's estimation of the distance between them. And as the number of links increases to include adjacent clusters, a distance overestimation tendency will tend to emerge. It is interesting also to note that Holdings found no significant effects of sex, thus indicating that both male and female individuals utilize a hierarchical form of representation of spatial data.

McNamara (1986) also had subjects learn locations within a subdivided layout. They were then asked to perform a direction judgment test, a distance estimation test, and a cue recognition test on the learned objects. The cue recognition test was used for priming purposes. It found that directional judgments were distorted in such a way that they corresponded with the superordinate spatial relationships represented in the hierarchy, and distance judgments were underestimated for within-cluster distances and overestimated for between-cluster distances (Figure 7.13). The priming experiment also indicated that cue names were recognized much faster when they were preceded by a cue in the same region (or same cluster) than if preceded by a cue in a different regional cluster.

In another clustering experiment, Merrill and Baird (1987) made the strong point that priming was far more effective when the environmental features used were both functionally and spatially related. In other words, locations generally found near each other in reality tend to fall close together in hierarchical representations.

7.10 Cognition and Behavior in Classical Models of City Structure

Zannaras (1973) undertook what is still a unique study in terms of examining the role that cognitive representations of the spatial structure of cities have on movements to the city center. Using the existing descriptive models of city structure discussed in Chapter 2 (i.e., concentric zonal, wedge and sector, and multiple nuclei),

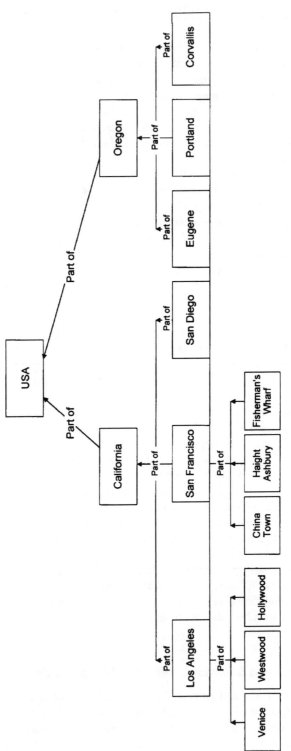

Figure 7.13. Hierarchical nonspatial arrangement of layout.

she found cities in Ohio that approximately fitted each model (Marion, Newark, and Columbus). Land use maps and scale land use models of each of their environments were constructed. Figure 7.14 illustrates how the base land use model was overlain with distance rings for calculating errors associated with wayfinding tasks. Her task was to see if changes in the physical structure of cities significantly influenced image building and consequent wayfinding behavior from the periphery to the city center. This study was unique because, whereas previous work on urban images had focused on deconstructing the city into its component imaged parts, there had been no previous (or later) attempts to relate organizational characteristics of elements in the urban image to the generalized models of the arrangement of city elements suggested in the geographic and planning literature.

Zannaras adopted a twofold approach that: (1) examined the difference in the importance attached by a sample of respondents to the same environmental cues as potential wayfinding aids in various urban structures (Figure 7.15); and (2) investigated differences in how the perception of distance might vary as the physical and built environment varied.

Two sets of hypotheses were proposed and were tested in the laboratory (us-

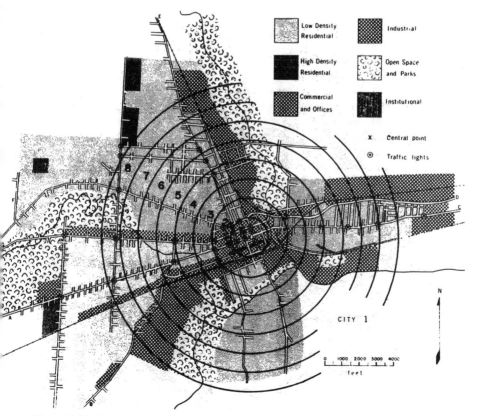

Figure 7.14. Sectoral city with land use and distance zones. *Source*: Zannaras, 1973.

How often do you notice the following features as you travel to the City Center? <u>Please circle the</u>
<u>appropriate number</u>.

1. Never   2. Rarely   3. Occasionally   4. Fairly often   5. Frequently

1 2 3 4 5	1. shopping centers
1 2 3 4 5	2. railroad tracks
1 2 3 4 5	3. direction signs
1 2 3 4 5	4. school buildings
1 2 3 4 5	5. banks
1 2 3 4 5	6. churches
1 2 3 4 5	7. movie theaters
1 2 3 4 5	8. restaurants
1 2 3 4 5	9. open space areas such as parks or green space
1 2 3 4 5	10. speed limit signs
1 2 3 4 5	11. the city skyline
1 2 3 4 5	12. traffic congestion
1 2 3 4 5	13. traffic lights
1 2 3 4 5	14. street width changes
1 2 3 4 5	15. billboards
1 2 3 4 5	16. bridges
1 2 3 4 5	17. neon lights in business areas
1 2 3 4 5	18. rivers
1 2 3 4 5	19. hills [b]
1 2 3 4 5	20. freeway systems
1 2 3 4 5	21. number and spacing of freeway exits
1 2 3 4 5	22. industrial buildings (factories)
1 2 3 4 5	23. public buildings
1 2 3 4 5	24. residential quality changes
1 2 3 4 5	25. residential density changes (spacing of houses)
1 2 3 4 5	26. smog
1 2 3 4 5	27. buildings become more numerous and closer together
1 2 3 4 5	28. major department stores
1 2 3 4 5	29. slums
1 2 3 4 5	30. construction work

[a]This figure includes the features which solicited the most responses.
[b]This feature had a mean of 2.83; all other features had means above 3.0.

Figure 7.15. Rating scales for environmental cues used in wayfinding. *Source*: Zannaras,
1973.

ing slides and scale models) and in the field. Additional hypotheses were proposed
to examine the effect of personal characteristics on cue selection and accuracy of
distance estimates.

The results showed that city structure significantly explained variations in the
mean importance assigned to environmental cues used for wayfinding (Figure 7.15)
(Zannaras, 1973). City structure also significantly explained variation in the mean

accuracy of responses associated with the slide and field trip experiments. Thus, it influenced subjects' ratings of wayfinding potential in each of the different settings, the type of urban features selected as cues to assist wayfinding, and the comprehension of the implicit interpoint distance relationship between selected features and the city center. She found that cities structured in terms of wedges and sectors better fostered an understanding of the spatial relationship between urban features and the city center. She also found that variations in city structure were more important than variations in personal characteristics in determining what features would be chosen, the accuracy of locating and sequencing such features, and the accuracy of using such features in wayfinding tasks from the periphery to the center.

Zannaras (1973) also found that navigational experience was a more important factor in successful task performance than length of residence in an urban environment. She hypothesized that residents who used the center on a regular basis (e.g., for work and shopping) would have more accurate distance perceptions and would be more confident wayfinders with respect to movement from the periphery to the city center; and these hypotheses were confirmed.

The critical finding of this study was that structure of the actual urban environment was of considerable importance in determining how the city was imaged and how behavior took place within it. She found also that real differences exist in the importance assigned to the same urban features as wayfinding aids as the structure of the environment changes (i.e., traffic lights and land use zones were more important in wedge- and sector-type structures than they were in concentric zonal or multinuclear-type structures).

Overall Zannaras (1973) found that the physical structure or layout of a city significantly explained variations in the accuracy of location tasks and on wayfinding tasks regardless of what mode (map, slide, model, or field trip) was used. She found also that the concentric structure model, apparently the simplest of all to understand, does not produce the best accuracy values for the range of tasks involved. Rather, the sectoral city proved to be the simplest structure for remembering locations, for arranging them, and for using them in wayfinding tasks.

7.11 Cognitive Distance

7.11.1 Distinguishing between Subjective and Objective Distance

In the cognitive mapping process, information is accumulated about relative locations and degrees of connectivity of environmental cues. The term *cognitive distance* has evolved to describe the relative spatial separation of objects in a cognitive map.

One of the earlier examples of applying the concept of cognitive distance occurred in the consumer behavior area (Thompson, 1963). In this context, *subjective distance* estimates were incorporated into conventional gravity models to help explain store choice. Along with this applied interest there developed a basic research interest in the concept of cognitive distance (Golledge et al., 1969; Briggs, 1969, 1972; Lowrey, 1970; Burroughs & Sadalla, 1979; Sadalla & Staplin, 1980a). In most cases a curvilinear relationship was seen to occur between subjective and objective distance (see Figure 7.16). Other authors, such as Cadwallader (1975, 1981), MacKay and Olshavsky (1975), and Pacione (1976), continued an applied interest in cognitive distance in the context of consumer behavior. In both basic and applied research it soon became obvious that the type of procedure used to estimate subjective distance, or to convert objective to subjective distance, was of critical importance (see Table 7.4). Some studies (Briggs, 1972; Lowry, 1973; Thorndyke, 1981) supported Stevens' (1956) contention that a power function was the most appropriate transformation. Both Day (1976) and Cadwallader (1979) argued that the type of function likely to be found was significantly affected by the measurement procedure used in obtaining subjective estimates. Features such as whether estimates of distance were obtained in metric or nonmetric units and whether the research question was specifically oriented toward spatial distance or temporal, social or functional, proximities or similarities, also proved to be critical.

Early investigations of cognitive distance produced some apparently contradictory results. Lee (1962) found that distances outward from the center of the Scottish city of Dundee were overestimated more than inward distances. Golledge et al. (1969), among others, found the distances toward downtown in U.S. cities were overestimated, whereas those toward the periphery were underestimated. Some reasons for the apparent differences were related to the size and functional complexity of the cities in question and also possibly to their internal structure (e.g., whether or not they had many regional shopping centers). Thus, differences in the nature of the local environment and its regional or cultural setting also proved to be a critical factor in the estimate of cognitive distance.

In discussions on the notion of cognitive distance there have been suggestions that differences occur due to sex (Lee, 1970), the nature and frequency of intervening barriers (Lowrey, 1970), and the nature of the endpoints between which cognitive distances are estimated (Sadalla & Staplin, 1980a, 1980b). It is accepted generally that cognitive distances may be asymmetric and that, in some cases, the distances may be interpreted in a functional, proximity, or similarity context rather than in a geometrical one. There have been speculations concerning the use of non-Euclidean metrics to represent cognitive distance (Spector, 1978; Richardson, 1979;

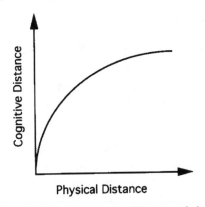

Figure 7.16. Relationship between cognitive distance and physical distance.

TABLE 7.4.
Relationship between Magnitude and Category Scales

Coefficients	Pearson Correlation Coefficients for Relationship between Magnitude Scale and Category Scale (Frequency)	Pearson Correlation Coefficients for Relationship between Logarithm of Magnitude and Category Scale (Frequency)
0.40 or less	0	0
0.41 to 0.60	7	7
0.61 to 0.80	32	35
0.81 to 1.00	26	23

Source: Cadwallader, 1979:568.

Golledge & Hubert, 1981; Tobler, 1976). Furthermore, Baird et al. (1982) have suggested that, given the asymmetric nature of cognitive distance, its representation may be impossible in any known geometric space.

7.11.2   Methods for Evaluating Cognitive Distance

Montello (1991) has defined cognitive distances as mental representations of distances encountered in large-scale environments that cannot be perceived from a single vantage point. Such distances often require movement through the environment before they can be properly apprehended. Alternatively, a scaled-down version may be developed from maps or diagrams. Montello argued that cognitive distance usually is defined using one of five methods: ratio scaling, interval or ordinal scaling, mapping, reproduction, or route following. *Cognitive distance* is thus different from *perceptual distance,* which can be defined as the estimate of distance between pairs of points or objects when both members of the pair are simultaneously viewed (Baird et al., 1982).

Of the five classes of methods used to estimate cognitive distance, Montello suggests that the first two represent traditional psychophysical scaling techniques. The last three methods are better suited for estimating distances rather than reproducing them. For example, mapping requires estimation or simultaneous representation of the relative locations of several places but at a scale that is reduced from the original environmental occurrences. Reproduction involves reproducing a traversed distance without a change in scale from the original environment in which the distance was traversed. For example, one may be asked to evaluate the distance between two buildings then, after ensuring that the buildings are occluded from sight, be asked to walk a distance equivalent to that which separated them. Alternatively, one can be walked over a specific length for a number of trials and then be requested to reproduce the length after some disorienting or distracting interval.

The psychophysical methods for estimating cognitive distance vary with respect to the degree of precision required. An *ordinal estimation* will require only suggesting which distances are shorter or longer than others; there is no need to say how much shorter or longer. *Interval measurement,* however, asks for quantitative measures of distances so that one can obtain an indication of how much longer or shorter one distance is than another. But, even in this case, one cannot describe the distance estimates so obtained as having the property of being proportional to another distance. This property belongs alone to *ratio scaling* where an absolute zero occurs on the distance scale, and meaningful distance ratios can be obtained.

Montello (1991) summarizes the four principal methods of ratio scaling as follows:

1. *Magnitude production*: Here an observer adjusts a test distance to a length that matches a number provided by an experimenter. The number is said to represent the length of a standard distance. The standard can be provided by the experimenter (e.g., 100 yards) or implicitly generated by the observer (e.g., equal to the length of three football fields).

2. *Magnitude estimation*: Here a test distance is provided with a number proportional to its perceived length in comparison to a standard distance assigned some other value. The standard distance and the number representing it (the modulus) again can be provided by either the experimenter or the observer.

3. *Ratio production*: Here a subject is given a standard distance and adjusts the length of a test distance until it appears to represent a prescribed ratio to the standard.

4. *Ratio estimation*: This is one of the most frequently used techniques and involves direct estimation of the ratio of the length of two distances. Alternatively, one may adjust one of a pair of lines so that the ratio between the two line lengths equals the ratio of a test distance to a standard distance. Examples in the geography literature include the use of magnitude estimation by Cadwallader (1979). Examples of ratio estimation include cognitive distance studies by Lowrey (1970, 1973), Briggs (1973, 1976), Byrne and Salter (1983), and MacEachren (1991).

Briggs (1973) used the ratio-scaling techniques to examine cognitive distances within a major urban area. In the geography literature, the standard line used to invoke the ratio-scaling procedure is often taken to represent a crow-fly distance between two well-known points (e.g., two landmarks). Depending on the scale at which the study is being undertaken, the standard distance may vary from a few hundred meters (for example, as might be the case when one is looking at estimates of distance across a campus or other institution) to a mile or more as would be the case if one were looking at interpoint distances at the level of an entire city. While these procedures are common in geography, Montello (1991) warns that there are difficulties with interpreting ratio-scaled data, particularly if different standards are used with different subjects and there is uncontrolled exposure to those standards.

Briggs (1973) found that the relationship between *subjective* and *objective* distance (i.e., cognized distance and physical distance) was best expressed by a power function. Often, interest is expressed in the value of the exponent in the power function. Where a value greater than 1 occurs, this indicates that subjective or cognitive distance changes more rapidly than does physical distance. Multiple repetitions of this finding have led to the generalization that shorter and well-known distances are often overestimated and longer and less-known distances are frequently underestimated. This process is sometimes referred to as "regressing toward the mean."

Even though magnitude and ratio scaling have been commonly used methods in geography and continue to be used in cognitive distance studies, Montello (1991) suggests that we should be aware of the following four difficulties associated with these practices:

1. *The ratio calculation problem.* Substantial differences may exist among individuals in their ability to determine the ratio between the length of a standard distance and the length of a test distance.

2. *The scale translation problem.* This might occur when the standard distance in an environment is represented by an arbitrarily scaled length of line to make the appropriate estimate; a subject must be able to perform the scale translation between the given standard and the real-world standard before a legitimate result can be achieved.

3. *The problem of bias.* Giving a standard distance with certain lengths provides no guarantee that all subjects will internally represent the standard as being of equivalent length.

4. *The orientation problem.* This may occur when the standard line is given in an orientation different to that normally experienced with respect to the usual frames of reference used in the environment (e.g., a north–south distance being represented by an east–west scale line). Vertical lines often appear longer than horizontal lines (i.e., vertical illusion). Aligning standards with real-world occurrences helps to reduce this type of problem.

Cognitive distances produced by interval- and ordinal-scaling methods are often collected using procedures such as paired comparisons, stimulus ranking, rating scales, proximity judgments, and similarity judgments.

Many of the difficulties associated with ratio-scaled data have been found to be relevant also for ordinal- and interval-scaled data (Montello, 1991). In addition, problems involving inconsistency of estimating procedures, asymmetries in paired comparison, and violations of transitivity when making judgments among all possible combinations of paired comparisons can produce problems and consequent difficulties of interpreting experimental results. For example, subjects might estimate differences differently depending on whether they are viewed as emanating from an anchorpoint or going to an anchorpoint. Similar problems may arise when one is estimating a distance between pairs in which one was an anchorpoint or landmark and the other was a lower-order feature or object. In this case, the cognized distance from the anchorpoint might be foreshortened, while the cognized distance to the anchorpoint may be enlarged.

Cognitive distances are often used to help an observer locate a place in an environment. To do this, classical surveying methods can be used. For example, *triangulation* involves drawing direction lines from two or more points to locate an object. When three or more locations are used, the intersection of these directional lines produces a triangle of error. Taking the smallest triangle produced by this method and then solving for its midpoint gives an estimate of the object's location. Another method is called *resection*, which uses the reverse procedure. Again a triangle of error is produced, and the centroid of the triangle is taken as an approximation of location. In the cognitive distance literature, triangulation is often done using a pointing procedure. For example, a subject may be asked to point to location "A," which is in some occluded place. Each time the subject points to the place from a different location, an angular measurement is made. The solution of the problem follows the same procedure in terms of finding the centroid of the triangle of error that results. This procedure is sometimes called projective convergence. Although such procedures work at local scales and in very familiar settings, as Golledge, Dougherty, and Bell (1995) show, people are not necessarily good at pointing toward distant places, particularly if they are not highly familiar with the place and the surrounding frame of reference is not well known.

Cadwallader (1973a, 1976) has compared ratio and magnitude estimation of cognitive distances in large-scale environments. This research, and later work by Montello (1991), isolates some of the problems associated with methods for measuring cognitive distances. These include:

1. Differences may be expected depending on whether subjects are instructed to estimate "over the route" distance or "crow-fly" distance.
2. Problems may arise when subjects confuse temporal separation between places with spatial separation.
3. Differences may occur depending on whether one is asked to estimate physical distance or perceived distance.
4. Serious questions arise concerning whether or not individual distances can be aggregated.

To summarize, although the *cognitive distance* concept has been shown to be a useful substitute for *physical distance* in models such as those predicting consumer behavior (Cadwallader, 1979), and cognitive distances are frequently used to provide estimates of the reliability of people's cognitive maps and sometimes their travel behaviors, undoubtedly there are problems associated with the collection and measurement of cognitive distance that must be addressed if the concept is to be widely accepted as a practical one that can be integrated into different predictive models. We have outlined some of the difficulties associated both with methods of collecting cognitive distance estimates and problems of using such measurements in explanatory contexts. As studies of cognitive mapping processes and cognitive maps generally become more widespread in geography and other disciplines, a solution to many of these problems becomes a more urgent research issue.

8

Activities in Time and Space

8.1 The Nature of Human Activities in Time and Space

All human activities occur coincidentally in *time* and *space*. Traditionally, human geographers have given considerable attention to the study of locational aspects of human activities. However, in the past three decades or so, the location of activities in time as well as in space has been given increasing significance in research. It is relatively easy to study the absolute spatial locations of activities and the locational outcomes of human decision making using GIS. But the absolute locations of activities can refer also to location in time that is derived from a clock or calendar. In this context clock time acts as a type of grid, and activities and decisions are put into it. We may consider the position of one activity in time in relation to the position of another activity in time or relative to some other location in clock time.

Parkes and Thrift (1980) emphasize the importance of analyzing human activities in a spatiotemporal framework:

> The separation of two items in space may be described by the distance between them and the separation of two items in time by the interval between. When spatial metrics such as meters or kilometers are used to measure distance, we have a measure of absolute distance. If temporal metrics such as hours or days are used to measure interval, we have a measure of absolute interval. However when an a spatial metric is used to indicate distance and an a temporal metric is used to indicate time, then distance and interval are being represented in relative terms, as relative distance and relative time. One of the most common relative space measures combines space with time, and distance with interval. Thus in everyday life we consider the time it takes to get somewhere. This notion of distance and interval in combination is now frequently referred to a as a time–space metric. The geographer's space time is not a new physical structure, as is the four-dimensional space time of Minkowski or Einstein; instead it is a technical convenience and a more realistic way of looking at the world. (p. 4)

We may think of human activities occurring within a context of locational space coordinates, with distance being separated within a space metric and all of these occurring within a time period, thus giving an overall space–time metric. In this way activities and events may be located on a time–space map.

This chapter discusses the nature of human activities in this spatiotemporal context. It begins with a review of how human geographers have incorporated the time dimension into the study of human activities in space through the concept of the time–space prism developed during the 1960s and 1970s. The nature of individual action spaces and activity spaces, particularly within large-scale urban environments, is discussed, including the evolution of an activity approach and early attempts to develop models of activity systems. The chapter concludes with a discussion on the collection of data to generate individual and household time–space budgets.

8.2   Time Geography

An innovative and instructive approach to the study of time, space, and human activities is what has become known as the *Lund time geography approach*. It is the result of attempts by geographers at Lund University in Sweden to develop a model of society in which constraints on behavior (activity) may be formulated in the physical terms "location in space, areal extension and duration in time" (Hägerstrand, 1970:11). Although known for more than two decades within Sweden, it was not until 1970 and publication of Hägerstrand's seminal paper "What About People in Regional Science?" that this approach first became more widely known.

8.2.1   Constraints, Paths, and Projects

At the heart of Hägerstrand's time geography as initially formulated was the notion that all of the actions and events that sequentially make up an individual's existence have both temporal and spatial attributes. Consequently, the biography of a person is always on the move and can be depicted diagrammatically at daily, weekly, annual or life-long scales of observation as an unbroken, continuous path through time–space. In the time-geographic approach, the first step is always to try and define a *bounded region of time and space* and to treat everything in it as a time–space flow of "organisms and artifacts" (Hägerstrand, 1970). The framework thus began with the environmental structure that surrounds every individual. It attempted to capture the complexity of interaction at the scale of the smallest unit. It permitted the physical boundaries of space and time to be evaluated as they impinge on intended activity programs. There were no direct references to motivational factors involved in interaction. Why we do things was an important factor in the understanding of human activities; but it was important also to know what stops us from doing certain things.

The approach suggested that there are a relatively smaller number of primary factors in everyday life that impinge upon all individuals and constrain their freedom to occupy certain space and time locations. It was hypothesized that when these *constraints* are identified, it is possible to deduce reasons as to why a particular individual follows one path rather than another. For Hägerstrand the logical place to begin the study of human geography was with the individual. A *project* in time-geographic terms is a set of linked tasks that are undertaken somewhere at some time within a constraining environment. Any *bounded region* will contain a population made up of individuals (organisms and artifacts) that may be represented, in a geographic context, as point objects. These describe trajectories or *paths* through time and over space, from the point when and where an individual comes into being (birth) to the point where and when he or she ceases to be (death). For the purpose of time-geographic analysis, such paths are followed until they make a permanent exit from the bounded space–time region that is under study or until they are transformed into some other entity, as in the case of many materials.

We may study point object populations in two ways: (1) one way is from a *demographic* point of view, where the concern is only with numbers and descriptive characteristics, such as age and gender; (2) the other is *geographically*, where the concern is to develop explanations of the ways in which different individuals and populations coexist in the same time–space frame or region.

The Lund time-geographic approach may be seen as capturing the spatial and temporal sequence and coexistence of events by using a dynamic map to represent the path of an individual in motion over space and through time. This is illustrated in Figure 8.1. Each *path* shown represents a particular *trajectory* of one individual. When the line of the path is vertical, there is no movement over space. When the line is sloped, *velocity* is registered. The shallower the slope, the greater the speed or velocity of movement, with less time being consumed to cover a given distance. The

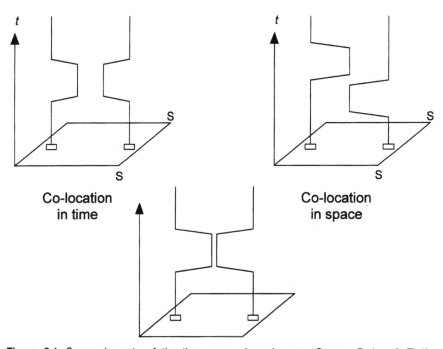

Figure 8.1. Some elements of the time–space dynamic map. *Source*: Parkes & Thrift, 1980:245.

opposite holds true when the slope of line is steep. If we record the movement of a number of individuals over a period, it is easy to see that we build a *web of interacting trajectories*. When this is repeated over the lifetime scale for each individual, we construct the geographic biography of the population within a "constrained environment" in which both space and time are limited in supply.

8.2.2 The Time–Space Prism

Time and space absolutely are inseparable from the intricacies of human behavior. Using time and space is contingent upon the constraints that operate on individuals. These constraints affect people's ability to influence their environment; thus, an individual's spatial reach is limited. This physical limit sets an accessibility field within the environment that constitutes a connected and continuous set of positions in space–time and has been called a *prism* (Lenntorp, 1976). The time–geographic approach may be described as physicalist, concrete observable realism of the location and movement of individuals, and not with individual experiences and intentions. However, this does not mean that there was no awareness of the significance of factors that underlie the conduct of human behavior. Motivations and intentions were seen as being elusive and, therefore, difficult to handle.

8.2.3 Physical Conditions

Hägerstrand (1975) summarized the physical conditions of existence that he considered necessary in the development of a time geography in the following terms:

1. There is an indivisibility of human beings and of many other objects. There are physical constraints or limits put on people by this environment.
2. There is a limited life-span of existence of all human and other physical entities.
3. There is a limited ability to participate in more than one task at a time.
4. All task are time demanding. A commitment to a particular task will diminish the finite time resources of the individual and ultimately, of the population.
5. Movement uses time.
6. Space has a limited capacity to accommodate events because no two physical objects can occupy the same place at the same time. Any physical object has a limit to its outer size and so limits the number of objects that can occupy a particular space. Thus, every space has a packing capacity, which will be defined by the types of objects to be packed into its area or volume.
7. Every physical object that has an existence will have a history or biography. Most nonhuman objects are sufficiently defined by their past alone. Humans have an ability to plan or permute the future and are sufficiently defined only by considering the past and the future.

Pred (1977) claimed that these time-geographic realities will always be true, regardless of any individual variations in the perception, conception, and measurement of space and time. They are facts of life, springing from the nature of physical being, and they are responsible for the local connectedness of existence with which time geography is concerned. In this way the web formed by the connectedness of individual trajectories, or *life paths*, is the outcome of collateral processes within bounded regions; or as described by Hägerstrand (1976:332): " . . . processes which cannot unfold freely as in a laboratory but have to accommodate themselves under the pressures and opportunities which follow from their common existence in terrestrial space and time."

8.2.4   The Nature of Constraints: Projects and Paths

The Lund time-geographic approach was based on the premise that individuals have goals. To attain goals, *projects* must be formulated. Projects are composed of a series of tasks and act as the vehicle for goal achievement. Thus, projects will include people, resources, space, and time. To complete a project, one must overcome the constraints that exist in the environment. The pursuit of projects involves events and actions that are incorporated into an individual's path. We have listed above the factors that are likely to limit these events and actions. The action and event sequences, or the behavioral choices, that accumulate along an individual's path on a day-to-day basis also may be thought of as being constrained by three major factors. These are:

1. *Capability constraints*, which circumscribe activity participation by demanding the allocation of large portions of time to physiological necessities such as sleeping, eating, and personal care. These constraints also limit the distance an individual can cover within a given time-span using the transport technology at his or her command.

2. *Coupling constraints*, which influence where, when, and for how long the individual must join other individuals or objects in order to form production, consumption, social, and miscellaneous activity bundles. Coupling constraints largely determine the pattern of the paths that occur within an individual's daily *prism*.

3. *Authority constraints*, which limit access to either space locations or time locations. They subsume those general rules, laws, economic barriers, and power relationships that determine who does nor does not have access to specific domains at specific times for either purposes. All environments are replete with control areas or domains of authority, the purpose for which seems to be to protect resources, according to Hägerstrand (1970:9). There is a hierarchy of domains of authority. These range from near absolute, regardless of individual attributes, to subordinate domains, which can be entered given social power of one kind or another.

According to Pred (1981):

> The composition of an individual's path will be circumscribed by the number and mix of specialized, independently existing, roles proffered up by the organizations and institutions found within any given bounded area (or by the activity system of a given area), and by the rules, competency requirements, and economic, class and other con-

straints that govern entrance to those roles. (The specialized roles embedded within organizations and institutions exist independently—until terminated or superseded—in the sense that when they are not filled by one person they sooner or later must be filled by another person.)

The paths of individuals may be regarded as dynamic maps consisting of *path space–time bundles* (Parkes & Thrift, 1980:250). Paths have different life-spans or durations. They meet at different *stations* as a result of different *tasks* and *projects*. Stations have different physical extent in time and space and are represented as tubes of varying size, according to the length of time for which they are in operation. The length of time a station is open may result from an authority constraint, such as a licensing law.

Pred (1981:236) has pointed out that the path perspective, with its emphasis on behavioral restrictions, can lead easily to the mistaken conclusion that the Lund time-geographic approach was merely a form of constraint analysis that is best suited for exercises in social engineering. This is a mistaken conclusion, but it is a reason why the time-geographic framework has not gained more widespread acceptance.

A *project* needs to be seen as consisting of "the entire series of tasks necessary to the completion of any goal-oriented behavior" (Pred, 1981:236). On the one hand, a project can be something as mundane as writing a letter. On the other hand, it can involve a chain of tasks as complicated as that necessary for organizing and carrying out production of goods and services. The tasks associated with either a simple or a complex project will have an internal logic of their own which requires that they be sequenced in a more or less specific order. We may view the logically sequenced component tasks in a project as synonymous with the formation of activity bundles, or with the convergence in time and space of the paths that are traced out either by two or more humans. Thus, "when an individual's *path* becomes wrapped up with the activity bundle(s) of a *project* defined by an organization or by some institution other than the family, there is a direct intersection between the external actions of the individual and the observable workings of society" (Pred, 1981: 236).

It does not matter whether a project has a one-time or a repetitive character, because the paths that must be synchronized will not normally be identical for every constituent activity bundle. Some paths necessarily will be tied up with all or most of the project's activity bundles, whereas other parts will have only an abbreviated association with the project. In order to for more complex projects to be completed, it may be required that several subprojects are carried out temporarily parallel to one another. For complex projects, it may also be necessary to develop subordinate and anticipated offshoot projects with activity-bundle sequences of their own.

Participation in a project places demands on the limited daily time resources of the individual. This is so because the daily time resources of individuals, groups, and entire populations are limited to 24 hours times their number. As projects must be carried out at one or more locations in space, any time–space regions will have a bounded capacity for accommodating the activity bundles that are generated by a project. During the period of time when the paths of people are committed to one activity bundle, they will not be able to simultaneously join an activity bundle belonging to another project. In this way, participating in a project will render individuals temporarily inaccessible to other individuals who might wish to join them for the purpose of doing some other thing at the same place, or at other locations. Thus, these individuals will opt to form their own project-related activity bundles at another time, and in turn they become inaccessible to other persons at that new time, possibly pushing aside or completely eliminating any chance that yet other project-related activity bundles can be developed. In this way, we can identify what is known as a spread effect in conjunction with carrying out projects, and a local connectedness is created between otherwise seemingly unrelated activities, events, and processes. We may view the whole operation of society as consisting of projects connected with economic production, consumption, and social interaction, plus acts of contemplation that are carried out in personal isolation. All these act in competition with each other as events and states for exclusive spaces and times.

8.2.5   Scales of Analysis

It is possible to discern three distinct scales of analysis in the time-geographic approach. These focus upon the individual, the station, and the population or activity system.

1. *The individual scale.* At this level the simple geographic device of the dynamic map can reveal the joint interplay of *path, location,* and *constraint.* It illustrates the fact that the choice of one task by an individual implies that there will be less time available to devote to alternative tasks. This choice may lead to a blocking effect, constraining other individuals from interacting or coupling with that individual. This can lead to a displacement (a knock-on effect) of those blocked activities into activities, and possibly lead to the formation of a queue. The family may be taken as an example of a unit consisting of individuals among whom there is a high degree of interrelatedness. Decisions on the part of one member of the family can have substantial effects on the others.

2. *The station scale.* At this level, which may include segments of the *paths* of individuals, the method of time geography involves mapping out the reach of each individual. In other words, we look at all the events in which an individual can participate in one way or another. These possible events, which must occur at stations, are specified by means of a *prism* or *activity* area, which is the volume of space–time in which it is possible to carry out a particular activity, or set of activities, at a given point at which the activity must start and finish. The *prism* is the *potential path area* (PPA), and its shape is described by the individual's speed of travel, and whether the station of origin is the same as the station of destination.

3. *The population or activity system scale.* At this level the time-geographic approach is built around the concepts of time supply and time demand. We need to distinguish between the population system and the activity system (see Figure 8.2). The population system consists of all individuals in a *bounded region of time–space.* A particular population's characteristics, such as its distribution of ages and skills, determines its dynamics. In addition, there are *capability constraints.* The total daily

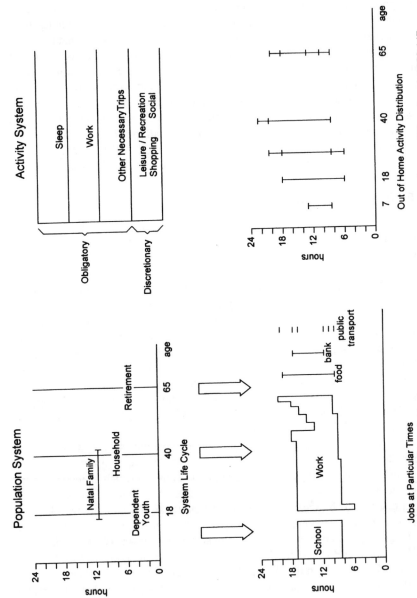

Figure 8.2. Mutual adjustments between population system and activity system. *Source:* Hägerstrand, 1972:147.

supply of time for the population system is 24 hours multiplied by the number of people. Time demand is the interplay between the population and activity system and is represented in the multitude of human projects that are conducted within various groups and organizations as part of the institutionalized activity system.

8.2.6 Applying the Space–Time Prism Concept within GIS

A study by Miller (1991) sought to apply the *space–time prism* concept developed by Hägerstrand (1970) and Pred (1977) but within a GIS, building on various attempts to apply the concept in spatial analysis and planning. Lenntorp's (1976, 1978) simulation analysis using data on the transportation characteristics of an area, the location and operating process of various urban functions and services, and a hypothetical activity schedule for households, has been used in locational analysis and transportation planning, but the costs incurred in calculating space–time prisms for a specific study area proved to be prohibitive. Although the prism does not require a large amount of data, it does require data at a detailed level of spatial resolution. In addition, there have been difficulties in manipulating and representing these data. However, the evolution of GIS technology has provided the potential to overcome these difficulties in operationalizing the space–time prism concept.

The space–time prism is an extension of the space–time path, but it does not trace the observed movements of an individual over an interval of time. What the prism shows are the proportions of space that are possible for an individual to be in at specific times. "The prism *potential path space* (PPS) delimits locations from which the probability of being included in an individual's space–time path is greater than zero" (Lenntorp, 1976:13). The PPS is a three-dimensional entity existing in a region of bounded space–time. The PPS is determined by an individual's time budget, or the time available for travel and participation in activities, the potential travel velocities in the environment, any time spent at unspecified locations to participate in an activity, and spatial constraints such as the need to be at a certain location at the beginning or at the end of a time period. But although the PPS is a revealing construct, "of more direct concern in travel modeling, locational analysis, transportation or other applications is the potential path area (PPA)" (Miller, 1991:291). The PPA is the projection of the three-dimensional PPS on a two-dimensional planar space, and it delimits the purely spatial extent or area within which the individual can travel. The PPA can be calculated directly without reference to the PPS, with any stationary stop time for activity participation netted from the overall time budget to reflect the reduced amount of time available for spatial movement (Lenntorp, 1976).

Miller (1991) proposed operationalizing the space–time prism in a GIS in the following manner:

1. The PPA is defined as continuous, two-dimensional space, but in reality a large proportion of the real space of an urban area is useless for travel or activity participation, because travel can only occur along paths (streets, public transport routes), and activities can only take place at specific locations (e.g., shopping can only occur where stores are located). Thus, the planar form of the PPA can be discarded in favor of more useful and easily calculated discrete spaces.

2. Within an operational GIS, two discrete space versions of the PPA can be derived. A PPA can be defined by locations at which activities occur in the study area, with this subset of spaces being distinguished as places feasible for activity participation by an individual—these represent an opportunity set of locations that are choices for participating in a particular activity. Travel in the environment between any two defined locations (such as a travel origin and an activity destination) can be calculated at a constant rate of space along a final line segment(s) connecting the pair. Thus, the PPA is defined by arcs (transport routes) that are feasible to travel between nodes (where activities are located).

3. It is possible to include variations in the possible travel velocity between different links in a transportation network due to variations in traffic congestion and differences in legal speed limits. Also, multiple paths can connect any two locations in a transportation network.

4. Across the urban transportation network, actual velocities attained during travel will vary during the day and over the week because of differing levels of congestion, particularly related to rush hours. Also, there may be different velocities of travel for each direction of travel through an area, related to time variations in congestion. Thus, travel times across the network fluctuate in accordance with time and direction of the trip.

Table 8.1 sets out the data requirements for PPAs based on a GIS. Miller (1991) proposes that:

> Using a GIS to derive space–time prism constructs such as the network based PPA can model individual accessibility within the format of a detailed and sophisticated representation of the total environment. A GIS can allow accessibility patterns within the

TABLE 8.1.
Data Requirements for Potential Path Areas Based on Geographic Information Systems

	Data characteristics	
Data requirement	Punctiform version	Network version
Travel origin/end locations	Point objects in planar space	Nearest nodes in a defined transportation network structure to the actual locations
Activity locations	Point objects in planar space	Nearest nodes in a defined transportation network structure to the actual locations
Travel environment	Direct paths between travel origin/end locations and activity locations. Each path characterized by a representative travel distance and a set of temporally dependent travel velocities	Arcs and nodes corresponding to linkages (streets, intersections) in the urban transportation network. Each arc characterized by the length of the linkage and a set of temporally dependent velocities. Each node characterized by a set of temporally dependent turn times

Source: Miller, 1991:295.

transportation network to be visualized. This alone provides a powerful perspective for the spatial analyst. The concept of accessibility is central to many theories and models of spatial behavior and location, and the ability to see this concept operating within its environmental context adds a valuable dimension to the intuition behind the construction of theories and models. (pp. 292–293)

The space–time prism in a GIS has direct applications in areas such as travel modeling, transport planning, and facility location such as retail stores. Individual accessibility can be modeled, and the GIS can combine individual constraints into aggregate measures of accessibility in a city. The identification of a choice set of activity locations enables modeling of spatial choice and travel behavior, including the trade-off between time spent undertaking activities and time spent traveling. The network PPA can be used to:

1. Derive a network based on accessibility that can evaluate the performance of a city's transportation infrastructure.
2. Identify variations in accessibility across the city.
3. Help inform policies and planning decisions to improve a network's efficiency and to alleviate spatial inequalities in the accessibility of specific disadvantaged groups to urban services by proposing optimal locations for new facilities.

Thus, Miller's (1991) research has provided a generic procedure for building network-based PPAs, utilizing the ARC/INFO NETWORK software package based on the space–time prism construct utilizing GIS technology.

8.3   Action Spaces and Activity Spaces

During the 1960s and 1970s behavioral geographers in the United States investigating the spatial behavior of individuals in large-scale urban environmental settings focused largely on the *perceived utility* of places in the operational milieu and on the delimitation of those places and spaces with which the individual was familiar and with which he or she interacted.

The concept of *action space* was developed to describe an individual's total interaction with and response to, his or her environment. It was hypothesized that people gather information about their environment and that they ascribe subjective values, or utilities, to various locations. In this way places are given a *place utility*. Wolpert (1965) denied that it was the nature and spatial extent of all our place utilities that comprised our action spaces. Thus, in a broad context, action space provides a framework within which individual or group spatial interaction can be viewed. Jakle et al. (1976) wrote that the action space: "specifically draws attention to the individual's relationships with his surrounding social and spatial environment and allows us to examine the patterns in which individuals interact in space. We can most effectively use the concept by dividing it into meaningful components—movement and communications" (p. 94). The components of *action spaces* and their nature require specific attention.

8.3.1 Individual Action Spaces

Although theoretically the individual has access to a broad range of environmental information, usually only a limited portion of the environment is relevant to his or her spatial behavior in any given context. Horton and Reynolds (1969:70–71) suggest that, even though an individual's *action space* is spatially limited, an examination of its formation led to the consideration of a wide range of spatial behaviors, such as the journey to work, school, shipping, visiting friends, and so on. One cannot ignore the individual's perception of the objective spatial structure of physical, economic, and social environments within which this behavior takes place. Because no two individuals perceive a given environment, such as a city, from exactly the same point simultaneously, and because each person bases his or her own interpretation of information obtained for the environment on past experience, action spaces vary from person to person. However, whereas perceptions and action spaces are, to a large degree, individualistic, there is reason to suggest that the formation of an individual's action space is affected by that person's group memberships, his or her position in social networks, his or her stage in the life cycle, and his or her spatial location relative to potential trip destinations. Generally, a person's residence is a primary node in that individual's action space. Although the perception of the urban environment was identified as an important determinant of an individual's action space, of equal significance was the perception of time and time preferences. In this context, *time preference* refers to an individual's weighting of time allocations for various types of activities. According to Horton and Reynolds (1969) it was: "appropriate to adopt a behavioral approach which examines the formation of the individual's action space as a function of his socio-economic characteristics, of his cognitive images of the urban environment and his preferences for travel" (pp. 70–71).

The individual's perception of his or her environment is not static, but changes via complex learning processes, as discussed in Chapters 5 and 7, Sections 5.4, 5.7, 5.8, and 7.2. Given socioeconomic constraints, and provided that an individual does not change his or her place of residence, we can expect individuals to modify their perception of the environment by moving within it and by communicating about it with their peers. Over time, an individual's behavior approaches a spatial equilibrium. However, this presupposes that the objective urban environment itself does not change; but in fact this is far from the case. The components of the built environment—such as its retail structure, the location of employment opportunities, and the quality of its residential areas—are constantly undergoing change. This results in the continuous reordering of perceived structures of urban space. In addition, technological change plays an important role in extending an individual's action space and in modifying its morphology. For example, the increased use of the private car for transportation has made it possible for an individual to extend the perimeter of his or her action space. At the same time the increased use of private transportation, particularly for the journey to work, has led to spatial distortions and place disutilities within the action space by yielding penalties in the form of congestion and its concomitant psychological and physiological effects.

An individual's time preferences also are likely to be important in affecting the morphology of his or her action space. It is well known that time preferences vary systematically throughout the life cycle of an individual. Other factors, such as income, education, occupation, and social status also can have direct and indirect influences. In addition, the location and spatial structuring of urban activities are important in understanding action spaces.

Thus, the formation of an individual's action space was formulated by Horton and Reynolds in the model reproduced as Figure 8.3. This represents one of several ways in which household action spaces can be examined in order to isolate factors that are important to their formation, spatial structure, and change.

8.3.2   Components of Action Spaces

One important part of action space is the collective movements of individuals. The movement component of an action space may be termed an *activity space*, which is defined as the subset of all locations within which an individual has direct contact as a result of his or her day-to-day activities. A second part of action space is defined as *communicating over space,* using interpersonal communication channels, such as the telephone, newspapers, magazines, radio, and television. Thus, activity spaces represent direct contact between individuals and their social and physical environments, whereas communication channels are indirect links with their environment.

The *activity space* for a typical individual will be dominated by three things: (1) movement within and near the home; (2) movement to and from regular activity locations, such as journeys to work, to shop, to socialize, and so on; and (3) movement in and around the locations where those activities occur. Furthermore, some individuals undertake extensive travel across broad areas of their city, its region, the nation, or even the world, as part of their work and leisure activities. As Jakle et al. (1976:93) have stated: "Activity spaces are an important manifestation of our everyday lives, and, in addition, represent an important process through which we gain information about and attach meaning to our environment."

Direct contact has an important influence on the way in which people define territories, both at the individual level and at the group level. Perceived as well as formally defined territories act to constrain and direct an individual's trips through space. People living in cities often avoid particular places in their spatial behaviors

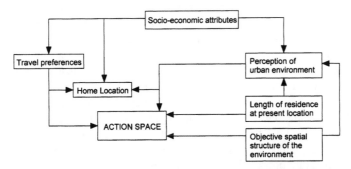

Figure 8.3. A conceptual model of action space. *Source*: Horton & Reynolds, 1969:73.

because of perceived hazards, such as areas with high crime rates. Formal boundaries, including international political boundaries and other administrative boundaries, also may act as constraints to movement. In this way, two important geographic concepts, *identification of territory* and *movement in the environment*, are closely linked.

Activity spaces also are closely linked to an individual's role within society. Both the temporal and the spatial aspects of an individual's activity space are the product of his or her defining a set of activities in which he or she desires to participate. The real-world behavior patterns of individuals, as defined by their activity spaces, thus may be seen in the context of an *activity system*. Through their experiences over time, people develop territorial familiarity that varies in degree according to the locations they interact with in space and the extent of cognitive reinforcement attached to those places. This occurs within their overall action space.

8.3.3 Temporal and Spatial Aspects of Activity Space

It is necessary to identify the *temporal* as well as the *spatial* aspects of activity space. Concerning the temporal aspects, movement to specific activity locations is related to the frequency and regularity with which an individual chooses to participate in a specified activity. The possible forms that such regularities might take are illustrated in Figure 8.4. Regularly scheduled activities, such as going to work, attending a club meeting, and going to church, have high probabilities of a trip being made at specific times and low probabilities of a trip being made during the intervening times (Figure 8.4a). Trips to purchase needed items that are consumed regularly have a probability distribution showing a gradual increase through time until the trip is made, after which the probability of going again is reduced to near zero. These types of trips, such as shopping trips, tend to be evenly spaced over time (Figure 8.4b). Trips to undertake time-contagious activities, such as playing a sport, show an increase in the probability of participation in the activity again soon with a probability that gradually decreases over time (Figure 8.4c). This type of activity and its associated trips tend to be clustered in time. Finally, Figure 8.4d illustrates trips to activities that occur randomly in time, such as trips in response to emergencies or accidents. Participation in this type of activity does not affect the probability of participating again.

In addition, we may view the temporal nature of activity spaces with respect to the overall mix of activities in which people participate. For example, early work done by Chapin (1965) took a planner's perspective in which activity spaces were viewed in terms of the time durations an individual, family, or household spends in different kinds of places. Such an approach is illustrated in Table 8.2. A simple classification for time budgeting is given in the left-hand column. With each type of activity, we can associate a trip type. At both the individual and household level, the progression of people and households through the life cycle means that particular types of activities and the associated trip types may occur and then disappear at key points. Thus, life-cycle stage has a direct impact upon the nature of activity spaces and the spatial nature of trips that are undertaken.

Examining the spatial aspects of activity spaces involves both the location of

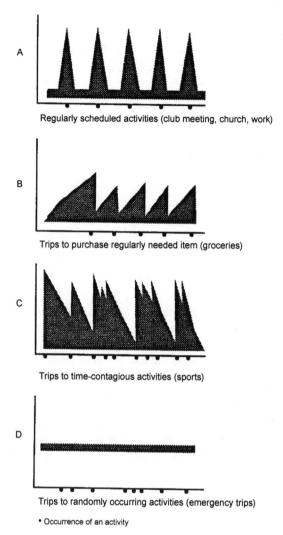

Figure 8.4. Relationships between time and the probabilities of trips to different activity locations. *Source*: Campbell, 1980.

activities (particularly relative to important nodes, such as the location of work) and distance decay. For planning purposes, it is important to map the pattern of trips and to record their timing, especially in forecasting transportation needs and helping to make decisions about facility locations. Certain spatial regularities have been noted in activity spaces. The simplest and most universal is that of *distance decay*, which is an aggregate concept that indicates a tendency for people to take trips most frequently to places nearby, with trips becoming less frequent as distances from the origin of the trip increase. At the individual level, however, the form of movement with respect to distance is typically much more complex, it being related, for exam-

TABLE 8.2.
Classification of Activity and Trip Types

Types of Activities	Types of Trips
Income producing (may only apply to household heads and other adults)	Journey to work Traveling salesmen
School (applies only to children)	Educational trips
Buying food and other goods	Grocery shopping Clothes shopping
Socializing with friends, relatives, etc.	Informal social trips
Community and club activities	Formal social trips Leisure activities
Outdoor recreation, entertainment	Recreational trips Tourist trips
Church, political activities	Cultural trips Participate in planning

Source: Modified from Chapin, 1965.

ple, to the specific location of a person's origin (such as home) and to the location of the other activities. Thus, significant variations among individuals often are apparent.

Linkages between the temporal and the spatial dimensions of activity spaces have been noted by many researchers. Activities with the greatest frequency of participation are generally located close to the home. An inverse relationship is found between the distance traveled to an activity and the frequency of participation in the activity, this being due in part to the greater time or monetary cost involved in making longer trips. Frequency of participation in an activity is related also to the spatial distribution of opportunities surrounding an individual's home. An individual is constrained by living where he or she does and becomes the victim of spatial inequalities in the distribution of locations where specific activities are available across the space of a city. The individual either can choose to incur the time and cost of travel to a place where the activity is available or can choose not to undertake that activity. As a result, some people have better *access opportunity* to activities than is the case for other people. It is useful to view an activity space map in two dimensions, thus introducing the idea of *directional bias*, which often is related to a partiality for a particular place over other equally distant places because of some perceived quality of the preferred place. A critical factor in any bias is the *distribution of opportunities*, but certain regularities can be seen in similar environments. For example, in most urban areas, similar distributions of activity locations exist. Traditionally in cities there is a dominant center, usually the CBD. Around and away from that center there tends to be decreasing population density with increasing distance and an associated decline in the number of activity locations. This simple pattern may be modified by important radial transport routes, secondary nodes of activity, and the particular physical characteristics in a given city. Nonetheless, urban areas tend to have common spatial structural characteristics. Because of this commonality in regularity, the activity spaces of people living in cities tend to have a common directional bias, usually in the form of a sector or wedge shape. People tend to move around in, and be familiar with, that part of the city that is on the same side of the city's center as their home location. It has been noted in many empirical studies that reasons for such biases include a resistance to travel through or around the center of large cities, the tendency for individuals to select activity locations nearest them when they are duplicated elsewhere, and the radial nature of transportation routes (see, for example, Adams, 1969; Zannaras, 1973). However, this bias may not hold for individuals living near the center of cities.

8.4 Activity Spaces, Trips, and Scale Differences

We may conceptualize *individual activity spaces* as being made up of a *hierarchy of movements*, as illustrated in Figure 8.5. At the top of the hierarchy is an individual's place of residence, which is a focal point in the network of his or her movements. Home is the place from which and to which the greatest number of trips are made. Focusing on the home location are short movements, which are often pedestrian, as well as the longer trips to and from activity locations using a variety of modes of transport. Other activity locations act as places around which additional pedestrian movements occur.

Short-distance trips involving pedestrian movements focus on and around the

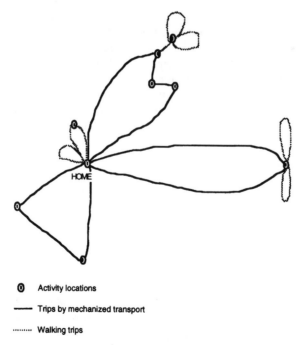

⓪   Activity locations

——   Trips by mechanized transport

········   Walking trips

Figure 8.5. Individual activity space: a hierarchy of movements. *Source*: Jakle et al., 1976:101.

home and other activity locations. Many built environments are planned and arranged in anticipation of certain types of walking behavior. For example, in large department stores, high-value-per-volume items, which are subject to impulse buying (such as women's jewelry), are placed on the ground floor near the main entrance, where the greatest pedestrian traffic flows occur. High-bulk items, which are more subject to purposeful buying, such as furniture and large appliances, are placed on upper floors where less traffic is found. In supermarkets, items with high profit margins often are placed in locations that would be encountered early in a shopper's trip through the store. Within the shopping center as a whole, walking behavior is an important consideration in design. For example, a typical large shopping center will place its two or three major department stores at polar locations with smaller specialty and comparison shops being strung out along the paths between them (see Chapter 10, Sections 10.5 and 10.6, for an elaboration of this point).

Much of the research on specific types of trips within activity spaces is focused on the journey to work. This is a segmented approach focusing on trip type. Trips for different purposes, such as work trips, social trips, or shopping trips, tend to exhibit contrasting time distributions. Work trips tend to peak at the traditional morning and evening rush hour times, whereas social trips tend to peak just after noon, in the evening, and at weekends.

We may distinguish further between trips made on the basis of the consistency of the activity location. Work trips usually go from home to a specific place and then back (sometimes with intermediate stops) on a regular basis, and usually they occur on five days per week. In contrast, social trips may have a regular component, such as the weekly participation in formal social organizations; or they may have an irregular component, such as visiting friends and going to parties. In the United States, an early study in Buffalo, New York showed that people make only one stop in two-thirds to three-quarters of all their trips. In large metropolitan cities, between 75% and 90% of all trips either began or ended at home, the remainder going between other activity locations. Thus, between 20% and 25% of trips were from one out-of-home activity to another. Apart from the home-based trips, approximately 40 other types of out-of-home activities were identified (Hemmens, 1970).

8.4.1 Work Trips

Work trips are viewed by planners as being the most important trips. Individual decisions determining the nature of the journey to work often are complicated by the decision of where to live. Thus, two separate processes are important in the journey to work; a residential location decision and a job location decision. These two processes in combination suggest two extreme situations in which individuals find themselves selecting a journey-to-work route. The first is one in which a householder does not want to move the residence, but a member of the household seeks a job location. An individual is likely to define an outer limit beyond which he or she will not be prepared to commute and an inner range within which the location of a job makes no difference. These limits are perceived distances, and they vary in accordance with individual travel preferences, the access of an individual to various modes of transportation, and the structure of the transport system available in the city. In between these two limits, distance may be considered as a trade-off with the attributes of the job. At the other extreme, an individual may acquire a job in a new city or state, and the household's selection of a residence determines the nature of the journey to work. A similar set of limits, this time around the new work place, may be set up within which distance is traded-off with other factors, such as the cost and size of the home, the type of neighborhood, and other site factors. As an additional complicating factor, the home location decision is often based on accessibility of other locations, such as shopping centers, schools, recreation areas and friends, in addition to the location of the person's job.

Thus, the trade-off between the location of home and job is not a simple one. In between these extremes many complicated situations occur. It may be that the location of a person's place of employment is moved by the employer to a place that is still within the commuting range of the person's home but that this results in a significantly increased commuting distance. This means the activity space now includes one very long daily trip. After tolerating such a stress for a time, a decision may be made by the household and the individual to reduce it either by moving the residential location closer to the work place, or by looking for a new job closer to the residential location.

8.4.2 Social Trips

Social trips contrast with work trips in a number of ways. In the United States a study by Wheeler and Stutz (1971) found that there was a strong relationship between social trips as a percentage of all trips and the family income. Also, household size affects the number and percentage of social trips, with larger households having a lower proportion of social trips. About 60% of social trips had a residential origin; 20% had an origin at a previous social activity; 10% had a work or business origin; and 5% had a shopping or recreation origin. It was evident also that different types of social trips exhibited different *distance decay* characteristics, with the highest distance decay being for trips to neighbors, with decreasing distance decay for social trips to friends, and with very little (if any) distance decay for social trips to relatives.

However, distance decay is only part of the explanation of the spatial characteristics of social trips. Another important concept is that of *social distance*. There is a tendency for social trips to occur between people, and hence neighborhoods, of a similar socioeconomic status. As individuals of similar status tend to live in proximate locations within cities this probably explains some of the *distance bias* that is seen in social trips. It explains also the numerous exceptions to the distance decay notion.

8.4.3 Trips to Other Activity Locations

Trips to other activity locations are many and varied. They may be taken either on a regular basis (such as trips for shopping, recreation, and participation in various cultural activities) or on an irregular basis. Such trips could be regarded as being scaled on a continuum from economically rational at one end to economically nonrational at the other. Shopping for groceries may reflect a genuine attempt to minimize costs (price of groceries plus transportation costs). In such cases, the distance traveled is not always the shortest, but it is minimized cost-wise in conjunction with the costs of items purchased. At the other end of the continuum are those trips that minimize neither cost nor distance but instead are made for aesthetic or hedonistic purposes. These include recreational trips, where the trip itself might be part of the recreation, such as a family going on an outing in the car that includes visiting a number of historic sites on a recreation trip.

8.4.4   Substitution Effect

Different trip types can be contrasted in terms of what is known as the *substitution effect*. In this case, one activity location, and its associated trip, is substituted for another; or an in-home activity is-substituted for a trip. Different activities are spaced in the environment in different densities. High density activity may be subject to more substitution because of a wide range of choices; but activities with only one or two locations in a city may not allow substitution. Thus, different activities and activity locations have varying degrees of flexibility in terms of the choices that people have to make.

8.4.5   Multipurpose Trips

Not all trips are for a single purpose. Some may focus on one location for several activities, whereas others may lead to several different locations for one or more purposes. A good example of the former are trips taken to major shopping centers and downtown areas for a variety of shopping purposes. Sometimes they combine social, recreational, business, and other activities. It is interesting to note that the demand for *multipurpose shopping trips* led to the development of higher order regional shopping centers providing a range of goods and services in one location. There has developed also a marked tendency for shoppers to bypass the nearest commercial center and to shop at more distant ones that offer a greater range of choices and functions. An early study by Looman (1969) in the city of Newark, Ohio, showed that 42% of all shopping trips were multipurpose trips. Of the remaining 58% that were single-purpose trips, 16% of them involved multistops (Recker et al., 1982; Pas, 1982).

The second type of multipurpose trip is one that includes numerous stops at separate locations. In a classic study of the city of Lansing, Michigan, Wheeler (1972) collected data on multipurpose trips that showed the correspondence of activity types and trip destinations with activities at trip origins (excluding going to and from home.) The data were analyzed from a matrix that showed the probability of particular types of trips being linked. Wheeler was able to identify the major linkages among various types of trips in multipurpose trip behavior as illustrated in Figure 8.6. He found that:

1. Many trip types were followed by trips to social and/or meal-eating destinations. These activities were quite important as part of multipurpose trips and were linked with a variety of other activities.

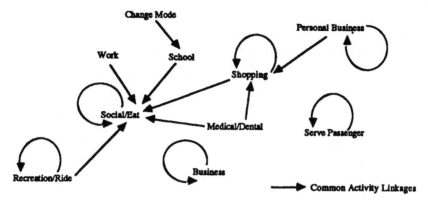

Figure 8.6. Activity types that are linked together by a probability of at least 20% in Lansing, Michigan. *Source*: Wheeler, 1972–645.

2. Certain activities often were linked to other activities of the same type. This was particularly so for personal business, social, and recreational trips, trips to serve a passenger, shopping, and business trips.
3. Work and medical/dental trips were rarely followed by a trip to a similar activity.

Intuitively these results would been expected.

Later research by Hanson (1980:247) sheds light on the reasons why an individual regularly may visit several establishments within one functional class in undertaking multipurpose trips. These are:

1. The individual's drive to reduce uncertainty by learning about available options.
2. The traveler's desire to spread risk by developing a portfolio of regularly visited destinations.
3. The variability in the spatial, temporal, and modal constraints found by the individual.
4. The individual's desire to reduce boredom by adding some variety to the travel pattern.
5. Differentiation of stores, or activity sites, within a given functional class means that different stores may meet different needs at different times.

For employed workers, the workplace is second in importance only to the home. Thus, within any given functional class the establishments chosen in connection with a journey to work may differ from those having the greatest utility when they are visited directly from home.

In a 1971 study of household travel conducted in Uppsala, Sweden, Hanson (1980) showed how individuals spread their stops for a particular purpose over several locations. She looked at four different land types to illustrate behavior for distinctly different purposes: convenience (food store) shopping; shopping for nonconvenience goods (clothes, etc.); conducting personal business (post office); and socializing/visiting (another person's home). She found that stops at food stores and other people's homes were made more frequently than were stops at clothing stores or at the post office. There was considerable interpersonal variability in the frequency with which a given land use type was visited, in the number of different (unique) locations visited per person over the 35-day study period, and in intrapersonal variation in the spatial diversification of destination. It was found that under 30% of food shoppers confined their shopping to one store, while 60% of post office users went to a single location. Thus, the purpose of the travel affected the number of destinations used. Furthermore, the study showed that for all land use types an average of one-half of the individual stops were made at the one place that was most frequently visited, this being 68% for food stops.

Hanson concluded that the results of the Uppsala study showed that although not everyone spread stops for a particular purpose among several different outlets, for three out of the four land use types studied (post office being the exception), more people spatially diversified by visiting more than one destination of a given type than did not; those who combined their visits to one location of a particular land use type were in the minority; thus it was "reasonable to assert that spatial diversification is an important component of spatial choice" (Hanson, 1980:251). She showed that there were groups, or bundles, of functions that were used frequently in combination on the same multipurpose trip. Principal components analysis was employed to show that there was a group of land uses that were linked to the home and to the work place, whereas other groups were linked most strongly to other urban functions. Even though the groups of land uses were not discrete sets, it was clear that bundles of functions or purposes frequently were combined on the same trip and that the composition of the bundle that contained a certain purpose might vary from trip to trip.

Finally, Hanson (1980) showed that there was little overlap between the set of destinations that individuals visited only on single-purpose trips and the set of places visited only on multipurpose trips. When all purpose travel was considered, the average number of different places visited per person on a single-purpose trip was 12; whereas the average number for locations contacted on a multipurpose trip was 24. However, an average of only 4.7 places per person were included in both sets. It was found also that more locations overall and within a given functional class (food stores) were contacted during multipurpose trips than on single-purpose trips, thus demonstrating the importance of the multipurpose trip in generating spatial diversification in destination choice. The particular bundle of purposes to be combined on a multipurpose trip is, therefore, important in deciding on the establishment chosen as a destination.

8.4.6 Summary of Factors Influencing Daily Activity Patterns

A number of general conclusions may be drawn from the foregoing discussion of the nature of human activities in time and space within the context of individual *action spaces* and *activity spaces*.

There appear to be three major components of daily activity patterns. These are: (1) the time of an activity; (2) the space over which the activity takes place; and (3) the type of activity. All of these are highly interconnected.

Time is taken into consideration in two ways: first, in terms of the *duration* of each activity; and second, in terms of the *time of occurrence* of the activity. The duration of each activity is a basic ingredient in the account of a day's activities. But not enough is known about this aspect of activity patterns. The time of day when activities occur also is important and may be critical in determining whether or not an activity will take place together with others that are needed during the daytime.

The *distribution of facilities in space* for particular activities appears to be of prime importance in determining whether or not a given activity will be performed and, perhaps more so, in determining the frequency with which an activity will be performed. We may hypothesize that activities will have a longer duration when access is close and easy, rather than when it is distant and/or difficult. Further, the distribution of potential places of activity is viewed by the individual from a certain perspective—that is, from where he or she is located. This means there is a need to relate activity patterns to the overall cognitive map that people have of a particular urban environment and, in particular, to identify orientation nodes. It is well known that an individual's view of opportunities available for interaction changes as she or he moves about within a city. The selection of any activity can be explained in terms of motivations, needs, wants, and the capabilities of individuals. The whole range of socioeconomic characteristics of individuals and household units thus become important factors in any attempt to explain the structure of various activity patterns. It is important also to consider the fact that activities are a function of preferences, tastes, information, habits, and financial circumstances.

8.5   An Activity Approach to Behavior in Space and Time

The time-geographic and the action–space and activity–space approaches to the study of human spatial behavior in a spatiotemporal context of large-scale environments have been integrated and extended in what is referred to as an *activity approach* to the analysis of individual and household activities and travel behavior. King and Golledge (1978) noted that an activity approach to urban analysis provides insights into the functioning of urban areas and to their spatial structure. Analysts of urban spatial behavior view the city as a collection of individual activities, actions, reactions, and interactions. They described urban places in terms of what is going on instead of in terms of quantities of land use of various types. In adopting this approach, the rationale is that if we know: (1) how people actually use an urban area, (2) how they respond in choice situations, (3) how they sequence their activities and the duration of their activities, and (4) the relationship of each of these things to changes in their own circumstances, then we will be in a better position to evaluate policies designed to change urban environments. And we will be in a better position to describe the city as it is used by the people living in it. Human activity analysis, guided by theory and time–budget records, offered the promise of supplying some conceptual guidelines for relating behavior patterns to the spatial organization of the city. As Cullen (1978) points out:

To understand human spatial behavior, rather than just monitor it, we must treat time explicitly as the path which orders events as a sequence, which separates cause from effect, which synchronizes and integrates . . . (in fact) the lack of an explicit treatment of time in behavioral studies, apart from the odd desultory venture into the field of predictive analysis. has been truly remarkable. (pp. 28–31)

8.5.1 Activities as Routines

Basic to the activity approach is the notion that activities are discrete episodes occurring at uniform intervals in the life of an individual or household. Their motivation is derived from a set of prior values. An activity occurs, produced by a choice mechanism, that has a spatial manifestation. Thus, activities are viewed in terms of routines. A *routine* is a recurring set of episodes in a given unit of time. A recurrence may not apply the same length of time for each activity, because different routines require different periods of operation. Activities may be modified on consecutive trials until feedback no longer alters substantially the activity and a routine. Thus, *habit* is formed. Almost all frequent routines will be daily, weekly, seasonal, or life cycle. It is the daily routine on which urban researchers and planners have concentrated most.

The main components, or episodes, of household activity patterns may be defined in terms of (1) how available time is budgeted and (2) where activities take place.

Also of importance is the mode of transport used to travel between the origins and the destinations that separate the activity episodes.

8.5.2 Activity Systems

The fundamental *activity systems* of a city will be those activity spaces of its residents. These collectively generate the need for, and set in operation, other entities that develop their own activity systems. An example will help illustrate. Activities associated with shopping, commuting, and recreation may progress to the stage where business and government activities will have to be ordered to cope with them. As a consequence, the environment in which activities take place will change. Schedules of individuals may have to be adapted to conform with schedules of institutional entities in matters such as working hours, transportation times, and so on.

8.5.3 Obligatory and Discretionary Acts

A further important general notion is that activity systems consist of obligatory and discretionary acts.

Obligatory acts include sleep, work, and school. *Discretionary acts* include recreation, some shopping activities, and leisure. They occur more or less in cycles that are regularly timed. Identifying these cycles is the first phase in being able to predict and plan for activities in cities. In a typical weekday up to 18 hours may be expended on obligatory activities such as sleep, work, home-making, and shopping.

Of the remaining six or more hours that are available for discretionary activities, most are spent at home. Thus, the importance of the home and its neighborhood is evident. This has implications for the allocation of public resources for leisure-time facilities. Discretionary acts will vary from household to household and produce some variation of behavior within groups; they may be deliberately selected by people, or they can stem from changing external circumstances.

8.6   Early Approaches to Modeling Human Activities

During the 1960s and 1970s, three basic modeling approaches were developed that sought to analyze and explain the nature of human activities in a space–time context:

1. *The transductive approach.* With this approach the focus is on solving urban planning problems through a better understanding of the living patterns of city residents, the way they allocate their time to different activities in the course of a day, the rhythm of these activities around the clock, and the locus of these pursuits in city space. It is an approach pursued by Chapin and his coresearchers at the University of North Carolina. Chapin (1974) suggests that there were variations in activity patterns among subsocietal segments in the population. As we study these variations, we may postulate antecedent ties that these patterns may have with felt needs and preferences.

2. *The routine and deliberated choice approach.* Cullen and Godson (1975) proposed that for any population it was the recurrent routine activities that structure the day. Thus the routines of an individual or household routines will be structured around a relatively small number of deliberate choices made rather infrequently. There is little demand for deliberate choice about the majority of activities in which people engage, and, in fact, there is usually neither time nor opportunity for alterations to the schedule. Cullen and Goodson (1978) claim that a few key times and activities will act as pegs around which the day is organized.

3. *The routine and culturally transmitted structure approach.* This approach was developed at the Martin Center for Architectural and Urban Studies at Cambridge University (UK). In this approach there is no explicit attempt to explain behavior, as evident through activity patterns, in terms of motivations or other aspects of psychological underpinnings of behavior. Instead, the approach postulates that time rhythms have a culturally transmitted structure. Shapcott and Steadman (1978) claim that there is a pattern of constraints that restricts and limits social and spatial behavior in such a way as to confer freedom and possibilities for a variable choice of patterns of personal activities. Thus a structuralist method is adopted that implicitly includes motivations as part of a determined and determinate social structure.

A full analysis of these and other early model approaches is given in Parkes and Thrift (1980: Chapter 5). What follows is a summary of the main aspects of those approaches.

8.6.1   The Transductive Approach

The approach developed in the 1960s by Stuart Chapin and his fellow workers at the University of North Carolina may be described as an aggregated, survey-based approach. In it, the population of a city is categorized on the basis of socioeconomic and demographic characteristics, and subgroups are seen to interact with the processes that produce the spatial organization of a city.

Chapin (1974:40) describes the model as *transductive* because it was concerned with both the individual and the household (whose behavior was regarded as being regulated primarily by biological time) coupled to learned behavior related to social time. The model sought explanation by assessing the relative importance of the role and personal background factors that precondition people from different population subsets to adopt particular activity patterns. In addition to looking at the relative significance of motivations and other attitudinal factors affecting the predisposition of a person to act in one way or another, predisposing factors are most closely associated with activity routines.

The approach uses human activity patterns as the means by which people satisfy needs and wants, of which two categories are identified:

1. *Subsistence needs*, such as sleep, food, shelter, clothing, and health care, plus activities that supply income to help satisfy these basic needs.
2. *Culturally, socially, and individually defined needs* that are met by the individual engaging in a wide range of social and leisure activities.

Chapin recognizes *obligatory* and *discretionary* activities. This distinction is very much related to the two types of needs outlined above. Chapin recognizes also that activities are the result of both learned behavior or acts of individuals and motivated behavior, which is on-going, initiated by an individual through wants, and directed to the end of satisfying these wants through the use of suitable elements in the environment. This is somewhat akin to operant conditioning in the tradition of Skinner's theory of learning. In this, "the environment not only provides the opportunity for the act, but also it conditions the act through its consequences" (Chapin, 1974:27).

In this context, Chapin's approach views activities as consisting of three components: (1) a *motivational component* (felt need or want directed toward a goal); (2) a *choice component* (alternatives are considered); and (3) an *outcome component* (observable action).

In later work Chapin (1978) developed a choice model of time allocation to daily activity patterns shown in Figure 8.7. This is in the form of a conditioned response model. Learned forms of behavior are seen to be shaped by four influences: propensity; an opportunity; an appropriate situation; and an environmental context.

The *propensity* is the motivational basis for the activity, conditioned by a person's specific constraints. A propensity element determined what activities are likely to fall in a person's realm of concern, thus defining the scope of choice. The *opportunity element* refers to the availability of a physical place or facility suited to the activity and to the congeniality of surroundings for engaging in an activity. The *sit-*

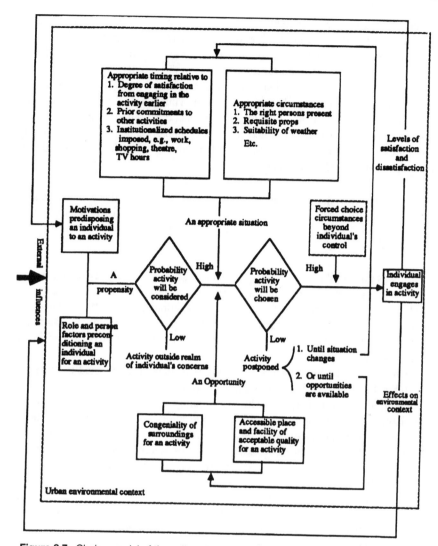

Figure 8.7. Choice model of time allocations to daily activities. *Source*: Chapin, 1978.

uational element refers to the appropriateness of timing and circumstances for the activity. The *environmental context* is a milieu within which choices are made, and it has everything of a nonphysiological nature influencing a person's behavior, including previous behavior.

In Figure 8.7 the choice process is represented by the flows on the diagram. The urban environmental context (the dashed outermost line) envelopes and influences the whole process of behavior choice. The motivations that predispose an individual to an activity combine with certain role and personal characteristics to precondition an individual for an activity. Predisposing and preconditioning factors are susceptible to inputs to the activity system, and they are influenced by the activities to which they give rise. Together these will initiate a propensity to consider an activity, and if the probability of engaging in an activity is high, then two additional factors will come into play before the activity is selected. An appropriate situation needs to exist, as must an opportunity to engage in the activity.

An important aspect of Chapin's approach concerns variation in the degree of postponability of an activity; that is, its *elasticity*. Thus there is an ordering of activities along a continuum from obligatory activities at one pole to discretionary activities at the other. It should be noted that the elements of the transductive choice model presented in Figure 8.7 are particularly pertinent to the discretionary activity mode. As Chapin (1974) indicates:

> The ordering suggests that on the average people have more latitude for choice at the discretionary end of the continuum and little or no latitude for choice at the obligatory end . . . . What is discretionary and what is obligatory are relative concepts. An activity is discretionary if there is a greater degree of choice than constraint. (pp. 37–38).

It was proposed that the basic physiological needs of everyday living, such as sleeping and eating, offered limited choice possibilities, were low in elasticity, and revealed little or no difference among populations drawn from different socioeconomic and demographic subgroups. In contrast, however, self-actualizing or discretionary needs were continuously confronting the variabilities incorporated in the transductive choice model.

Chapin applied his model to the analysis of data sets collected in 1968 and 1971 in Washington, DC. Two communities were sampled; an inner-city, black community and a central, transitional white community. High- and low-income groups were distinguished, as were male and female groupings, and the degree of full or part-time employment. Activity patterns according to 11 activity categories were generated for an average spring weekday. The emphasis of activities was on discretionary pursuits, which were particularly amenable to explanation by the transductive choice model.

By distinguishing between male and female groups, those working full-time or part-time, the sample could be related to the causal factors of precondition, and predisposition, which the transductive choice model suggested were activity pattern-determining factors. It would be assumed that situational and opportunity components of the model were satisfied because the results referred to activities that were actually current. Furthermore, because only a single day (i.e., a 24-hour period) was involved, the urban environmental context was constant, having little or no influence on activity choices that were made during that period. Thus there was no feedback of new information to alter predisposition or precondition, and there were no changes in the timing and spacing of activities. Satisfaction of needs were based on what has been learned from the previous day and on experience of persistence over a longer period of time.

Chapin summarizes the influence of motivational factors with respect to men working full-time and men working part-time. This is done according to rank scores from a stepwise multiple regression analysis.

The component *esteem* (i.e., the need for status) represented by the declared importance of having neighbors of similar socioeconomic background may be taken to illustrate the kind of interpretation that Chapin's transductive model permitted. To the men in the black community, apparently this was less important than for men in the white community, especially as a predisposing factor before they engaged in social activities. The transitional white community was, in fact, about 3 miles further from the city center than was the black community, and during the summer, in which the survey was taken, the black community had experienced considerable violence in the streets in their neighborhood. The concern about violence in the streets became a very significant predisposing factor, generated by the unsettled urban environmental context in the days preceding the study. Social activities of all kinds that involved leaving home were thus affected, especially in terms of their time location. In-house activities were adjusted in consequence. Given preconditioning factors, such as stage in the life cycle or health status, both study groups had their activity patterns determined to a similar degree. The health status factors were marginally more significant in the determination of activity choice in the white community than was the case in the black community.

From this early work by Chapin, it is evident that there is considerable value in having a theoretical structure upon which to base interpretations. According to Parkes and Thrift (1980) and Chapin's transductive model:

> The theory guides the selection of empirically derived relations. It should be noted that by emphasizing choice factors in activity participation this becomes an ideal tool for planners. When making decisions, they are better able to accommodate aggregated choices assessed through a model which is initially sensitive to the motivations which pre-dispose action; to the role in person characteristics which pre-conditions action and to the perceived availability and quality characteristics of facilities in time and over space. (p. 217)

8.6.2   The Routine and Deliberated Choice Approach

In a study of urban networks and the structure of activity patterns, Cullen and Godson (1975:7) asked the question: "How do we find out more about the way in which the individual's time–space (space–time) dimensions are structured?" Basic to their approach are:

1. The role of motivational and psychological factors.
2. The proposition that activity patterns are highly structured and routinized and largely beyond the control of the individual except in terms of a small number of significant decisions made during a lifetime.
3. The proposition that motivational and other psychological factors, such as stress, levels of expectation and degree of commitment, combine to influence these significant, if infrequent, decisions or deliberated choices.

Cullen and Godson (1975:8) propose the following generalization: "Activities to which an individual is strongly committed and which are both space and time fixed (that is to say they must occur at a particular space location and at a particular time), tend to act as pegs around which the ordering of other activities is arranged and shuffled according to their flexibility ratings." They hypothesize that, while not consistently rational, behavior contains *highly organized episodes* that are structured by physical patterns and needs with priorities based on considerations such as the financial importance of the activity, the presence of certain other participants, the order in which activities are planned to occur, and the likes and dislikes an individual has for specific activities. *Constraints* were seen to operate on choice, and when related to priorities, any single activity choice would be subject to a fixity rating according to the degree of commitment of a person to that activity. Cullen and his fellow workers attached great importance to the *fixity ratings of activities*, suggesting that these enabled scheduling to facilitate *synchronization* of activities and movements and to consume limited time.

The amount of time allocated to an activity is taken to be a measure of its *utility*. Time is seen also as a linking medium; thus time becomes a path that orders events as a sequence, separating cause from effect, always synchronizing and integrating activities. In this context, the time-budget approach is interpreted as an accounting medium for the analysis of activities. A basic problem relates to how activities may be classified in terms of their functional characteristics, as is the case in the transductive approach discussed previously. This makes it harder to identify motivational factors involved in activity choice. As a partial solution to this problem, Cullen and Godson (1975) suggest it is more appropriate to classify activities directly according to motivation or degree of commitment, or of *deliberated premeditation*. This enhances the meaning given to human activities differentiated and classified on a functional basis. For example, if we were to take the activity of shopping, various subcategories of this classification are evident according to whether it is premeditated or impulsive, planned at some specific time in advance, or forced by circumstances, and so on.

We may regard activities in terms of whether they are: (1) planned or unplanned; (2) deliberated or spontaneous; (3) expected or unexpected; and (4) fixed in time or not fixed in time. In studying daily activity patterns of individuals and households, Cullen (1978) notes that many activities are deliberated quite carefully, yet the profile of activities in a normal day is swamped very much by a dominant pattern of repetition and routine. Thus, there are a few deliberated choices concerning only a small number of key events in life that act as pegs around which other activities become scheduled. Cullen cites key events—such as marriage, the job one has, the school one is sent to, whether one has children or not—as being the primary structuring agents for the daily activity routine. When the day begins, its program is already printed, and the stage is set for another performance much like yesterday's. To Cullen (1976:407), action is seen as a "response to the means which are attached to and define the social and spatial institutions against which action is set, and adaptation was a process, whereby long-term deliberated choices abut where to live, what job to take, and which clubs to join are translated into a pattern of daily and weekly routines."

This notion of *routine activity* in a long-term choice context is illustrated in Figure 8.8. In this way, Cullen suggests that each day is a product of a mixture of choices and constraints. In the long term, we deliberate seriously over only limited aspects of our situation, and we accept others that are simply beyond our control.

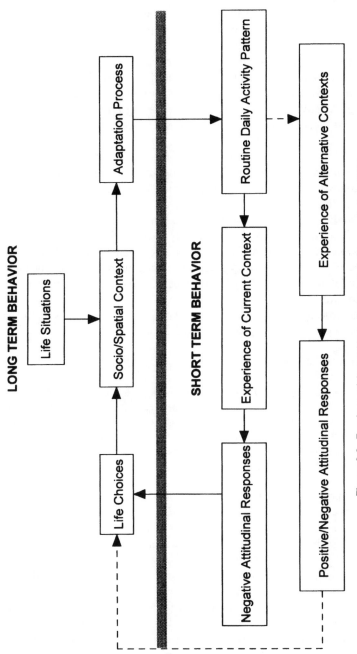

Figure 8.8. Routine activity and long-term choice. *Source*: Cullen, 1978.

297

Time enters the cycle in the form of the dotted line in Figure 8.8, and it helps overcome a paradox whereby actions that are dependent on motives and attitudes are, in turn, the initiators of motives and attitudes through experience gained in action. The cycle is broken with explicit reference to time, thus enabling us to differentiate the long-term attitude behavior relationships from those that operate on a day-to-day basis.

Cullen claims that, to a greater or lesser degree, every activity is fixed in time or space—and sometimes in both. He identifies four categories of activities that affect the subjective rating of *fixity* people may ascribe to activities. They relate also to existing objective fixity factors, like the opening times and closing times for shops and the locations of recreation facilities. The four subjective factors of fixity are:

1. *Arranged activities*, involving joint activities with others, which may be more or less difficult to arrange at some other time or some other place depending on the status of those involved.
2. *Routine activities*, which are undertaken with sufficient regularity and frequency that they may become highly fixed, particularly in time.
3. *Planned activities*, which need not involve others but will be assigned a level of subjective fixity according to commitment. Generally, the nearer in future time an activity is, the harder it is to adjust its planned space–time location.
4. *Unexpected activities*, which may impose fixity ratings onto existing arranged, routine or planned activities.

Cullen (1978) argues that:

An understanding of long term behavior must be based upon a logically prior understanding of everyday activity patterns for one very simple reason. Whilst the important choices which determine the way people fashion their own environments, and which obviate the need for choice in the short-term, are infrequent and oriented to distant time horizons, their implications are experienced day-to-day as discrete moments of pleasure, indifference or pain. (p. 33)

In empirically testing the routine and deliberated choice approach thus developed, Cullen and his colleagues focus on the *day-to-day diary technique* as the core element among their survey instruments. This is appropriate as the aim was to explain choices in the long-term, thus the model requires monitoring not only of overt behavior but also of the way attitudes develop against the backdrop of everyday activities. This can be achieved only if it is ensured that the environment, the activity, and the attitude are each recorded with a device that had minimum impact on the way they are experienced. The diary approach is a device that recalls a sequence of events from the recent past. Although it is far from perfect, it categorizes, precodes, and punctuates everyday behavior in a variety of more or less arbitrary ways.

Analysis of data collected from diaries involves frequency counts using *activity episodes* as units. This allows cross-tabulations to be produced, for instance, of prearrangement and subjective fixity. Time allocation analysis involves statistical description of the time budget, based on computed mean durations and other statistical moments around the mean, with activities further classified according to fixity, deliberation commitment, and so on.

An example from Cullen and Godson (1975) is given in Table 8.3. Means and variances for the duration of time allocated to activities, classified by function, premeditated choice, and so-on, were computed. It was possible then to calculate intercorrelations among all activity classes and subsequently to use factor analysis to determine whether there is any underlying structure in the day. Furthermore, time series analysis was used to analyze relative time location; that is, when something happens in relation to some other event. Using these approaches, it was possible to answer questions such as: "Which activities tend to lead and which tend to lag other activities?" and "How does one activity tend to lead or lag another activity?"

The sequencing of activities is an important aspect of this analytic approach. Computing transitional probabilities and using time series analysis are two methods that clarify the degree to which there is internalized structuring of activity patterns, independent of time allocation and durations. The object was to investigate and determine the probability of occurrence of given sequences of activities.

Cullen and Godson (1975) used time series to study survey data on activities and their various attributes collected for half-hour time periods throughout the whole day over 14 consecutive days. They divided the period of 17-hour time interval from 0700 hours to midnight (the period during which most people are awake) into 100 equal 10-minute intervals, and for each time interval they were able to

TABLE 8.3.
Allocation of Time by Degree of Prearrangement and Fixity (Mean Duration Figures in Minutes)

	Mean number of episodes per day	Mean percentage of episodes	Mean time in minutes/day	Mean percentage of time	Mean duration of episodes
Prearrangment					
Arranged	2.8	8.8	133	9.2	47.0
Planned	5.4	16.8	225	15.6	41.7
Routine	9.8	30.6	301	20.9	30.6
Unexpected	6.8	21.1	196	13.6	29.0
Time–space fixity					
Not able to do anything else at that time	7.1	24.7	324	22.6	45.6
Not able to do activity at any other time	9.2	28.5	375	26.1	40.8
Not able to do activity elsewhere	10.1	49.8	534	37.2	52.9
Could not have been elsewhere at that time	9.1	29.7	351	24.4	38.6
Generalized fixity	3.6	11.2	203	14.1	56.3

Source: Cullen & Godson, 1975:21.

show the number of people engaged in a particular activity. Also they determined the number of people who described the activity in which they were engaged as being of a certain fixity or deliberated choice type, such as a routine, planned alone, fixed in space, fixed in time, and so on. Thus it was possible to develop an indication of the changing shape or structure of the activity and its perceived constraints during the day. A total of 19 activities, or attributes of activities, were studied in detail. Three of these are discussed here to illustrate the approach. The results are reproduced in the graphs in Figure 8.9. In interpreting these graphs, it must be remembered that the data were collected from an institutional population, namely students at a British university. The graph shows the characteristics somewhat akin to a normal curve, peaking around the noon hour. There are considerable oscillations in the second graph. The two marked peaks in the third graph are probably different from what would be expected in a normal community and reflect the institutional community on which the survey was based.

Cullen and Philps (1975) also looked at stress as an explicit feature of the routine and deliberated choice approach. They studied 50 married couples in Hackney, London. They concluded that although the structure of the normal day was highly routinized, the subjects did note the stresses of everyday life and were more prepared to talk about them than they were to talk about less stressful highlights of the day. Stresses have to be coped with, and they form part of the adaptation process that operates to produce long-term behavior. Stresses may be graphed in a way similar to that shown in Figure 8.9.

As might be expected, the graphs thus indicated that the number of husbands who reported feeling under stress at various times of the day showed lags of about 4 hours related to work distribution peaks. Although less than 30% of the working day was spent at work, over 70% of all stress was experienced during that time. However, the cumulative effect of stress was marked; sensitivity increased as morning became afternoon and as afternoon became evening.

From the routine and deliberated choice approach it seems that it is possible to structure the analysis of human activities in a manner allowing very positive contribution to be made toward explaining some of the basic characteristics of human activities. Parkes and Thrift (1980:229) suggest that the possibility of coupling the transductive and the deliberated choice approaches is a fruitful area for research, particularly in transport choice studies.

8.6.3 The Routine and Culturally Transmitted Behavior Approach

The aim in this approach is to estimate the distribution of aggregates of people in a three-dimensional frame of *location*, *time*, and *activity*. Environmental constraints may be built into an elementary entropy-maximizing method. Sets of equations have been developed allowing the prediction of population aggregate distributions in relation to the three principal dimensions of location, time, and activity. This method is outlined in Thomlinson (1969). A simple diagrammatic scheme (Figure 8.10) was used to illustrate the various normal constraints operating on daily behavior. The sum of the cells formed by the intersection of the planes of time, location,

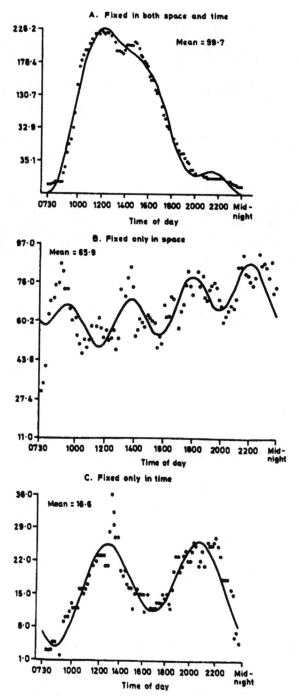

Figure 8.9. Daily activity structure with time and space fixity. Redrawn from Cullen & Godson, 1975.

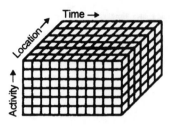

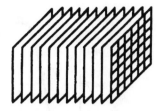

(a) A three-dimensional array of cells representing activity, time and location

(b) The total population in cells in each activity/location plane must not exceed the population constraint for the cell

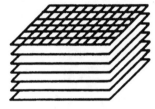

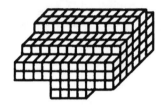

(c) The total population in cells in each location/time plane is restricted by the time budgeted for that activity at a location

(d) Some activity/location combinations are not available, nor are some activity/time combinations

Figure 8.10. A three-dimensional time, location, and activity matrix. *Source*: Bullock *et al.*, 1974.

and activity accommodate all the people engaged in various activities. The modeling problem involves allocating population groups to activities in time and space in accordance with times allocated to similarly classified activities derived from time budgets. Parameters of the dimensions have to be estimated, subject to restrictions on the availability of locations. Figure 8.10a, an interrupted matrix is shown. This indicates that certain cells are not available for occupation because of time, location, or activity restrictions. The problem to solve is to find the most probable permutation of the permissible restrictions. The result is the distribution of the number of people engaged in any activity for each time period in each location over the day (Bullock, Dickens, Shapcott, & Steadman, 1974:48).

This modeling approach is based on a number of propositions. Shapcott and Steadman (1978) suggest that "a rhythmic pattern of behavior is established, and is transmitted culturally in the form of a known timetable of activities." This is a routine rather than erratic conduct that is taken to be the primary factor in structuring an activity pattern for an aggregated population. No attempt is made to examine patterns of activity from the standpoint of the individual as a decision maker. Rather, the time budget is taken as the starting point for the modeling of activity patterns in time and space, and the concern is with the overall pattern of behavior of groups of people who are identified by similar, easily measured characteristics. The approach has similarities with Chapin's transductive approach discussed earlier. The approach has no underlying individual behavior model on which it is based. It is similar to the deliberated choice model in that significance is assigned to the structuring powers of routinized behavior, and it is hypothesized that there is stability of activity patterns over a long period of time. However, there is an absence of a motivational, premeditative component in the model.

A key to the model is the definition of the *constraints* and *opportunities* that apply in a society or culture as a whole. Shapcott and Steadman (1978) claim that:

> Constraints (and the opportunities which they reveal) appear as rhythmic structures
> . . . for example, the fixed hours of work might be imagined as a series of clusters of
> tubes appearing regularly between the hours of 0900 and 1700, the location of these
> tubes in the horizontal plane corresponding to the relevant spatial positions of the
> workplaces in the city or region in question . . . an instantaneous time slice cut
> through this picture will reveal the effective map of the spatial area in question, as it
> is defined in activity terms at the chosen time of day. (pp. 52–54)

A further interesting notion is that activity patterns have a *culturally transmitted* structure. The previous approaches discussed above placed human activity patterns in terms of interaction between the real physical environment of the city and the subjective environment of individuals. In this approach, a third world mediates between these two environments. This is the *world of objective knowledge* (Popper, 1972), which is a *cultural product*. In the study of human activity, it provides a perspective between that of the wholly subjective, psychological point of view, and the wholly rigid, cultural and environmental deterministic point of view. At any single point in time, the social structure of temporal regularities is "a given so far as the individual is concerned" (Shapcott & Steadman, 1978:73). There have been a number of major studies that have applied this approach. We refer to one of these as an example of the application of this approach.

A study involving 450 residents was conducted by Shapcott and Steadman (1978) in the city of Reading in southeast England. It was possible to calculate the amount of time, averaged over 7 days of the week, these residents spent on various activities. The population was split into three categories: men, working women, and nonworking women. The study was compared with a time-budget study that had been undertaken in Reading in 1961 by the BBC to assess long-term stability of population activity structure. The argument made was that the changes in culturally transmitted structure between 1961 and 1973 would have initiated significant differences in the structure of the daily routine. Comparison of the 1961 and 1973 figures showed that during a period when real incomes increased by about 30% there was an increase in the proportion of women in the work force from 30% to 42% (Shapcott & Steadman, 1978:67). The researchers concluded that:

> . . . decisions about the timing of most of the major kinds of activity are not made on a
> daily basis at all. Instead, individuals and households collectively commit themselves

over a much longer time scale by a number of life decisions—marriage, taking particular jobs, choosing a particular place to live, sending their children to a particular school—to a series of rhythmic constraints which then govern the greater part of their daily routine and around which and within which their optional and variable activities which do occur at the day-to-day level must be fitted in. (p. 67)

The links with the routine and deliberated choice approach need to be emphasized.

Thus, it would seem that routine and constraint enable people to cope with the complexity of an urban environment that is faced on a day-to-day basis, thereby eliminating the necessity of considering anything but a small fraction of the true variety of that environment.

8.7 Collecting Data on Activities in Time and Space

8.7.1 Time Budgets

Time budgets and time–space budgets are devices used to investigate the ways in which human activities occur coincidentally in time and space. They provide us with a greater understanding of the patterning of human activities, both at the individual and at the aggregated societal levels, and they are required to operationalize many of the models that have been developed to analyze and simulate individual and household activities.

A *time budget* is "a systematic record of a person's use of time over a given period. It describes the sequence, timing, and duration of the person's activities, typically for a short period ranging from a single day to a week." A logical extension of this type of record is a *time–space budget* which "includes the spatial coordinates of activity location" (J. Anderson, 1971:353).

Studies using these devices do not constitute a unified research field. Time budgets are used widely in market research for the mass media and in nonspatial ecological studies of life-styles and leisure. During the 1960s and 1970s considerable impetus was given to spatially oriented research on daily activity patterns. Since then travel diaries have been widely used for transport-planning purposes.

It is interesting to consider the notion that time itself is a quantitative measure highly appropriate for studying the social organization of society. Time is a scarce commodity, and it is feasible to suggest that it is the main rival to money in our society. In may situations *time is money*. It is a quantity to be spent, saved, or wasted. In addition we have to keep to time deadlines to order our activities in particular sequences as well as to take account of the amounts of time required to get from place to place and from one activity to the other. Its main advantage (unlike money) is that everyone who remains alive within a given period has exactly the same amount of time to allocate. As we have seen earlier in this chapter, the allocation of time to compete in activities involves a *tradeoff*.

Detailed activity records can describe just the whole-time use pattern, or preselected types of activity of individuals, such as travel, in a given area over a particular period of time. Whether data are collected in questionnaire or in diary form, activity records are classified by J. Anderson (1971:356) into four categories, namely:

- *Time budgets.*
- *Space–time budgets.*
- *Contact records,* which cover face-to-face and other preselected types of interpersonal communication.
- *Travel records,* which typically describe origins, destinations, purposes, modes, and types of trips.

In addition, there is a further requirement:

- The *activity location,* if the purpose is to generate a time–space budget.

In time budget surveys, as in other surveys, decisions need to be made about the size and structure of a sample. Other important questions relate to factors such as:

- Should the data be collected in the presence of the subject?
- How much explanation of the object of the study should be given to the subject?
- What should be done about nonresponse?
- How is an inadequate interview or diary to be determined?

These are general problems associated with social survey design, but they are compounded in time-budget surveys because of the inevitable demand for a very high level of cooperation from subjects and because of the relatively long periods of time required to complete the survey. Furthermore, these surveys are expensive. Often it is necessary to survey individuals or households before and after a residential relocation, or before and after some important change in the operational environment, such as changes in public transport modes and schedules or the construction of a new road.

8.7.2   Methods of Collecting Time-Budget Data

There are generally four methods of time-budget recording. These are:

1. *Recall method (1),* where the activities of some specified period in the past are recalled with as much precision, with regard to time and location, as is possible for the objectives of the study. A major problem here is the *ex post facto problem,* and it is important to limit the period of recall to as short a time period as possible.

2. *Recall method (2),* where the activities of some normal period are recalled.

The basic problem here is the interpretation of the term "normal" in relation to time location, duration, and space location of each activity.

3. The *diary method*, where subjects keep a diary for a specified period of time. They record activities either according to predetermined time locations, or in a free-formated manner. This may or may not require interviewer intervention, depending on the degree of availability of money to conduct the diary survey. An ex post facto evaluation of a diary with the subject present is desirable as a check on reliability. The free-formated approach is fraught with problems, since the subject is given little guidance in terms of specification of activities and locations, and so forth. The free-formated style of diary is also extremely difficult to code.

4. *Game-based methods*, of which there are various types. There are many reasons why this approach is used, but it is usually used in association with some other method, often in the context of a "post diary" or postinterview situation. Game-based methods are basically aimed toward investigating changes in the contingencies of the decision-making environment.

A major problem to be solved in time-budget studies concerns the *classification of activities*. There are an almost unlimited number of classification schemes that are available to the researcher conducting time-budget studies. Some standardization is essential in order to make comparative studies possible. It is necessary to generate theory. We can use the frequency, duration, and sequence position of activities relative to some marker as a basis for activity classification. Activities can be classified as well in terms of their tendency or probability to be predecessors or successors. They can also be classified according to whether they are undertaken by the individual alone or in association with other individuals. Furthermore, they may be classified in terms of whether they are obligatory or discretionary. An activity pattern is a dependent variable, dependent upon the predisposing and preconditioning attributes of the subpopulation that it represents. Thus, the activities that make up an activity pattern require classification. Chapin (1974:21) defines activities as "classified acts or behavior of persons or households which, used as building blocks, permit us to study the living patterns or life ways of socially cohesive segments of society."

Chapin's group at the University of North Carolina developed a classification scheme based on a dictionary of about 230 activity codes that are grouped into activity classes at two levels of *aggregation*. A less complex system of classification was that used in the *multinational time-budget study* by Szalai, Converse, Feldheim, Scheuch, and Stone (1972). An extract is provided in Table 8.4 that gives an example of the scheme and the towns that were involved in the multinational study. Here the full 24 hours of the day were broken down either into the original category code of 96 categories or into a more manageable 37 activity codes obtained by collapsing certain of the detailed codes. The 37 activities were reducible to nine subtotals, such as work, and so on. Activities are classified as primary and secondary based on the respondent's description given in a diary. A particular activity may, on some occasions, be described as primary and on other occasions as secondary, thus presenting a problem. It should be noted that the classification scheme, when applied to all activities, whether primary or secondary, has a two-digit code. The first digit divides

TABLE 8.4.
Multinational Time Budget Classification of Activities (an Extract)

	Belgium	Kazanlik, Bulgaria	Olomouc, Czechoslovakia	Six cities, France	100 electoral districts Fed. Rep. Germany	Osnabruck, Fed. Rep. Germany	Hoyerswerda, German Dem. Rep.	Gyor, Hungary	Lima-Callao, Peru	Torun, Poland	Jackson, USA	Forty-four cities, USA	Pskov, USSR	Kragujevac, Yugoslavia	Maribor, Yugoslavia
Sample size	2077	2096	2193	2805	1500	978	1650	1994	782	2759	1243	778	2891	2125	1995

00 regular work	25 outdoor playing	50 attend school	76 party, meals
01 work at home	26 child health	51 other classes	77 cafe, pubs
02 overtime	27 other, babysit	52 special lecture	78 other social
03 travel for job		53 political courses	79 travel, social
04 waiting delays	28 blank,	54 homework	
05 second job		55 read to learn	
06 meals at work	29 travel with child	56 other study	
07 at work other			
08 travel to break		57 blank	
09 travel to job	30 marketing	58 blank	
	31 shopping		80 active sports
	32 personal care		81 fishing, hiking
	33 medical care	59 travel, study	82 taking a walk
	34 administrative service		83 hobbies
10 prepare food	35 repair service	60 union, politics	84 ladies hobbies
11 meal cleanup	36 waiting in line	61 work as officer	85 art work
12 clean house	37 other service	62 other participation	86 making music
13 outdoor chores		63 civic activities	87 parlour games
14 laundry, ironing	38 blank	64 religious organization	88 other pastime
15 clothes upkeep		65 religious practice	89 travel, pastime
16 other upkeep	39 travel, service	66 factory council	
17 garden, animal care		67 misc. organization	
18 heat, water	40 personal hygiene	68 other organization	90 radio
19 other duties	41 personal medical	69 travel organization	91 TV
	42 care to adults		92 play records
	43 meals, snacks		93 read book
	44 restaurant meals	70 sports event	94 read magazine
20 baby care	45 night sleep	71 mass culture	95 read paper
21 child care	46 daytime sleep	72 movies	96 conversation
22 help on homework	47 resting	73 theater	97 letters, private
23 talk to children	48 private, other	74 museums	98 relax, think
24 indoor playing	49 travel, personal	75 visiting with friends	99 travel, leisure

Source: Szalai et al., 1972:576–577.

activities into 10 main groups (0 for work, 1 for housework, etc.). The second digit elaborates on these main groups; for example, something coded as work may be further coded as 00 regular work, 01 work at home, 02 overtime, and so on. Locational data may be coded either on a functional basis, such as using home or place of work, or by a geogrid coding system. Graphs can be produced from diary data collected and classified according to this scheme to show, for instance, the proportions of people from different social or demographic subgroups of the population that are undertaking certain activities in certain locations at different times during the day.

8.7.3 Duration, Frequency, and Sequence of Activities

Duration, frequency, and sequence of activities are important aspects of time budgets:

1. *Duration* of an activity may be evaluated simply in terms of the amount of time spent on that activity. It may be considered either as a ratio of the total amount of time in a given period spent on that activity or as a ratio of time spent on some other activity over a given period of time. Many simple statistical measures can be used to conduct such analyses, including calculation of the mean, variance, standard deviation, coefficient of variation, and correlation and regression statistics. The ratio of the standard deviation to the mean of an activity duration has been widely used in many studies as a measure of group elasticity of the activity. In this way it is possible to compare the activity elasticities between social and demographic subgroups of the population. The elasticity of the activity is greater as the size of the coefficient increases.

2. *Frequency* of an activity may be considered in two ways, either as the frequency of occurrence of an activity for an individual within a specified period of time such as a day, or as the frequency of occurrence of an activity across individuals in a population or sample within a defined period. Frequency with which an activity occurs can be used as an indicator of the importance of the activity.

3. *Sequence* of an activity is a difficult aspect of time-budget analysis. Sequence may be approached by the use of time series analysis to describe sequence of activities over a period such as a day, using autocorrelation methods by determining the fixity of activities in terms of space and time locations. In this way the structure of a period in terms of a single activity or set of activities can be determined. We may also determine the degree to which activities tend to lead or lag behind others by using cross-lag-correlation approaches. Use of transitional probabilities can give an indication of the probability that an episode of a certain type will be immediately followed by one or another type in any person's sequence of behavior.

9

Activity Analysis in Travel and Transportation Modeling

9.1 The Evolution of Transportation and Travel Models

It is in the field of modeling travel behavior and in the analysis of travel patterns, especially in the context of transportation planning, that the behavioral approach to investigating human activity patterns has been said to have great potential application. Time–space budgets have been used extensively as a means of data collection to investigate destination choice in transport planning. The strength of the behavioral approach to travel modeling is that it is formulated on a set of hypotheses related to the decision-making unit such as the household; the models developed have adopted the decision-making unit as the unit of analysis, even though most individual choice models seek a set of parameters aimed at describing and predicting the aggregate behaviors of groups and populations. These approaches have been concerned mainly with modeling recurrent travel behavior, for it is this behavior that is most relevant in the transport planning context.

This chapter discusses traditional urban transportation planning models and their limitations, and it traces the development of disaggregated discrete-choice models that focus on utility in choice behavior, including the multinomial logit model. Significant developments up to the early 1980s focused on the application of activity time–space budget studies in estimating the nature of aggregate travel behavior, particularly in evaluating changes in travel mode and transportation system characteristics. Recent models place activity choice analysis within a cognitive framework and, in particular, focus on activity scheduling using computational process models. These integrate human activities, transportation systems, and land use in a GIS framework and in a regional setting to simulate activity behavior and model the impact of urban growth and transportation system changes.

9.1.1 The Shortcomings of the Traditional Urban Transportation-Planning System

The majority of transportation-planning studies in countries such as the United States have used a five-step procedure known as the *urban transportation-planning system* (UTPS). This was developed in the 1960s and since has been refined incrementally. UTPS uses aggregate data for sub-regions (zones) of a city to estimate travel on present and proposed transportation networks. Figure 9.1 represents the standard travel-forecasting process used in the UTPS. The process has been used internationally, and it became "institutionalized," despite widespread criticism and controversy.

Supernak (1983) noted that with the exception of disaggregated discrete mode-choice models, very little progress had been made in travel demand-modeling over 30 years. The UTPS continues to be used widely despite its lack of behavioral content. As a result the UTPS "prevents the analyst from evaluating alternative policies that are unrelated to investment proposals for major facilities. These problems, always apparent and annoying, have taken on a new significance as (in the United States and or in other nations) the Clean Air Act Amendments of 1990 and ISTEA of 1991 place greater modeling and analysis burdens on States and MPOs" (Stopher, Hartgen, & Li, 1993:1). The UTPS model also lacks feedback between model elements, thus preventing the system from reaching equilibrium in forecasting environments. For example, output speeds generally are lower than those assumed in the model, resulting in overestimation of travel demand for new transportation proposals. Goodwin (1992) has found in the greater London area that consumers were observed to be much more flexible in response to increasing congestion than was originally envisaged, indicating that there is a range of behavioral adaptations that people can and are prepared to make that is far greater than what can be modeled using the UTPS approach.

There are further pressing reasons why activity-based approaches of the type discussed in Chapter 8 should be used to develop more sophisticated models. As Axhausen and Gärling (1992:323) have noted: "The growth of car ownership outdistances the growth of road capacity in most urban areas around the world, forcing planners to consider a new mix of policy instruments to maintain minimum service standards in the face of further increasing congestion. An important part of such a policy mix are measures to manage traffic demand." Two main instruments have been identified for traffic or *travel demand management*:

1. The restraint of traffic through costs and road-pricing policies and the regulation of parking, access to particular areas of cities, and limits on speed.
2. The use of information technology to provide both planners and the traveler with more and better information about traveling conditions before and during trips to guide route selection and parking choices.

However, fundamental issues for the development of policy instruments are the degree to which travelers are prepared to substitute out-of-home with in-home activities and to reschedule out-of-home activities, including switching the location of activities and/or travel mode (Axhausen & Gärling, 1992:323).

To address these issues it is necessary to understand fully the interrelations between decisions of individuals and households in undertaking activities in space and time that involve travel, including: when, where, and for how long activities take place; why particular modes of travel and routes are chosen; and with whom

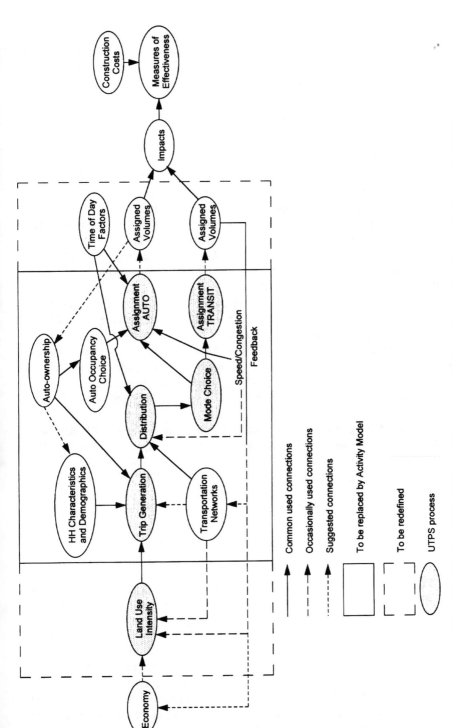

Figure 9.1. The standard travel-forecasting process: UTPS modelling system. *Source:* Stopher et al., 1993:3.

people participate in those activities. Thus, it is important that transport planning not be restricted to the use of the traditional UTPS approach and that policies and planning be based on a fuller understanding of the behavioral basis of decision choices of households and individuals undertaking travel in cities.

9.1.2 Discrete Choice Models

Disaggregated behavioral approaches to the study of travel in an activity context have been dominated by *discrete choice models*. These models are based on the economic theory of consumer choice and theories of how households allocate time (Becker, 1965; Winston, 1982). The latter assumed (questionably) that people allocate time according to the principle of utility maximization (Prelec, 1982). These models basically assume that, "within the time and cost constraints imposed by their budgets, people choose to spend time in activities which is proportional to their (process and/or goal achievement) utilities" (Axhausen & Gärling, 1992:326). Over time, people may change their activities, and utilities are assumed to be dependent on duration, frequency, and time of day.

Discrete choice modeling techniques have been used in transportation research to model the choice of mode, departure times, or vehicle type (Ben-Akiva & Lerman, 1985; Bunch, Bradley, Golob, & Kitamura, 1993). A significant shortcoming is that "they are limited to the modeling of how single trip choices are related to properties of the choice alternatives and characteristics of the group to which individuals making the choices belong" (Gärling, Kwan, & Golledge, 1994:1).

However, these models have provided a considerable understanding of:

1. The characteristics of trip chaining (Damm & Lerman, 1981; Kitamura, Kazuo, & Goulias, 1990); choice of activity participation and duration (Kitamura, 1984).
2. Choice of activity pattern (Adler & Ben-Akiva, 1979; Recker et al., 1986a, 1986b).
3. The utility of time (Winston, 1982, 1987).
4. The nature and structure of activity patterns taking account of spatial, temporal and interpersonal constraints (Hanson & Huff, 1986, 1988; Pas, 1988; Pas & Koppelman, 1987).

But one of the problems with most of the utility-maximizing models is that they have failed to produce a conceptual framework on how people actually manage or do not manage to schedule activities so that utility is maximized.

To overcome this difficulty, recent research has seen the development of more eclectic conceptual models, which do not rely on the assumption of utility maximization. These models have tried to build on knowledge about how people actually make decisions (Abelson & Levy, 1985) and the capacity of people to process information (Simon, 1990). The focus of these models has been on conceptualizing how a given set of activities are scheduled, assuming that choices of participation in a set of activities have already occurred, and computer simulation is envisaged as a viable modeling tool.

9.2   Early Behavioral Approaches

At a general level it is possible to distinguish two early approaches to modeling activity and travel behavior in cities:

1. *Economic–psychological-based models* that assumed utility maximization in which travel is modeled as a choice process in isolation from a wider set of human activities.
2. *Behavioral-based models* that, while also employing a type of utility-maximizing assumption, assumed that travel is only one of a range of complementary and competing activities operating in a continuous pattern or sequence of events in time and space.

9.2.1   Multinomial Logit Models

In the latter context, travel was seen as the procedure by which individuals trade time to move location in space in order to undertake successive activities (Hensher & Stopher, 1979:12). In terms of the development of models, two broad approaches were adopted: first, those concerned with attempts to improve the *multinomial logit (MNL) models* that formulate individual choice; and second, those concerned with attempts to examine alternative model structures of travel behavior, in particular, Markov and semi-Markov processes, and threshold models.

The MNL model is a discrete choice model in which the fundamental notion is that an individual chooses an alternative that is perceived to have the greatest utility. (The model is discussed in detail in Chapter 10, Section 10.3; so we do not elaborate on the basis of the model and its formulation here.)

The MNL model of choice assumes that the ratio of probabilities of choosing one alternative over another, where both alternatives have a nonzero probability of choice, is unaffected by the presence or absence of any additional alternatives in a set. This results in two major problems. First, a failure to recognize that all alternatives are distinct can lead to a biased estimation of model parameters. Second, prescribed market share changes may occur when any existing alternative is altered or the set of alternatives is changed by addition or deletion (Hensher & Stopher, 1979:13). But in recent years there have been many improvements to the MNL models. In particular, researchers have sought alternative formulations of individual choice models that are capable of producing information to assess the degree of similarity among choice alternatives. There has also been considerable progress in using information theory as a basis for assessing MNL models.

9.2.2   Markov Process Models

Markov process models, and particularly *semi-Markov process models*, provide an alternative form of transportation models. In these models it is necessary to define a set of states that an individual might occupy at various points in time. The process then defines either the probability of an individual being in a given state at time T, or, given the total number of individuals in various states at time (T-1), the model estimates the number in each state at time T, based on a transition probability matrix

(Hensher & Stopher, 1979:18). These Markov models have been attractive because the concept is one of a dynamic model. But they do have some drawbacks. Foremost is assumption of independence of previous states occupied by the individual; this in turn needs an assumption of stability in the transition probabilities. There is also a problem of defining the state that an individual occupies at any given time.

9.2.3 Threshold Models

A further approach used by early behavioral researchers was to employ *threshold models*. While these operate within the utility-maximizing assumption, an important departure from the logit (discrete choice) model approach is that an individual is not seen as reacting in a continuous fashion to attribute change but, rather, is seen as reacting only when the changes are sufficiently large to cross a threshold. This could be a threshold of awareness or a threshold of acceptability. The notion of threshold is very much related to the notion of man as a *satisficer*. An extension of the threshold approach was the *elimination by aspects (EBA) model*, which postulated a hierarchy of attitudes or aspects, each of which must meet a threshold value in an alternative before that alternative can be considered or chosen (Tversky, 1972).

9.2.4 Notion of Indifference

Inherent in models relating to thresholds and elimination by aspects is the notion of *indifference*, which relates to the way a traveler trades off between the generalized costs of alternative modes of travel. The notion of what is the *just noticeable difference* on an individual's choice of mode of travel will depend upon his or her indifference surface, which comprises probability bands of indifference around levels of utility for alternative modes. This approach incorporates a classical economics indifference surface analysis proposed by Quandt (1956). The action that is required to move from a lower to a higher utility level involves a utility cost that is not compensated for by the gain in utility. This effect is analogous to the threshold concept, suggesting a net utility band of indifferences.

9.2.5 Early Time–Space Activity Models

The 1970s saw the evolution of travel and transportation models with a broader theoretical behavioral base. Their objectives were to isolate the mechanisms by which actual and optional interaction of individuals in time and space occurred. Their focus was on:

- The nature of choice and constraints.
- The sequential courses of events at both an individual and an aggregate level.
- How events are distributed in coherent blocks of space–time.

This time–space approach to the study of human activities as spatial behaviors within a choice framework provided the basis for subsequent development of models incorporating the activity time–space concept. The unit of analysis in this approach became the individual or household carrying out a typical regime of daily activities. Activities were defined by: performance time for an individual on an origin/destination point basis; the spatial locations in which activities occurred were mapped in time and space; and the trips taken on a multimodal transport system were modeled. This approach was exemplified by the PESASP *simulation model* developed by Lenntorp (1978), which evaluated alternative paths to show the number of possible ways in which a given activity program can be carried out in a particular environment.

9.3   The Human Activity Approach

A *human activity* approach to modeling travel behavior has been conducted within the *time–space prism* context that was discussed in Chapter 8. As explained by Brög and Erl (1981)

> By the mid 1970s, an international discontent had arisen with models generally used for transportation planning. New, behaviorally-oriented approaches started to dominate the discussion among people involved in transport research. While the standard five-stage transportation demand models had depicted the locational change of the individual as the basic transport situation, and statistically combined these into trip patterns, the new approach focused on the individual as being embedded in a complex social and material environment. The notion that out-of-house mobility is not an end in itself, but rather serves as a means for individuals to participate in activities which cannot or should not take place in their direct neighborhoods, was of far reaching consequence for further research. It made it necessary to change from a mobility or trip survey to a method of determining out-of-house activity patterns. Since the analysis of activity patterns quickly showed that an individual's activity pattern is not only influenced by his or her own autonomous decisions, but also by the other members of the household within which an individual lives, it became necessary to include entire households in empirical surveys. (p. 1)

In discussing the need for the development of models based on the human activity approach, and in identifying the problems inherent in the discrete choice-based models used in the 1960s and 1970s in transport planning, Burnett and Thrift (1979:116–119) list the following:

1. A major defect of choice-based models was that they paid lip service to the notion of travel as derived demand.
2. The majority of models were based on the assumption of two-stop, single-purpose journeys, which represented only about 60% of individual journeys within cities, with multistop journey sequences constituting the remainder.
3. Many of the models assumed static equilibrium conditions. The MNL model of mode or destination choice assumed that there was a single allocation of trips between choice alternatives by a given population group, and this allocation presumably arose from stable individual utility functions. The allocation of trips

reached a new equilibrium following a shift in the socioeconomic characteristics of travelers or a change in mode or destination attributes.

4. There are important long-run changes in utility functions that many models do not allow for. Long-run alternatives in travel needs or constraints can generate alterations in the individual's daily activity set and hence in purpose, type of frequency, timing, destination, and route choice.

5. Choice models derived from economics and psychology, and that were presented in a utility-maximizing framework, do not really explain individual travel behavior per se. The individual is considered atomistically, that is, out of the context of the individual's household of which he is a member.

The human activity approach provided a theoretical framework that did not suffer from these limitations. The approach linked urban travel behavior to the fulfillment of the needs of individuals in households through the formation and accomplishment of activity sets. Travel was handled as a *derived demand*. Daily allocation of trips could be described and explained. Single- and multipurpose journey generations could be understood. Short-run variations in demand for public transport facilities could be related to factors such as diurnal, weekly, or monthly changes in activity sets, as well as to changes in learning about destinations, modes, or routes. Long-run variations in demand could be associated with changes in socially determined needs, with changes in time and space constraints through alterations, and with changes in socioeconomic status, technology, or characteristics of the physical environment. Travel was seen as the outcome of all of these elements.

9.3.1 The Activity Concept

In using the human activity approach as a basis for modeling in transport planning, three time–space-related concepts were used to transform the general theoretical approach into a rigorous and precise framework with many potential applications:

- Travel choice options constrained in space–time.
- Trip generation as an outcome of activity choices.
- Constant time budgets.

Because transportation planning ultimately is concerned with traffic flows and volumes, it was necessary to use the individual trip as the basis for an activity-oriented study. However, the trip should not be viewed in isolation. Rather, the manner in which an activity is related to a specific trip, how this activity fits into the overall activity program of the individual and the household, and which out-of-house activities result needed to be considered. Brög and Erl (1981) presented this highly complex set of interrelationships in the diagrammatic form reproduced in Figure 9.2, which illustrates the *activity patterns* of the individuals in a household. A husband is seen to drive his car to work. A wife goes shopping twice; once she walks, and once she uses a bicycle. Later in the day, she uses public transport to go and meet her friend. A son uses his own car to drive to the university, and on the way home, he stops at a supermarket to do some shopping for the family. A grandmother

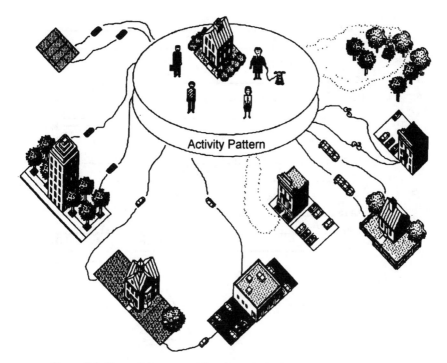

Figure 9.2. The activity concept for a household. *Source:* Brög & Erl, 1981.

takes the dog for a walk. The household members have undertaken eight out-of-house activities during the day. These activities may be classified as work, education, shopping, and recreation trips. The household members actualize their activity patterns differently. In total, members of the household have left the house seven times in order to achieve their activity patterns, and they make a total of 15 trips using three modes of transport: individual modes (7); public modes (2); and nonmotorized modes (6).

Another way of representing travel and activities of individuals in households is shown in Figure 9.3. Duration of travel between activities is presented on a time axis in Figure 9.3A. The activity time perspective is added to the travel perspective in Figure 9.3B. Note that at some locations multiple activities take place, and travel itself is an activity, classified as A3. A further concept is illustrated in Figure 9.3. As it takes time to travel through space, there is a limit to the range of destinations that may be visited in a given period of time. Travel by a faster mode of transport will increase the distance range and therefore the number of practical opportunities. This concept, formalized by Hägerstrand (1970), is the *space–time prism* discussed in Chapter 8, Section 8.2.2. In Figure 9.3C, an individual is committed to take part in activity A_1, at location l_0 (say home), from t_0 to t_1, and to engage in activity A_2, at l_1 from t_2 to t_3. He or she has uncommitted time from t_2 to t_1, but the use that can be made of it depends on the speed with which travel from 2 to 1 can be undertaken. On foot it might take $(t_2 - t_1)$, leaving no time for an alternative primary activity.

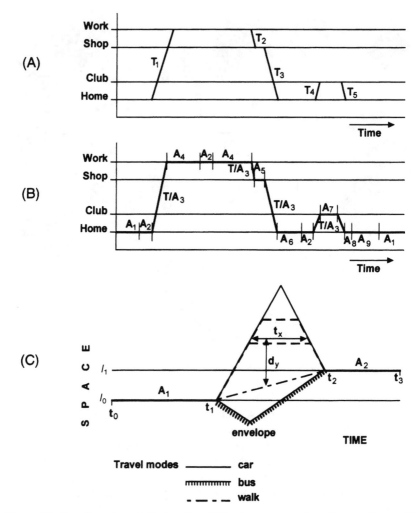

Figure 9.3. Travel mode, activity travel, and the time–space prism. *Source*: Jones, 1979: 63–64.

Faster modes of travel will enable the person to reach an enlarged area of space, and within this envelope, some trade-off distance between intermediate destinations can take place. Public transport waiting times effectively shrink the prism and reduce the range of destination choice and the time that may be spent there. Space–time prisms will thus define a potential range of choice.

In the transport-planning context, trip generation may be regarded as a "trip/no trip" travel option, and this will be a function of a number of elements, including socioeconomic characteristics of the potential traveler, the level of service offered by public transport systems, and so on. Within an activity framework, Jones (1979:65) says that the "trip/no trip" dichotomy can be replaced by an examination of activity choices that lie between participating in activities at the decision maker's current location, involving "no trip," or in other activities that make use of facilities elsewhere, hence requiring travel. This decision has three components:

- The range of possible activities.
- The set of destinations offering suitable facilities for activity participation.
- The characteristics of the available transport system.

Individual choice is a function of the attributes of these three factors and is constrained in space–time for the appropriate *prism*. In this way, travel can be handled as a derived demand. Activities utilize a finite limit of daily time budgets (24 hours). They also require facilities to be available at a limited number of locations in space (often only at home). Further, locations operate at fixed times or between certain hours of the day. Other activities are fixed in timing and/or location to a varying degree, and they provide fixed points in space–time around which more optimal or flexible activities have to be analyzed. The degree of freedom of choice at such times is limited absolutely by the appropriate space–time prism, and often by financial constraints, which are determined by the overall availability of transportation modes. Choices lie between combinations of activities and locations within the prism that are perceived as options by the individual. At an aggregate level, factors operate that tend to control overall time allocations to activity and travel time, placing a further limitation on activity and travel patterns (Jones, 1979:66). The role of travel is particularly important within this activity approach, because it can both be diminished and enhanced through the role of time. Travel takes time, thus diminishing the number of activities that can be undertaken. But travel is also an enhancing agent because it smooths the operation of daily activity patterns in space and time.

9.3.2   A Model of Individual Travel Behavior

An early example of attempts to model individual travel behavior was proposed by Kutter (1973), who sought to explain cause–effect relationships in people's movements in urban areas by studying the behavior of a sample of households in Germany. He proposed that people determine the individual structure of role behavior. In turn, this implies certain time sequences of place-related activities. The sequence of spatially distributed activities becomes the basis for seeking explanation of cause–effect relationships in individual travel behavior. He focused on the day's schedule of activities for one individual. The individual (P_i) has at his or her disposal an urban system composed of communication facilities (G_k), and the communication means and paths (W_k). In the course of one day, the individual (P_i) carries out activities (A_{ij}), corresponding to the individual structure of role behavior, with the parameters of the communication facilities (G_k) and the parameters of time and duration of the activity itself. The spatial distribution of facilities induces changes of place (F_{ij}) (for $j = 2, n$) with the parameters of the activities related to these changes of place. These relationships are illustrated in Figure 9.4A.

   Kutter defined the *individual activity pattern* as the sum of all activities carried out by one person in the course of one day, plus the movements required for these activities. The individual movement pattern only included movements within the scope of an activity. The *activity system* of an urban area was defined as the sum of all individual activity patterns for persons within that urban area. As such activities were elements, and the movements were relations in a behavioral system.

Total urban person movements (which include all changes of place, even walks to a shop or rides on a bicycle) are the sum of the individual movement patterns caused by the superimposition of the activity system as a behavioral system on the urban environment as the material system. This is represented in Figure 9.4B. Thus the individual characteristics of people in the behavioral system and the characteristics of the existing material system influence the nature of regularities of urban person movements in an urban area.

(A)

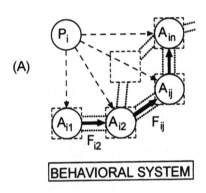

BEHAVIORAL SYSTEM

(B)

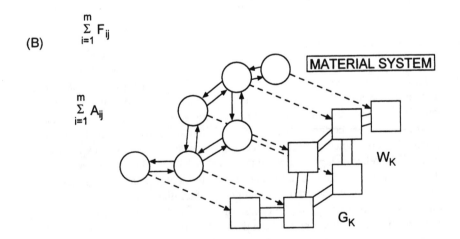

A = An individual activity pattern
B = Superimposition of a behavioral system on a material system

Figure 9.4. Kutter's activity pattern individual travel behavior model: (a) behavioral system; (b) material system. *Source*: After Kutter, 1973:241.

Kutter limited his analysis to what actually exists in terms of structures in the urban environment, and he hypothesized that spatial and temporal structures of urban person movements would be determined by the concurrence of individual activity patterns. In this way he was able to assign typical behavior categories to these individual behavioral patterns. The empirical data necessary to conduct such an analysis are time budgets of out-of-home activities, and required the:

1. A priori grouping of individual populations according to individual characteristics of persons (socioeconomic status, stage in life cycle, car ownership).
2. Detection of components of both activity patterns and movement patterns for these defined groups of people.
3. Examination of a priori grouping with regard to its significance for activity patterns and movement patterns (work, school, shopping, personal, business, social, recreation activities).

An *activity index* was defined as the quotient of persons being out of home and of all persons in a given population. In this way Kutter was able to divide the activity profile of a population into activity profiles of single activities shown in Figure 9.5.

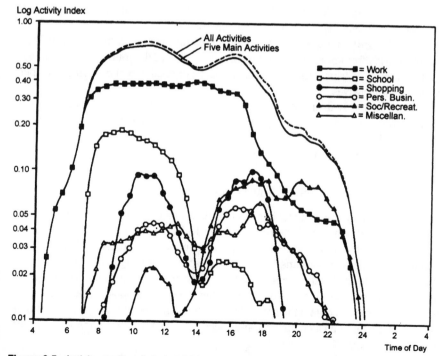

Figure 9.5. Activity profile of study of Tokyo household, divided into the activity profiles of main activity. *Source:* After Kutter, 1973:241.

The main activities depicted accounted for about 80% of all activity trips and accounted for 93% of the duration of all activities in the activity pattern. The duration of activities was a determining factor for the requirement of an individual time potential, and an activity model was defined as the matrix of duration of the five main activities applied to a number of urban zones and to a number of time intervals in the day. In this way, the movement indices for any population or any group in a population could be specified according to the definition of activity indices. For any group of the population, it was possible to show the distribution of specified time period trip generations that are influenced by the duration of main activities, such as work, school, and shopping.

A further aspect of individual travel behavior analyzed by Kutter was to determine whether certain activity categories and their time and space ties were relevant to certain groups of the population. The determining factor in the structure of individual activity patterns was the occurrence of work over a day's period. The time ties of work also determined other activities of employed persons. Considerable importance was attached to the fact that car ownership in no way affected these individual movement patterns. Although car ownership might increase recreation trips, such an increase in trips is of little relevance to the travel patterns that are of interest to travel planners. Three broad conclusions were drawn by Kutter (1973:252–253) from his study:

1. The total number of daily trips is related, to some degree, to the age of a person and role behavior. It is evident that trip reduction is an insignificant value to differentiate human behavior in an urban environment.

2. A significant grouping of urban populations results from the hourly distribution of trips from home and back (first and last leg of the journey). According to these home-based trips, urban populations can be divided into pupils, employed persons, and housewives.

3. The hourly distribution of changes of place is determined by the temporal distribution of activities over the period of a day. As individual activity patterns are constituted by simple journeys, activity profiles are indeed significant criteria of the resulting movement models.

9.3.3   A Demand Model for Travel Using Queuing Theory

A study using a queuing model approach to investigate demand for travel in an activity framework was conducted in Japan by Kobayashi (1976, 1979). Using the analogy of transportation systems on queuing models, a representation of transportation and activity systems by *serial queues* was investigated. The model assumes that time required for transportation and various activities is a key factor to determine the number of trips for users within the study area. A cost-effectiveness function was used to obtain the optimal trip patterns of users. In this way, an optimal visiting rate could be calculated for any specification of transportation and activity time.

The basis of the model is shown in Figure 9.6. A route from an origin area A to a destination area B in Figure 9.6a provides the structure of the activity model.

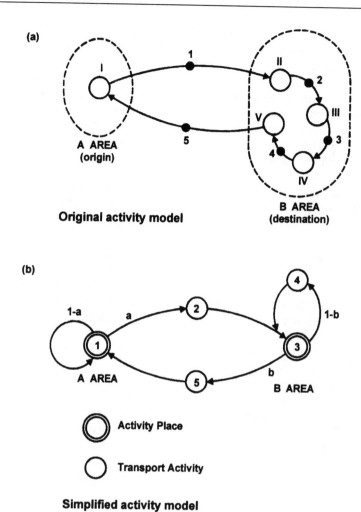

Figure 9.6. An activity model for the analysis of trips. (a) Original activity model. (b) Simplified activity model. *Source*: Kobayashi, 1979:111.

Starting from the origin (I) in area A, an individual travels using a commuting transport system (1) to the first destination (II) in area B. After spending some time at II, he or she leaves for the next destination (III), using one of the area's transportation systems (2) in area B. The same approach is applied to the rest of the journeys within the area's transport system. An objective for the model was to predict the number of visits to the destination places in area B and the number of journeys to the area B from the origin area A. These numbers of trips are usually considered to be a function of effectiveness of both the transportation system within the area B, and between A and B areas. Area B may be regarded as the destination cluster of similar functions such as shops, leisure facilities, office blocks, and so on. Its probability characteristics are postulated as being equal. The most random case is when the probability distribution function of the required times to be consumed at destinations is expressed in an exponential form. Figure 9.6a can then be simplified as Figure 9.6b, and the destination places (II, III, IV, and V) are combined into one queuing stage 3. The area trips 2, 3, and 4 are also combined into one stage 4. In Figure 9.6b, five queuing stages are shown to represent both the activity and the transportation systems. Queues 1 and 3 have a double circle to show the activities at the origin and destination areas respectively, and queues 2, 4, and 5 have a single circle to show the transport activities within and between areas A and B.

Kobayashi made a number of assumptions about the queuing process.

- The service times of each queue stage are exponentially distributed.
- The interarrival times at queue 2 are also exponentially distributed.
- The number of servers for each queue is infinite.

The interdependence of each queuing stage was finite, and waiting time of users could be calculated. A transition probability matrix was calculated between origins and destinations, and a specifically conceptualized cost-effectiveness function was developed to calculate optional visiting rates at activity locations. Optional trip patterns under various transport circumstances were calculated.

Kobayashi tested the model empirically using data collected in October, 1976 in the Japanese cities of Osaka and Takatsuki from samples of 1,000 in each city. He collected data to identify the utilization time for various mixes of journey purposes. Interviews were conducted in all cases at the time of the trip making. The most important data to be collected were allocations of time. In addition, socioeconomic characteristics of the interviewees were noted. The time associated with each trip and activity and a record of all trips undertaken were collected from respondents.

Activity place codes were used in coding the sequence of activities of journeys. Thus, a place trace code of 51345 would represent a journey from home (5) to office (1), to leisure (3), to private business (4), and home (5). The time trace for each activity and travel phase could also be recorded, for example, 30(150)20(50)15(40)30. This represents 30 minutes travel time from home (5) to office (1), 150 minutes spent at the office, followed by 20 minutes travel from the office to a place of leisure activity (3) where the person stayed for 50 minutes. This was followed by a 15-minute trip to complete private business that took 40 minutes, and finally the individual traveled to the final destination, home (5), which took 30 minutes.

In performing a validation test, actual visiting rates and the calculated visiting rate were compared using the activity data, including the times associated with the use of the commuting transport system, the time associated with activity places, and the time spent using the area transport system. Actual visiting rates for any segment were calculated. The relationship between the average number of visits predicted from the model and that present in the population could thus be calculated for various values of differences between actual and calculated visits. It was found that about 75% of the individual observations were predicted by the model to within 20% of the actual number of visits researched. The best agreement occurred with nonstudent shopping journeys on a Sunday and home-based daily shopping journeys. Commuter and office-to-home journeys produced good levels of agreement. Home-based daily shopping journeys and office-based, noncompulsory journeys showed lesser levels of agreement. The correlation between the actual number of visits and the calculated number of visits for each segment, and for the total sample, overall was 0.84. This suggests that the calculated optimal visiting rate was strongly related in a positive way to the actual visiting rate. Kobayashi concluded that the model was validated by real-world data, at least for noncompulsory journeys (Kobayashi, 1979:107–110).

9.3.4   The Household Activity–Travel Simulator Model

A further example illustrating the human activity approach in the development of individual-oriented behavioral models in transport planning is the *household activity–travel simulator* (HATS) model developed at the Transport Studies Unit at Oxford University in England. This is an interactive gaming device that can be used in the context of a household interview situation. It enables members of the household to create a physical representation of their daily activity–travel patterns, which can then be modified under guidance from the interviewer to explore responses to different hypothetical external changes, such as revised public transport schedules or the introduction of staggered working hours.

In the HATS model, each member of the household was provided with a board on which he or she set out his or her activity–travel patterns. The lower part of the board comprised three horizontal, parallel grooved sections marked with 15-minute time periods relating to away-from-home, traveling, or in-home activities. The upper part of the board was in the form of a map display for the environment in which the person lives. On it were marked the locations at which activities take place and on which the travel routes can be indicated. Colored blocks of different sizes were provided that fit into grooved sections on the board to represent the time periods spent on different types of activity. Up to 10 activity groups were distinguished by color, and up to six different modes of travel could be represented.

In explaining how the HATS game works, Jones (1979:7) tells how once each member of the household has set the board to show his or her activity locations, the duration of those activities, and the route and mode of travel used, then it is possible to look at changes that will be made by each individual member of the household in response to some given external space–time modification. The household member who is directly affected first by the external modification will alter his or her board and then consider the likely secondary effects that will result throughout the day. Other members of the household then consider any indirect impacts they might experience. In this way, an interactive procedure is used to arrive at a solution that is acceptable to all. The game encourages discussion among all individuals in the household at all stages of the procedure, and a tape record is kept for subsequent analysis. The use of the gaming board provides a clear visual display of daily behavior and draws attention to the structure of the individual activity patterns among household members. It enables the identification of any anomalies that may arise from any tentative adjustment, such as being in two places at once or having unaccounted periods of time.

The model was used to study the impact of a proposed school hours change in West Oxfordshire in England (Jones, 1979). In this study, the model identified three forms of repercussion: (1) changes in the school journey; (2) changes in other travel patterns and associated out-of-home activities; and (3) changes in in-home activity patterns.

The experiment worked well, and household response was generally favorable. The use of the HATS game was able to improve considerably on intuitive forecasts about the impact of the proposed change in school hours.

9.3.5 A Situational Approach to Modeling Individual Travel Behavior

An interesting extension of the disaggregate behavioral models used in transport planning derived from the human activity approach was what Brög and Erl (1981) described as the *situational approach*. This approach assumed that the individual has a certain number of options in the operational environment and that these form an objective situation. The options that an individual has were determined by:

1. The material supply of the transport infrastructure.
2. The constraints and options of the individual and his or her household (which can be sociodemographically deduced).
3. The social values, norms, and opinions relevant to transport behavior.

The situational approach assumes that individuals and their behavioral situations determine individual options and that decisions concerning spatial behavior are made using a subjective logic, which is quite different from that of the transport planner or politician. This does not necessarily mean that individuals act in a totally irrational way, only that rationality would be subjective. The approach has as its basis the activity concept illustrated earlier in Figure 9.2, which showed a highly complex set of interrelationships among members of a household. The activity and travel patterns can be considerably altered when, for example, a new transportation policy is instituted. As Brög and Erl (1981:5) state: "this can lead to a reorganization of individual activity patterns, substitutions in destinations, persons and/or times. These substitutions can go hand in hand with a mode switch or they can take place using the same modes which were previously used."

Let us assume that there is a change in the activity patterns of the household. Assume further that there has been an improvement in the public transport system in the city in which the household lives, which permits easy access to a spouse's place of work, who in this case is the husband and father in the household. He switches from using the family car to using public transport to get to work. He continues to participate in his recreational activities, but now these can take place elsewhere, such as at a location within walking distance from his office. In the evening after school, the son picks up his father from his recreational activity, and they drive home together in the son's car. The wife and mother now has use of the family car, and she uses it to do the shopping that her son had previously done on his way home from the university. She no longer uses public transport to meet her friend and now can take her friend along when she drives to the store. Shopping is done by her as previously, but the grandmother now also does some shopping, since the wife and mother is now occupied elsewhere with the car. She stops at a store that she passes each day while walking the dog.

The transport planner, who is only interested in the number of trips made by public transport, is overlooking an important consideration; namely, the number of additional passengers either being attracted to or being lost from public transport. In the example given by Brög and Erl (1981), after the introduction of the new public transport route there are significant changes in trip modes by members of the household before and after this change as depicted in Table 9.1. It is interesting to compare the before and after activity patterns and number of trips and modes of trips by individuals in the household. The activity patterns of all household members have been affected, even though the activity program of the household remains the same. Brög and Erl refer to a major study in West Berlin in Germany that showed the effect of lengthening an already existing subway line led to changes in 11% of all trips, but these changes affected 19% of the activity patterns, 28% of the households, and 35% of the mobile persons included in the study. The effect that the new stretch of subway had on out-of-house activity patterns could not have been identified if one had simply looked at the number of trips made.

This type of model was used in the West German city of Hannover, a city of approximately 1.1 million inhabitants (Brög & Erl, 1981). The aim was to forecast the likely impact of an urban rapid train system. The planning authority was interested in ascertaining how the number of passengers using public transport would increase if public transport time were decreased by 10%, 20%, and 40% . In addition, the authority wished to forecast the likely impact of increases in gasoline prices by 50%, 100%, and 150% . The study looked only at trips made on weekdays. The study was a large one, with over 10,000 persons participating.

TABLE 9.1
Activity Patterns for a Four-Member Household Before and After a Change in Public Transport

| | | Number of times home is left | Number of Trips | | | | Activity Pattern |
			Total	Individual Mode	Public Transport	NM*	
Husband	Before	2	4	4	—	—	W-A-W-F$_1$-W
	After	1	3	1	1	1	W-A-F$_1$-W
Son	Before	1	3	3	—	—	W-B-F$_3$-W
	After	1	3	3	—	—	W-B-S*-W
Wife	Before	3	6	—	2	4	W-E$_1$-W-E$_2$-W-E$_2$-W
	After	3	6	4	—	2	W-E$_1$-W-F$_2$-E$_3$-S*-W
Grandmother	Before	1	2	—	—	2	W-F$_3$-W
	After	1	3	—	—	3	W-F$_3$-E$_2$-W
Total	Before	7	15	7	2	6	
	After	6	15	8	1	6	

NM*, nonmotorized trips; S*, service.
Source: Brög & Erl (1981), p. 7.

In aggregate terms, it was found that for each driver almost every second car trip could have been made on public transport. However, when one looked at the actual options for the individual (that is, when one identified those persons who were actually oriented toward, or amenable to, the use of public transport in their choice decision making), the proportion of trips that could have been made with public transport was very much smaller. The study identified eight situational groups that related to trip categories, not persons. They were determined either by objective reasons (such as no alternative mode of travel), by personal reasons (car needed at work), by subjective reasons (persons insufficiently informed), or people disliking a specific mode.

Brög and Erl (1981) showed that it was possible to determine the maximum potential for change and estimate the equivalent responsiveness in making changes in public transport in a city. Reducing travel time resulted in a much greater potential for change than increasing the price of gasoline. Elasticity coefficients showed that, when using the car became increasingly difficult, the number of trips made by public transport increased considerably. However, when the impact of increased cost of gasoline was further examined, it was shown that many adjustments could be made to given activity patterns for all situational groups, and changing to public transport was only one (albeit a major) way of reacting. The potential for change to public transport use from current car–driver trips in the Hannover study increased when persons were better informed, demonstrating the scope of possible adaptability of persons in the learning process when a planned policy is introduced. From the status quo, where the maximum potential gain is a 5% conversion to public transport, this might be increased to a maximum of 35% with improved information and long-term attitude changes resulting through equalizing the perceived values of the time dimensions, cost, and level of service.

9.4   A Cognitive Framework for Analyzing Activity Choice

The intent in some of the modeling approaches discussed above was to conceptualize, analyze, and predict repetitive travel behavior. But the approaches used were devoid of the cognitive spatial learning and choice paradigms discussed in Chapters 5 and 7. While traditional theories of urban and economic geography and their assumptions about repetitive travel behavior had yielded uniform accounts of the behavior of consumers and producers, no cognitive–behavioral equivalents of those theories had been produced. To incorporate cognition in an analysis of repetitive travel behavior, two strands of work in behavioral geography are relevant: *spatial choice* and *cognitive mapping*. These two paradigms have quite different approaches:

1. In the *spatial choice* paradigm, the structure of trips as revealed behavior is accepted, as is the dichotomy of dependence of choice and independent attribute variables. Explanation of trips hinges on inferred choices, subjective utilities, and preferences. Empirical studies have furnished a considerable increase in the knowledge of cognitive dimensions of trip destinations, and have established statistical re-

lationships between attitudes to sites and trip frequencies to them. In the area of discretionary travel behavior, the choice paradigm has had a pervasive effect, especially in studies of shopping and recreational trip behavior.

2. In the *cognitive mapping* paradigm, an individual's travel patterns, within the notion of activity space, provide considerable insight into the way people develop cognitive images of their environment. However, this cognitive approach has had little impact in overt travel behavior studies.

A synthesis of the choice and cognitive paradigms was seen by Pipkin (1981:171) as being essential if there was to be a significant development in the analysis of repetitive travel behavior in a cognitive spatial-choice paradigm. Developments over the last decade or so have addressed this concern.

9.4.1 Destination Choice in Trip Behavior

In the earlier discussion of individual-oriented behavioral models in transport planning and the analysis of travel, a central factor was the notion of choice by individuals, especially choice relating to mode, and particularly for discretionary activity and travel. *Discretionary* activity and travel, such as some shopping and recreation trips, have tended to receive most of the emphasis in studies by behavioral geographers; whereas the work trip, which is an *obligatory* trip, has been the dominant concern in disaggregate travel demand modeling. The emphasis in cognitive studies on choice in discretionary travel supposed:

1. A priori segmentation of trips from individual behavior.
2. The separation of point destinations from their objective context in real space and from their associative context in cognitive schemata.
3. A choice base model of explanation that, at best, drastically simplifies the account of behavior provided in cognitive psychology.

The approaches developed have sought explanation from cognitive, attitudinal, and other psychological transformations of traditional predictors of travel behavior, such as attributes of sites, distance, and locational variables of activity destinations, and relevant individual characteristics (Pipkin, 1981:148). In particular, the focus has been on destination choice and the perceptions of sites, rather than on size or price, as being salient in preferences in destination choice for discretionary behaviors such as shopping. A problem here is that the attributes cited as being important in preference and choice often are incompatible across different types of activities and their finer subcomponents. Furthermore, in an activity such as shopping there is the major problem of disentangling store attributes from attributes of shopping centers and the retail clusters in which they are found. The multipurpose trip phenomenon is a further compounding issue.

We saw in Chapter 5 the importance of distance perception in behavioral geography. However, it has been relatively rare for distance perception studies to be conducted specifically in terms of destination choice. Schuler (1979) recognized explicitly that cognitive transformations of distance are uniformly implicated as important variables in destination choice, and Spencer (1980) also has provided evidence that accessibility is confounded with site attributes in a single cognitive dimension. The mode choice literature seems to indicate that time spent waiting is perceived as more costly than is time in motion (Stopher, 1977).

In both the conventional and the cognitive models of destination choice studies, it has been common practice to control individual characteristics, usually on the basis of factors such as socioeconomic and demographic characteristics of respondents, car ownership, length of residence, and the location of residence. There is considerable empirical evidence that points to differences in the way mean distance traveled to undertake some activities, such as shopping trips, correlates with socioeconomic status. Also, cognitive-rating models predict consumer behavior for lower-income groups more accurately than they predict behavior for high-income groups. But these attempts at social stratification do not provide a psychological-based behavioral explanation for such differences. The problem of segmentation on the basis of socioeconomic and demographic phenomena and by location is a particularly important issue. The role of attitudes, preferences, intentions, and perceived constraints has been recognized for a long time in travel demand models as being as important, if not more important, than social and age segmentation (Dobson, Dunbar, Smith, Reibstein, & Lovelock, 1978).

9.4.2 Problems in the Spatial Choice Paradigm

Two major problems exist in the use of destination choice models for studying repetitive travel behavior. These involve the divergence of recorded and revealed preferences. Possibly this divergence is due to inconsistencies in individual behavior, as well as to deficiencies of questionnaire techniques used to elicit the true attitudes of respondents to surveys. Furthermore, predictors identified as being related to preference formation and choice behavior may well differ in their salience for different individuals, including within groups that are homogeneous on the basis of socioeconomic and demographic characteristics. Probably it is incorrect to assume that choice actually exists for other than very few groups of people differentiated on the basis of socioeconomic status. Spencer (1980) has shown that, although attitudinal and preference scaling procedures might elicit an individual's true preferences, because of the constraints on income and mobility, these cannot be attained accurately. Thus, desired but unattainable choices will never be observed in actual behavior. Hensher and Louviere (1979) and Dobson et al. (1978) have suggested further that observed discrepancies between actual behavior and attitude may represent a lack of adjustment among structures that evolve at different rates and that there will be different behavioral intentions among people toward specific alternatives in the choice set. According to Pipkin (1981a:158), "These perspectives on the relationship between attitudes and behavior indicate clearly that an understanding of the cognitive transformation of site and location attributes, no matter how detailed, is insufficient in explaining repetitive travel."

Considerable literature has evolved that criticizes the choice paradigm approach to the study of repetitive travel behavior. This is the result of a number of factors:

1. Conventional choice models are not really behavioral. The findings from the time–space approaches to human activity (discussed in Chapter 8, Sections 8.2 and 8.7) have shown that the spatial structure of activity patterns in the behavioral context of travel may be explained to a high degree through the notion of constraints that people face in their day-to-day activities.

2. There is the problem of multipurpose travel, and probably it is more appropriate to focus on studying the structure of trips in their own right, particularly placing emphasis on the strength of linkages between various types of trips, activities, and their locations. Within the context of multipurpose travel, decision making should be recognized as an extremely complex phenomenon, in which potential stops acquire opportunistic utility by virtue of their proximity to other sites. Hanson (1980) provided some interesting insight into modifications that need to be made to choice models in order to address this problem. She suggested that the frequency of visits and inferred preferences for sites within a functional category do not vary on single and multipurpose trips, with some functions actually possessing stronger trip linkages to the workplace than to the home or to other establishments. Traditional categories of data, and their modes of collection, have tended to minimize the importance of such trips. She makes a strong case for the collection of detailed longitudinal information to distinguish between stages of journeys to study the time and space aspects of multipurpose trips.

3. In most geographic studies of trip choice destination, choice sites are characterized by their location relative to the respondent's home. In shopping trips, for example, proximity to other sites, such as banks, drug stores, and discount stores, may be equally important components in the attractiveness of grocery shopping (Schuler, 1979). Furthermore, multipurpose trips need to be studied in both the temporal and behavioral context.

9.4.3 The Cognitive Paradigm and Repetitive Travel

Pipkin (1981:162) claims that there is a fundamental discontinuity between the cognitive approach seeking psychological transformations in predictor variables with respect to trips and the goals originally formulated for cognitive behavioral geography. A distinctive feature of a cognitive paradigm is the structuring of behavior, not by observable, spatial, or topological properties but by inferred motives and intents. Behavior needs are seen from the viewpoint of the primacy of inferred cognitive structures, as represented through cognitive maps. This approach sought to develop postulates of behavior that are general and cognitive and, in some sense, spatial (Cox & Golledge, 1969). It is difficult to reconcile this cognitive paradigm with work on repetitive travel behavior. As Pipkin (1981) states:

> A simple dichotomy of "habitual" or routinized and genuine or innovative choices is inadequate to understand the behavioral context of repetitive travel, since even habitual trips are likely to be linked to stress eliciting choices concerning timing, car availability, income constraints and the like. Thus, in the specific context of decision making, and in the more general task of showing relationships between ongoing mental activity and relevant behaviors, geography has failed to satisfy the requirements of cognitive explanation. (p. 164)

Until the 1970s, little was understood about the formation and characteristics of cognitive maps and the cognitive mapping process that has become the central focus of concern within the cognitive paradigm in behavioral geography. The constructs of cognitive maps had not been integrated properly into behavioral studies of repetitive travel destination choice studies; nor, for that matter, had they been integrated into studies of overt travel behavior in general. One of the problems was that studies of destination choice, particularly in discretionary travel behavior, such as shopping trips, had emphasized the *affective* aspects of urban image, particularly those relating to attitudes toward price, convenience, travel time, and so on. Destination choice studies were concerned with point or specific location site characteristics, which could be characterized only in a few geometric measures (such as distance from home or distance from origin of the trip), with contextual measures relating to the locational setting of retail clusters and shopping centers. But cognitive maps are represented by configurations in a more holistic form. In travel studies, attitudinal and other cognitive variables usually had been applied as independent or explanatory variables in relation to dependent choices or the preference rankings of individuals. In studies of cognitive maps, the contents of images usually were treated as dependent variables in relation to factors such as socioeconomic status, length of residence, mobility characteristics, and the activity patterns of people (Pipkin, 1981:164–165).

Little evidence was available to demonstrate the existence of a general axiomatic system for travel behaviors that was strictly cognitive and distinctively geographic. Even the approach used by Burnett and Hanson (1979), in which single or multipurpose trips were studied in a time–space context, was basically noncognitive (in the sense that the cognitive paradigm is concerned with viewing behavior as a continuum of motivated acts in which various goals may be pursued simultaneously). It was apparent that, in order to bring together the choice and the cognitive paradigms, it was necessary to concentrate on the cognitive processes of decision making rather than on the structure of schemata as such.

9.5 Computational Process Modeling of Household Activity Scheduling

Significant advances toward addressing the issues raised above have been made in recent years through a renewed focusing of research on the *scheduling* of individual and household activities involving the development of *computational process models* (CPM) of travel choice as a "valuable alternative or complementary tool to mathematical–statistical techniques of discrete choice modeling (Gärling et al., 1994:1).

The scheduling of activities is a process consisting of independent choices where each involves the acquisition and storage of information, memory retrieval, accuracy–effort tradeoffs, and conflict resolution (Gärling et al., 1994). It is useful to consider this as analogous to a production system and to develop a model based on how people think when they solve problems as an alternative (or at least a complement), to discrete choice models. A production system is conceived as "being realized in a cognitive architecture featuring a perceptual parser, a limited-capacity evoking memory, a permanent memory, and an effector memory" (Gärling et al., 1994:6). An operational CPM is a production system implemented as a computer program. A CPM can provide a more detailed description of the individual choice process than is the case in discrete choice models. But there is a requirement for more than travel diary data, which often record only spatiotemporal actions of individuals and households and which rarely consider the cognitive processes underlying decision making and choice behavior.

New techniques of data collection that trace the cognitive processes preceding overt choices need to be developed, and this is occurring in areas where CPMs are being used (for example, Ericson & Simon, 1984; Ericson & Oliver, 1988). The objective is to use CPMs to reveal principles that will apply generally. This involves aggregating detailed descriptions of individual choice processes if we are to use CPM to forecast the consequences of travel and transportation policy measures in cities. However, CPM have significant advantages over the discrete choice model where the goal is to test theory, whereas a disaggregated approach is preferable in principle because it is more sensitive to possible violations of assumptions.

9.5.1 Modeling Travel Route Choice

For 20 years researchers have been attempting to implement a conceptualization of travel choices in a computer program aimed at emulating how people make choices. Probably the first attempt to develop a CPM of travel choices was Kuipers' (1978) TOUR model, which sought to represent an individual's cognition of the environment, how it was acquired, and how it was used to choose routes. TOUR was successful in many respects, but it was not based on the extensive empirical research on people's cognitive maps, their spatial orientations, and how they go about wayfinding (as discussed in Chapters 5 and 7). Recently, work by Gopal et al. (1989) and Gopal and Smith (1990) has built on the empirical work of Golledge et al. (1985) investigating children's acquisition of spatial knowledge. A program called NAVIGATOR has been produced that models route planning by means of various heuristics. If information for making a route choice is lacking, then "making a random turn at an intersection," or "moving in the same general heading," are examples that are implemented in the model.

Route planning in a static environment has been modeled by Leiser and Zilberschatz (1989) in their TRAVELER model, which makes the assumption that the relative locations of origins and destinations are known. An unknown route from origin to destination is then constructed through the process of search starting from both the origin and the destination.

A model of route planning in a dynamic environment (ELMER) has been proposed by McCalla, Reid, and Schneider (1982) in which routes are conceived as sequences of instructions for how to travel. When one is navigating an environment, routes are retrieved as the need arises; thus, planning is interwoven with the execution of the plan.

But these models neglected the dependencies that exist between different travel choices and between travel and activity choices.

9.5.2   STARCHILD: Identifying and Choosing Representative Daily Schedules

Root and Recker (1983) have proposed a comprehensive conceptual framework for the analysis of the daily scheduling process shown in Figure 9.7. This is a two-stage model:

1. In the *pretravel stage*, an individual schedules activities that satisfy his or her needs and desires.
2. The execution of the schedule is monitored during the *travel stage*, with revisions, additions, and deletions being taken in response to unforeseen consequences of the schedule, new demands, or unexpected incidents, such as a road accident.

The model assumes that people select an optimal sequence of activities in terms of their durations, locations, and modes, so as to maximize the total utility of the set of activities while maintaining the flexibility of the schedule, minimizing time spent away from home, and minimizing the coordination effort involved in the scheduling of activities.

In operationalizing the STARCHILD model, a computer program developed by Recker et al. (1986a, 1986b) makes provision for assigning utilities to activities and their associated wait and travel times depending on whether the activities are planned out-of-home activities (such as work), unplanned out-of-home activities (such as shopping), or discretionary in-home activities (such as watching TV). The utilities of planned out-of-home activities are assumed to be proportional to their perceived importance, multiplied by an objective index of probability of performance.

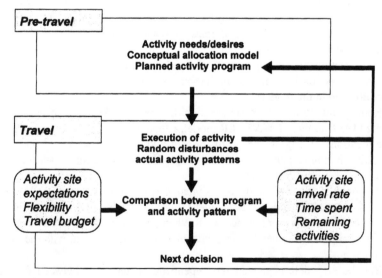

Figure 9.7. A dynamic utility-maximizing framework for choosing between representative activity schedules. *Source*: Axhausen & Gärling, 1992:328.

But the utilities of wait and travel time are assumed to be inversely proportional to the actual wait and travel times, with a slope that decreases with the importance of the activity. Similar methods are used to assign utilities to the potential of individuals to participate in unplanned out-of-home activities. This is achieved by multiplying perceived importance by an index of the probability of performing the unplanned activities at available locations. The utility is made dependent on the time duration since the activity was last performed. Also, the expected disutilities of the additional travel time are taken into account. In-home activities are a problem in calibrating the model because no information was available from travel diary data used in the study; thus, all such activities are assumed to be discretionary, with their utilities dependent on duration at home and the number of other household members being at home.

Axhausen and Gärling (1992:328–329) note that STARCHILD provides an estimated model for the choice between alternative activity schedules; a utility-maximizing framework is used in which the "utilities of alternative schedules are computed by surmising utilities assigned to the activities, wait times and travel times according to rules which differ depending on whether activities are planned out-of-home, unplanned out-of-home, or discretionary in-home activities." In the models, expected utility formulation is similar to the subjective expected utility theory proposed long ago by Edwards (1954) and the more recent theory of planned behavior proposed by Ajzen (1985). But, as its authors acknowledge, the STARCHILD model has a weakness in that it is probably implausible that people consider many different alternative activity schedules. An important issue is how such models reduce the number of alternatives from all the feasible alternatives that people might consider. Furthermore, in this and similar models it is assumed that the time horizon is a day or shorter, whereas this may differ for different activity types. "Other relevant time horizons are conceivable, such as, for instance, the weekly commitments, the time until the next major holiday and the time until the next vacation. Because these time horizons may be similar for a large part of the population, it becomes essential to include them in any conceptual framework and empirical analysis" (Axhausen & Gärling, 1992:329). Also, an individual will respond to the claims of other people with whom he or she interacts, and it is unclear to what extent moral obligations (for example, accompanying a child to a sporting event) are subject to narrow benefit–cost analysis. Heap (1989) has pointed out how societal rules governing choices between activities often dominate decision making. All this serves to demonstrate just how complex it is to model daily schedules of individuals and of households.

9.5.3   SCHEDULER: Generating an Activity Schedule

Gärling et al. (1989) have proposed a conceptual framework that addresses some of the shortcomings of the STARCHILD model. In part it is based on the cognitive model of planning proposed by Hayes-Roth and Hayes-Roth (1979) in which activities are defined as a means by which the environment offers individuals or households an opportunity to attain their goals. "Individuals have preferences for these means based on their beliefs about how instrumental they are and how important the goals are. Choice of participation in activities is determined by these preferences in conjunction with prior commitments" (Axhausen & Gärling, 1992:330).

The concept of *scheduling* is defined as the process of deciding how to implement a set of activity choices during a defined time cycle. This involves an interrelated set of decisions made by an individual, interactively with others, concerning who would participate in what activities, when, where, for how long, and how that person would travel between locations where the activities can be performed. A primary goal is the feasibility of the schedule, and a secondary goal is the minimization of travel. This approach draws heavily on contemporary utility theory (Ajzen, 1985; Fishbein & Ajzen, 1975) and is based on the notions that:

- Commitments (appointments, obligations) are factors determining activity choices.
- Preferences are assumed to be enduring but may change depending on changes in the salience of different goals.
- There is no specified time horizon for commitments.
- Scheduling is assumed to have a limited time horizon of a day or shorter.

Figure 9.8 illustrates the components of the household scheduling model proposed by Gärling, Säisä, Böök, and Lindberg (1986). Rather than assuming that all alternative schedules are generated, it is assumed that a set of activities is chosen from that part of an individual's memory (called the long-term calendar) that stores information about activities. Selection of activities is an order of priority, with cut-off points dependent on how the priority changes and a judgment of the number of activities possible to perform in the time slot being considered. Most people select activities that are few but have a high priority to be performed during that day. The

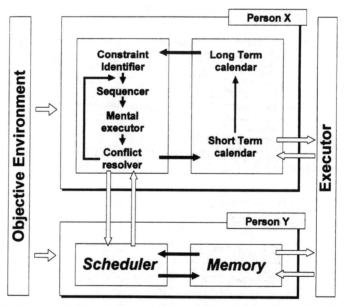

Figure 9.8. Components of a household scheduling model. *Source*: Axhausen & Gärling, 1992:331.

SCHEDULER in Figure 9.8 retrieves memorized information about the environment and identifies when and where each activity can be performed. Gärling and Golledge (1989) have noted that this identification is not trivial because of the many possible biases in how an individual remembers the environment.

The SCHEDULER model takes account of temporal constraints (such as retail opening hours) that impact the sequencing of activities. Activities are required also to minimize total travel, using a variant of the *nearest neighbor* heuristic (Gärling, 1989). Starting with the first activity to be executed, a more detailed schedule is then "mentally executed," with the aim being to evaluate its feasibility under given circumstances. At this stage decisions are made about travel modes. More detailed information is sought also about activity durations, departure times, travel times, and wait times. "Mental execution" will depend on the extent to which this information is accessible. There is an upper limit on how many activities are "mentally executed." Conflicts threatening the feasibility of the schedule, such as overlapping end and start times, are resolved if possible, either by changing the order between the activities in conflict, or by replacing a chosen activity with another activity of lower priority. When the schedule has been formed for the highest-priority activities, attempts are then made to find lower-priority activities in the longer-term calendar that fit into the remaining time slots.

The SCHEDULER model does not assume that all scheduling is accomplished at a certain point in time but that scheduling of activities may continue as the schedule is being executed. What is called the "short-term calendar" in Figure 9.8 works out a skeletal or detailed schedule for a few activities, and, once this is executed, other activities are added. This permits an individual to retain a degree of preferred flexibility in scheduling activities. Also, information may be lacking, thus not permitting the schedule to be formed.

The model framework assumes that some activities may require the participation of two or more members of the household (e.g., family outings), whereas others may be performed alone by the individual (e.g., work). This means that the scheduling of an individual's activities needs to take account of the schedules of other members of the household and also often of other individuals with whom interaction occurs. Thus, information on the time and the location of those interdependent activities is required. In this way, SCHEDULER represents a "rudimentary representation of social interactions between household members while scheduling [sic] their daily activities" (Axhausen & Gärling, 1992:331).

9.5.4 Data Requirements and Operational Complexity of the Models

The complexity of the conceptual framework underpinning activity-based approaches to travel analysis, as illustrated in the STARCHILD and the SCHEDULER models discussed above, highlights the rigorous and detailed nature of empirical data required to operationalize models concerned with the scheduling of activities as distinct from the more straightforward time-allocation approach. "Their behavioral base needs to be strengthened, but that reflects our lack of knowledge in many areas which these frameworks address. Two particular issues are important: the handling of changing and the handling of routines" (Axhausen & Gärling, 1992:332).

The current development of theory in cognitive psychology and cognitive science is shedding some light on how deliberate decisions are assumed to be replaced by the execution of scripts or routines over time, but "the implication may not be tenable that the automation of activity scheduling is identical to the automation of many motoric activities, such as cycling or car driving. The repetition observed could be the result of the avoidance of the work involved in scheduling itself, of the repeated execution of a script superior to others, or the lack of knowledge about alternatives, and the lack of alternatives" (Axhausen & Gärling, 1992:332).

It is not surprising, then, that relatively few models of activity-scheduled choice have been calibrated. But over the last 10–15 years, empirical studies have provided a number of insights into individual and household behaviors:

1. The STARCHILD model, applied to a study of activity patterns in Winceham in 1979, revealed how subjects weighted travel times independently of the importance of trip purposes but that they allowed also for extra slack times to ensure the performance of important activities. In addition, subjects valued positively the potential to engage in unplanned activities, and they more strongly preferred to spend time at home when all other members of the household were present (Recker et al., 1986a, 1986b).

2. In a study of sequential analysis of decisions to participate in different activities conducted in Munchen, Germany, a multinominal logit model was used to estimate every choice of a new activity during a day. Socioeconomic characteristics of the household, and the degree of previous participation in the same type of activities by individuals and other household members were included, and the model estimated separately for working and nonworking subjects. The results demonstrated an influence of previous participation in the activity by the same subject in almost every case, whereas the influence of the degree of participation by other household members was less visible. Weights associated with degree of previous participation increased during the day, indicating a nonlinear effect and a time-of-day dependent influence of this factor (Mentz, 1984).

3. A study of activity types in Baltimore in 1977 proposed a logit model of how choices of activity types—including personal business, social and recreational, shopping, escorting passengers, and return home (on a nonhome basis)—depend on socioeconomic characteristics, number of cars, time of day, and degree of previous activities. Both home and non-home-based trips were estimated. For home-based trips, the choices of activity type were influenced by time of day and the degree of previous enjoyment in the same activity. For non-home-based trips, it was found, too, that choices were influenced by time of day and by the degree of previous participation in the activity, but to a lesser extent. Also, the duration of out-of-home activities and the number of journeys strongly influenced the timing of the return home, suggesting either a long and intensive or a very restrained and short participation (Kitamura & Kermanshah, 1984).

4. A study in 1983 in Israel used a complex nested-logit model of shopping participation with the whole week as the planning horizon. It included data on the socioeconomic characteristics of subjects, household composition, and available time for shopping before and after midday. Not unexpectedly, there were differences

for the days when shops were closed or shopping hours were restricted. But, the interaction of shopping activity across the week was captured by budget variables—the frequency of shopping and the number of earlier shopping trips—and by the logsum term of a nested scheduling model focusing on the timing of shopping (Hirsh, Prashker, & Ben-Akiva, 1986).

These and other empirical studies indicate that:

1. Choices of activity participation and duration are dependent on the position of the day in the week and the time of day.
2. Activity choices are influenced by previous and planned activity participation, and as a result are not made independently.
3. Social interaction within households is an important factor (Axhausen and Gärling, 1992:335). The models have weaknesses related both to their reliance on utility maximizing principles and their reliance on observations of actual choices or revealed preferences. As a result it is difficult to infer casual relationships.

Axhausen and Gärling (1992:335–336) have highlighted the need for research to use a "combination of naturalistic observations and scheduling through diaries or similar techniques, known as experience sampling, which in the past has been used for the performance of activities; observations under simulated but fairly realistic conditions, and observations under experimental conditions using many of the techniques developed in psychology, such as analysis of performance errors, chronometry and, possibly, complete modeling of think aloud' protocols."

According to Axhausen and Gärling (1992), other questions that need resolution include:

1. How fast, and to what degree of accuracy people acquire or update knowledge about destinations, travel times, and travel costs?
2. What is the precise role knowledge plays in activity scheduling?
3. How are scheduling decisions organized over time?
4. To what degree are decisions made simultaneously?
5. What time horizons are used?
6. Why, when, and how are scheduling decisions later revised?
7. How are different spatiotemporal constraints identified and taken into account to achieve an aspired-to degree of realism?
8. What costs are minimized and how?
9. To what extent are activities performed in response to demand by other people within or outside the household?
10. How do utilities or priorities change over time?

9.6 Planning Applications: A Simulation Model for Activities, Resources, and Travel

The evolution of behavioral-based models to analyze individual and household activities in a time–space context represents a significant paradigm shift from the old, but still widely used, UTPS models outlined at the beginning of this chapter. A challenge is how to incorporate these disaggregated behavioral models within broader model frameworks that relate our increasing understanding of the cognitive basis of choice in activity travel with transportation network and services and land use planning in a metropolitan setting. This would provide a powerful policy and planning tool.

Such an approach has been stated by Stopher et al. (1993) through their *Simulation Model for Activities, Resources, and Travel* (SMART). It seeks to link, in a behavioral way, the three major dimensions of the transportation problem: (1) households, resources and decisions; (2) transportation systems; (3) land use while providing the key output required to link volumes and flows on transportation systems.

The model attempts to draw together a range of approaches discussed earlier in this chapter and previously in Chapter 8, including: activity simulation; time–space paths; role allocation and adaptation; activity structure and household decision making; and time–space budgets and activity systems.

9.6.1   The Characteristics of the SMART Model

The model makes a number of crucial assumptions about the structure of travel (Stopher et al., 1993:2–3):

1. The household, through its individual members, is the primary decision-making unit. Individuals, by engaging in activities, carry out these decisions.
2. The requirement to engage in activities that satisfy household needs is the driving force behind decisions that result in travel.
3. The decisions to participate in activities (and who will participate) are the result of negotiation and role allocation within households.
4. Most travel is habitual, depending for form on infrequently made decisions. Commonly, travel patterns do not shift.
5. Travel is, generally, an incidental item to the process that results from the requirement for activities outside the home. Modeling the travel process correctly and realistically requires that activities, *not* travel, be the center of attention.
6. In response to sufficient large external or internal stimuli, such as changes in household structure, resources, transportation investment, or land use change, households adapt incrementally by changing activity patterns and hence travel, within time-budget constraints.
7. Over time, land uses and transportation systems also adapt to serve evolving household activity requirements. Congestion and accessibility serve as one, but not the only or the most important, influence of land use and activity-site selection."

The elements of SMART are shown in Figure 9.9. The model consists of a number of major elements. Land uses are identified by the amount of activity participation in subareas of a city. Transportation system access is described primarily

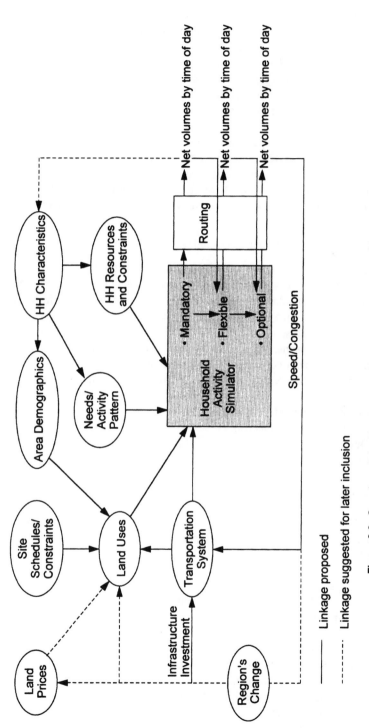

Figure 9.9. Overview of the SMART model structure. *Source:* Stopher et al., 1993:5.

— Linkage proposed

---- Linkage suggested for later inclusion

by intraregional separations with speed feedbacks influencing the system. Household characteristics, resources, and constraints are tied to each individual. Household needs are described as a set of activities that must be performed by each member of the household. A *household activity simulator* generates a specified set of activity patterns that meet time constraints and have a minimum generalized travel cost given the resources and needs of the household.

A GIS format is used to assign trips stochastically to routes via an integrated routing procedure. Speed feedback mechanisms between traffic congestion on route by time of day and activity specification permit claiming, spreading, time shifting, and other policy options to be modeled. Simulation of overall region effects on household activities can be summed to give micro to macro effects as well as permitting macro- to micro-level effects to be analyzed. In this way, the model integrates household decision making and travel with regional land use change.

9.6.2   The Household Activity Simulator

At the center of the SMART model is the household activity simulator, in which household activity patterns are simulated by closely "pegging" mandatory activities in time and space, then adding flexible and optional activities around those requirements. Travel mode choice is viewed primarily as the outcome of the car allocation process (using both personal and household vehicles if available) not primarily of network cost comparisons. Public transport use is limited to no-car-available (i.e., captive) and lower-travel-cost (i.e., choice) situations.

Household and individual activities are categorized by six typologies:

- Mandatory, flexible, or optimal.
- Personal or group specific.
- Duration.
- Location.
- Frequency of occurrence.
- Required start/end times and ranges.

The SMART model requires the collection of household based data involving the characteristics and resources for each person and the needs (activity) patterns of households. Data should cover:

- Age, gender, relationships.
- Cars/vehicles assigned or owned; drivers licensed.
- Activities required (mandatory).
- Frequency of engagement.
- Hours required and locations of mandatory activities.
- Activities that are flexible or operational.
- Frequency of engagement.
- Total household income; lifestyle/stage.
- Travel time budget.
- Total cost budget.

Activity needs data are required for each member of the household to show what activities are assigned or optional for each person, including duration required to perform the activity and the range in duration, start time and end time. The model works by either (1) simulating a set of observed activities for a household, given a background pattern of household characteristics and required activities, against a pattern of land uses (activity sites) and transportation access; or (2) estimating a new pattern of household activities in response to changes in household characteristics, land use, or transportation.

"Model calibration is achieved by choosing an estimated activity pattern set against a set of observed patterns for each household" (Stopher et al., 1993:9). Activities are analyzed according to whether they are mandatory, flexible or optional, and are dealt with in that order. This means that the mandatory activities are identified as quickly as possible before a move to the flexible optimal activities.

9.6.3 A GIS for SMART

The SMART model utilizes GIS technology for the management of both land use (i.e., macroscale) and household activity (i.e., microscale) data. In order to operationalize the GIS, activities are defined so that:

1. In the household context, an activity is any one of a set of goal-satisfying behaviors in which a household member participates.
2. In the land use context, activities at sites are taken to be the number of units of allowable activity, either in space or by the participating individual or group of individuals, that are built or are available at a site.

Figure 9.10 illustrates the GIS view and the data view of a city region from which the GIS is built to operate the SMART model. Each zone in the GIS is, in effect, a "slice" of the activity-trait matrix for the region, and within each "slice" activity traits are information items that describe the "rules" for activity engagement at that site. These include data on: the number of units of activity (e.g., employees, commercial floor space by type, number of households by income); a schedule of allowable activities (i.e., the times open); density and underdeveloped/developed area ratios; total and component land area; built-out or maximum land areas for each activity (which is a function of area and zoning).

Activity participation rates are defined as the per-unit participation rates for activity sites, and these correspond basically to trip rates, expressed as the expected or average utilization of sites, per unit activity, area allowable, per day. As the region grows or changes, this matrix will grow or change.

Household activities are located in the GIS describing the urban spatial data and are treated as a special "activity pattern" layer. Forecasts are made by incrementally changing patterns, land use, household characteristics, or activities to use new services or to fit new circumstances.

Obviously the model requires the collection of comprehensive and complex survey data from all individuals in households using sample survey techniques as well as a comprehensive GIS for a city region. Large budgets are required for such

GIS View   Data View

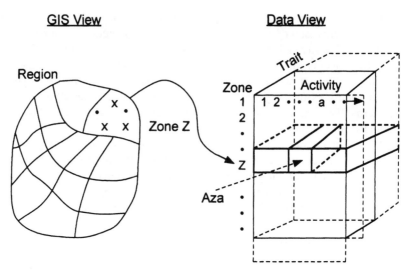

Figure 9.10. GIS and data views of a region for the SMART model. *Source*: Stopher et al., 1993:7.

data collections, and great care needs to be given to research design and data-collection procedures to ensure that valid and reliable data are collected to minimize error and bias. The necessity to use diary methods of survey data collection adds to both cost and data reliability and validity problems in these studies.

9.6.4   Operationalizing the SMART Model

Stopher et al. (1993) outline 11 steps for operationalizing the SMART model using the HAS and the GIS:

1. *Locate the household's activity sites.* Present sites used are located in space using zones or coordinate data that places them in a zone. Activities are classified as: mandatory, flexible, or optional.

2. *Set time blocks for activities.* Fixed times for mandatory activity engagements are specified in the household activity researching system by individual.

3. *Allocate activities to person.* For mandatory activities, person and times are assigned directly, thus "pegging" people to locations at particular times for specified durations.

4. *Provide vehicles.* Vehicles owned by a specific person are assigned for that individual's use, and remaining vehicles are allocated to other household members in priority order.

5. *Estimate travel times.* Spatial separation of activities and the transport network are used to compute generalized costs of travel by mode for zone pairs. The "congested network" is used to estimate travel costs for a particular time of the day when travel occurs.

6. *Choose mode.* Cars are assigned to eligible household members prioritiz-

ing mandatory activities (or pooling is used if possible). Then nonsolo car modes for lower-priority activities are used. A nested logit is used to determine probability of travel for each nonsolo mode.

7. *Choose departure time.* For out-of-home activities, departure times are selected so as to achieve expected arrival time at mandatory activities. Partial time–space diagrams may be drawn for each mandatory activity and its required time.

8. *Cumulative budgets.* Required travel times for each person are cumulated to check that they do not exceed the range of arrival/departure/travel times.

9. *Compute disruption index.* For each person in the household, a set of mandatory activities is available, thus a "disruption index" may be constructed representing the amount of disagreement/difference between estimated and actual mandatory activity patterns.

10. *Network routing and speed feedback.* To ensure that network congestion and site activity are accounted for in analyzing activities, all mandatory activities are first located in the region. Travel is generated corresponding to the travel movement between zones. A set of travel records is computed for the entire data set for each sequential pair of activities for which travel is required. Route choice is left to the individual traveler. This requires partitioning of activity-based travel patterns by time of day (actual start time of trips), by hour, including feedback times under congestion.

11. *Flexible and optimal activities.* Flexible activities are those that have a required fixed location or time, but not both (e.g., shopping, personal, business). Optional activities are those in which time, location, and the event itself are not fixed. An "expert system" routine may be used to determine who in the household is a possible participant in each flexible or optional activity, in which the typical allowable duties of household members are arranged. These activities must take place inside the time and space "windows" created by the mandatory activities. As shown in Figure 9.11, for a hypothetical three-member household, the only time available for total family activities is a "window" in the evening. A minimum-path algorithm can be used to determine travel times and distances to each feasible site for an activity, and travel time and cost are accumulated for each member of the household. In this way, the algorithm allows for the "real world" type of interactions that go on between mandatory and nonmandatory activity choice on the basis of constrained paths from the mandatory model.

9.6.5 Applying the Model

Stopher et al. (1993:19–21) have applied this model to households surveyed in a 1985 home interview travel survey in Charlotte, North Carolina. By way of example, Figure 9.12 shows for two households, each containing five persons and two cars, how each household's travel pattern may be represented as a series of times on the zone map, with locations of home, work, and other activities shown. Each line represents a directed trip (a "desire line") between origin and destination. The plot is in ARC/INFO, using a coordinate system for the Charlotte zone structure.

The SMART model can be used in numerous transportation-planning situations. For example, Stopher et al. (1993:19–22) illustrate how a "random sample enumeration" method of forecasting could be used to model sequentially each

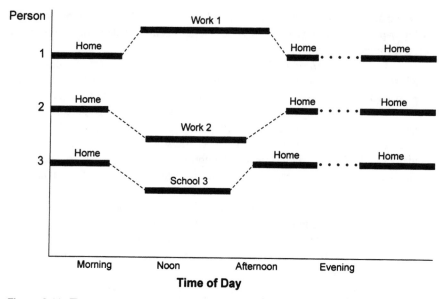

Figure 9.11. Time–space diagram of mandatory activities and windows for flexible/optional activities. *Source*: Stopher et al., 1993:18.

household's time–space activity record and then "expand" the resulting trip pattern by factoring it up to a projected regional total of households. If a new road or public transport service were being considered, the model would analyze each household record in sequence, and only three households that are affected by the proposal would show activity changes, with other patterns not changing. The revised and the unrevised travel patterns would then be expanded to the regional total, and trips would then be assigned to the transportation networks.

If the interest were in changes in household structure or resources, such as adding to the size of households or providing an additional car, then the model would proceed by specifying in advance which households were to get the modified characteristics by altering the appropriate variables in the HAS. The model would then recalculate activity patterns to serve the changed needs of households, or to exploit the new resource, and the resulting trips would then be expanded.

Changes in land use in a region can be viewed at in a variety of ways. For example:

1. If additional services, such as new stores or commercial sites are added, a modified land use pattern would be entered with the land use matrices of the GIS. Use of these new facilities by customers would then occur as households close to the new location consider its effects on their activity need sets.

2. If additional households are added through residential expansion, then the new households would need to generate new activity using a simulation effect derived from specified mixes of household sizes and characteristics to

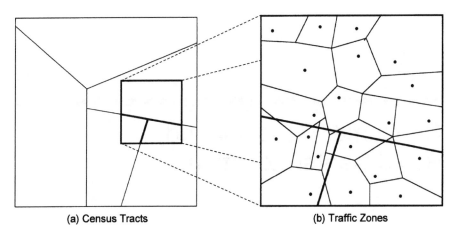

(a) Census Tracts   (b) Traffic Zones

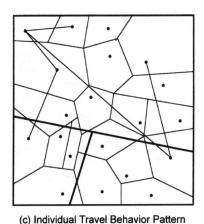

(c) Individual Travel Behavior Pattern

Figure 9.12. Example activity patterns: (a) census tracts; (b) traffic zones; (c) individual travel behavior pattern.

create activity records typical of the household types to which they belonged.

3. For work sites, new linkages between present households, differentiated by factors such as income range, and the site could be developed so that the new work site attracts employment.

Stopher et al. (1993:21) note how "the entire model could be "looped" in 5 year increments so that land-use patterns could adapt to increases in the number of households or changes in transportation supply, and vice-versa."

Thus the SMART model, utilizing the HAS, has a wide variety of potential applications.

10

Consumer Behavior and Retail Center Location

10.1 The Evolution of Consumer Behavior Models

Consumer behavior traditionally has been a fertile area of research in geography. Because of the inherent spatial nature of consumer shopping activity—where the consumer is likely to shop directly affects the location and organization of the retail market system, and vice versa—many researchers have been concerned with examining the relationships between consumer behavior and the spatial structure of the retail environment. These were introduced briefly in Chapter 2.

Among the earliest approaches is that of Reilly (1931), who was concerned with defining market areas using a *law of retail gravitation*. Central place theory developed by Christaller (1933) was also relevant to understanding consumer behavior. Concepts such as the *range of a good* and *threshold demand* requirements were particularly appropriate.

In early geographic research into consumer behavior, two parallel streams developed. One was largely *non behavioral*, having its antecedents in the standard *social gravity* model (Lakshmanen & Hansen, 1965) and the *entropy-based model* developed by Wilson (1969). The other consisted of *behavioral* models of consumer choice and decision making, which departed radically from the conventional formats. Based on innovative articles by Thompson (1963) and D. L. Huff (1962, 1964), this second stream examined the importance of variables such as subjective distance in the decision-making act, or following Huff, replaced deterministic allocations of people to places with *probability surfaces* describing the potential for interacting with sets of places.

Although D. L. Huff's probability model was based on objective variables (for example floor space and travel time), his 1964 paper, conceptualizing the consumer decision process, provided an innovative schema upon which many geographers began to build. For example, Rushton (1965) developed a gravity model-type format that substituted empirically based *revealed preferences* for town attractiveness and town-distance trade-off functions for conventional objective variables. Thus, consumer activities were conceptualized in terms of preferences for places rather than strict objective allocation problems. Similarly, Golledge (1965) conceptualized the market decision process in a *dynamic choice* framework and illustrated the importance of learning activities in the development of habitual and stereotyped behavior. Golledge (1967) also developed a locational choice schema from both the consumer and the producer points of view that was based on continuous behavioral change. Golledge and Brown (1967) emphasized the importance of both search and learning in the market decision process. *Dynamic Markov models* were used in both studies as the process mechanism for showing the change of choice proportions amongst alternatives as information about their relative attractiveness accumulated over time.

Shortly after these preliminary attempts, R. M. Downs (1970a) introduced yet another idea into behaviorally oriented research on consumer behavior. He emphasized that shopping centers had both an *objective physical existence* and a *subjective* or *cognitive existence*. He outlined procedures for defining the cognitive structure of a shopping center and for determining the significance of a wide range of attitudinal, attributional, preference, and evaluative variables, all from the point of view of potential patrons of the centers. His paper on the cognitive structure of a shopping center is indeed a landmark and has been one of the most influential and widely quoted papers relating to consumer spatial behavior, consumer preference, and consumer choice.

Briggs (1969) pioneered the use of *multidimensional scaling* in an attempt to construct subjective utility measures relating to the choice of competing shopping centers by randomly selected sets of consumers. The potential use of multidimensional scaling for examining a variety of spatial problems (e.g., cognitive maps, consumer behavior, migration behavior, etc.) was summarized by Golledge and Rushton (1972). Shortly afterwards, an article by Burnett (1973) showed the feasibility of using multidimensional scaling procedures to explain and to predict consumer choices of shopping centers for both convenience and shopping goods.

By the early 1970s, therefore, a complete arsenal of methods, concepts, and emerging theories was available for the geographer interested in both subjective and objective approaches to the analysis of consumer spatial behavior. Using a variety of methods, Golledge et al. (1966), Clark (1969), and Clark and Rushton (1970) had attacked central place theory from the consumer and behavioral points of view, showing empirically that in many different contexts consumers did not exhibit the single-center–least-effort syndrome required by both Christaller (1933) and Lösch (1954) and that a wide range of apparently "irrational" economic and spatial behaviors existed. This evidence, along with the emergence of new conceptual bases and methodologies, helped to modify existing approaches designed to explain consumer spatial behavior.

10.2 Gravity Model Approaches: From Deterministic to Probabilistic Analysis

10.2.1 Classical Retail Gravity Models

Early investigations of consumer behavior in the context of market area definition were related to purpose of trip, trip frequency, distance to centers at different levels in the retail hierarchy, and general spatial interaction principles (McCabe, 1974:10).

In 1931 William J. Reilly made a significant application of market area analysis with his *Law of Retail Gravitation*. Reilly's Law states that two centers attract trade from intermediate places approximately in direct proportion to the size of the centers and in inverse proportion to the square of the distance from these two centers to any intermediate place.

Mathematically this law was expressed as:

$$\frac{T_a}{T_b} = \left(\frac{P_a}{P_b}\right)^N \cdot \left(\frac{D_b}{D_a}\right)^n$$

where:

T_a = the proportion of the retail business from an intermediate town attracted by city A

T_b = the proportion of the retail business from an intermediate town attracted by city B

P_a = the population of city A

P_b = the population of city B

D_a = the distance from the intermediate town to city A

D_b = the distance from the intermediate town to city B

N, n = empirically derived constants

This model was tested empirically, and P. D. Converse (1949) is credited with a significant modification of Reilly's original formula that made it possible to calculate the approximate point between two competing centers where the trading influence of each was equal. This was the *breaking point*, expressed by:

$$D_a = \frac{D_{ab}}{1 + \sqrt{\frac{P_a}{P_b}}}$$

where:

D_b = the break-even point between city A and city B, (in kilometers from B)

D_{ab} = the distance separating city A from city B

P_b = the population of city B

P_a = the population of city A

These early efforts provided a basis for systematic estimation of retail trading areas. Reilly's empirical tests suggested an exponent of 2 for the mitigating effect of distance, while it suggested that the attractiveness of population was linear with size (i.e., $N = 1$). This value was observed to have a range from 1.5 to 2.75. Some of his observations for specific goods led to power functions being derived that ranged from 0 to as high as 12.5.

There were, however, several conceptual and operational limitations associated with the use of the Reilly-type retail gravity models. Perhaps most important was the break-even-point formula, which was incapable of giving graduated estimates above or below the break-even position between two competing centers. This made it impossible to calculate the actual total demand for either center's services. In addition, when several centers were included, invariably there was overlapping of trade areas. Often some areas would not be allocated to any center's trade area. But, more importantly, these models assumed rational economic, and spatially invariant, individual behavior on the part of the consumer.

10.2.2   Probabilistic Retail Gravity Models

To overcome some of the limitations of the Reilly-type retail gravity models, D. L. Huff (1962) proposed an alternative probabilistic retail gravity model. This model incorporated the realistic notion that customers do not always select one center for exclusive shopping. Lack of consistency in choice is apparent where there are a large number of centers of varying size and functional complexity within a reasonable distance. This is the case in large urban areas. Thus, his model focused on the customer rather than the retailer. After all, it is the customer who is the primary agent affecting the trade area of retail centers. Huff's model describes the process by which potential customers choose from among acceptable alternative retail centers to obtain specific goods and services. Huff suggested that areas of competition between shopping centers overlap, thus shopping trips must be apportioned among competing accessible centers in a probabilistic manner.

A formal expression of Huff's *probabilistic retail gravity model* is:

$$P_{ij} = \frac{\dfrac{S_j}{T_{ij}^{\gamma}}}{\displaystyle\sum_{j=1}^{n} \dfrac{S_j}{T_{ij}^{\gamma}}}$$

where:

P_{ij} = The probability of a consumer at a given point of origin i traveling to a particular shopping center j

S_j = the size of a shopping center j (measured in terms of the square footage devoted to the sale of a particular class of goods)

T_{ij} = the travel time involved in getting from a consumer's travel base i to a given shopping center j

λ = a parameter that is to be estimated empirically to reflect the effect of travel time on various kinds of shopping trips

The *expected number of consumers* at a given place of origin i who shop at a particular shopping center j equals the number of consumers at i multiplied by the probability that a consumer at i will select j for shopping. That is:

$$E_{ij} = C_i \cdot P_{ij}$$

where:

E_{ij} = the expected number of consumers at i who are likely to travel to shopping center j

C_i = the number of consumers at i

P_{ij} = probability of a consumer at location i choosing to shop at location j

Huff goes on to draw some general conclusions about the nature and scope of trading areas associated with retail centers:

1. A trading area represents a demand surface containing potential customers for a specific product(s) or service(s) of a particular distribution center.
2. A distribution center may be a single firm or an agglomeration of firms.
3. A demand surface consists of a series of demand gradients or zones reflecting varying customer-sales potentials. An exception to the condition of demand gradients would be location in a unique geographical setting, thus presenting an absolute monopoly in providing products and/or services that are necessary. Under these conditions, no gradients would exist, but, rather, a single homogeneous demand plane occurs.
4. Except in the case of monopoly, demand gradients are of a probabilistic nature (i.e., $0 < P < 1$).
5. The total potential customers within a distribution center's demand surface (trading area) is the sum of the expected number of consumers from each of the demand gradients.
6. Demand gradients of competing firms overlap; and where gradients of like probability intersect, a spatial competitive equilibrium position is reached.

Thus, a *trading area* may be defined as:

$$T_j = \sum_{i=1}^{n} P_{ij} \cdot C_i$$

where:

T_j = the trading area of a particular firm or agglomeration of firms; that is, the total expected number of consumers within a given region who are likely to patronize j for a specific class of products or services

P_{ij} = the probability of an individual consumer residing within a given gradient i shopping at center j

C_i = the number of consumers residing within a given gradient i

A typical outcome of the application of a Huff-type probability gravity model at the intraurban scale uses census tract centroids, weighted by households or populations for each tract, as a surrogate for demand, and investigates probability trade area surfaces for a selection of competing retail centers. The model has been widely used to investigate the potential trade area impact of a proposed shopping center development, especially its likely effects on competing centers. Figure 10.1 gives an example from Adelaide, South Australia (Stimson, 1985), for a shopping center development at Port Adelaide in the northwest sector of the city. The probability surface is shown. Figure 10.2 illustrates the potential trade area impact on competing centers of varying size if the proposed Port Adelaide shopping center were to be increased in size, holding population constant.

10.2.3   Refinements of the Huff Model

A refinement of the Huff formulation is the Lakshmanen and Hansen (1965) model. The main parameters of this model refer to the mass/attractive force measurement (α) and the distance or interaction decay measure (β). The exponent may be a

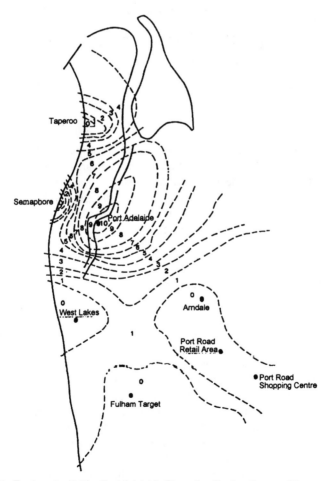

Figure 10.1. Trade potential for Port Adelaide Shopping Center. *Source*: Stimson, 1985.

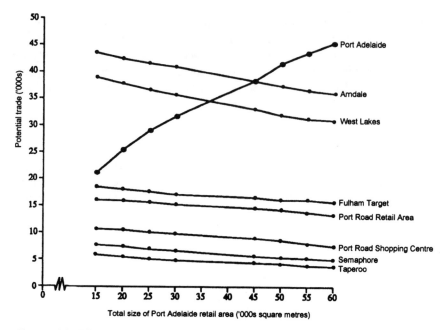

Figure 10.2. Effects of change in total retail area in Port Adelaide. *Source*: Stimson, 1985.

weighting factor reflecting center attributes, such as car-parking facilities, amenity value, and range of functional units. It may be a multivariate mass-weighting exponent. The exponent is identical to the distance exponent in the Huff model. Stanley and Sewall (1976) have added a retail chain image measure to the Huff probability retail gravity model that is empirically derived from survey data on store and center image using multidimensional scaling techniques.

10.3   Disaggregate Discrete Choice Models

10.3.1   Types of Discrete Choice Models

Many geographers are interested in studying how humans make spatially relevant decisions, such as where to go shopping or where to go to school. In studying this type of human, or *consumer*, decision making, geographers often use mathematically based models. Many of these have been "borrowed" and adapted from other disciplines, such as economics and psychology. Perhaps the most widely discussed and applied models of consumer choice are those belonging to the family of discrete choice models. These models are also known as random utility models, and they stem from research on microeconomic theory, psychological judgment theory, and developments in statistical analysis of categorical data.

Historically, these *discrete choice models* have been used in studying questions relating to marketing and transportation problems. More recently, they have been used to study problems of spatial choice behavior, such as the choice of a grocery store, a place to shop, where to go to school, and the like. It is interesting to note that discrete choice models were developed because the previous microeconomic, or marginalist, models were inadequate for investigating spatially relevant decisions. Marginalist models are inadequate because the assumptions upon which they are based do not accurately reflect the decision-making and choice processes of humans.

The family of discrete choice models includes the *multinomial logit* (MNL), the *nested logit*, the *dogit*, the *generalized extreme value* (GEV), and the *multinomial probit* (these models are ordered in terms of their complexity and comprehensiveness). The nested logit, dogit, and GEV models are commonly called "halfway house" models, as they are more realistic than the multinomial logit model, but less general than the multinomial probit. Although the multinomial probit is the most theoretically appealing, it has rarely been applied in choice contexts with more than two alternatives.

Of all the discrete choice models, the MNL is the simplest and most computationally tractable. It has received the widest application in the study of spatially relevant consumer choices.

Many important decisions made by individuals involve selection (or choice) from a limited, constrained set of discrete alternatives such as the choice of a house, a car, an occupation, a college, a shopping destination, and so on. The assumptions upon which a model concerning prediction of choice is based are therefore important. To derive and successfully apply the MNL model, six assumptions must be satisfied:

1. Each *individual* decision maker is assumed to be faced with a *discrete* set of choice alternatives. Proper definition of the choice set is a critical issue when models of the discrete choice family are applied. It is desirable that all individuals in the sample have access to the same alternatives. For convenience, we label these discrete *choices* as A_{li} through A_{ji}, where i denotes a given person, and j denotes a given choice. Thus,

$$A_i = (A_{li}, \ldots A_{ri} \ldots A_{ji})$$

2. The choice rule for discrete choice models is *utility maximization*. In other words, an individual will choose that alternative yielding the highest utility or (U). Therefore, alternative r will be selected if and only if,

$$U_{ri} > U_{gi} \quad \text{for} \quad g \neq r, g = 1, \ldots, j$$

Specifically, the utility (U) of choice r for person i is greater than choice g, where g stands for all choices other than r.

3. The MNL is a model of *probabilistic* choice. In other words, an individual will have a certain likelihood of choosing different alternatives, such as a 50% chance of shopping at store A, a 40% chance of shopping at store B, and a 10%

chance of patronizing store C. In the event that the condition of competing alternatives holds, it will occur with the probability

$$P_{ri} = \text{prob} \left[U_{ri} > U_{gi} \quad \text{for} \quad g \neq r, g = 1, \ldots, j \right]$$

where P_{ri} is the probability of individual i choosing alternative r.

4. An individual's utility for each alternative is divided into two components,

$$U_{ri} = V_{ri} + E_{ri}$$

where:

V_{ri} = the *observed* or systematic component of utility

E_{ri} = an *error* or disturbance term that reflects the unobserved attribute of the choice alternative.

The systematic element is defined in terms of observed attributes of the choice alternatives and/or the decision makers. The stochastic (random) component enters the utility function because certain choice-relevant attributes are unobserved and because the valuation of observed attributes varies from individual to individual.

5. Given the above definition of utility, the probability of choosing the rth alternative can now be written as:

$$P_{ri} = \text{Prob} \left[V_{ri} + E_{ri} > V_{gi} + E_{ri} \quad \text{for} \quad g \neq r, g = 1, \ldots, j \right],$$

which can be rearranged to express the following:

$$P_{ri} = \text{Prob} \left[E_{gi} - E_{ri} < V_{gi} - V_{ri} \text{ for } g \neq r, g = 1, \ldots, j \right],$$

This is known as the *random utility model*. When the partitioned expression of utility is incorporated into the probability model described above, the random utility model of choice can be derived. The equation states that the probability that individual i will choose alternative r can be defined in terms of the difference between the random utilities of alternative g and alternative r being less than the difference between the observed utility levels of alternative g and alternative r. To recapitulate, the random element attempts to account for the inability of the analyst to fully represent the dimensions determining preference in the utility function.

6. The MNL model is derived by making assumptions regarding the random component of utility. The E_{ri} is *independently and identically distributed* (IID). The statistical probability distribution of E_{ri} is Type 1 extreme value (double exponential). This leads to a computationally tractable form of the model:

$$\text{Pr} | j = \frac{e^{V(Z_{ri}, S_i)}}{\sum_{g=1}^{j} e^{V(Z_{ri}, S_i)}}$$

The above equation represents the most common form of the MNL model.

10.3.2 Discrete Choice Models: An Example

Suppose that a survey of consumers has been undertaken to understand their shopping habits and to determine the percentage (i.e., the market share) of shoppers a new store might attract. Three existing stores (stores A, B, C) and one proposed store (D) were rated on the following dimensions: (1) variety of goods; (2) quality of goods; (3) parking availability; and (4) value for money. Given these variables, an MNL model can be used to estimate the relative importance of each attribute and to derive market-share predictions with and without the new store.

The average attributes obtained from the sample are set out in Table 10.1. The importance weights (a_d) were estimated by fitting the shopper's choices of existing stores to their ratings using the MNL model.

The attractiveness or utility of each store can be derived as follows:

$$U_r = \sum_{d=1}^{4} (a_d \cdot X_{rd})$$

where:

U_r is the utility of store r
a_d is the importance weight for dimension d
X_{rd} is the rating for store r on dimension d

As an illustration, the utility of store A is

$$U(\text{Store A}) = (2.0 \cdot 0.7) + (1.7 \cdot 0.5) + (1.3 \cdot 0.7) + (2.2 \cdot 0.7) = 4.70$$

The results of the logit model analysis of this simple hypothetical situation are given in Table 10.2.

As an illustration, the market share estimate for store A (including store D) can be obtained by:

$$\Pr(\text{Store A}) = \frac{e^{U(\text{Store A})}}{e^{U(\text{Store A})} + e^{U(\text{Store B})} + e^{U(\text{Store C})} + e^{U(\text{Store D})}}$$

$$= 109.0/(109.0 + 27.1 + 77.5 + 55.7)$$

$$= .407$$

TABLE 10.1
Attribute Ratings

Store	Variety	Quality	Parking	Value for Money
A	0.7	0.5	0.7	0.7
B	0.3	0.4	0.2	0.8
C	0.6	0.8	0.7	0.5
D	0.6	0.4	0.8	0.5
Importance weight (a_d)	2.0	1.7	1.3	2.2

TABLE 10.2
Sample Logit Model Analysis

Store	Utility $(U_r = \Sigma a_d X_{rd})$ (a)	e^{U_r} (b)	Share Estimate Without New Store (c)	Share Estimate With New Store (d)	Draw (e = c − d)
A	4.70	109.0	0.512	0.407	0.105
B	3.30	27.1	0.126	0.100	0.026
C	4.35	77.5	0.362	0.287	0.075
D	4.02	55.7		0.206	

TABLE 10.3
Predicted Market Penetration

Store	e^{U_r}	Share Estimate
A	109.9	0.376
B	27.1	0.092
C	77.5	0.266
D	77.5	0.266

A major drawback of the MNL model is the "independence from irrelevant alternatives" property. It implies that a new alternative entering a choice set will compete equally with each existing alternative and will obtain a share of the market by drawing from the existing alternatives in direct proportion to the original shares of the market held by these existing alternatives. This is demonstrated in column e of Table 10.2, where the draw is proportional to share. If it were decided to build store D right next to store C, and each store were perceived to have the same utility, the predictions in Table 10.3 would be yielded.

These predictions indicate that store D has drawn shares proportionally from each of the other stores. However, since store D is perceived to be identical to store C, it is intuitively more plausible that the market share for store C would be "cannibalized" to a much greater degree than the other stores.

10.4   Information-Processing Models

10.4.1   From Rationality to Satisficing and Information Processing

Many of the early models of consumer behavior required assumptions not only of economic and spatial rationality but also a set of assumptions related to psychological rationality. For example, decision makers were:

1. Required to have consistent and fixed preferences.
2. Assumed to be unable to separate preferences and beliefs (their preferences and ultimate choices were assumed to be one and the same).
3. Assumed capable of computing optima of one sort or another.

Just as assumptions about economic and spatial rationality were assaulted by both empirical evidence and theoretical probing in the 1960s and 1970s, so too did the necessity of these psychological assumptions become questioned during the same period. For example, Tversky and Kahneman (1974) claimed that the majority of individuals have rather limited information-processing capabilities, and because of these limitations they were unlikely to be able to perform or to make decisions according to some a priori computed optimal function. This echoed the earlier thoughts of Simon (1957), who had argued strongly for the replacement of the economically rational being with assumptions more relevant to rationality or satisficing capability.

10.4.2   Information Integration

Louviere (1974) introduced an information-processing paradigm into geographic research on consumer behavior. In particular, this new approach viewed choice as a process rather than a final act, it considered choices to be purposive with the decision maker being an active manipulator of information about the choice environment, and it tried to model consumers as selective information processors in which general decision-making criteria are filtered down to specific choice instances. Known as *information integration theory,* the approaches also echoed earlier suggestions that consumer decision making necessarily involved feedback from each purchase act and could therefore be considered a type of learning process. Although some of these earlier approaches had decided limitations—for example, they did not cover a wide range of decision-making heuristics, they only sketchily covered the influence of task attributes, and were somewhat incomplete in their descriptions of information processing itself—their pursuit continued. While much of the work is still at a conceptual stage, it has led to the development of a major tie between consumer behavior models and computational process models of decision making and choice. These latter are now most commonly found in the cognitive science and artificial intelligence literature (Smith et al., 1982).

One approach that has enjoyed a considerable measure of success, despite the complexity of its task situation and general propensity to operationalize the model in laboratory rather than field settings, is information integration theory (sometimes also appearing in the literature under the heading of *functional measurement theory*).

In this type of model, individuals are assumed to be able to discern key attributes of products or situations about which they have to make choices. They are then assumed to have the capability of developing trade off functions among different quantities of each of the attributes, and to combine all attribute and trade off information or factors into a single numerical judgment—that is, they perform an information integration task prior to making a choice (Louviere & Henley, 1977).

Basically, it is assumed that individuals can supply direct numerical estimates of their degree of preference for quantities of a particular attribute or stimulus and that this degree of preference can be represented on an interval scale. Thus weight and scale parameters can be developed and insights gained into the types of combination rules that an individual appears to be using when making a choice. In general, the term functional measurement refers to the evaluation of what the individual sees as being "functional" in the decision-making process. This perception of different functional degrees is translated into relative weights that are then incorporated into a trade-off function that is accessed during a particular decision-making and choice process.

10.5   Imagery and Consumer Behavior

10.5.1   Distance-Related Studies

One of the earliest tenets of behavioral research in geography was that cognized, or subjective, variables were likely to prove more important than objective variables for explaining of consumer behavior. It was not difficult to accept this reasoning for individually based models. When put in a context of population aggregates, however, important questions were raised concerning the definition and meaning of cognized variables, such as shopping center or store image, subjective distance, familiarity, preference, and so on. Attempts to define these concepts and to incorporate them into operational models of consumer behavior have been at the forefront of much recent behavioral research.

Most studies looked at a different component of what is generally called an individual's *cognitive map* (see Chapter 7, Section 7.3). This contains the keys to explaining individual behavior. If a place is cognized as being difficult to get to or dangerous, its probability of patronage is reduced. If a shop or a shopping center is imaged as being glamorous, attractive, and exciting, there is a tendency for customers to patronize it more frequently than any of its competitors.

Golledge et al. (1969) used arguments such as these in their study of configurations of distance in intraurban space to help explain the decline of patronage of CBDs, whose declining image encouraged patrons to turn their trade toward suburban shopping centers. In other words, the argument was advanced that if a consumer's cognitive map is incomplete, distorted, or fragmented, or has within it substantial *distance, directional,* or *temporal biases,* the individual may behave "irrationally" in terms of choosing possible destinations from an available opportunity set. Such behavior might include acts such as choosing routes or time paths that were not close to minimal path solutions, or patronizing stores or centers that were much further away than more obvious, closer ones, even when the stores were from the same chain or the shopping centers had approximately the same functions in them.

One of the most fundamental components of a cognitive map is distance. Cadwallader (1975) tested a hypothesis that spatial rationality is more closely linked to cognitive than objective Euclidean distance. He found that a greater proportion of sampled shoppers believed that they patronized the nearest supermarket rather than did actually patronize the closest one. He suggested that more distant shopping centers had exaggerated attractiveness image components, and in a later stud, Cadwallader (1981) demonstrated that goodness of fit between model choice and actual choice was improved by substituting cognitive rather than objective distance in the model. In a similar vein, MacKay and Olshavsky (1975) investigated the significance of cognitive time and cognitive distance in consumer behavior situations. They also examined the differences that exist between *preference* (an ideal state) and *choice* (an actual state severely constrained by a host of economic, environmental, and other factors), showing there was no one-to-one mapping between the two.

10.5.2 Search and Learning

Behavioral work has emphasized that consumer choice is not a repetitive invariant act. As D. S. Rogers (1970) had pointed out, consumer behavior, even with respect to frequent purchases such as groceries, took 6 to 8 weeks to form. Prior to that, considerable search activity and experimentation were undertaken, and over time the funneling of behavior took place (Figure 10.3). Funneling simply involved successive elimination of unfavorable alternatives consequent to evaluation of a particular choice act such that over time the average number of places visited per chosen episodic interval first decreased and then became relatively stable. This confirmed the theoretical suppositions made by Golledge (1967), who had developed a model of the consumer behavior (or market-decision process) in the form of a dynamic Markov chain, which eliminated a number of alternatives after feedback had shown that reward expectancy thresholds were not reached.

Further studies by Potter (1967, 1977, 1978, 1979) have indicated that consumers undergo some type of spatial search process before sufficient information is accumulated about retail opportunities to begin the process of winnowing out the less favorable opportunities. He developed the notions of usage and information fields. His concepts paralleled the notion of action and activity space discussed in Chapter 8, Section 8.3. As with other studies, Potter found that information fields and the opportunity sets contained therein were sectoral in nature and, given the context of a dominant CBD, the sectors were frequently stretched from a home location in the direction of the CBD. Potter also provided evidence that higher-income

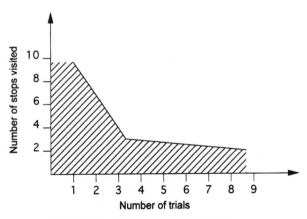

Figure 10.3. The process of funneling.

groups were aware of more opportunities than lower-income groups and that they had a tendency to travel greater distances to a greater variety of retail outlets than did the lower-income groups.

The importance of the information component was emphasized by Hanson (1978, 1984), whose work on households in Sweden demonstrated that the majority of spatial opportunities theoretically available to a given population subset were very poorly known. Only a few members of what might otherwise be regarded as a feasible opportunity set were recognized, and there was a pronounced distance decay effect found with respect to familiarity with retail and other shopping opportunities. The distance effect decayed gently up to 2 miles from the home base but beyond that it tapered off quite sharply. This and other work lend strong support to the original behavioral hypotheses that assumptions of complete information (or perfect knowledge) were far too strong for most real-world situations, even when the spatial scale was small.

10.5.3 Store and Center Image

While this work on subjective distance, information integration, and search and learning proceeded apace in the 1970s, a gradually increasing body of parallel work focused on the characteristics of store and center image. Using a variety of multivariate techniques on subjectively collected data sets, researchers tried to uncover which dimensions influenced consumer choice processes. Whereas the studies quoted in the previous section emphasized the friction effect of subjective time and distance in reducing the size of opportunity sets, store and shopping center image research focused more on the attractiveness component and how it conceivably influenced consumer choice.

10.5.4 Multidimensional Scaling

An indirect method of recovering image concepts uses a technique known as *multidimensional scaling* (MDS). Here data are collected from subjects who evaluate similarity or dissimilarity among pairs of alternatives (e.g., pairs of stores or pairs of shopping centers). The similarity measures yield a set of independent dimensions that represent the structural relationships among the alternatives. The resulting MDS configuration of stimuli reveal the space of minimum dimensionality in which all the stimuli can be mapped. The number of dimension, then, represents the number of critical variables used by subjects in the reasoning process (i.e., when making judgments about such stimuli).

Identification of dimensions can be undertaken in two ways. If some prior data are collected on a number of dimensions hypothesized to have an effect on consumer choice, then the different factors can be correlated with the projections of the stimuli on each dimension in turn, thus identifying which factor is most highly related to which dimension. This property fitting procedure, PROFIT (Carroll & Chang, 1970) can be used with subjectively or objectively defined factors. Alternatively, once the MDS configuration has been obtained and a space of minimum dimensionality determined, a posteriori PROFITs can take place (in a manner similar to that used in factor analysis) to identify dimensions. In either case, however, testing is done to see whether or not the hypothesized factors relate to the discovered dimensions.

Examples of consumer behavior studies using MDS procedures include Burnett's (1973) study of women's choices of both convenience stores (e.g., groceries) and shopping goods (e.g., women's apparel), and studies by Singson (1975), Lloyd and Jennings (1978), Spencer (1978, 1980), Blommestein, Nijkamp, and Veenendal (1980), and Hudson (1981). Retail consumer behavior has proven to be one of the more popular and productive areas of consumer behavior research. The benefits of recovering the latent dimensional structure underlying the choice act and using identifiable components of this structure to define store and center image-attractiveness components have been consistently demonstrated (Table 10.4).

There continues to be advocacy of each of the many techniques developed to examine store images (e.g., semantic differential or Likert scaling, repertory grid analysis, and MDS procedures), and the relative advantages and disadvantages of the different methods are discussed in Timmermans, Van der Heijden, and Westerveld (1982a, 1982b).

10.5.5 The Retailer's Cognition of Store and Shopping Center Environments

As pointed out by Golledge (1970b), the way producers imaged their consumers' behaviors will influence how the product is presented to the potential buying population, the location chosen for the display and sale of the product, the reaction toward competition, the dollars spent on advertising, and so on. For example, the retailer who assumes that the buying populace are Marshellian calculating machines, thus adopting rigid economic principles to guide purchase acts, will focus only on getting a message to the consumer that that retailer's prices are lowest. Regardless of where the store is located, the retailer would argue that the lower prices will attract all the customers needed to surpass that threshold population required for profitable store operation. The retailer who believes clients are primarily Freudian consumers may invest heavily in advertising a product's personal appeal, and packaging becomes the dominant theme of the marketing project. Those who assume customers are basically Veblenian social animals will emphasize locations in prominent busy shopping places, such as CBDs or regional shopping centers, where purchasing the product will be part of a social interaction process.

The types of locational decisions made by different retailers, therefore, give some insights into how they image their potential customers and what forces drive their marketing actions. But retailers are also influenced to a substantial degree by the quality of the local environment in which they develop their service. As pointed out by Golledge, Halperin, Gale, and Hubert (1983) in a study of retailers in Melbourne, Australia, a wide range of attitudes are considered to be important (see Table 10.5).

TABLE 10.4

Sample of Field Survey Data for Women's Wear Retailers, Sydney, Australia: Selected Centers

Center Characteristic	Double Bay	Eastwood	Gordon	Top Ryde	Warringah Mall	West Pymble
Distance*	10.1 mi. 43 min.	4.3 mi. 15 min.	0.3 mi. 5 min.	4.3 mi. 10 min.	7.5 mi. 30 min.	0.2 mi. 3 min.
Barriers on route	Harbour Bridge C.B.D.; congested major roads	Cross two main roads; railway; social divide; off main rte.	None	Cross one main road	Pleasant drive	None
Parking	0.75–1 hr. curbside, 2 parks; full	0.5–1 hr. curbside, 1 park; full	Extensive off-st. parks	Planned parks; half full	Space near shops	Always space
Women's wear shops	49 shops 3 branches high-class chains	10 shops 1 department store	15 shops 1 department store	3 shops 1 department store	3 shops 1 department store	1 shop
Shop price rating[†]	1 (Max. Price one dress on display A$200)	3	2	2	2	1
Shop quality	1	4	2	3	2	2
Shop service rating[†]	2	4	1	3	2	3
Within-shop range[†]	4	4 4 (McDowells)	4 2 (Farmers)	4 3 (Grace Bros)	4 2 (David Jones)	4
Center size status[‡]	Regional, fashionable, high income, nucleated	195 shops including 1 planned center	112 shops	Planned regional center	Planned regional center	Neighborhood
Center environment[§]	2 (includes planned development)	5 (older strip)	4 (strip center)	3	1	2
Special features	6 specialty import shops; 1 fashion house	Railway divides center	Main road through center	None	High standard amenities	Modern planned unit

*Straight-line distance in miles and shortest driving time from the nearest major route intersection to the samples' customers.

[†]Rating on a five-point scale; 1 very high or very good; 2 high or good; 3 moderate; 4 low or poor; 5 very low or very poor. The age, income, and activities of the sampled consumers were borne in mind.

[‡]Classification by National Cash Registers, *Shopping Centres in Australia* (continuous broadsheets), where available.

[§]Rating on the same five-point scale as in [†], taking into account the ease and safety of shopping within the center, the external appearance of the shops, and the presence of amenities such as childminding centers, air conditioning, rest areas, etc.

Source: Burnett, 1973:193.

TABLE 10.5
Influences Affecting Retailer Location Choice

Attributes of Retail Environments
A. Layout of shopping center
B. Shopping center appearance
C. Access for delivery and loading
D. Number and size of shops
E. Business mix
F. Accessibility to main roads
G. Access within center
H. Location relative to competing center
I. Customer base
J. Turnover in types of business
K. Overhead costs
L. Property costs
M. Labor turnover and absenteeism
N. Town council and planning controls
O. Diverse activities in town helping trade
P. Future of the town
Q. Town draw from wide area

Retailer choice as to whether to locate in an existing business district, or in a dispersed suburban location, might depend on:

1. Their image of the rigidity of the local board of supervisors or town planners.
2. Their perception of the cleanliness, orderliness, or degree of control exerted over the shopping environment.
3. Their image of how accessible their place of business is to available on-street and off-street parking facilities.
4. Or it may simply be a result of where they believe the hub of shopping activity is currently focused.

Fears of competition, not only from other retailing specialists, but from other competitive shopping centers, appear to play an important role in deciding whether to continue leasing a given location or to undertake a major move. Similarly, image of turnover rates in, say, a suburban planned shopping center environment may be quite critical in terms of influencing retailers to retain an established business district location or vice versa.

10.6   The Development of Planned Shopping Centers in the United States

10.6.1   The Evolution and Proliferation of Shopping Centers

The concept of a center for shopping and exchange of products is not new. Across the world for thousands of years, intermittent and permanent, stationary and periodic markets have been set up to serve the same purpose—that is, to provide the public with a single place in which products can be obtained via exchange or purchase. In cities the location of these centers has changed constantly, from the periphery to the center and back to the periphery again. As cities grew, more and more distinct and separate areas for shopping developed and began to accumulate labels such as the CBD, the *major district center*, the *strip center*, the *regional center*, and, of course, the small *isolated dispersed store*.

Although the first recognizable out-of-town center was opened in the United States in 1923—"The Country Club Plaza" in Kansas—the shopping center as we know it today (as distinct from the commercial strip or the CBD) emerged around 1950 in the United States. Prior to this time retailing was highly concentrated in town centers, emphasizing the nucleated nature of commercial districts. Accessibility by public as well as private transportation was encouraged and catered for. Since the 1930s the private motor vehicle has become a major visual and physical intrusion into the urban landscape, and the resulting problems of congestion, pollution, and lack of parking have offset the relative economic success of the traditional centrally located business districts. The 1920s and 1930s saw the beginnings of a massive population expansion into suburban areas, and this residential expansion was shortly followed by decentralized office and business activity and the location of extensive "shopping parades" close to new housing districts. These shopping parades can be regarded as the precursors of today's out-of-town and other decentralized shopping centers (Golledge, 1985).

The shopping center came into prominence in North America as the retail industry's answer to population decentralization. It was a response to the dramatic shift of the nation's population from central cities to the suburbs. Although at first it was largely a response to the demand for local convenience goods by the suburban population, nowadays shopping centers provide nearly all desired retail products (except those representing the most highly specialized of a shopper's needs). From 1957 to 1965, the number of shopping centers in the United States more than tripled from about 2,200 to over 8,000, while the aggregate space they represented increased fivefold to more than 1 billion square feet. This was accompanied by a decline in retail sales registered in CBDs (see Figure 10.4). This decline appears to have leveled off, but dispersed shopping centers now produce more than 70% of all sales among the nation's largest retailers. Recently, however, central business areas are revitalizing, and a slight increment in their total retailing activity has been occurring.

In the United States and Canada, by 1957 there were approximately 2,500 shopping centers. This had increased to about 11,000 by 1967, and to 14,000 by 1970. Seventeen thousand shopping centers existed in 1973, and this number has continued to increase to around 30,000 in 1982 and to approximately 40,000 in the 1990s (Figure 10.5). The number of shopping centers in countries other than the United States and Canada, however, has been comparatively small. Up to 1973, the leading nation in the rest of the world as far as the number of shopping centers was concerned was Australia, but at that time it had fewer than 100 centers. Since 1973 shopping centers have been opening in the United States and Canada at the rate of about 1,000 per year. Of these approximately 10% are generally regarded as being

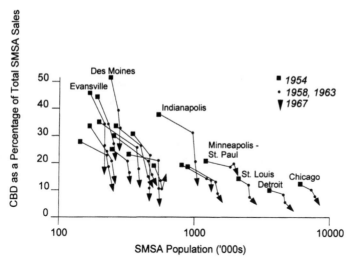

Figure 10.4. Decline in CBD sales. *Source:* Berry, 1981:31.

at the major regional level and others are of a lower order such as community, neighborhood, or street-corner centers.

After 1975, in the United States and Canada there was a slowdown of the building of new centers, particularly the super regional centers, partly as a result of saturation of the market and partly due to intervention by state and local governments. This governmental intervention has been brought about by an increasing concern for the environmental impacts and ecological effects of shopping center development and by an increasing trend toward integrated regional planning with rationalization of the number of regional level centers located in each area. Inflation and the rapidly increasing costs of services and utilities, together with some uncertainty about energy futures have also contributed to the slowdown. Currently there appears to be adequate potential for continued development of the medium- and smaller-sized centers, both in well-developed urban areas and in smaller places. Whereas in much of the West, Southwest, and in the South these smaller centers continue to develop as strip centers, in much of the central, northern and eastern parts of the United States, even smaller centers are adopting the enclosed mall design so that heating and air-conditioning can potentially increase trade during the winter and summer seasons. This convention in turn is supposed to offset or neutralize additional construction and development costs.

10.6.2   Tenant Mixes and Functional Make-up

The backbone of many of the earlier shopping centers consisted of food stores, such as supermarkets. An anchor (such as a junior department or variety store) often was located at the opposite end of the center. As the technology of merchandising and marketing changed, and as consumer tastes and preferences and travel behavior changed, *regional centers* emerged, which were generally harnessed to the drawing

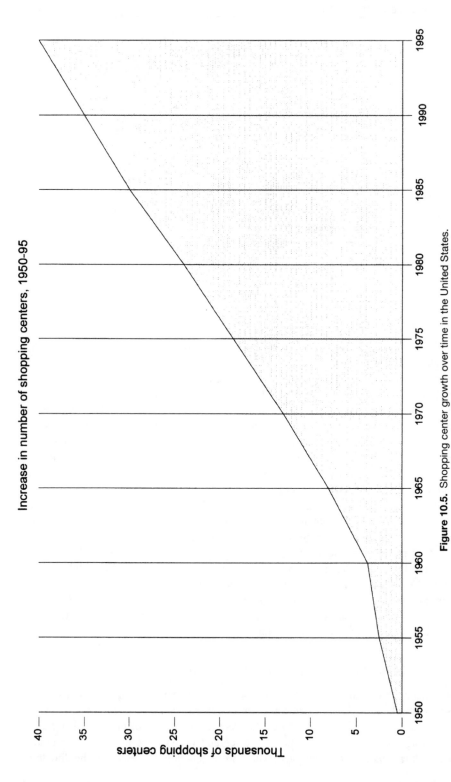

Figure 10.5. Shopping center growth over time in the United States.

power of a major department store (Table 10.6). There are now in existence a number of "super regional" centers with several major department stores and large numbers of ancillary tenants, but the superstore/supercenters are still relatively few in number and confined to the heavily populated areas of the country. Even so, the largest supercenter in the world is located in Edmonton, Alberta, where the West Edmonton Mall covers 5.2 million square feet—the size of 115 football fields! Malls in the United States produce $830.2 billion in annual sales. About 70% of the shopping population patronizes planned centers or malls.

Since the early development of the decentralized shopping center, a considerable functional change has taken place. Rather than acting as a merchandise mart, many of the higher order centers today act as *community centers* and as meeting places. They provide recreational facilities, incorporate entertainment and dining-out facilities, and have become a focal point for the social activities of part of the population within their trade area. In the larger shopping centers, spiraling costs of land combined with various objections to large, sterile paved areas have led to multistoried shopping areas and multistoried parking garages and parking decks.

What has been happening in the *central city* areas while this decentralization and regionalization of shopping has been occurring? There have been dramatic attempts to try and maintain the viability of inner-city areas, particularly specialized areas such as Chinatown in San Francisco, which has maintained its viability because of a particular mix of characteristics catering to a local ethnic population and to the tourist trade. The buildings, except for the ornaments, are not very different from the buildings found in just about any major western city.

In other areas in the inner city (such as the Vineyard in San Jose and the Cannery in San Francisco), development has taken place by refurbishing old warehouses or major buildings and turning them into enclosed malls with a wide variety of

TABLE 10.6
Characteristics of Shopping Centers

Center Type	Leading Tenant (Basis for Classification)	Typical GLA	General Range in GLA	Usual Minimum Site Area	Minimum Support Required
Neighborhood center	Supermarket or drugstore	50,000 sq. ft.	30,000–100,000 sq. ft.	3 acres	2,500–40,000 people
Community center	Variety, discount, or junior department	150,000 sq. ft.	100,000–300,000 sq. ft.	10 acres or more	40,000–150,000 people
Regional center	One or more full-line department stores of at least 100,000 sq. ft. of GLA*	400,000 sq. ft.	300,000–1,000,000 sq. ft. or more*	30–50 acres or more	150,000 or more people

*Centers with more than 750,000 sq. ft. GLA usually include three or more department stores and hence are superregionals.

Source: Urban Land Institute, 1981.

shopping facilities. These can be quite attractive, have interesting architectural designs, and provide a place for people to meet even if the shops are not open.

Many of the contemporary *planned shopping centers* are dominated by tenants representing national chains. This trend had become well established by 1981, when national chains comprised 46% of tenants in neighborhood centers, 62% of tenants in community centers, and 72% of tenants in regional centers (see Table 10.7 for representation of other types of ownership). National chains have continued to dominate all levels of planned shopping centers in terms of sales and rental income. Well established by the 1980s, this tendency is even more pronounced today. In many regional centers and inner-city malls, for example, rentals are extremely high, and local entrepreneurs and local chains often have difficulty in surviving. The death rate of small firms in large shopping centers is high; space has a high proportion of turnover; and only the most determined or most unusual local entrepreneurs manage to survive.

Shopping centers usually can be differentiated by the types of tenants that dominate them and are represented most frequently in their *tenant structure*. Table 10.8 gives examples of the most frequently found tenants in regional, community, and neighborhood shopping centers. For example, in *regional centers*, general merchandise represents over 30% of sales, whereas clothing and shoes represent 26% or more of sales. In *superregional centers*, clothing and shoes become more important, occupying about 35% of sales, whereas general merchandise produces only about

TABLE 10.7
Examples of Tenant Mixes in Different Types of Shopping Centers

	Center Type					
	Neighborhood		Community		Regional	
Tenant Type	1972	1981	1972	1981	1972	1981
A. Tenants (%)						
National chain	20		27		37	
Independent	58		48		35	
Local chain	22		25		28	
B. GLA (%)						
National chain	44	46	54	62	63	72
Independent	31	31	19	17	14	10
Local chain	25	23	27	21	23	19
C. Sales (%)						
National chain	55	58	57	67	62	67
Independent	20	17	15	13	11	11
Local chain	25	215	28	21	27	22
D. Rent income(%)						
National chain	50	37	48	52	72	59
Independent	27	38	25	24	11	18
Local chain	23	25	27	24	17	23

Source: Urban Land Institute (1981).

TABLE 10.8
Tenant Mix of Traditional Free-Standing Shopping Centers

	Percentage of GLA			Percentage of Sales		
Tenant Type	Neighborhood	Community	Regional	Neighborhood	Community	Regional
Retail						
Food	26.5	15.7	5.5	43.8	33.2	7.9
General merchandise	16.3	35.8	53.4	8.9	30.1	48.5
Clothing and shoes	5.5	7.8	16.9	4.2	7.6	20.4
Dry goods	3.8	3.0	11.4	1.6	2.0	1.5
Furniture	4.8	3.7	1.4	1.7	3.6	1.7
Other retail	16.9	14.5	10.1	13.4	15.5	13.3
Total retail	73.8	80.5	88.7	73.6	92.0	93.3
Services						
Food service	5.1	3.7	2.9	3.2	3.8	3.6
Personal	5.6	3.6	1.2	2.5	1.8	1.1
Financial	3.0	2.5	1.6	0	0	0
Office	4.7	2.2	0.4	0	0	1.1
Total service	18.4	12.0	6.1	5.7	6.6	5.8
Other tenants	4.7	4.3	3.0	20.7	2.4	2.0
Vacant	3.1	3.3	2.2	0	0	0

Source: Urban Land Institute (1975).

22%. In both cases, food shopping is not important, comprising only 6% of sales in regional centers and 4% of sales in superregional centers.

In comparison, the leading tenants in *community centers* are food and food services representing approximately 20% of the total sales (Table 10.9 (a–d). General merchandise closely follows with 18–20%. Large and small appliances are of similar importance, as are financial services such as banks and insurance agents. In community centers, general merchandise represents about 30% of total sales, food and food services about 36%, other retailing activity about 15%, and clothing and shoes 7%.

At the *neighborhood* level, food and food services again represent the most frequently found tenants with ladies' clothing and shoes, large and small appliances, hardware, and dry goods each being significant. Over 40% of sales come from food. About 20% are derived from services, 13% from other retail activity, and the balance from a mix of functions. Clothing and shoes represent only about 4.2%.

10.6.3 Retail Center Policy Trends in the United States

Local zoning approvals are perhaps the most troublesome procedures required prior to commencement of construction on shopping centers. In recent years, developers have run into a myriad of federal, state, and local environmental regulations, all implemented by separate agencies. Other regulations are blossoming at all levels of government that will require review and approval prior to commencement of construction of new shopping centers and that will govern and force compliance during operation of the shopping center.

TABLE 10.9
Twenty Most Frequently Found Tenants

(a) Neighborhood Centers		(b) Community Centers		(c) Regional Centers		(d) Super-Regional Centers	
Tenant Classification	Rank	Tenant Classification	Rank	Tenant Classification	Rank	Tenant Classification	Rank
Food and Food Service		Food and Service		Food and Food Service		Food and Food Service	
Supermarket	2	Supermarket	2	Candy and nuts	9	Candy and nuts	10
Restaurant without liquor	8	Restaurant without liquor	17	Restaurant without liquor	12	Restaurant without liquor	13
Ice cream parlor	20	Restaurant with liquor	20	Fast food/carryout	17	Restaurant with liquor	19
						Fast-food/carryout	9
General Merchandise		General Merchandise		General Merchandise		General Merchandise	
Variety store	14	Junior department store	6	Department store	14	Department store	18
		Variety store	13	Variety store	18		
Clothing and Shoes		Clothing and Shoes		Clothing and Shoes		Clothing and Shoes	
Ladies' specialty	15	Ladies' specialty	14	Ladies' specialty	6	Ladies' specialty	3
Ladies' ready-to-wear	7	Ladies' ready-to-wear	1	Ladies' ready-to-wear	1	Ladies' ready-to-wear	1
		Men's wear	7	Men's wear	2	Men's wear	2
		Family shoe	5	Family wear	11	Family wear	15
				Family shoe	3	Family shoe	6
				Ladies' shoe	7	Ladies' shoe	5
				Men's and boys' shoe	20	Men's and boys' shoe	8
						Unisex/jeans shop	17
Dry Goods		Dry Goods		Dry Goods		Dry Goods	
Yard goods	18	Yard goods	11	Yard goods	13	Yard goods	11

372

Furniture
Radio, TV, hi-fi — 19

Other Retail
Hardware — 16
Drugs — 4
Cards and gifts — 10
Liquor and wine — 12

Financial
Banks — 13

Offices
Medical and dental — 3
Real estate — 9

Other Offices
Medical and dental — 3

Services
Beauty shop — 1
Barber shop — 6
Cleaners and dyers — 5
Coin laundries — 11
Service station — 17

Furniture
Radio, TV, hi-fi — 19

Other Retail
Drugs — 8
Jewelry — 12
Cards and gifts — 10

Financial
Bank — 16
Insurance — 18

Services
Beauty shop — 4
Barber shop — 9
Cleaners and dyers — 15

Other Retail
Books and stationery — 8
Drugs — 19
Jewelry — 5
Cards and gifts — 4

Financial
Banks — 15

Offices
Medical and dental — 16

Services
Beauty shop — 10

Other Retail
Books and stationery — 12
Jewelry — 7
Cosmetics — 4

Financial
Banks — 14

Services
Beauty shop — 16
Optometrist — 20

Source: Urban Land Institute (1981).

National Historic Preservation Act	1966
National Environmental Policy Act	1969
Federal Clean Air Act	1970
Vermont Act 250	1970
Land Use and Environmental Law	1973
California Environmental Quality Act	1970
Coastal Zone Management Act	1970
Florida Development of Regional Impact Act	1970
American Folklife Preservation Act	1976
National Environmental Policy Act Amendment (A-95 Guidelines)	1979
Community Conservation Guidelines (Repealed 1981)	1979

Figure 10.6. Relevant U.S. impact acts.

Within the United States generally, directions to each of the individual states have normally come from federal legislation such as the National Environmental Protection Act of 1969 or the Coastal Zone Management Act of 1972 (Figure 10.6). Currently there are on-going attempts to adopt further federal land use regulation policies, which would give the states even greater control not only over shopping center development but over all land use development. Figure 10.7 illustrates the impacts that a new shopping center can have in an area and why different types of impact analyses are required prior to development.

State and local land use legislation is usually one of two types: The state either sets up a mechanism for reviewing the environmental impact of a facility, or it establishes an agency to designate minimum and maximum types of desired uses on land. This latter mechanism seems to be an ultimate goal for much of the current land use legislation.

For proposals that are deemed to be of regional or more than regional importance, the application process requires the applicant to analyze the impact of the shopping center on:

1. The environment and natural resources of the region including air quality, water quality, solid waste, noise, radiation, topography, natural vegetation, animal life, aquatic life, soils.
2. Historical and archaeological sites.
3. Parks and recreation areas.
4. Geology.
5. The economy of the region including employment and user characteristics.
6. Public facilities of the region including sanitary sewers, storm-water disposal, water supply, solid waste, power supply.
7. Public transportation of the region.
8. Housing of the region.

In most instances developers are required to use different consultants at a cost, which often exceeds $100,000 per DRI application.

As the participation rate by women in the work force rises, as the proportion of working families changes, and as the shopping needs of working singles get ex-

Economic Impact		Environmental Impact		Social Impact	
Positive	Negative	Positive	Negative	Positive	Negative
Adds new stock	Reduces old stock	Modernizes outworn areas	Changes traditional character	Allows for efficient shoppers	May favor car-borne shoppers
Accommodates larger modern stores	Discriminates against small independents	Reduces land-use conflicts	Creates new points of congestion	Provides new shopping opportunities	May limit choice to stereotypes
Increases rates and revenues	Increases monopoly powers	Scope for new design standards	Intrusive effects on older townscapes	Provides more safety	Creates new stress factors from crowds
Creates new employment	Changes structure of employment	Provides weather protection	Creates artificial atmosphere	Provides more comfort and amenities	Attracts delinquents and vandals
Improves trade on adjacent streets	Reduces trade on peripheral streets	Leads to upgrading of some streets	Causes blight on other streets	Concentrates shopping in one area	Breaks up old shopping linkages
Enhances status of central area	Affects status of surrounding centers	Integrates new transport	Causes pressure on existing infrastructures	Potentially greater social interaction	Becomes dead area at night

Figure 10.7. Impacts of a new shopping center. *Source:* Golledge, 1985: 105.

pressed, there has been a shift in shopping habits, particularly in terms of an increased demand for more evening or weekend shopping hours to accommodate changes in family and household habits. There also has been a shift in demand for consumer goods, including dramatic increases in demand for recreational and leisure-type activity, thus changing the number and type of retail functions available in both old and new shopping centers.

10.6.4 Environmental and Other Public Policies

Environmental protection has become a fact of life. In North America, the Clean Air Act of 1970 established federal control over air quality including indirect pollutant sources such as shopping centers and large parking facilities. The Federal Water Pollution Control Act Amendments 1972, the Coastal Zone Management Act (passed in 1972 and amended in 1976), and repeated attempts to pass a federal land use bill are examples of the series of federal regulations that limit unrestricted development at both state and local government levels.

While these additional regulations and controls have produced a slowdown in commercial center development and have increased the time required for approval of projects, responses to them have brought innovation and refinements to many phases of retail development. So, there are both positive and negative effects of any direct or indirect control measures. Although the increased delay has in turn increased construction costs and operation costs (which are passed on through increased occupancy costs to tenants and hence via pass-through charges to customers), there appear to be noticeable benefits in terms of the general public good from such controls.

Critical features that are part of the public policy arena include drainage, solid waste disposal, parking, traffic, and aesthetics.

10.6.5 Locations and Characteristics of Shopping Centers

We have seen in the previous sections of this chapter that the planned shopping center has emerged as a significant component of the commercial structure of most of today's economies and societies. We focus now more on the nature and structure of such centers and discuss the processes involved in their location. These processes are examined both from the developer and consumer point of view.

Just as we see that today's urban systems exhibit a hierarchical structure ranging from mega metropolitan region, to metropolitan area, to city, to town, village, and hamlet, so too has there emerged a hierarchy in the commercial structures found within today's cities. Although decreasing in relative dominance over the past 40 years, the CBD is still an important feature in most commercial structures. In many of today's larger cities, there have been significant developments of other centers structured similarly to the traditional CBD that often rival it in importance (e.g., as a financial district). However, most CBDs still contain a significant retailing component and are still designed to serve entire cities rather than sections of it. The superregional shopping center has evolved as a decentralized retailing center. Often these include several millions of square feet of gross leasable area (GLA) and a variety of principal tenants that usually include several types of department stores. Such superregional centers are, however, relatively few in number and are associated more with urban conurbations or incipient megalopolis (i.e., where two or more significantly large cities anchor continuous, built-up intervening areas).

Shopping center location must consider a substantial number of critical dimensions. Timmermans (1982) has identified some of the most critical of these. They include:

- The number of shops involved.
- The quality of parking facilities.
- Distance to or from a home location (i.e., proximity to customers).
- The quality of the shopping center.
- The range and quality of stock.
- Sets of service attributes.
- Sets of environmental attributes.
- Sets of stock attributes.
- Sets of accessibility attributes.
- Prices.
- Ambiance of the center.

Although locational decisions for shopping centers are still made using traditional models, such as Huff's probability model, it is far more usual to find decisions being made today via *location-allocation models*. These models are iterative computer programs that incorporate the ideas of location and trade area definition. They allow the decision maker to assess a potential market and its market penetrability, then to estimate potential sales. Figure 10.8 details the steps that are involved in the more traditional market area analysis. These steps similarly guide the application of location-allocation models, but different emphasis is placed on the significance of each step.

A process known as *malling* has become widespread. This is in a sense the development of a shopping center within an existing shopping district. Examples include the Eaton Center in Toronto and the inner-city Horton's Plaza mall in San Diego. There are several different types of malls that have been developed:

1. *Infill centers* usually comprise up to about 20 units built on a redeveloped site within an established shopping district.
2. *Extension centers* are add-ons to existing retail centers to improve the quantity and quality of products offered and to meet growing retail needs. This is often associated with urban practices such as renewal, redevelopment, and gentrification.
3. *Core replacement centers* have the primary purpose of replacing structurally and functionally obsolete retail property in prime retail locations with new design space allowing for contemporary management and sales techniques.
4. *Multiuse centers* are usually complex developments in which retailing is only one of a number of activities mixed within a single building complex,

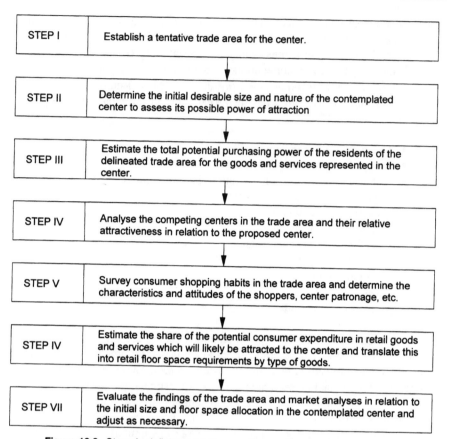

STEP I	Establish a tentative trade area for the center.
STEP II	Determine the initial desirable size and nature of the contemplated center to assess its possible power of attraction
STEP III	Estimate the total potential purchasing power of the residents of the delineated trade area for the goods and services represented in the center.
STEP IV	Analyse the competing centers in the trade area and their relative attractiveness in relation to the proposed center.
STEP V	Survey consumer shopping habits in the trade area and determine the characteristics and attitudes of the shoppers, center patronage, etc.
STEP IV	Estimate the share of the potential consumer expenditure in retail goods and services which will likely be attracted to the center and translate this into retail floor space requirements by type of goods.
STEP VII	Evaluate the findings of the trade area and market analyses in relation to the initial size and floor space allocation in the contemplated center and adjust as necessary.

Figure 10.8. Steps in defining a trade area. *Source:* From McCabe, 1971:21.

such as the Renaissance Center on the Detroit waterfront, which is basically an 80-story hotel with 30,000 square meters of retailing in a single megastructure; the Trump Tower in New York; or the Omni Center in Atlanta.

The traditional shopping center, however, is feeling some competition from new merchandising methods. These include mail solicitation by limited distribution complexes (offering fashion goods, music entertainment, specialized clothing and jewelry, and so on). Whereas some of the larger and traditional catalog sellers (e.g., Sears) have reduced or eliminated this phase of their activities, catalog buying via mail supports many smaller industries these days. Some simply mail out their catalog and hope that the variety of product and the convenience of ordering from home and having it delivered will attract buyers. Others induce catalog buying by offering surprise gift packages or enroll the purchaser in sweepstakes with prizes of cash or goods.

In addition to this, special consumer channels are now available through cable television networks. While some of these emphasize particular products such as jewelry, others simply offer products that are not offered for sale in regular retail outlets. These may range from collectors' items to recreational or fitness equipment, instructional books and videos (e.g., aerobics or tennis lessons), to safety and theft protection devices for vehicles or homes. Some experimentation (e.g., in Chicago) has also taken place with respect to getting access through home computers to lists of goods offered at different shopping places. The individual simply makes a choice from the list, and the products are packaged to await pickup.

All these changes in retailing reflect changes in the attitudes of consumers toward the purchase act and the attitude of producers toward providing their product. In other words significant behavioral changes have produced structural changes in the entire industry. The simple normative assumptions discussed in Chapter 2 rarely dominate in today's market decision-making processes. Instead, critical assumptions include:

- Minimizing time of shopping (e.g., by being able to do it from the house via computer or television).
- Convenience (home delivery rather than an out-of-home shopping trip).
- Satisficing as opposed to the traditional economic assumption of utility maximizing by minimizing costs.
- Least effort in terms of the spatial domain.

In many larger North American cities, shopping center developers and locators must take into consideration housing preferences of the new higher-income, young home owners not involved or pressed by childrearing and with two working household members, one or both of whom may be professional. Such individuals seek neighborhoods in the inner city with geographic clusters of housing structures capable of yielding high-quality environments. They demand also a variety of public amenities within safe walking distance such as scenic waterfronts, parks, museums, art galleries, universities, distinct architecture, historic landmarks, and convenient, high-quality shopping. In addition, they require an increasing amount of out-of-home food services including both fast-food and high-cost, menu item specialty restaurants.

10.7 The Role of Government in Shopping Center Locations

Shopping centers today are constrained in terms of locational decisions by federal, state, and local policies. For example, prior to initiating construction, developers must go through lengthy processes of seeking permission from local authorities to develop a site. The local authorities, in turn, have the responsibility of ensuring that state and federal controls or suspected impacts are properly evaluated. Today, invariably, those at the predevelopment stage must have performed an *economic impact*, an *environmental impact*, and/or a *social impact assessment* (see Figure 10.9). This is so because of concerns for the environment, concerns for equity and social justice, and concerns for economic viability, as well as to some extent a felt need to protect the infrastructure and tax base of downtown business areas. For example, the infrastructure investment and tax bases in most downtowns, their centrality, their accessibility, and their function as a hub for local mass transit facilities all provide reasons for protection from unrestricted competition. Such areas are seen to serve

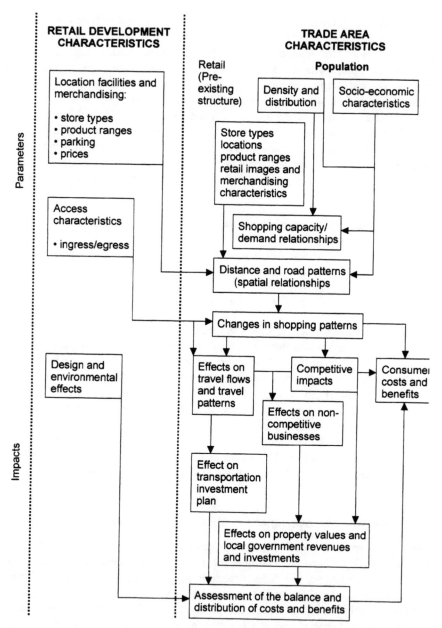

Figure 10.9. Procedures for environmental impact analysis. *Source:* Rodgers, 1979:399.

the entire city in many different ways, and attacks on their viability from sectors such as retailing are often closely monitored.

Within the last two decades, *economic impact analysis* has become a major tool of government. In some cases this tool has become political, for its primary function is to preserve the viability of existing commercial areas from competitive intrusion by new shopping centers. These existing commercial areas include both CBDs and the existing business hierarchy both planned and unplanned in a given political district. Such a use appears quite rational if the existing physical facilities can be modernized, if examination of the market area potential calls for preservation and upgrading facilities, and if existing lease terms and conditions allow tenant changes to be made. Otherwise insistence on detailed and time-consuming impact studies prior to receiving development permission have been used as a political tool to protect existing investments.

Throughout most of the history of shopping center development in the United States, adherence to zoning controls was virtually the only approval required before submission of a plan to build. However, changes in merchandising and marketing methods have produced more sophisticated consumers who have transferred to the regional center development context their concern for negative environmental impact or degradation as the result of shopping center location. Since the National Environmental Policy Act of 1969, there has been an increased tendency for various concerned groups of private citizens and state and federal governments to request not only environmental but also economic or social and cultural impacts relating to the development of new shopping centers.

Different attitudes toward this need change as national political power changes. In 1979 and 1980, for example, President Carter's Community Conservation Guidelines were in operation, and the need for economic impact was greatly emphasized. In 1980 President Reagan took office, established a Task Force on Regulatory Relief, and repealed the Carter guidelines. During their existence, however, the Carter guidelines had been used to delay some suburban developments and to prohibit them in some areas. They also encouraged developers and planners to see suburban center development as part of a general metropolitan system of centers that both complemented and competed with each other. While they were in operation the guidelines generated a number of conflicting legal positions depending at least in part on the strength of local environmental feeling. This perhaps reached its epitome in California with the development of the Coastal Commission and the California Environmental Quality Act.

The Carter guidelines reflected a felt need to effectively, efficiently, and equitably manage scarce federal resources. Their essence was to direct efforts toward examining externality effects of federal policies such as subsidy or support policies for redevelopment as well as attempting to anticipate the future effects of development or growth (Figure 10.10). This resulted in more urban impact analyses. In 1979, an amendment to the National Environmental Policy Act (the so-called A-95 Guidelines) was passed. These required urban impact analysis to be completed within 45 days of the submission of plans for development associated with federal assistance. Underlying this act was the feeling that Congress, the administration, and the American people in general have the right to expect an evaluation of the po-

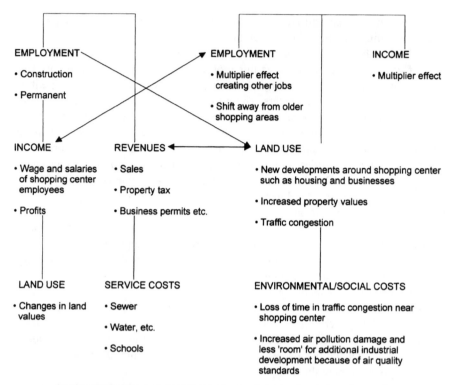

Figure 10.10. Direct and indirect impacts of new regional shopping center.

tential impact of federal actions before they occur. Alternatives then may be developed to mitigate the most serious negative consequences.

Many critics of impact analysis in urban areas argue that the responsibility for the development of federal and state policy can be laid directly at the doorstep of developers of regional malls. They argue that the latter were the chief culprits for a host of urban ills including damage to drainage systems, runoff, overloading of utilities, traffic congestion, and increased problems of safety.

10.7.1 Economic Impacts

The traditionally accepted definition of *economic impact* is that it "is an attempt to assess the economic efficiency and fiscal impacts of (alternative) land use proposals by local agencies, as an aid in evaluating public and private land proposals" (Public Resources Code USA Section 2115, 1979).

By tying environmental impact to the notion of *economic efficiency*, however, this definition avoids questions of equity, disguises the importance of externality effects, and leaves the door open for ambiguity. Ambiguity arises from different definitions of economic efficiency. For example, it can be defined in terms obtaining a product at the lowest cost, or it can refer to providing a maximum quantity of service or goods at a given cost.

Equity, in this sense, is a criterion for assessing "who should bear the cost" of a particular activity. Should it be the developer? The local government? The state government? The businessman? The consumer? Local area residents? Or should all of these bear the brunt in some prorated manner?

Economic impact analyses are a means for allowing governments to anticipate the effect of their decisions and to avoid problems resulting from those decisions. Such analyses also allow governments to anticipate the effect of private sector development on their own resource base and to avoid problems resulting from those decisions. A need to secure timely information with respect to the impact of private or public actions should be obvious. What is less frequently appreciated is that the standard economic impact analysis is analogous to studies that some responsible merchants and developers initiate prior to beginning work anyway. However, what the impact policies do is to take the results of such studies out of the private domain of the developers and put the results firmly before the public eye. Also, this allows the public adequate time to absorb relevant information about the advantages and disadvantages and short- and long-term consequences of particular acts.

Thus the primary objective is to provide federal, state, and local action on the basis of an objective evaluation process that will enable agencies to distinguish clearly between costs and benefits of any development. They will be able to evaluate the net benefits of revitalization of existing areas or to develop *tradeoff* functions between loss of positive features associated with economic, physical, or social environments after development with the convenience, cost, and other benefits obtained from the actions. If the impact analysis demonstrates that significant negative externalities or direct negative consequences will result from a given action, then the appropriate decision-making agency usually either requires modification of the development plan or refuses to allow permission for the development to take place.

Thus, economic impact assessment becomes the vehicle by which decision makers and planners work out what is economically efficient. But in addition to this, information is often gathered in terms of environmental impact and sociocultural impacts.

10.7.2 Externalities

Externalities arise when the action of one person or group brings unanticipated costs or benefits to another. These costs are usually those that the initiators of the actions do not have to bear. The benefits are those that may inadvertently or indirectly accrue because of an action. Thus, externalities are either positive or negative.

- A *negative externality* might result from the discharge of waste or pollution into a water body or the atmosphere, thus affecting the health or safety or welfare of others.
- A *positive externality* might result from a retail development that brings additional utilities such as water and sewage to a residential area that had re-

lied on wells or septic tanks. Other positive externalities might be an unanticipated increase in employment, the provision of better community services such as child care, an increase in land values, or improvements to road systems or public transportation systems.

Sometimes impact assessments are accepted or rejected on the basis of how well these externalities are highlighted and understood.

10.7.3 Social Justice and Equity

The provision of services usually is judged to be efficient when their locational pattern provides *maximum profits* to the entrepreneur within a *competitive* framework in a system. A loss of efficiency occurs when locational advantages change via increased number of retailers or shops in the support base, such that profits are effectively reduced.

When attempting to obtain an efficient solution, most location analysts welcome the process of *minimizing system costs*. These include maximizing accessibility of the retailer to surrounding consumers. Thus an efficient retailing system would *maximize sales* while universally *minimizing the travel costs, time*, or *effort spent* by potential consumers.

In such a system there is no objective evaluation of the distribution of benefits among individuals, and an efficient system can grossly violate conditions of *equity* and/or *equality*. These principles were combined by Harvey (1973) into the concept of *social justice*. Social justice is sometimes violated when a system is developed purely on the basis of economic efficiency or when actions are taken to preserve the status quo even though the beliefs of the community or the values of a society may be violated.

Equity in the spatial sense is defined as justice with respect to location. This notion is absent from most theories and models of location theory, which tend to stress economic efficiency. The idea of providing some type of equity arises when society feels that some of its members may not be receiving a fair share of the available goods and services. Inequity can result from inefficiency in the economic order or it might exist in spite of or because of the selection of efficiency as a goal criterion. It also can directly result from inequities in the distribution of wealth in the system and the consequent differences in the range of goods and services designed to service that unevenly distributed wealth.

In spatial terms, *systems equity* sometimes is defined in terms of maximal average travel time or travel distance. Thus a system average concept provides a weak measure of equity that is sometimes used in location-allocation models. This interprets efficiency in a social sense more so than in an economic sense. More typically, equity is measured by means of a socially imposed minimum standard. With respect to a locational pattern of services for example, one may consider "equitable" to be no more than some acceptably small proportion of people who are more than a critical distance from a place at which goods and services are located. This maximum distance constraint often is built into today's location-allocation models. Inequity arises in the retailing center when a distributional pattern emerges that unduly penalizes particular segments of the population. Much of the value of a decentralized, free-standing planned shopping center is related to the question of improving the well-being of the mobile, suburban homeowner in that no longer did these exurbanites have to travel far distances over congested roads to "undesirable" inner-city locations to purchase a reasonable array of required goods and services. But even these decentralized centers usually are served only by individual transportation modes, namely the private car. Access via public mass transportation is poor. This means that other populations distributed throughout an urban area are denied access to them.

This concept of social justice has been tied to the notion of a more equitable distribution of services and facilities and includes: (1) raising the average level of access for people to sets of facilities (e.g., more facilities located closer together); (2) reducing the proportion of people more than some critical distance from a facility (i.e., changing the locational pattern of services); and (3) reducing the variability of access or distance traveled to facilities.

Questions of social justice include whether the costs obtained in meeting the above criteria should be absorbed by the provider of the service rather than be passed on to the consumer, or whether costs are absorbed by some governing agency or decision-making unit.

10.8   Future Trends

In this chapter we have progressed in scale from the individual consumer to the decision-making and regulatory activities of governments at national, state, and local levels. Throughout we have stressed that critical components include the assumptions being made by the decision maker and the rules or strategies that guide the decision-making process. We have seen that often these are incompatible when we compare individuals and institutions. These incompatibilities are rationalized differently within different economies, different societies, and different ideologies. But underlying all of this, we have found sets of principles that are becoming universally acceptable at both the individual (*micro*) and institutional (*macro*) levels.

In making any statement of policy, the decision-making authority must be aware of what appear to be the dominant future trends in the area for which policy is made. For example the vitality of a shopping center depends on the continuance of the demand for its products. Besides creating new centers, the shopping center industry has to be involved in the careful management and operation of its old centers. This requires not only maintenance of each center's productivity and efficiency, but also recommendations for internal adjustments as characteristics of a center's trade area change and as consumer tastes and preference change. It is recognized that any constraining policies have to be dynamic and must lend themselves to change also. Three factors, then, appear as critical forces for the future:

1. Characteristics of the trade area population's buying power and patronage patterns.
2. The continuing need to build or alter centers to meet changes in consumer demand.

3. The financial situation of operators and developers and changes in systems costs that would be reflected in new building costs.

Changing characteristics of trade area population include discernible trends in population composition, distribution, income levels, social structure, age structure, ethnic structure, and tastes and preferences for goods and services, as well as willingness and ability to travel. Other forces include changes in public policy, particularly environmental-related policy, and changes in the patterns of energy use in transportation. The final critical issue concerns trends in financing, including such things as variations in interest rates.

11

Place and Space

11.1 Views of Place and Space

Some people think of space as just one characteristic of objects in an environment (Hart & Conn, 1991). This emphasis is most dominant in fields such as environmental psychology, spatial cognition, and behavioral geography.

In these circumstances, space often is viewed in the *Kantian* sense of being a container in which human action takes place after the processing and manipulation of information sensed in the spatial container. Spatial cognition is seen to focus more on spatial relationships and the interdependencies among objects and attributes in the spatial domain. Hart and Conn (1991) argue that it is more appropriate to substitute the term *place* for that of *space*. This would, they claim, bring one closer to meaning and action, for place is seen as the focus of human intentions. Adopting this point of view readily allows incorporation of affective components into the spatial domain, for attributes such as thinking, feeling, and acting are presumed to be relevant to places in an environment rather than to abstract spatial structure.

This approach is *action oriented*. Generally in the social sciences, action is defined as goal-directed behavior. In spatial cognition and in much of behavioral geography, movement is seen as the outcome of a choice act that is undertaken in the process of achieving objective goals. Hart's (1981, 1984) previous research with children has supported the Piagetan concept that, at least in terms of infants, action can come before representation; that is, representation emerges from nonintentional actions, which eventually lead to action schemas. Consequently Hart and Conn (1991) suggest that many of the actions of adults are rationalized after completion as being goal-directed behaviors, when they may have been proceeding simultaneously with the processes of representation and evaluation. In pursuing this line of thought, a greater importance must be placed on the social environment rather than the physical environment. In essence actions are thus presumed to be the result of transactions between people and environments.

11.2 Affect and Emotion

Significant influences are exerted on the nature and tone of the human–environment relationship by affective appraisals and the resulting emotions that are tied to those appraisals. Amedeo (1993) suggests that there are two premises underlying work on the relationship among human activity, experience, and environment:

- The first is that activities and experiences take place in environmental configurations (e.g., settings, places, landscapes).
- The second assumes that humans are capable of interpreting, praising, evaluating, and perhaps adapting to the environments in which they live.

These premises assume that environments consist of things external to mind. But they suggest also that people transact and interact with the information sensed from these external sources and in doing so, ascribe meaning to that information. They imply that the way in which the information is experienced and the value and belief systems that in part control the way it is processed will influence heavily the process of interpretation and addition of meaning. The ways that our emotions color interpretation and become strongly tied to environmental representations are of particular importance in terms of ecological evaluation and assessment.

11.2.1 Difficulties of Definition

Amedeo (1993) points out that *emotion* and *affect* themselves are difficult concepts to define. He selects a definition by Kleinginna and Kleinginna (1981), which is reproduced in Strongman (1987):

> Emotion is a complex set of interactions among subjective and objective factors, mediated by neural/hormonal systems which can (a) give rise to affective experiences such as feelings of arousal, pleasure/displeasure; (b) generate cognitive processes such as emotionally relevant perceptual affects, appraisals, labeling processes; (c) activate widespread physiological adjustments to the arousing conditions; and (d) lead to behavior that is often, but not always, expressive, goal-directed, and adaptive. (p. 3).

Such a definition includes the idea that emotional experiences must be understood both in terms of how those experiencing them comprehend them and how others recognize expressions and translate the meaning of those expressions into social or cultural situations that activate responses. Examples of *emotions* are given in Figure 11.1. Thus, in practice, feelings, emotions, and moods are easily considered to be components of an emotional category.

In terms of landscape assessment and evaluation, the standard list of emotions has been expanded to include preference and aesthetic evaluation. The term *affect* is, according to Amedeo, much more useful in describing this larger domain. Thus, he suggests that affect may be used to refer to both feelings and the emotions commonly and jointly associated with those feelings.

Happy	Sad
Hopeful	Hopeless
Like	Dislike
Love	Hate
Excitement	Dullness
Active	Passive
Placidity	Anxiety
Courage	Fear
Calmness	Anger

Figure 11.1. A list of emotions and feelings.

11.2.2   Critical Components of Emotions

Amedeo (1993) identifies a number of critical components or themes that reoccur in studies on emotions themselves or emotional responses to different environmental situations. These include:

- A physiological component.
- A cognitive component.
- The integration of physiological and cognitive components.

These components combine to facilitate a common way of viewing emotion in the spatial context. Amedeo suggests that places and spaces as well as people evoke emotional responses. Places that evoke such responses are not just the beautiful, mysterious, ugly, or degraded natural landscapes, but the rooms, houses, neighborhoods, and cities in which we live our lives and experience the environment. For example, particular districts and cities can be labeled friendly, threatening, fearful, frustrating, or beautiful, and they can induce feelings of desire, hate, concern, and fear.

Mehrabian and Russell (1974) suggest that the emotional states generated by an individual in a particular environmental or behavior setting will be influenced by the nature of the setting and the initial emotional state of the perceiver and that they will be due to various personality traits of the individual. In other words individuals have *emotional dispositions* that are brought out to a greater or lesser extent by the settings with which they interact. They claim that the critical emotional dimensions are *pleasure, arousal,* and *dominance*. One shortcoming of this approach, however, is that it makes no provision for cognitive appraisals of behavior settings and the environmental information obtained from those settings. It fails also to account for the difference in affective experiences that a single person may have of the same setting, or indeed that different individuals may have for the same setting. Thus, from a given viewpoint overlooking a heavily wooded natural area, one individual may be dominated by ecstatic feelings of joy and happiness or pleasure with the natural harmony of the scene, whereas a neighbor may be dominated by feelings of avarice in terms of assessing the scene for its potential for providing marketable timber.

11.2.3   Hierarchical Levels of Affective Response

Ulrich (1983) provides an alternate conceptualization postulating that affective components are the first that are aroused when visual perception of a given environment or setting takes place. He suggests there may be several *hierarchical levels of affective response*, the first involving broad categories such as liking or disliking, interest or lack of interest. It is response relating to the development of a preference, not primarily to the processing of information. More detailed affective responses are generated as experience with the many dimensions of the setting or environment occur. He does suggest, however, that there is probably a linked hierarchical component in the affective response and that the initial impression and the emotion generated in that initial impression are likely to control the search and explication of other feelings and emotions. In other words, there is a multileveled, hierarchical arrangement in which, once having entered the hierarchy at one general level, there is a greater tendency to look for more detailed feelings of that general type rather than to cross over to another cluster in the hierarchy.

Since the result is a strong person–environment–behavior tie, Ulrich hypothesizes that effective reactions to natural settings serve an adaptive function in the way they effect behavior. As quoted in Amedeo (1993:96), Ulrich suggests that "An individual's affective reaction motivates, or serves as an action impulse for adaptive behavior or functioning." Actions or behaviors motivated by effective reactions may include search and exploration, approach or avoidance, attention or misdirection, focusing or scanning.

11.2.4   An Interactional Constructivist Model

Moore and Golledge (1976) adopt a *person–environment–behavior paradigm* that suggests that human behavior and experience can be fully understood only when viewed in terms of the circumstances in which they occur. Thus an environment constitutes the surroundings or a medium in which behavior and action take places in response to the sensing of information. In other words, people involve themselves with the production of internal representations (cognitive maps) by means of which they acquire, synthesize, store, recall, and use environmental information. In this process of storing such information in long-term memory, often the environment provides the context that adds meaning to behavior and provides some insights into the nature of behavioral reactions and responses. In interacting or transacting with environments, individuals are guided by prior experience, in that this prior experience is a prerequisite for action as much as it is a basis for identification and attachment of meaning to processed information. But since the information that individuals obtain about an environment is filtered through human sensors, what is being apprehended has little meaning until it is related to the purposes, motivations, or goals and objectives of sensing individuals.

Thus, Moore and Golledge suggest (1976) there is an *interactional constructivist* position arguing that knowing about an environment is related to an ongoing process of using information obtained via the transaction process but adjusting, altering, or revising it according to constructs such as anticipations, expectancies, and a host of other templates already stored in memory. Instead of a pure transactional approach, they offer an interactional–constructivist model that emphasizes more the role of previous experience in apprehending the environment. Information stored in memory can be internally manipulated by thinking, or it can be modified by the process of inputting new experienced information from previously unexperienced settings or environments. Modifications to an internal knowledge structure and the affective responses tied to components of that structure are seen to be the categorical rules that guide the individual's awareness and selection of relevant information from the mass of information generally available in any given environment or setting. Concepts such as cognitive maps, prototypes, scenes, images, and schemata have been suggested as the mechanisms for transforming sensation into knowledge.

11.2.5   The Individual's Assessment of Setting

Amedeo (1993) suggests that *emotion* is reflective of an individual's *assessment of a setting*. In other words, an emotion reflects the personal significance given by the subject to a setting. The emotional response embodies both the reaction to context, to previously experienced circumstances, and to the social and cultural filters through which a setting is viewed at any particular point in time. Amedeo and York (1984), for example, have examined the reaction of different population groups to settings ranging from run-down, inner-city neighborhoods, to suburban household environments, and large, apparently unbounded, natural settings complete with flora and fauna. They suggested that particularly intense feelings and emotions—components of a general class of responses called *affective*—influence one's experience of different physical environments. Whereas cognitive responses include thinking, information processing and manipulation, coding, decoding, representation, and recall, affective responses consist of the range of feelings and emotions, beliefs and values resident in long-term memory. For example, it is argued that affective responses influence verbal and nonverbal relationships with environmental settings. Amedeo and York suggest that there are *emotional norms* that underlie affective responses to physical surroundings. To examine this hypothesis they analyzed the response patterns of 79 subjects to three different environments: inner-city urban street scenes, pictures of a social gathering inside a suburban residence, and a scene of a natural environment.

Amedeo and York (1984) claimed their results support the common knowledge that people's responses do not always conform with their beliefs. They suggested that people appear to be equipped with emotional norms and therefore are assumed to invoke rules of feeling in their affective responses to environments. For some environments, these emotional norms are multiple and distinctive; for others they are universal. The "feeling rules" hypothesized to exist are said to be "schemas"—that is generic, long-term integrations that can be accessed and used in different problem situations. Amedeo and York suggest it appears that the inner-city subjects felt safe and secure even with scenes of dilapidated inner-city areas, but they felt fear and concern when faced with views of geese and animals in a natural environment.

11.3 Landscape Perception

Landscape has been a focus of geographic interest since the inception of the discipline. Aitken et al. (1989) suggest that *landscape* invariably is seen through the eyes of the beholder and is interpreted through a filter of cultural values and personal beliefs. Landscape is mostly viewed holistically and as a *Gestalt*. When it is broken down into isolated segments or components, it loses its landscape quality and becomes a mere collection of geographic layers.

Always considered a part of cultural geography, landscape became enmeshed in behavioral research under the rubric of environmental perception. Innovative research using this paradigm included that by Lowenthal (1967) and Lowenthal and Prince (1965). But despite being an extremely popular term in the geographer's vocabulary, landscape proved to be one of the most difficult to define. Sell, Taylor, and Zube (1984) have suggested that in landscape perception the terms *landscape* and *environment* are often used synonymously. For some researchers, however, these two terms connote different things. Both can refer to natural and built scenes, but landscape perception invariably focuses on the tangible, visual/perceptual elements of environments. Particular emphasis is placed on features and characteristics of landscapes that express social and cultural trays.

Aitken et al. (1989) argue that no single convenient taxonomy can be applied to the range of modified environments that are open to human study. Convenient descriptors include rural landscapes, natural landscapes, wilderness areas, park lands, and natural settings. Some researchers argue that the term should not be used in an urban context or in other human designed and architecturally diverse spaces. Even though these latter cityscapes are accepted as special types of cultural landscape, there is still some hesitation in treating them in the same way that we treat natural environments or human-modified biogeographical settings. Zube (1984) and Mitchell (1984) suggest that the emphasis on natural and rural environments ties strongly to areas of natural resource management.

11.3.1 Landscapes and Cultural Values

Lowenthal (1975) examined the relationship between the human values embedded in different cultures that shaped the landscapes in which they lived. Interpretation of landscape requires giving it meaning in terms of cultural experience and cultural values. This research recognized that landscape experience influences and, in turn, is influenced by *culture*. Cultural values serve to condition the way people interpret, regard, and interact with landscapes. Often landscapes are interpreted not just in terms of the mix of physical resources that occur within a setting but in terms of the way that cultures symbolically interpret, rearrange, or otherwise impress themselves on the arrangement of landscape features.

Aitken et al. (1989) claim that landscape assessment is strongly tied to historical practice and cultural values. They suggest also that environmental symbolism is often captured not just in the real world but in fictional novels and stories in factual treatments of space and place and in art and in other literature such as in poetry, travelogs, guidebooks, photographs, cinema, and television (Zonn, 1984; Aitken,

1994). Aitken et al. (1993) further suggest that landscape perception lends itself to study by the methods of phenomenology, particularly by stressing the inward, personal nature of environmental experience; to support this argument they point to the work of Rowles (1978) and Hill (1985).

According to Aitken et al. (1993), Meinig (1979) has interpreted landscape as "text" and uses historical method and the analysis of discourse to unpack not only the physical information contained in landscapes, but also the beliefs, values, shared habits, and preferences that different cultures exhibit in the ways their landscapes are arranged. More recently different perspectives on landscape have emerged from the tradition of Marxist and socialist geography. Often these have focused on the built environment (Harvey, 1985; Blaut, 1983)—those settings that were somewhat neglected by traditional landscape analysts.

11.3.2 Place

As Aitken et al. (1989) interpret it, *place* implies a location and an integration of society, culture, and nature. It generates strong psychological and emotional links between people and places. These links are dependent on the range of experience that people have with places. Strong arguments have been put forward that people develop and respond to a sense of place, and it is this sense of place that identifies the felt coherence of features in a setting, as well as the feelings and emotions that the place generates. The sense of place incorporates not only the concepts of location and pattern but feelings of belonging, invasion, mystery, beauty, and fear. As time changes, cultures alter settings, and the sense of place may change. This implies that the perception or cognition of the place, its symbolism, its meaning, its cultural significance, and even its boundaries, may change. In many ways this way of looking at landscape captures the essence of geography, for it provides clear evidence that places differ. Even if they are similar, people may experience them differently and allocate different meanings and symbolism to them.

The importance of the relationship with the past has been expressed in many different ways by Lowenthal (1975). Lowenthal argued that there is a systematic relationship between individual experience and landscape components. Thus his approach is embedded firmly in the experiential perspective. He suggests that relationships among landscape components have isolated factors such as coherence and legibility. These factors rate high in understanding human preferences for different landscapes for different purposes.

In early treatises on landscape evaluation, Appleton (1975a, 1975b) argued that the development of an acceptable and reliable system for landscape evaluation would not evolve until the theoretical vacuum existing at that time was overcome. He argued there was a scattering of bits and pieces of theory from perceptual psychology, architecture, design, forestry, outdoor recreation, geography, and environmental studies. But no commonly accepted or overarching framework existed. Later, Sell et al. (1984) argued that four distinct paradigms of what they called "landscape perception" had emerged:

1. Pragmatic applications designed for use by managers and planners (the *expert* paradigm).
2. The search for meaning in the perception of landscapes (the *psychophysical* paradigm).
3. The search for meaning in landscape in human terms (the *cognitive* paradigm).
4. The search for meaning of human landscape interactions (the *experiential* paradigm).

In the *expert paradigm*, trained professionals undertake objective surveys of landscapes and measure and record landscape features and characteristics in assessments of characteristics such as scenic beauty. Much of the work in this area has evolved in assessment and evaluation that is commonly part of environmental impact studies. The overwhelming sense of this paradigm, however, is that it is the individual expert only who is capable of seeing the nuances of the environmental gestalt.

The *psychophysical* paradigm assumes that landscape has value, and its value is related to its potential as a stimulus. It is external to the individual, relatively invariant over time, and is perceivable through visualization. This paradigm has led to the development of measurement techniques such as signal detection modeling as a procedure for assessing scenic value of forests or estimating air quality or neighborhood quality. It also allows us to deal with problems of environmental degradation caused by the intrusion of human activity or even perhaps at times as a result of natural hazards. Much of the recreational landscape perception literature falls into the psychophysical paradigm.

The *cognitive paradigm* focuses on landscape quality as a construct of the human mind. Visualization of landscape is the normal input modality. Here aesthetic preferences, adaptation levels, and arousal theories all contribute. Appleton (1975b) suggests that this approach can draw on information from history, philosophy, and ethnology to provide an evolutionary-based appreciation and perception of landscape qualities. Within this paradigm also is housed a considerable amount of cross-cultural research on landscape perception and evaluation including that by Sonnenfeld (1969), and Zube and Pitt (1981). Sonnenfeld compared Alaskan and Delaware residents' perceptions of landscapes, and Zube and Pitt undertook cross-cultural comparisons between culture groups comparing professional and general public attitudes toward and perceptions of landscapes. Within this paradigm also lie most of the studies comparing natural versus built landscapes.

The *experiential paradigm* focuses on the phenomenon of human environment interactions. It is relied upon in art, historical, and literary traditions. It suggests that the aesthetic quality of landscape lies both in the natural environment and in the meaning of that environment to people. It is almost exclusively undertaken using unstructured phenomenological exploration, literature review, or art criticism.

These paradigms appear to be pervasive and lead to the conclusion "that landscape perception research efforts are marked by separation: separation by paradigm; separation by theoretical origin" (Saarinen, Seamon, & Sell, 1984:68). However, these authors then go on to offer an overarching schema for landscape research that incorporates the four paradigms. In their schema landscape perception is considered a function of interaction between humans and landscapes and the outcome of perceptual interaction feedback that modifies human, landscape, and interaction components. In this model the human component refers to the implicit or explicit assumptions about the nature of humanity. This includes statements concerning the level of skill of a landscape observer, the degree to which individuals act as information processors or filters, the degree to which they are active or passive in the interactive process, the degree to which past experience influences what is being perceived, the prior knowledge state about the landscape that is stored in long-term memory, the expectations and aspirations set by anticipated exposure to a landscape, and the personality and sociocultural context of the individuals or indeed the landscape itself.

Landscape properties include both the *individual* elements and the *entirety* or *Gestalt* of landscapes as entities. Both tangible and intangible relationships are assumed to be critically involved in the perceptual interaction process. Landscape features may be given salience based on characteristics of art, design, or indeed ecology or cultural value. Often it is assumed that landscape properties are the ones capable of manipulation through different types of management processes. The tangible elements might include factors such as approximal relations of vegetation and water, whereas the intangible elements might include difficult concepts such as mystery, beauty, or faithfulness (Figure 11.2).

The key to this paradigm is the *interaction component*. Thus, work focuses on the human landscape interaction at either the individual or group level, as well as the interaction between and among humans within particular landscape contexts. Interaction processes, according to Sell et al. (1984), are divided into components such as active or passive, purposeful or accidental, unique or habitual. Outcomes of the human landscape interaction process can include tangible ones of physical change and intangible ones such as alteration of feelings, emotions, values, beliefs, or states of mind. This model basically is a model of *transactionalism*, in which values given by humans to different landscapes evolve from the transactional process of mutual influence. Using this paradigm, Sell et al. (1984) argue that human landscape interaction is grounded in the following properties:

Tangible	Intangible
Human-Natural Mix	Beauty
Economic Value	Mystery
Landuse Mix	Serenity
Slope	Discordant
Vegetative Covers	Gestalt
Water	Cohesion
Flora	Placelessness
Fauna	Aesthetic Value
Economic Value	

Figure 11.2. Tangible and intangible components of human–environmental interaction.

1. *Landscapes surround*: In this way they permit movement, search, and exploration, and thus the participant becomes part of the landscape.
2. *Landscapes are multimodal*: In other words, multisensory input is processed simultaneously in assessing and evaluating them.
3. *Landscapes involve both central and peripheral information*: They use all the senses' information simultaneously received from all points of the compass or from the entirety of the surrounding landscape.
4. *Access information*: The mass of messages and information emanating from landscapes is redundant and is far too massive to be wholly perceived or understood.
5. *Action*: Landscapes cannot be passively observed but immediately involve the observer even if only in terms of the act of looking or sensing; landscapes call forth purposeful actions.
6. *Landscapes* always have *ambiance*.

These attributes were suggested first by Ittelson (1973). Like his contemporary, Jack Wohlwill, who argued that the environment is "not in the head," Ittelson argues that environments and the landscapes embedded in them cannot be studied in the same way that one studies objects in a laboratory. In other words, part of the exploration of landscape and the development of understanding involve human experience. The ways that different humans and human groups undertake transactions with the environments in which they live are reflective of the moral and ethical values of culture, society, and economy. An essential premise of this approach, then, is that landscape perception cannot be restricted to the aesthetics of looking at scenery. It is embedded in the larger set of values that influence meanings, and those meanings can vary considerably depending on the human or group landscape transactions that are involved. Sell et al. (1984) suggest that research in the area commonly identified as landscape perception can be integrated into a single overarching framework. The approach illustrated in Figure 11.3 suggests a combination of research paradigms in which concern with landscape features and resources merges with compositional characteristics such as mystery and hazard to present a wholistic view of landscape. This appears to be a paradigm that is becoming more accepted by both expert and public groups concerned with environmental use and management and with protectional preservation.

11.4 Historical Preservation

One area where landscape assessment, evaluation, and indeed perception have spanned the gap between academic and applied work is the area of *historical preservation*. Typical research activity in this area is focused on the association of economic impacts of preservation. These include situations such as gentrification and renewal or processes of reversing inner-city decline, or focusing on the economic benefits of outdoor recreation as a compliment to the historical use of natural resources in different areas.

Ford (1979) examined the place of historic preservation in the process of altering both the social and physical character of U.S. cities and towns. He concentrat-

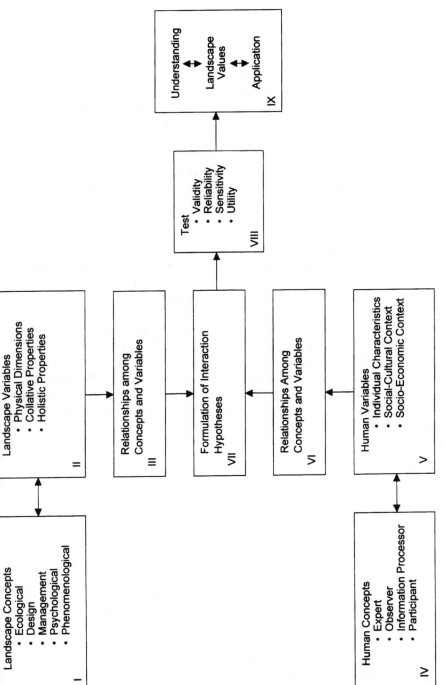

Figure11.3. Integrated framework for landscape perception research. *Source:* Sell et al., 1984:81.

ed on areas identified in the traditional concentric zonal model of urban structure—urban cores, zones of discard, and older lower-socioeconomic-status neighborhoods. Historical preservation was seen as part of the way to identify and maintain a sense of place.

Datel and Dingemans (1980) suggest that there are a number of underlying dimensions behind the concept of preservation. These include the use of literary and historical sources to describe national or universal attributes to the past and judgments of those attributes worth preserving. Lowenthal and Prince (1965) have examined historical artifacts in the United States and Britain to determine how people perceive, experience, and manipulate the past. They see the landscape as a historical source itself, and by viewing landscape, one can find evidence of a melange of social, political, religious, and aesthetic thought and experience that have been instrumental in preserving or altering landscapes over time. By examining literary and artistic records, they are able to show how attitudes toward landscape are altered from the past to the present. This approach has also been instrumental in characterizing the current concern with landscape restoration, preservation, and reconstruction as deriving from particular views of national heritage and responsibility.

Lowenthal and Prince (1965) contrast perceptions of the past and historical preservation practices in the United States and the United Kingdom. They argue that American history has adopted the practice of isolating, purifying, and ossifying (i.e., museumizing) in an attempt to preserve social messages about American values and virtues. It is no secret that Americans have not hesitated to use history and historical values for economic gain. This is particularly true with respect to the gentrification and "yuppie" movements attracted to revitalized inner-city areas. Many of these inner-city areas had been purchased at low value by major financial institutions such as banks, insurance companies, and mutual funds, and revitalization was simply a means of turning such acquisitions into profitable venues. Similar actions have been taking place in inner-city areas, particularly in cities with waterfront frontage where warehouses and storage sheds that for many years simply blighted the visual landscape have been turned into minimalls, tourist attractions, and large but reasonably priced residential rentals. Lowenthal and Prince argue that the English have preferred their historical places to look lived in. Traditionally they have been integrated into the working landscape but are emphasized as landmarks or places of historic interest or importance. There is some argument that the English are more reluctant to exploit their historical places for commercial purposes, but there is no doubt that this has become an increasingly important role for such places as economic conditions make it difficult to retain them in their pure forms. Lowenthal and Prince suggest that American preservationism in more recent years, however, had less to do with reconstruction for the sake of gaining tourist dollars and an increasing concern for preservation as a record of past existence, beliefs, and values.

As an example of this, Datel and Dingemans (1984) report on research comparing historic districts in five metropolitan areas—Philadelphia, Washington, San Francisco, London, and Paris. They note the growing similarities among preservation tastes in each of these places. Emphasis is placed on maintaining features and characteristics of civic cause that are often the center point of public meetings and preserving historically surviving residential quarters that often have aristocratic associations. Datel and Dingemans (1984) also emphasize the preservation of ancient villages that may have been originally outside a historic city but have been absorbed later by suburban development. Other places in these cities that are subjects for preservation include early industrial sites, canals, institutional complexes with innovative design or historical importance, heritage homes or houses and grounds of the rich or upper class, gardens, parks, cemeteries, and land adjacent to streams, rivers, or ponds. Datel and Dingemans (1984) argue that differences among the different metropolitan areas in terms of the way they decide on preservation can sometimes be explained by differences in the cities' resource base or its administrative, political, and economic structure. They give examples. Among London's 404 conservation areas, 175 can be classed as suburban residential developments constructed between the end of the 19th and the early part of the 20th centuries. In Paris, few such areas are designated as being worth preserving. Pennsylvania's village settlement pattern has been reflected in the form and structure of Philadelphia. The plantation system of Virginia is embodied in a significant part of Washington's historic districts. Despite these differences, Datel and Dingemans argue that the same qualities of human scale, antique texture, harmonious architecture, mature landscaping, romantic associations, and traditional appearance are themes that commonly and repeatedly occur in recognizing places as being worth preserving for historic reasons.

However, the five cities studied also give different preservation powers to government authorities and provide incentives to owners or residents to protect and enhance the historical value of dwellings or other buildings. In the United States, tax incentives for certified restorations are common. Europeans often use the process of restricting property rights in historic areas. Datel and Dingemans claim that the central cities of Philadelphia and San Francisco have not advanced far in the local designation and regulation of historical districts, but Washington and both London and Paris, with their much stronger planning traditions and tougher regulations on preservation, are far more advanced. All of these areas provide examples of historic districts that have decayed; some are being restored, while some have remained in a stable and decayed manner. In the latter case in particular, new development rather than wholesale restoration is permitted but is tightly controlled. Within the United States this often has taken the form of restoring the original appearance of places, and this is equaled only perhaps by the attention given to the restoration process by the French. In all cities, factors associated with restoration and preservation have had the effects of displacing residents, often lower-socioeconomic and ethnic minorities, who have succeeded in establishing themselves in the form of high-density tenantries in buildings formerly located in the inner-city and occupied by single families.

Datel and Dingemans (1984:135) have derived the notion *sense of space* to refer to "the complex bundle of meanings, symbols, and qualities that a person or group associates (consciously and unconsciously) with a particular locality or region. This bundle derives not only from direct contact with the place, but also from many other experiences and from literary and artistic portrayals." They go on to argue that those interested in sense of place must of necessity unpack the historical preservation bundle, for this embodies the societies' and cultures' traditions about what aspects of place are deemed to be worth perpetuating. Tuan (1974, 1977), perhaps more than anyone else, has examined the union of space and place and has extended this concept beyond the traditional landscape bounds to include the sense of place that is celebrated in architectural beauty, historical events, the characteristic structure of neighborhoods, and the awareness that particular places, and the way they are preserved, carry with them the spirit, tastes, preferences, values, and beliefs of generations. Similar sentiments have been expressed by Lynch (1960) in terms of identifying imageable components of different cities.

11.5 Evaluating Places and Assessing Landscapes

It makes little sense to speak of "the environment." There are *many environments*, each with its own sets of attributes and characteristics, rules and regulations, feature contents, perceptions, and mysteries. Physical and built environments mean many things to different people, depending on how they view them, experience them, or use them. One cannot assume that the same environment means the same thing to all people.

There is a need, therefore, to find out if there are a fundamental set of characteristics that are commonly perceived or known about environments. This becomes of particular importance to activities such as planning for new uses, planning to preserve all uses, or the establishment of policy with respect to how interaction with an environment can take place. Previously in Chapter 6 (Section 6.8.2) we quoted the work of Nasar (1984), who showed that perceptions of the built environment in Knoxville, Tennessee varied considerably depending on whether one was a permanent resident or a visitor to the city. Other cross-cultural and intracultural studies have shown how different subgroups differentially weight different features and attributes in specific environments resulting in the designation of different anchorpoints and some noticeable difference in the cognitive maps of those environments.

Why would one need to assess or evaluate an environment? Typically *environmental assessment* would take place as part of the preparation of an environmental impact report prior to any translation of the natural to a built environment. In the United States *environmental impact* reports are mandated prior to the beginning of construction for new residential developments, shopping centers, hazardous waste or nuclear waste disposal, for historic preservation, or for leisure and recreation. The former process is usually carried out by professionals and has the following components:

- A survey of what is there in physical reality.
- Designation of sets of attributes whose presence or attributes can be measured in some way.
- A selection of the indicators of the attributes.

11.5.1 Describing What Is There

A description of what exists at a particular place is often made from both experiential and secondary materials:

1. *Experientially* the assessor can travel through the environment observing the multiplicity of features and relative frequency of occurrence of key

characteristics (e.g., endangered species, unique vegetation, geological faults, geomorphic features such as soil slips or sinkholes, and so on).

2. *Secondary sources* often involve the interpretation of aerial photographs, satellite images, land use maps, and prior evaluations of the utility or usefulness of the area for specific activities.

In the course of this description, *classificatory schemes* for cataloging features must be used. These can be developed idiosyncratically, but it is more likely that a standardized system would be used, such as have been produced by the Bureau of Land Resource Management in the United States and the Countryside Commission for different parts of the United Kingdom. Applying these schemas to the local inventory process allows the development of an ordered and detailed listing of the identifiable components of the local area. Overviewing a categorized list facilitates the building of a holistic description of the environment. Often such schemes are represented on maps. These maps can be the traditional paper and pencil maps of the cartographer or, as is more frequently occurring today, the visual representations of phenomena on a computer screen as presented by a GIS. This inventorying process provides a set of base information against which comparisons of worth and value or value and belief system can be checked. The critical assumption here is that once an attribute of the physical environment can be identified, it can be measured numerically (e.g., the number of trees, the quantity of area devoted to certain types of natural vegetation, the area of crops, building heights and shapes, amount of planned roads or highways, etc.). However, sometimes the most significant attributes are not directly measurable. Rather they may reflect an aesthetic valuation produced by a particular mélange of features. Such a characteristic may be the beauty perceived in certain views or certain scenes; the mystery enshrouding places at certain times of the day or night; the joyful pleasure experienced by people anticipating or experiencing the environment (e.g., the rush of pleasure associated with free exploration of a natural area unconstrained by the paths, utilities, buildings, or activity normally found with a built environment).

It is possible sometimes to capture some of these aesthetic qualities pictorially in slides, photographs, videos, or movies. A classic example is the Peter Weir movie *Picnic at Hanging Rock*, where story, acting, and imaginative photography were interwoven to produce a wonderful mixture of beauty, mystery, and suspense. At times, however, it is the sense or experience of being at a place, the feeling or emotion that is generated by interacting, that is the most significant attribute. These attributes are the most difficult to convert to some type of numerical form. In many cases one can merely suggest that they are present or absent, representing them only at a nominal level. It is apparent that lists of land uses and measures of their amounts of occurrence are not likely to produce an understanding or appreciation of these aesthetic values or feelings aroused by an environment, nor are they likely to allow one to assess the folkloric, religious, cultural, or philosophical properties of a place.

When a set of significant attributes concerning a place are determined, there remains still the problem of deciding what indicators of their occurrence best represent them. This is a measurement problem. Such a statement often produces intense negative reactions from many who argue that feelings, emotions, aesthetics, values, and beliefs cannot be measured. But here we must return to the discussion of types of measurement in the earlier chapters of this book. A measurement may be as simple as recording the presence or absence of an object, feature, feeling, value, or belief. Such a recording can be made in a variety of situations such as in the presence of the place, in the presence of an excellent representation of the place, or by asking people to imagine that they are at the place. Thus simple measurements can be collected in the field or in the laboratory. It is standard practice in many spatial cognition studies to expose people to environmental layout, then ask them to imagine that they are standing at some place in the layout and point to the location of other features. This type of imagination experiment reveals the degree to which survey-level understanding of an environment has been achieved.

When we are dealing with feelings or emotions and when it is not possible to place a person in an actual setting, then pictorial representations of the setting (e.g., slides, videos) frequently are used. The imagined condition again is potentially useful in addition to these. The information generally collected, however, is in the form of verbal descriptions rather than numerical enumeration of attributes.

In many of today's environmental impact requirements it is mandated that both objective and subjective indicators be developed. Thus, some indication of the location and quantity of vegetation in an area might be required plus some indication of its assessed value in terms of concepts, such as the beauty, enjoyment, satisfaction, order, or danger that is associated with such places. Thus this first descriptive stage of an environmental assessment can be compiled using different types of direct viewing or imaginable conditions and objective and subjective counts or ratings of the presence or absence of particular attributes. These often provide the first level of input in evaluating land use plans, management decisions, or as input to public hearings soliciting impressions of proposed developments.

11.5.2 Specification of Attributes

The specification of attributes of a particular place for which an environmental assessment is being made requires a mix of protocols. As mentioned in the previous section, exhaustive lists of features may be developed, and, after examination, the principal attribute dimensions of the physical reality may be determined. Such dimensions may include degree of roughness, quantity of water, types and condition of vegetation, wildlife varieties, and endangered species. A quite different set of attributes may develop from asking for people's opinions and obtaining their descriptions of places. Written or recorded descriptions may be examined using techniques such as item analysis to identify critical verbal cues that anchor attribute dimensions. In this way, indications of qualities such as peacefulness, pleasure, noisiness, safety, mystery, beauty, and so on, may be elicited. Sometimes these are represented in bipolar terms (e.g., beauty/ugliness) and with the individual being asked to rate a given place somewhere along the continuum anchored by bipolar endpoints. This rating procedure can be undertaken by experts (e.g., planners, architects, designers, artists) or by the ordinary person.

Rachel Kaplan (1991) points out that environmental assessment has many hidden assumptions. She argues that description is based on categories, yet the choice of categories often is not explicit. The acceptance of categorization often leads to the neglect of psychological dimensions, particularly affective components. The allocation of rating scores to features by experts is often dependent on their purposes, and these purposes are frequently insufficiently specified. Kaplan also argues that assessment is often guided by what is countable. This carries with it the assumption that "if you can count it, it's likely to be important." Unfortunately this leads to the position that if it can be counted, it is likely to be included in a model whether or not it is important for the purpose at hand. The evaluator or assessor has to be constantly aware to include what matters as well as what can be counted.

This raises a particular set of dangers. In the 1960s and 1970s, geographers dealing with large quantities of data (e.g., from censuses) embraced the technique of *factor analysis*. This is a data reduction technique that collapses sets of variables to common dimensions, sometimes challenging the researcher to find appropriate anchors to describe the dimensions. Some argue that the GIS is the equivalent of factor analysis in today's world. Their comprehensiveness and versatility in both representation and analytical modes allow one to input a huge quantity of recorded information without necessarily knowing whether or not it matters. Using extremely versatile GIS functionality, information can be represented as layers, then overlain or decomposed, compressed or expanded, analyzed, or visually represented in such ways to allow description. To avoid traveling the same path as was traveled with factor analysis, it is important to ensure that the purpose for which GIS is used in the environmental assessment context can be clearly and unambiguously stated.

Purposes themselves may be selected without sufficient attention to the sets of assumptions underlying them. A commonly accepted purpose behind environmental assessment is economic development. This assumption frequently is not questioned, although it is now common for environmentally aware groups to challenge the appropriateness of development. This challenge might take place with respect to proposed change in what is seen to be highly valued land. It occurs also in those circumstances where the use value of a piece of natural environment is low, but the primary criterion simply is to keep it in its natural state and preserve what is there.

11.5.3   Selection of Indicators

In most environmental assessment the purpose is not simply to produce a pictorial or symbolic record of what exists, but in some way to assign values to the individual and configurational content of a place. To describe what is there, we have seen how recourse often is made to existing descriptive or classification systems that can provide a way to attach a quantity to an attribute. It is assumed that once indicators have been defined, data can then be collected about them, and some type of analysis can proceed. Of course, if such indicators are capable of being quantified (however weakly or however strongly), one has input into management decision-making simulations that may be able to provide some idea of what would happen if particular environmental components were manipulated. For example, if a unit of area were an integral part of a scene, then removing that unit and replacing it with something different (e.g., a housing tract) might destroy completely the values associated with the holistic imagery of the original scene. This could well impact not only the way that people view the environment but also affect their potential behaviors and interactions with respect to it. Typical indicators used in environmental assessment are the classes listed in Figure 11.4.

Assessment of an environment thus involves *description*, *evaluation*, the development of *indicators*, and attempts at *explanation*. We emphasize here that the

1. **Physical attributes**	landforms landcovers atmospheric quality light ruggedness height soil type water quantity water quality
2. **Landscape attributes**	variety of land uses color texture form line uniformity variability shape
3. **Subjective attributes**	perceived complexity diversity harmony coherence dominance of visible forms uniqueness, mystery beauty enjoyment satisfaction aesthetic value traditional use value historical significance memories
4. **Human environment interactions**	familiarity travel frequency land use types expected or planned uses scale configurational structure/partition visual perspective

Figure 11.4. Sample of indicators used in environmental assessment.

first phase (description of the physical reality) is not the whole product. It is needed to complement the vast quantities of remotely sensed and image-processed data from satellites that can now be manipulated and stored in many a GIS with mapped or visual images and perceptions, memories, values, beliefs, ethics, and morality. Thus, assessment must be conscious of both the physical reality and the transformations of that reality that are stored in memory as part of the cognitive mapping process. There is some evidence that views of attributes such as exploitable, endangered, risky, mysterious, beautiful, and safe have been perpetuated and provide norms in different societies and cultures for evaluating novel environments. The question that remains unanswered, however, is what represents the "core" of such designations in different regions, in different cultures, in different societies, and in different economies? If such "cores" can be determined, then a step has been made toward providing a minimal set of attributes that must appear in the environmental assessment undertaken within a particular economy, society, or culture.

11.6 Another View of Space and Spaces

According to Beck (1967) the delineation of forms in space is a basic part of the process of differentiating the environment into discrete and heterogeneous parts, percepts, and concepts. He argues, for example, that the six cardinal directions are not endowed with equivalent meaning for humans. Rather, orientation epitomized in spatial prepositions—such as up and down, front and back, left and right, and inside and outside—has particular meanings because humans are bilaterally symmetrical terrestrial animals. For a psychologist, Beck adopts a rather supportive geographic view that the study of spatial distributions, both real and perceived, is the cornerstone of any investigation of the relationship between humans and the environments in which they live.

Beck suggests that there are at least three basic kinds of space:

1. *Objective space* is that of physics and mathematics. Its dimensions are measured by universal standards such as distance, size, shape, volume, length, and surface.
2. *Ego space* is the individual's adaptation of observed or objective space. Its purpose is to present a consistent view of the components of physical space such as the size, shapes, and distances.
3. *Immanent space* is subjective, the space of the unconscious, or dreams, or fantasy. It includes the space in which individuals embed their cultural notation system that is embedded in cultures. It is fundamental because it is a space that is tied to the anatomy of our bodies.

Beck suggests that spatial typologies can be broken down into a set of dichotomous descriptors:

1. *Diffused space* versus *dense space*: These represent dispersion versus compactness.

2. *Delineated* versus *open space*: These refer to bounded versus unrestricted freedom.
3. *Verticality* versus *horizontality*: These represent the dominant factors in different spatial planes.
4. *Right* and *left* in the *horizontal plane*: These are vectors of the horizontal plane.
5. *Up* and *down* in the *vertical plane*: Two vectors of the vertical plane.

Beck argues that personal spatial systems may be observed by eliciting preferences and determining which symbols embedded in these preferences represent the spatial parameters used by an individual. He developed a spatial symbols test in which pairs of figures representing simple geometrical shapes such as points and lines represented the five variables described above. The task given to subjects was to choose which pair of symbol figures they preferred. The complete paired comparison design involving 67 pairs of symbols was administered to 611 subjects. An example of the figures used in this spatial symbols test is provided in Figure 11.5. Beck found that there was a developmental evolution in the number of dimensions underlying the comparative test that increased from the youngest age group (5- to 6-year-olds) to the older children (13- year-olds). The clusters appearing at the earlier-stage group contained items from all five spatial scales. He suggests that this means that beliefs are undifferentiated according to dimension at this age group and infused into a single system. Of the 32 items composing the two clusters (factor groups), 22 represented the spatial planes of left–right, up–down, and horizontal–vertical. By the age of 19 years, the number of clusters (factors) increased to three. Spatial differentiation, particularly along the diffuse–dense dimension emerges strongly at this point. In 13- to 14-year-olds, five clusters appear, each composed of items from a separate grouping of the variables mentioned earlier. Beck concludes from this that spatial experience leads to differentiation in spatial meaning.

11.7 The Language of Space

Whereas visual perception offers one way of sensing and interpreting space and place, *language* is a very different modality and is often used as an agent of spatial description. And although *visual perception* is important in object recognition, *language, touch*, and *hearing* modalities are also important for *sensing* objects, relationships, structures, patterns, behaviors, and meanings. People who are blind have not the valuable experience of vision to help them sense an environment in all its complexity. But *audition, haptics, smell*, and *taste*, along with the senses of *flow* and *motion* provide them with the means of immediate contact with that environment and, like vision, can act as filters for the many to whom it may concern messages emanating from that environment.

Couclelis (1991) argues that *language* is a higher level faculty than *perception*. She suggests that many animals have visual perceptual skills equal to, if not better than, those found in humans, but we have not been able to establish clearly

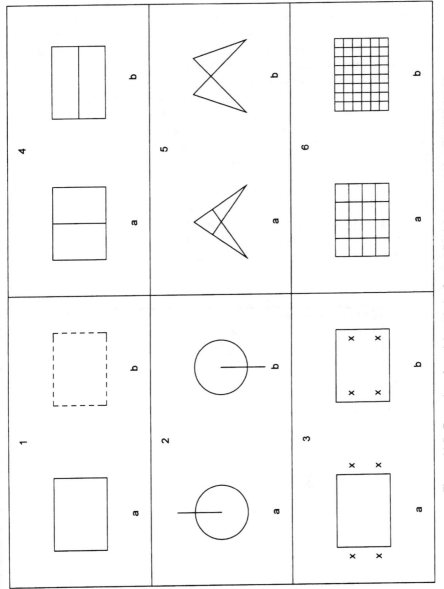

Figure 11.5. Examples of spatial symbols. *Source:* Beck in Lowenthal et al., 1967:37–38.

that those animals communicate via language. However, one must at this point be aware of Wittgenstein's admonition that even if a lion could talk, we could not understand it! This implies that lion language, lion concepts, lion grammar, and lion meaning would be so alien to those developed by humans that we may have no common base for developing translations.

Lakoff (1990) suggests that *imagery* is the connection between language and visual perception. However, the output of visual perception differs from the memory of say a perceived or imagined scene, or even the stored memory or knowledge of a given scene whose features may have to be assembled from felt, remembered, and imagined experiences. In fact, some argue that visual scenes cannot be understood until they are converted into mental imagery.

Talmy (1983) suggests a strong connection between *language* and *spatial imagery*. He argues that language is organized to direct the listener's attention toward sets of fundamental parameters or categories related to conceptualization of space. He distinguishes three imaging systems in language that are used for understanding and structuring space: schematic structure, viewpoint, and distribution of attention (Figure 11.6).

1. *Schematic structure* is imparted by prepositions assisted by demonstratives, such as "this" and "that," and nouns such as the use of "timber" versus the term "forest" to describe trees and water as opposed to a pond to define an amount or type of liquid. Such structures draw also on geometric and other idealizations to convey dif-

Schematic Structure	-	far
		among
		across
		along
		into
		this
		that
Viewpoint	-	global
		local
		relational
		perspective
Imaging Systems	-	figure/reference
		ground/target
		closure
		focus
		dominance
		against
		along
		at
		near
		close to

Figure 11.6. Imagery systems used to structure space.

ferent aspects of spatial relations or activities, such as motion, or indeed the spatial properties of objects such as size and boundedness. Unfortunately, the prepositions and demonstratives are not scale invariant in terms of their interpretations, and they are often used in wildly different concepts—the snail crawled "across" the car, and the Englishman walked "across" the Sahara. Such concepts may also be neutral with respect to shape, such as we may find in the phrases "in the cup" or "in the valley." A further problem relating to spatial prepositions and demonstratives is that relating to closure, as may be found with the use of the preposition "inside."

2. *Viewpoint* is the second of Talmy's imaging systems. He distinguishes between a static global viewpoint such as might be found in the statement "There are a few children in the playground," and a sequential, dynamic one with a more localized scope such as "One can occasionally see a child or two in the playground."

3. *Distribution of attention* is the third component used to structure space. It is similar to the selection scene of visual perception and is made obvious in the figure–ground conundrum frequently illustrated in visual perception. In language terms, however, the figure–ground concept can be expressed as a difference between a statement such as "The entrance is near the kiosk," as opposed to "The kiosk is near the entrance." This may indicate that proximity in language is not symmetric as it is in mathematics and geometry. Instead it involves distinguishing between reference objects and target objects. Sets of spatial prepositions that produce the same complex situations include against, along, at, near, close to, and many others.

Thus, language reflects the common sense world. This naive or common sense or human view of the physical world is often incomplete, fuzzy, distorted, schematized, abstract, and summarized. Many of the locative expressions in our language may appear perfectly logical and comprehensible in a world of points, lines, areas, and surfaces that are more or less Euclidean in nature. However, they often fall down when used in the world of subjective reality.

Landau and Jackendoff (1993) suggest that the thing that sets humans apart from other species is an ability to express spatial experience by means of language. They explore the language of objects and places, in particular examining the nature of nouns and spatial prepositions in the English language. They argue that when objects are named as objects, the nouns incorporate some geometric properties relating to them, such as shape (axis, solid and hollow volumes, surfaces, and parts). In contrast to this, when features or things are seen to have the role of figure or ground in a locational expression, only very fuzzy or coarse geometric properties are represented. They argue further that spatial prepositions tend to be nonmetric and unless further qualified, provide only loose and fuzzy spatial information. Thus, although designed to clarify and enrich our understanding of features, objects, places, and relationships in space, the language that we use often adds uncertainty, imprecision, and multiple meanings.

11.7.1 Nouns

Landau and Jackendoff (1993) claim that the average adult has a vocabulary of about 10,000 names for things. They define these as "count nouns" that label different objects and features. For many people, object categories shape is one of the most important criteria used in the identification process for deciding what label should be attached to it. Thus categories of things with the same or similar shapes are often labeled the same. The use of shape is so universal that even very small children can identify categories of things from simple black-and-white line drawings. For example, when an aspect is represented by shape, which is then described by a noun (e.g., "this is a box"), objects that are similar in nature may be generally referred to as "box-like." There is, therefore, a tendency to develop shape bias in classificatory terminology. The number or variety of shape categories that can be discriminated depends on the versatility of the language for the purpose of differentiating among slightly different shapes. Examples of language categories developed to deal with shape bias are seen in Figure 11.7.

Landau and Jackendoff (1993) argue that in order to name objects and their parts, spatial representation of them by *shape* is a rich descriptive system that contains attributes that can be generalized and used to differentiate among object categories.

11.7.2   Defining Where Things Are

The language used to describe *where* often focuses on geometric characteristics (Figure 11.8). Landau and Jackendorf (1993) argue that the standard linguistic representation of an object's place requires three elements:

Shape Categories and Names

Whole objects (e.g., simple 3-dimensional components such as cubes or cylinders).

Names for object parts (e.g., trunks, handles, fingers, foot, ceiling).

Names for spatial parts (such as axis and axial parts, e.g., front, back, top, bottom); Implicit in such labeling are orientation and direction axis that help to differentiate and relate labels such as front, back, and sides.

Names for objects best described as surfaces (e.g., 3-dimensional phenomena that are often represented as 2-dimensional irregular polygons such as lakes, regions, and buildings represented by their facades).

Names for negative object parts (e.g., a ridge as opposed to a valley, where the valley is interpreted as a negative part chiseled out from an object rather than adding to it).

Names for containers and related objects (e.g., containers such as depressions, bowls, administrative districts, and a variety of solid or hollow volumes).

Figure 11.7. Shape categories and names.

point	off
line	behind
area	in front of
region	on top of
landmarks	beneath
reference node	behind
route	place
to the right	location
to the left	far
turn	from
community	above
neighborhood	higher than
adjacent	lower than
connection	away
network	across
hierarchy	along
on	in

Figure 11.8. Examples of language of "where."

1. The object to be located or what can be called the *figure*.
2. The reference object or background or environment in which the object's place is to be determined, often called the *ground*.
3. A *relationship* between the *figure* and *ground*, usually expressed by means of a spatial preposition.

When we say "The house is in the suburbs," then "house" is the figure, "suburbs" is the ground, and "in" represents the relationship between the two. Relationships can be expressed in a simple form by spatial prepositions or by verbs that incorporate spatial relations—such as "going" (e.g., "go toward," "go across").

Using *spatial prepositions* often produces confusion, since different individuals represent relationships in asymmetric form. For example, one person may say "The book is on the table," whereas another says "The table is under the book." Convention usually suggests that the larger and more immobile object is the reference or ground, and the smaller or more mobile object is the figure. Conventionally, then, we would expect to hear that "The book is on the table" rather than "The table is under the book." Confusion in verbal communication often results from mixing figure–ground relationships in descriptive expressions. This characteristic becomes important in developing cognitive maps of environments. For example, important landmarks become reference nodes, and other environmental features are related to them using spatial prepositions such as "near," "beyond," "to the left," and so on. These reference nodes give much more accuracy or precision when used as the ground in a locative statement (e.g., "My house is near the Mission").

Some of the most frustrating uses of spatial prepositions occur when an object's shape carries with it information relevant to the meaning of a selected preposition. The terms "along" and "across" provide excellent examples. "Along" implies that an object has a principal access of significant elongation so one can travel "along" a road, but one generally does not travel "along" a desert or "along" a round table. On the other hand, one may talk about traveling "across" a desert. Convention also guides the use of other prepositions such as "on" and "in" ("on" the lake as opposed to "in" the lake), and "inside" and "in" (one may be "in" a pool, but not "inside" a pool) (see Figure 11.9).

Thus, in verbal communication about features in space and relationship among those features, the choice of nouns and prepositions conveys conventional information that is often fuzzy or inaccurate. In many cases the language implies an underlying geometry, and unconventional use of a relational term may cause considerable confusion. But since humans apparently do not develop incredibly precise representations of phenomena, the language that has been developed is *loose* enough to convey meaning but *meaningful* enough to be commonly interpreted and acted on. In the absence of a more technical expression of place and space that provides locational and relational precision, our knowledge of and our communication about the environments in which we live are necessarily *fuzzy, incomplete, distorted, generalized*, and at times *confusing* (e.g., to my "left" is to your "right").

If spatial language used to identify place and spatial relationships is instrumental in the encoding of sensed information, it carries the imprecision that is found in most cognitive maps. Object and place recognition by humans, therefore, may often be *descriptively rich* but *spatially inaccurate*. The nouns provide the richness, and the prepositions provide the fuzziness and inaccuracies. For example,

Spatial Prepositions		Temporal Prepositions
near to	downward	before
on	foward	after
in front of	left	ago
under	right	during
in	there	despite
out	here	like
under	east	into
above	south	out of
apart	north	
across	west	
along		

Figure 11.9. Spatial and temporal prepositions.

Landau and Jackendoff (1993) point to the concept of relative distance as an example of this richness/inaccuracy conundrum. They suggest that English prepositions embody four levels of description:

1. Location in the region *interior to* the reference object, described by terms such as "in" or "inside."
2. Location in the region *exterior to* the reference object but in contact with it, described by terms such as "on" or "against."
3. Location in the region *proximate to* the reference object described by terms such as "near" or "close to."
4. Location *distance from* the reference object described by terms such as "far" and "beyond."

They suggest also that some additional precision can be obtained by adding a negative—such as the house is "not" near the corner or the landmark is "not" visible from the store.

When discussing direction, Landau and Jackendoff (1993) suggest that the appropriate spatial prepositions are tied to concepts of gravitational attraction. Thus, "over," "above," "under," and "below" refer to a *vertical* dimension or the primary direction of gravitational pull (Figure 11.10). Other prepositions, such as "near,"

Object Geometry -	Figure Object Geometry	Relation of Region to Figure	Paths
in on near at inside on top of behind along across between among	along across around allover throughout	interior (in, inside, throughout) contact (on, all over) proximal (near, around) distal (far) negatives (out, off, far) vertical direction (over, above, under, beneath) horizontal directions (by, along, next to, close to, besides) front to back (ahead of, behind, beyond, before) axial (on top of, in front of) visibility and occlusion (on to of, underneath, before	earth oriented (up,down, east, west, north, south) figure object axis oriented (forward, ahead, backward, left, right)
		(adapted from Landau and Jackendorf (1992)	

Figure 11.10. Reference object geometry.

"by," and "beside" reflect relationships that are orthogonal to gravitational pull and are expressed primarily in the *horizontal* domain. More generally, spatial prepositions are seen to be of significance when identifying:

- *Relative distance* (e.g., "on," "against," "near," "far").
- *Direction* (e.g., "above," "beneath," "next to," "far from," "in front of").
- *Distance–direction combinations* (e.g., "on top of," "in back of").
- *Visibility* and *occlusion* (e.g., "behind," "underneath," "on top of").
- *Paths (e.g., "forward," "sideways," "right," "north," "west," "to," "from").*
- *Functional situations* (e.g., hitting a ball "to" someone vs. hitting a ball "at" someone).

Thus, the concepts of *figure, ground, region,* and *relationship* provide differing quantities of precision in spatial language. Descriptions of the *what* and *where* of places become more or less interpretable as people choose appropriate nouns and prepositions to formalize relationships. Descriptions of place can be like an artist's rendering of a painting, interpreted idiosyncratically by everyone who looks at it; or they can be dramatically and spatially correct—that is, properly encode all the relevant information about the spatial domain in which features or objects are placed and related one to another.

11.8 The Utility of Place

One talks of a *sense of place* as something that is of significance to people. Much of the literature on sense of place refers to intangible characteristics or some unmeasurable but comprehensible mix of place characteristics that makes it attractive or unattractive, but that influences people positively or negatively toward it. There is also a clear implication that a sense of place influences people's behavior.

The behaviors may be economic, social, political, or be undertaken for other purposes. This sense of place previously has been interpreted by geographers (Wolpert, 1964) using an economic paradigm called *place utility.* Place utility referred to the value that individuals allocate to a place or the levels of satisfaction they would achieve by interacting with the place in some way—ranging from looking at it, to living on it, or exploiting it. In a sense, place utility was indicated by the amount of time, effort, or resources, that an individual, group, or society would expend in preserving or using the place.

Beyond the initial place utility concept, sense of place was reinterpreted in terms of *social transactions* involving place and people. Again, this was sometimes elaborated in a cost and benefit scenario, but sometimes its elaboration went beyond the strictly economic to the social, the ethical, and the philosophical. Where defined in economic or social terms, however, there was general agreement that maintaining sense of place depended in part on its natural physiographic characteristics, its cultural values, and its historical setting and significance. Under these circumstances, often the benefits were not expressed in monetary terms but in terms of stability of expectations or the personal security of knowing that a place treasured because of some peculiar sets of characteristics was being maintained and not degraded or developed.

People often derive a great deal of pleasure from living at a place. The pleasure so derived comes from locational characteristics of the place, the community setting in which it is embedded, and the social and cultural relationships that accrue to a person by living at that place. At times these pleasures have been expressed in simple economic terms, such as the value of land or housing, or the total quantity and quality of public and private goods available at that particular place. Often differences are pointed out between the sense of place developed by residents and the sense of place developed by nonresidents such as tourists or recreationers.

In several other chapters (5, 6, 7, 12, 13) we have incorporated explicitly the notion of a sense of place into our discussions of human behavior. At times this has been contained within concepts of preference and patronage. Empirically we have discussed activities such as migration and labor force changes, all the time stressing intra- and interregional variations in expected and actual compensations associated with incorporating a sense of place into a person's decision-making structure.

However, it is difficult to develop objective indicators of the utility associated with place—or the *value* that is incorporated in a sense of place. Almost invariably this has had to emerge via an individual expressing a preference, revealing an attitude, or undertaking an assessment or evaluation. Bolton (1989) asks us to consider a region in which there is high turnover of population but in which the demographic and socioeconomic characteristics of the population do not change. This might be the case if in the aggregate the profile of out-migrants were matched by the profile of in-migrants. However, in this circumstance there could be an erosion of sense of place. While being demographically or socioeconomically equivalent, the values and beliefs that the in-migrants bring to the community may be substantially different from the original ones, thus causing a change in the way that both individuals and the community assess their sense of place.

Often the question of sense of place is tied to *scale*. Tuan (1974) defines *topophilia* as an *affective bond* between *people* and *place* or setting and suggests that this bond usually is developed between people and relatively small areas. Perhaps this is why economists have expressed little interest, for the most part, in trying to characterize, observe, and measure sense of place. Thus, if it is difficult to think of a sense of place developing for a large region, then it can be argued that the concept is not made operational at a scale of interest to economists generally. More likely it is the fact that even if a sense of place can be associated with a large region (e.g., in the United States, the "South," the "East Coast"), the boundaries would be so diffuse and so difficult to define that comparison of variables within and without such regions may be difficult.

Bolton suggests four reasons why economists have not been interested in a sense of place:

1. Because the concept is so intangible.
2. Place means many different things to many different people.
3. Many scholars are reluctant either on methodological or philosophical grounds to offer objective descriptions or measures of a sense of place.

4. There is today a tendency to focus extensively on the loss of a sense of place rather than to specify what a sense of place really consists of.

The last point is one that reflects the attitudes of many people in today's economy and society who argue that technological change and development have resulted in excessive destruction of important natural and historical assets and erosion of local differences. This extends into the homogenization of information via mass media and uniform product design and marketing strategies of national and multinational corporations, as well as reflecting a growing tide of genuine concern about future livability on this planet.

Bolton (1989) argues that it is possible to analyze a sense of place as an asset—a durable but intangible asset similar to commodities such as good will and reputation that are often associated with the transfer of businesses to new owners. He suggests a number of different ways of regarding place as an asset:

1. As *capital*: that is, as location-specific infrastructure that cannot be moved out of a place.
2. As an *investment*: In the sense of maintaining a sense of place as an investment designed to uphold the market value of the area in which this sense is grounded.
3. As an *economic return*: that is, an asset that can be accumulated and expanded and generally can be considered a public good.

In order to interpret sense of place from this economic point of view, a number of assumptions are necessary.

1. A sense of place is a human creation. Thus, it must be created by investment either by individuals or by communities. Real income is sacrificed in order to receive returns from the asset.
2. A sense of place must be maintained. Here one must be able to identify the acts of needed maintenance (e.g., preservation or rehabilitation).

From these two assumptions Bolton (1989) is able to raise a number of questions of interest to an economist, such as:

1. What is the relative efficiency of different kinds of transactions in creating a sense of place?
2. How do market transactions compare to philanthropic activity in creating a sense of place?
3. How important are investment and public goods that identify and preserve a sense of place compared to investment in private goods that can otherwise benefit a community?
4. What sort of returns must be obtained from a sense of place before there is private and public realization that it is a characteristic to be preserved?

To examine these questions requires different forms of measurement of relevant variables. Many geographers and others would take issue with Bolton that such questions are irrelevant, and it is impossible to objectify and measure the variables required to answer them. Bolton (1989) suggests, however, that one way of dealing with this problem is to use Lynch's (1960) concept of *legibility*. As we saw in Chapter 7, Section 7.7, Lynch has suggested that city images can be recorded in terms of paths, edges, districts, nodes, and landmarks. Bolton suggests that a region may be defined as having a strong sense of place if it contains a number of these legible forms. He suggests also that some index of accuracy or error of people's cognitive maps of different environments will reveal which places at various scales in the environment are legible. His argument is that to develop sense of place there must be the form of legibility that can be defined and quantified.

However, as many geographers, ecologists, and natural resource scholars have suggested, the intangibility factor associated with the sense of place may be the dominant one. Thus, fuzziness and lack of clarity may be of more value in defining a sense of place than precision and legibility. We would suggest, however, that regardless of whether the tangible or intangible position was taken with respect to examining the sense of place, it should be possible to develop either a subjective or an objective scale (or some combination of the two) that captures the essence of a place. Thus researchers such as Stephen and Rachel Kaplan (1976) talk of the mystery or beauty characteristics of a place as being of significance, while at the same time they argue that these features themselves are multidimensional and psychological and difficult to define clearly. As mentioned in Chapter 1, Section 1.6.4, great strides have taken place in the development of rigorously based qualitative research methods, and we suspect that it will be in careful consideration and rigorous application of these qualitative methods that a final solution will be found to capturing a sense of place in such a way that it can be accepted, validated by others, and have meaning to many.

11.9 Environments for Recreation and Leisure.

The expanding role of *leisure* in postindustrial society has produced a dramatic growth in tourist activity at both the domestic and international levels making tourism a major element in many economic development plans at the urban, regional, state, and national levels (Stough & Feldman, 1982). For a place to attract tourists, it needs to possess qualities and attributes that make it attractive, desired, and enjoyable.

Typically tourist areas are dynamic. They emerge, evolve, expand, and change over time. Butler (1980:5) cites several factors that are responsible for this, including "changes in the preferences and needs of visitors, the gradual diversification, and possible replacement of physical plant and facilities, and the change (even disappearance) of the original national and cultural attractions which are responsible for the initial popularity of the area." According to Stough and Feldman (1982: 25–26), of special significance is the observation that "the type and volume of tourists, most cultural changes, community changes, and resident reaction to tourists vary in a predictable manner with stages of tourist attraction development."

Some of these relationships are set out in Figure 11.11, and their implications are discussed in what follows.

11.9.1 Product Cycle and Typology of Tourists

Numerous authors (for example, Butler, 1980 and Stansfield, 1978) have observed how tourist attractions appear to pass through what seems to be fairly predictable stages that are akin to the product cycle concept. The notion is that different types of visitors are drawn to particular tourism spaces and places at different stages in the availability and the development of tourism attractions. As suggested in Figure 11.11, an underlying dimension of these typologies may be described as being anchored at one end by the visitor seeking "strangeness" or "novelty" and on the other end by the visitor who desires to travel but wishes to do so within the context of "familiar" surroundings. Stough and Feldman (1982) show how the "unique" and the "strange" can be experienced best during what is the "nascent" or "exploration" stage of a tourist area's development, whereas a "familiar" environment is experienced more readily once an area has become well established and developed as a tourist resort, that is, when it has achieved "consolidated" development (Butler, 1980).

Attraction Development Stages

Undeveloped				Fully Developed
Discovery	Local Response and Initiative		Institutionalized	
Exploration	Involvement	Development	Consolidation	Stagnation

Tourist Typologies

Seeking the Strange or Novel				Seeking the Familiar
Adaptive (Seeking Unique Experience)				Passive, Demanding
Drifter	Explorer	Individual Mass Tourist		Organized Mass Tourist
Allocentric	Near Allocentric	Mid-Centric	Near Psychocentric	Psychocentric
Explorer	Elite	Off-Beat	Unusual	Incipient Mass	Mass	Charter

Community/Cultural Changes

Low Impact		High Impact
Self-sufficient		Exchange Economy
Subsistence		Urban/Technologies
Agrarian		Society
	Positive Impacts	Negative Impacts
	(Employment, Income,	(Social, Environ-
	Amenity Expansion)	mental, Economic
Euphoria	Apathy	Irritation	Antagonism

Figure 11.11. Attraction development states, tourist typologies, and community cultural changes. *Source:* Stough and Feldman, 1982: 25.

Plog (1974) provides an interesting typology of tourists and visitors that pro-vides elaboration of the relationship among the stages of tourist area development, the type of visitor, and the community and cultural impact on an area. He identifies five tourist types who are strung along a population curve of psychological groups shown in Figure 11.12:

1. *Allocentric individual*: This person is akin to a "drifter" or "explorer," with the tourist seeking visitation environments that are totally different from his or her normal living environment. It involves wishing to see the tourist environment and to experience it as it is, for example, a wilderness area or an undeveloped natural fea-ture. These tourists are "discoverers," and they rarely return to an area for a second visit. Reime and Hawkins (1979) claim they constitute about 6% of the population.

2. *Near-allocentric individual*: This person is similar to the former, in that he or she seeks to see and interact with an area as it is, but is different in that some ac-commodations are desired, such as lodgings and restaurants intended to serve the needs of the tourist. Such people tend to travel often, and work actively to discover new places, as evidenced by the type of person who engages in beachlanding and no tourism activities. They comprise about 15% of the population.

3. *Midcentric individual*: This person belongs to the dominant group of peo-ple who seek "security" and "peer approval" and as a result tend to be oriented to tourist areas with a variety of commercial attractions that typify a tourist resort.

4. *Near-psychocentric individual*: This person is the tourist who requires a high density commercial environment and seeks tourist areas with an abundance of hotel chains, a wide range of restaurants, souvenir shops, and so on, as found in re-sorts such as Miami. Typically such people seek attractions that are easy to reach by car or air and that are domestic rather than foreign.

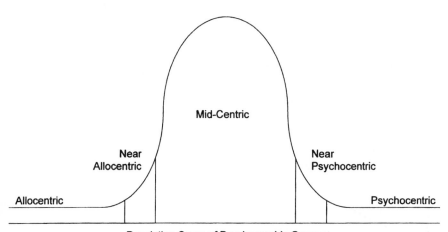

Population Curve of Psychographic Groups

Figure 11.12. Population curve of psychographic groups. *Source:* S. C. Plog, Why destina-tion areas rise and fall in popularity. Unpublished paper presented to the Southern California Travel Research Association, 1972.

5. *Psychocentric individual*: This person seeks highly familiar areas and tends to enjoy resort areas, amusement parks, family restaurants, shopping malls, and souvenir shops. Such people require that the attraction environment is constructed to meet their interest, thus artificial tourist facilities, such as theme parties, prevail as attractions. This type of tourist tends to be a repeat visitor.

Stough and Feldman (1982) and others have shown how Butler's (1980) model of evolution of tourist regions and Plog's classification of tourists interact to produce a variety of characteristics in the nature and complexity of tourism regions. Figure 11.13 provides a hypothetical representation of the evolution of a tourism area. Those in the "exploration" stage tend to have relatively simple tourist infrastructures with visitors coming in small numbers, interacting with local people, and using local facilities and services. The physical, economic, social, and cultural makeup of the region is unchanged by tourists. In Australia, the Cairns region was like this up to the 1970s.

In the "involvement" stage, the tourist area begins to produce services for visitors and to market itself. Tourist seasons emerge as tourism becomes more organized. There are changes in the social and living patterns of many people in the region. Additional income and employment opportunities arise, transportation connections with tourist source areas expand, and specialist tourist facilities begin to be planned and developed. In Australia, the Gold Coast entered this phase in the late 1950s and Cairns in the Far North Queensland region in the late 1970s.

The "development" stage sees the region emerge as a major tourist market with a tourist population at peak periods that may be larger than the local population, and the region becomes dependent on extensive advertising and extension of

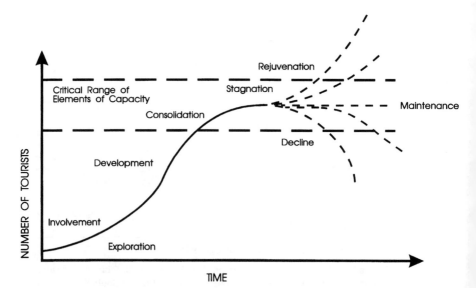

Figure 11.13. Hypothetical evolution of a tourist area. *Source:* Butler, 1980.

its tourist catchment. Local involvement and control in developing and producing tourist services declines in relative importance, with modern facilities being produced by externally generated and controlled capital. Artificial man-made tourism attractions are introduced that become as important as, or even more important than, the natural amenities of the region. Significant economic benefits accrue to local residents, but major environmental, social, and cultural impacts occur as the development stage unfolds. Local residents tend to be segregated, often by choice, from the tourist populations, and more complex forms emerge. In Australia the Gold Coast area of southeastern Queensland experienced this development stage in the 1960s and 1970s.

The "consolidation" stage sees the rate of visitor increase slowing but with peak tourist numbers well exceeding the local population as is the case with Lake Havasu on the Colorado River on the Arizona–California border. The economic impact of tourism becomes a major part of the region's economic base, and the marketing and operating of tourism are now highly institutionalized. "Stagnation" can be reached when the region's attraction has peaked. There are now few opportunities for new attractive developments, the image of the region may have become tarnished, and it becomes a less fashionable place to visit. Surplus capacity can occur. The Gold Coast in Australia reached this stage in the late 1970s.

"Poststagnation" stages see the tourism area adjust through rejuvenation, maintenance, or decline. Miami Beach is an example with renewed appreciation of its "kitschy" 1930s art deco hotels. "Decline" is frequent as the region loses it competitive advantage to other places both nationally and internationally. "Rejuvenation" can occur through the addition of new artificial attractions and the development of previously undeveloped natural attractions. The tourism area taps new tourist sources as a result. The emergence of golf as the focus for integrated resorts and the Japanese tourist market in the Gold Coast of Australia in the 1980s is typical of rejuvenation. To "maintain" itself, the tourist region requires extensive physical and institutional reinvestment to repair and modernize, and often this requires national and multinational corporations to become dominant players. The Gold Coast tourist core area has been experiencing this reinvestment and rejuvenation since about the mid 1980s.

11.9.2   Some Public Policy Implications

There are clearly some public policy implications of the above with regard to the development of tourism areas, particularly relating to the local planning of tourist attractions, facilities, and their funding that render, maintain, or enhance the "space and place attractions" of the tourism area for tourists. Stough and Feldman (1982) identify issues such as:

- Avoiding decline or managing it whenever it cannot be avoided.
- Rejuvenating an area's attraction once growth or stagnation ensues.
- Diversifying tourism products to cater for niche tourist groups.
- Addressing the balance between marginal benefits to the tourist and the marginal cost of expansion to the local community.

They provide a model (Figure 11.14) that shows the relationships among the components of an area's tourism industry in terms of managing social, economic, and environmental impacts in providing the tourist attractions and infrastructures needed to maintain the area's image and attractions as a tourist destination. This involves policies and strategies directed toward attraction, maintenance, and development; promotion and marketing; and local and regional transportation network development. Public sector agencies need to provide leadership with respect to setting rules, proprietary arrangements, partnerships with the private sector, and the development of supportive management tools.

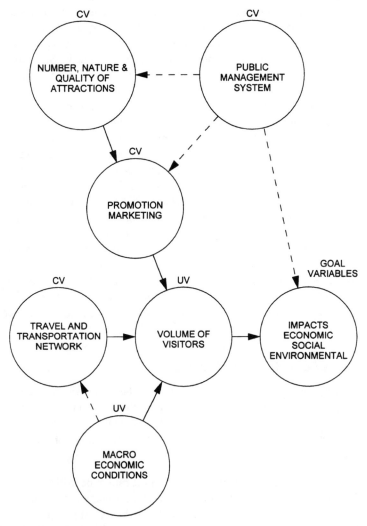

Figure 11.14. Relationships between the components of an area's Tourism industry. *Source:* Stough & Feldman, 1982:35.

Stough and Feldman (1982) refer to the experience of Charleston, South Carolina, which experienced little economic growth for a century following the Civil War. But when historic renovations became popular in the 1960s, much of the city's downtown became a target for preservation. By the late 1970s, visitor volumes had grown substantially, and the city was then host to the Spoleto Festival. The key to its success has been the strategy to preserve the historic core of about 100 blocks of the city bounded on three sides by water. The mayor of the city formed a Tourism Management Strategy Committee in 1979, and a Visitor Reception Transportation Center was designed. Plans were introduced to reduce pressure on the old historic downtown through preservation areas and restricting vehicle access. A tourism management strategy was implemented with a major focus on public relations, crowd control, and ordinance enforcement oriented toward minimizing the negative impact of tourism on the community.

Thus it is important in the development of tourist destinations and the maintenance of them as desired destinations for strategies to be formulated and for actions to be put into practice to ensure that they continue to exhibit those attributes that give them that distinctive sense of place as tourist attractions and for them to have place utility for tourists. This requires attention to be given to both the landscape attributes and the way they are perceived by users (i.e., visitors as well as locals) and the conditions necessary to attract the investment required to provide the tourism products as economically going concerns. Addressing these issues may be assisted through a consideration by civic planners and decision makers of the place and space utility, and perception and use considerations discussed in this chapter.

12

The Causes and Nature of Migration

12.1 A Complex Phenomenon

As a general economic and social process, migration is an important event for the person or household making the move, for the places of origin and destination of migration flows, and for society as a whole. Migration has been studied for a long time, and a plethora of theories to explain migration and to model migration flows have been developed. However, our theoretical knowledge of migration is far from complete.

It is well established that migration is an outcome of a range of factors including social, economic, cultural, political, institutional, psychological, and physical, considerations that independently or together may influence the way in which a migration decision is made or enacted.

Migration is a complex phenomenon. In spatial terms, it may be short or long distance; it may involve the crossing of boundaries within countries and across international boundaries; it may involve areal units that are communities, counties, states, nations, and cultures; it involves individuals, households and groups; and it may be voluntary and involuntary. The motivating factors underlying migration also are complex, based on the social organization of the migrants, their family ties, and the role of individuals in society. Migration may be the result of predominantly socioeconomic factors that manifest themselves in a push–pull manner.

But recognizing that migration is complex is insufficient. It is necessary to unravel the complexity in order to explain trends, patterns, flows, behaviors, and attitudes that characterize and underpin the migration decision and residential choice process of individuals, households, and groups so that we may predict and plan for the impact of migration flows on communities that are both destinations and origins for migration. In this chapter we consider some aggregate and disaggregate models of migration. In Chapter 13 we focus specifically on mobility within cities and the residential search and choice processes underlying those relocations.

12.2 The Nature of Migration and Migration Studies

Migration is the fundamental process that redistributes people over the surface of the earth. It has profound implications both for governments at all levels in providing infrastructure and human services, and for the private sector in affecting the size and nature of markets for goods and services. It can be studied at various scales, both *spatial* and *social*. At the spatial scale, migration can involve movement over long distances or over very short distances. At the social level, it can involve individuals, households, or groups. Migration involves the movement from one location to another, plus the setting up of a permanent or semipermanent residence at the new location. It is self-evident that migration is an important spatial decision that people make, often several times during their lives. For example, in Australia people move on average 11 times during their life.

12.2.1 Data Sources

Typically census data are used in national and regional migration studies. This has limitations in that population movement is treated largely in a cross-sectional manner, with migration being measured as a transition across statistical or administrative boundaries over a specified period of time, usually an intercensal period of 5 or 10 years. In some nations the census collects data on migration over the previous year.

International migration studies also are reliant on secondary data sources and, in particular, on statistical sources developed by government agencies that collect data on the movement of persons in and out of a nation. In addition, nations have immigration policies and regulations, and data are available on the number of immigrants by category of immigration and by countries of source of immigrants.

Secondary data sources on migration are available from statistics collected by government agencies and are available through the population and household census. But these data provide no information on the *motivations* underpinning migration flows, and they provide only limited data on trends and changes in *migration flows* over time. Multivariate statistical modeling techniques have been used to analyze the relationship between aggregate migration flows between regions and other social and economic data, such as the structure of economic activity and employment at origins and destinations.

However, to study the *behavioral* basis of *decision making* by individuals and households in undertaking migration, it is necessary to use survey research techniques to collect factual, attitudinal, and opinion data relating to the decision to move. Usually this involves what is known as *ex post facto* recall by the individual or members of a household, both of reasons why they moved from the place of origin and why they chose the destination.

Typically, migration is measured at an aggregate level in either absolute terms or as a rate. *Absolute measures* usually describe migrations in terms of distributions by age, sex, occupation, and so on, and may be used in analyzing population attributes of migration flows. *Rate measures* are used to examine the rate at which geographic areas gain and lose migrants with respect to their socioeconomic characteristics.

12.2.2 Approaches to the Study of Migration

Various lines of inquiry are evident in migration research. Six of these may be identified in geographic studies, and these may be summarized as follows (after Shaw, 1975):

1. *Migration level activity and differential studies*, in which the principal explanatory variables are age, gender, family status, education, occupation, career, and life-style.
2. *Economic aspects of migration*, in which the principal explanatory variables are wages, salaries, and employment opportunities and in which cost–benefit and factor allocation techniques are used.
3. *Studies of behavioral aspects of the decision to migrate*, in which migration is related to place utilities, preferences, stressors, and residential complaints.
4. *Migration probability studies and mover/stayer continuum studies*, in which migration expectancy is examined in the analysis of intra- and interregional flows.
5. *Stochastic process studies*, in which migration histories and cumulative inertia are investigated.
6. *Studies of spatial aspects of migration*, in which the explanatory variables are distance, directional bias and information flows and in which intervening opportunities and gravity models are used.

These approaches may be classified further into *macro studies*, which are concerned with flows at the aggregate level between origins and destinations, and *micro studies*, which seek the behavioral explanations of the residential location decision process. But, these approaches are complementary. As stated by Clark (1981):

> Knowing why an individual decides to move, and the impact on the system of his decision to change residences, can lead us to a better understanding of why some districts are areas of high in and out migration and why some areas are more stable. While the microanalytic or behavioral approach focuses on the decision to move, on the role of differential access to sources of information in shaping that decision, and on the spatial patterns of search, the macro analytic is still largely concerned with the spatial regularities in migration streams and the interrelationship between areas of in and out migration. (p. 187)

In addition to the *macro* and the *micro* approaches, Green (1990:1335) has suggested that the time has come to explore a "new perspective," which she suggests could be labeled a *meso analysis* approach, that would combine the collective experiences of migrants to quantitative flow data in a way that provides behavioral insights into migration flows. And I. R. Gordon (1992) has pursued another perspective, which addresses the implicit incongruence between the macro and micro dichotomy by suggesting that migration research has three functions. He claims it should:

1. Be "anticipatory" so as to allow preparation for future changes.
2. Be "problem oriented," analyzing trends and being illustrative of issues underlying changing migratory behavior.
3. "Illuminate ways in which space and place are used within a changing economic and social order" (I. R. Gordon, 1992:131).

All modes of analysis are important, and one can provide information that is relevant to the others. All yield insights into the nature and processes of migration, mobility, and residential relocation processes.

12.2.3 Migration and Mobility

It is important to distinguish between *migration* and *mobility*.

1. *Migration* refers to the relocation process that involves an individual or household shifting geographic location from an origin to a destination. The term also is used in a wider sense to refer to aggregate levels of migration flows from county to county, city to city, and so on.
2. *Mobility* is used in the context of migration studies to refer to the propensity of an individual or household to change residential location in a given period of time. Sometimes it refers to the actual number of moves, relative to some benchmark, by an individual or household over a given period of time, regardless of scale. In ratio or rate form, it is also used to represent the proportion of households in a given geographic area (e.g., nation, state, city, or suburb) that change their residential location in a given period of time.

A further distinction can be drawn between migration and mobility depending on the nature and scale of moves. Often mobility refers to residential shifts within the same general geographic area (e.g., within a city), whereas migration refers to residential shifts involving movement across or between geographic areas, such as countries or states.

Yet a further meaning often is attached to the term *mobility* when it is used to describe the movement of individuals or households up or down some type of socioeconomic status scale.

12.2.4 Behavioral Approaches

A *behavioral approach* focuses on the *migration decision* and the *residential location search and choice process*. Before migration, an individual or household must decide whether to move or to stay put, and if the decision is to move, then where to move. These decisions are not easily separated. Different households have different propensities to move and different degrees of freedom in their migration and residential location decisions. For example, in the United States many wealthy households virtually are free to move or stay as they wish. Many poor and ethnic or minority groups have been forced to move because of government-imposed urban renewal projects, freeways, and other land use change programs. Ecological invasion and succession and residential segregation theories have shown clearly that some groups often are forced to move for a wide variety of reasons. In making their migration and residential location decisions, higher-income groups may exercise choice over a wide spectrum of regions and of housing markets within a city. In contrast, underprivileged minority and poverty groups often are severely constrained in their choices by economic, social, and discrimination factors. Differences in the *relative freedom to exercise choice* in these two basic migration decisions have been observed at various levels of scale, ranging from black poverty groups living in ghettos in the United States to the totally enforced migration of refugees and displaced persons.

The majority of migration decisions appear to be voluntary, occurring in the context of individual or household *place utility*, *preference*, *choice*, and *cognitive* frameworks. A person's place utility forms the framework within which he or she makes a decision to move and where to choose a residential location. The evaluation of alternative locations within his or her space preference surface evolves over time, and the choice made reflects relative abilities to satisfy needs and fulfill aspirations at both the large-area scale (such as a region or a city) and the site scale (such as the area of a chosen city and the dwellings within that area).

12.3　Types of Migration and Mobility

It is convenient to distinguish among various types of migration and to point to the nature of spatial segregation and assimilation that often characterize larger migration flows from a specific origin to a destination region or city.

12.3.1　Long-Distance Migration and Residential Mobility

It is useful to distinguish between *long-distance migration* and *residential mobility* as they have little in common apart from a change in address or geographic location occurring between two points in time. The magnitude, motivation, and implications of these two forms of movement of people in space are quite different, and they are best approached as separate phenomena even though they are elements in a continuum of movement that incorporates:

1. *International migration*, involving movement of people from one nation to another across national borders.
2. *Internal migration*, involving movement of people between regions within a nation.
3. *Intraurban migration*, involving movement of people from suburb to suburb within a city, the simple changing of a dwelling by people within the one suburb or town.

The term *residential mobility* is used to distinguish *intraurban movement* from the *migration* by people over *longer distances*. Thus, the difference is conceptual or theoretical as well as involving a difference of scale, with the motivations of moves being different at the different scales.

12.3.2   Partial Displacement Migration or Mobility

The vast majority of migrations are of a *partial displacement* nature, in which the moving unit stays within the same general area, such as a city, but relocates to a new residence. This type of movement thus takes place within the household's action space as discussed in Chapter 8, Section 8.3, and the area searched will be determined by the combined effects of the activity space, the location of friends, and the hierarchical structure of information sources. Sometimes it is referred to as *mobility*. Typically it is biased to those areas that are most accurately defined in the individual's cognitive map of the city, and it is influenced also by the spatial biases of the media and agent information sources that are consulted. The current residential location of an individual or household is the base from which search occurs in partial displacement migration.

12.3.3   Total Displacement Migration

These migrations are to a new major geographic area, such as from one city to another or from one nation to another. But they do not occur within a directly searchable action space, and thus they are related to different information-gathering and evaluation processes. It is usual for the search phase at the *total displacement level* to be influenced by a *distance decay of general information flow*. However, the urban hierarchy of a region or nation strongly influences the directions and volumes of such flows. Also, interpersonal communication ties between particular areas tend to cut across this distance decay prediction. A good example of this was the immigration of southern European settlers to Australia in the post-World-War-II period, where flows from countries such as Italy and Greece were typified by so-called *chain migration,* in which people from a specific region (of, say, Calabria in Italy) tended to concentrate within Melbourne and in a well-defined inner section of that city. The factors underlying total displacement migrations usually are less complex than is the case for partial displacement migration, and they tend to relate to things such as job opportunities (both perceived and enforced transfers) and retirement resettlement to "sun belt" locations. Total displacement migrations often are made by migrants moving back to areas with which they had previous ties (reverse migration), but over time and with the sequence of moves, the place utility for previously experienced locations tends to decrease.

It is not unusual for total displacement migration to match household types with places, so that within the United States, for example, white and black populations and rich and poor groups are not clearly differentiated at a regional scale. However, in the past many immigrants from overseas have tended to develop regional concentrations, such as Scandinavians in the upper-Midwest of the United States, and Italians in some of the larger cities of the Northeast. Retirement resettlement to "sun belt" areas, such as Florida, California, and Arizona in the United States and to the Gold Coast in Australia, are examples of the former general rule.

It is significant to note that many migrant streams display a marked degree of spatial segregation demonstrated in the choice of new residential locations.

12.3.4   Relating Geographic, Social, and Occupational Mobility

It is interesting to conceptualize mobility not only in spatial terms but also in social and occupational terms. Figure 12.1 shows how social and occupational mobility either can occur or not occur in association with residential mobility. Thus, social space moves can occur without migration. However, it is common for partial displacements to be related to changes in reference groups. Total displacements, however, are linked strongly to occupational mobility, and, by definition, they involve substitution of old social contacts for new ones.

12.3.5   Necessary or Obligatory Moves versus Moves Caused by Needs

It is instructive to consider the *motives* behind migration. In this context we can consider the influence of *push–pull* factors. At an international level, we may think of migrations as moves caused by necessity or obligations versus moves caused by needs. *Necessary* or *obligatory* moves are related to push factors, such as *forced* or *semiforced* moves from an area of origin because of political, religious, or other push factors. In contrast, moves caused by *needs* are a result of *both push* and *pull* factors, associated with economic forces pushing people from areas of origin accompanied by economic pull factors at the destination. For example, Gould and Prothero (1975) documented these factors as being important in migration moves within African populations.

W. Peterson (1958) proposed that migration is either a decision by individuals or groups to improve things at a destination over their condition at the origin or an attempt to retain what they have in response to changing conditions. He classified migrations into four categories:

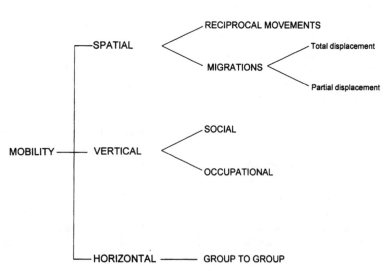

Figure 12.1. Relationship among geographic, social, and occupational mobility. *Source:* Jakle et al., 1976:155.

1. *Primitive migrations*, where innovating and conservative movements are largely attributable to man's inability to cope with natural forces.
2. *Forced or impelled migrations*, where the primary cause is some state or some functionally equivalent social institution.
3. *Free migrations*, where the will of the migrant to either achieve the new or retain what they have is the decisive factor in the decision to move.
4. *Mass migrations*, where innovating or conservative movements become a style or an established or semiautomatic behavior.

12.3.6   Circulatory Migration

In this type of migration it is typical for members of a household, usually the husband or grown sons, to move from one place, such as a rural village, to another place, such as a rapidly expanding city, in order to seek work in the service or manufacturing sector of the economy, with that member returning on a periodic or seasonal basis to the permanent place of residence. Thus a *circulatory* pattern is established that is repeated over time. Examples occur in the rapidly changing nations of the developing world, such as Indonesia, where circulatory movement takes place between rural villages and big cities like Jakarta.

A similar type of circulatory migration pattern, but at a different level of scale, is evident in the movement of unskilled and semiskilled workers from southern and Southeast Asian countries on a seasonal or annual basis to the oil-producing countries of the Middle East. In developed countries, circulatory migration also occurs for reasons of regional labor shortage. For example, in Europe, unskilled laborers from Turkey and other Eastern Mediterranean countries have moved seasonally and on a contract basis to West Germany in response to shortages of unskilled labor in specific industries.

12.3.7   Migration and Assimilation

For migrants who are members of a minority ethnic or racial group and who enter an area dominated by a larger homogeneous population, a specific issue faced is that of *assimilation*. This is about the process whereby the minority group is gradually absorbed by the host population. Assimilation has both *social* and *spatial implications*.

From the *social perspective*, members of the minority migrant group gradually assimilate into the host society, and numerous changes take place during this process. Hanson and Simmons (1968) suggested that individuals gradually change their roles with respect to the host society, as well as with respect to people within their minority group. In this way, the migrant adopts a role path that is a movement through social space that may correspond to some types of movement through geographic space. M. M. Gordon (1964) referred to these changes as *acculturation*. In this process the minority group accepts certain cultural patterns of the host population, such as language, cultural norms, religious beliefs, and practices. As its members enter the host society's groups and institutions the minority group begins to lose some of its ethnic character. During this process there is both an elimination of prejudicial attitudes on the part of the host society and an elimination of overdis-criminatory behavior. The two groups must eliminate power and value conflicts that are due to ethnic differences between the migrant and the host population. This process is clearly evident in the Northern American city, particularly in the North-east, which from the late 19th century to the early 20th century received wave after wave of eastern and southern European migrant minority groups who, over time, became acculturated within the American urban society. However, some more re-cent urban migrant groups, such as Puerto Ricans, Mexicans, and Asian nation mi-grants, have made less progress toward assimilation in these terms.

Assimilation also has a *spatial* dimension. Minority racial and ethnic migrant groups moving into a city tend to exhibit a high degree of *residential segregation* from the remainder of the population by settling in well-defined residential areas of a city, often leading to the development of ghettos. Over time, as acculturation oc-curs with the assimilation of the migrant minority group through a gradual move-ment in social space, spatial assimilation is likely to occur through the *residential diffusion* of the migrant group households out of the ghetto into the suburbs. This spatial assimilation process is illustrated in the model in Figure 12.2.

12.4   Modeling Aggregate Migration Flows

12.4.1   Laws of Migration

Ravenstein (1885) is usually credited as offering the first comprehensive theoretical structure to explain migration behavior. From a study of British census data in the late 19th century, he inferred some general principles or patterns that have been condensed into seven *laws of migration* to provide the basis for the development of migration studies (E. S. Lee, 1969):

1. Migration numbers and distance traveled are inversely related.
2. There is a greater propensity for rural people, rather than urban dwellers, to migrate.
3. Migration flows are strongest from rural areas to urban centers, but the process may not always be conducted in one step.
4. Each main "stream" of migration produces a (usually smaller) counter-stream.
5. Numerically, female individuals appear to favor short-distance migration.
6. Technological advancement has positively influenced migration rates, and centers of increasing technological investment tend to attract migrants.
7. While economic and social oppression influence out-migration, the strongest migratory motivational force appears to be the in-migrant's indi-vidual perception of enhanced personal gain.

In subsequent research two broad approaches have emerged. One is an *empir-ical approach*, the aim of which is to refine the description of migration distance by fitting a suitable curve to data on migration volume by distance. Such a formula simply represents the migration field in a more concentrated and convenient form

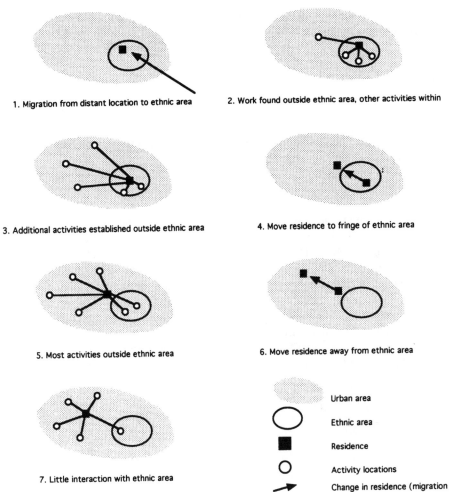

1. Migration from distant location to ethnic area

2. Work found outside ethnic area, other activities within

3. Additional activities established outside ethnic area

4. Move residence to fringe of ethnic area

5. Most activities outside ethnic area

6. Move residence away from ethnic area

7. Little interaction with ethnic area

Urban area

Ethnic area

Residence

Activity locations

Change in residence (migration

Daily activity trips

Figure 12.2. Spatial assimilation model showing spatial behavior at different levels of assimilation. *Source:* Jakle et al., 1976:161.

than does the original map or table. The other is a *deductive approach*, in which researchers begin with a set of plausible basic assumptions and then elaborate hypotheses of migration from them. These hypotheses may be given mathematical formulation and then tested using empirical data.

12.4.2 The Empirical Approach: Migration Fields

Analysis of the migrations to or from a certain place looks at the volumes of movements that come from specified distance zones. Thus, for a given destination, such as a city receiving migrants, we may plot the number of migrants coming from all points of origin according to the distance of those points from the destination. In this way, we may fit a Pareto-type curve to such data. This will take the form:

$$y = ax^{-b}$$

where:

y = the number of migrants coming to a destination from an origin
x = the distance of an origin from the destination
a = the y intercept
$-b$ = slope of the distance decay curve

The migration field of the city receiving migrants is thus plotted, as demonstrated in Figure 12.3. Note that in the above equation the distance decay variable (exponent b) is negative. The negative b exponent will become smaller as the distance from which volumes of migrants are attracted to the city increases.

Using this approach, we may define the *migration field* of a city, which is the area that *sends* major migration flows to a given place. Conversely, it may be the area that *receives* major flows from a place. Thus, we may have an *in*-migration field and an *out*-migration field for a given place.

An important study of the migration fields of towns was undertaken in Sweden by Hägerstrand (1957). In studying the migration fields of various communes

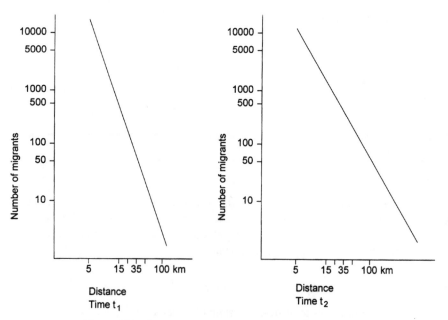

Figure 12.3. The migration field of a city: Increase over time.

over the period from 1860 to 1930, he found that there was a decrease in the negative slope exponent, which defined the migration field of a typical commune. He concluded that there was an increase over time in the distance from which communes attracted migrants, with an increasing proportion of total migrants coming from increasingly distant places. This coincided with improvements in transport methods and reduction in the relative cost of transport.

It is possible to draw a number of general conclusions about the migration field of cities. In the United States, for example, cities have had a significant level of in-migration from nearby metropolitan areas and from larger metropolitan areas at greater distances. Jakle et al. (1976:168–171) have commented on the influence of the urban hierarchy on migration flows that make up the in-migration field of any city. The urban hierarchy structures the flow of information through the media, thus having an important impact on people's mental maps of the United States. As major cities play a predominant role in the media, they are likely to be prominent areas in people's mental maps. In addition, there is a greater probability of obtaining specific information about jobs and amenities in a given metropolitan area than similar information about a small city. Thus *hierarchical migration* occurs, with the larger places being likely to provide larger numbers of migrants than smaller places.

Most major cities in the United States have had significant in-migration from nonmetropolitan areas in their own state and adjacent states. This is called *hinterland migration*. Much of the origin area is a region that is influenced, to a considerable degree, by the metropolitan functions of that city. However, the spatial pattern of hinterland migration usually is characterized by a *distance decay* effect. It is influenced also by the flow of information patterns that relate to the structure of the lower levels of the urban system, and thus to the lower levels of the media hierarchy. In addition, it is influenced by the degree to which the city is incorporated within the activity spaces of people within the hinterland, thus affecting the degree to which the place figures prominently in the mental maps of people in the hinterland.

A final form of migration related to the migration field of a city may be termed *channelized migration*. This refers to the fact that many cities have at least one significant source of migrants from a nonmetropolitan area at some distance from the hinterland. Interpersonal contacts are important here. This can lead to a form of *chain migration* developing.

12.4.3 The Gravity Model and the Intervening Opportunities Model

Ravenstein's laws of migration have fashioned the way the *gravity model of migration* has evolved. Over the ensuing century Redford (1926) and Thomas (1938) concentrated on the age differentials of migration flows, and Jerome (1926) looked at economic affects on gravitational flows. Later Stewart (1948) explored a social physics analogy by seeking to find a flow matrix to explain trend patterns in migration.

In its most basic form, the gravity model is defined as:

$$I_{ij} = f(P_i \, P_j / d_{ij}^m)^N$$

where:

I_{ij} = an index of interaction between places i and j
P_i = the population in place i
P_j = he population in place j
d_{ij} = the distance separating places i and j
n, N are empirically determined exponents

As gravity studies focus on the spatial aspect of interactions, highlighting the effects of migration volume over space, they aggregate quantitative statistical data. The gravity model has been used widely to predict potential levels of interaction between cities and towns. Implicit in the gravity approach is that people all have similar motives for migrating, probably because many population researchers have tended not to explore the various qualities displayed by each migrant, or migrants in general. Similarly, the basic gravity model considers distance alone as a limiting factor on migration. Thus the gravity approach allowed "professional geographers [to unadvisedly] discount the role of human agency" (Earle et al., 1989:179).

Stouffer (1940) questioned the implicit certainty of the gravity approach in seeking to answer the problem why all migrants out of rural areas in the United States did not seek an urban life-style. He hypothesized that physical and social variables exist that intervene in the migration decision-making process. Stouffer's assumption was that it was not distance as such that limited migration flows; rather it was the available social opportunities at the places that were alternatives. Thus, in some regions migrants may only have to travel a relatively short distance to attain their desired objectives, whereas in other (possibly less populated) areas the distance may be considerably longer. Implicit in Stouffer's hypothesis was that people do not necessarily wish to go to the place offering the maximum number of opportunities, but rather there are a variety of places that appealed to certain migrants. Stouffer also developed the concept of *intervening opportunities* that might siphon off migration streams. This intervening opportunities model took the form:

$$I_{ij} = \frac{P_i P_j}{Q}$$

where:

I_{ij} = some index of interaction between places i and j
P_i = the population of place i
P_j = the population of place j
Q = some measure of intervening opportunities between i and j

Stouffer's model thus proposed that the number of people moving a given distance is directly proportional to the number of opportunities at that distance and inversely proportional to the number of intervening opportunities. It was not the intervening distances between places per se, but rather the amount of intervening opportunities for interaction that retarded interaction. As an example, take the case of the United States cities of Philadelphia and Boston in the Northeast and the cities of Dallas and Houston in Texas. Boston and Philadelphia are similar in size to Dallas and Houston. However, there is a considerably greater amount of movement between Dallas and Houston and Houston and Dallas than is the case between Boston and Philadelphia and Philadelphia and Boston. This should not be surprising because there are many more opportunities for migrants between Boston and Philadelphia (e.g., cities in New York and New Jersey) than is the case between Dallas and Houston. Because there are fewer intervening opportunities between Dallas and Houston, there is a greater degree of migration both ways between those cities than is the case between Boston and Philadelphia.

Hägerstrand (1957) sought to test the intervening opportunities model in Sweden using in- and out-migration data for every parish for 1946 to 1950. He drew 10 kilometer distance zones around each parish. Using the Stouffer model, he calculated the expected migration against the observed levels of migration. Using six randomly selected parishes, Hägerstrand found that the agreement between observed and expected frequencies of migration was poor. One of the problems might have arisen because he treated every member of a migratory group (such as a family) equally. This presented some flaws, since two identical opportunities at a given place can be filled in one case by a single person and in another by a large family. Hägerstrand conducted a second test using groups of migrants instead of individuals. He found, however, that agreement between expected and observed migrations was no better. It is worth noting also that a study by Anderson (1955) in the United States found that nothing is gained by substituting geographic distance with intervening opportunities in investigating migration flows.

The gravity model and intervening opportunities models were astutely judged by Hägerstrand (1957): "Each formula which in some way makes the frequency of migrations directly proportionate to the population and at the same time introduces a reverse proportion to distance, is somehow able to be brought into line with actual observations, yet the ultimate conformity is missing."

Interest in these concepts has continued. E. S. Lee (1969) revised Ravenstein's "laws," taking into account important theoretical contributions by geographers and sociologists up to that time. He suggested that there were qualitative factors at both the origin and destination that encourage and/or discourage migration decision making and that the perception of these factors as being either positive or negative was largely in the eye of the beholder (Lee, 1969:50). He added that even if a prospective migration decision were to be positive, there are intervening variables that may alter substantially the initial locational decision. These intervening variables might be either institutional, such as a second competing alternative location, or personal, such as issues of family ties; and distance may be a constraint. Migrant behavior continued to be seen as a response to the perceived qualities flowing from the institutional arrangements in the originating and alternative locations, but it could be enhanced, or conversely checked, by variables not directly associated with the institutional structure. Figure 12.4 represents these concepts.

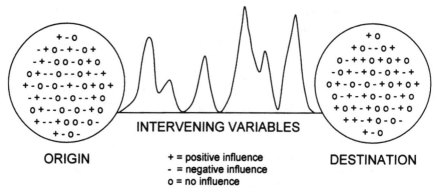

ORIGIN   INTERVENING VARIABLES   DESTINATION

+ = positive influence
- = negative influence
o = no influence

Figure 12.4. Origin, destination, and intervening variables in migration. *Source:* Lee, 1969:50.

12.4.4 Push–Pull Models

A fruitful approach to the study of migration has been to regard it as a response to *push* and *pull* factors. The *push* factors were seen as those situations that give rise to dissatisfaction with the present location of an individual or household; *pull* factors are attributes attracting people to a destination. They have been analyzed using two basic approaches:

1. The first uses a *regression* format in which migration distance is an independent or explanatory variable, whereas migration volume is dependent upon a range of push and pull factors relating to the place of origin and the place of destination.
2. The second approach takes a *structural* view of push and pull factors. In this way, both distance and push–pull factors are incorporated within a gravity-type model.

The multiple regression model was used by Olsson (1965) in a study of migration in Sweden. On the basis of an exhaustive analysis of a number of empirical migration studies, Olsson selected a number of independent variables that he considered relevant in explaining variations in migration distances. These related to measures of income and unemployment at the place of origin and destination, age and income of the intended migrant, plus the traditional variables relating to size of origin and destination that were included in the gravity model. Olsson tested the relevance of these variables in explaining variations in the migration distances in a randomly selected group of Swedish population. This was done using a *stepwise multiple regression* technique. This method of analysis was designed to include one variable at a time, and at each step, the variable accounting for most of the nonexplained variation in the dependent variable was entered. This meant that the variable with the highest correlation coefficient with respect to the dependent variable was the first one to be included in the model. Partial correlation coefficients were then derived for the remaining variations in the dependent variable. The variable explaining most of the remaining variation was included next. Subsequent steps were taken by adding new variables and repeating the process. Olsson found that:

1. Length of migration is positively related to level of income at the place of origin.
2. Length of migration is positively related to the degree of unemployment, both in the place of origin and the place of destination.
3. Length of migration is positively related to the population size of both the place of origin and the place of destination.
4. Length of migration is negatively related to the age of the migrant.
5. Length of migration is positively related to the migrant family's income.

The test conducted by Olsson was encouraging, with a multiple correlation coefficient exceeding 0.885, indicating that the model would explain 78% of the variance in migration distance. Individual migration distances were found to be positively related to increases in the level of income of the migrant, the degree of unemployment, and the population size in both the places of origin and destination. On the other hand, the variables representing personal characteristics of the migrant did not contribute significantly to the power of the model.

A consideration of *push–pull* factors as determinants in explaining migration habits has roots in Pavlov's (1927) *reinforcement theory* and Skinner's (1938) *stimulus–response theory*, and a number of early behavioral studies of migration developed quite elaborate analyses of their nature. The notion was that *push* factors at the origin were perceived and/or assumed to have a negative influence on quality of life, whereas *pull* factors at the destination were positive factors drawing prospective migrants to the destination (Bogue, 1969).

Table 12.1 lists Bogue's principal push and pull factors. In general, what is viewed as negative in one place will have a positive opposite in another location. But when Bogue's factors have been tested in research using survey methods to elicit the reasons for migrating, significant correlation between sets of push or pull factors and migration has not necessarily been found (e.g., Skinner, Rampa, & Spriggs, 1983). Shuval (1982:678) has suggested that in such structural methodologies there will always be a "symmetrical relationship between factors that motivate people to

TABLE 12.1
Bogue's *Push* and *Pull* Factors in Migration

Push Factors	*Pull* Factors
• Decline in regional income, causing localized recession.	• Perceived superior career opportunities in another location
• Loss of employment (from causes other than recession, e.g., mechanization).	• Greater income in another place.
• Political, religious, ethnic, and/or other forms of oppression or discrimination.	• Personal growth opportunities, such as better education, group association, etc.
• Little or no pathway to increase personal development in structures such as marriage, status, or career.	• Preferable environment, such as climate, housing, schools, and/or other institutional facilities.
• Catastrophe, for example, floods, fire, earthquake, war, etc.	• Desire to be with kin, or other favorable people, in another place.
	• Lure of different social or physical activities in another place.

Source: Summarized from Bogue, 1969:753–754.

leave a permanent residence and those that attract them to another specific locus." Rather than provide explanation of migration behavior, this *stimulus–response* determinism has enabled the formation of a set of differentiated physical and social characteristics for various locations largely, it seems, based on abstracted supposition. In effect, these differentiated characteristics have given enhanced credibility to the gravity approach because they identified factors that either repelled or attracted migrants, thus suggesting some value for the idealized functions of institutionalized spatial structures (e.g., Campbell & Krieger, 1981).

12.4.5 An Optimizing Gravity Model

By providing qualitative examples of push and pull factors to illustrate preferred social functioning in cities that attract large numbers of migrants, some migration researchers have amended the gravity approach to account for the hypothesized variations in place attributes. For example, Dorigo and Tobler (1983) proposed an optimizing model based on the basic gravity computational model but with a capacity to incorporate a *shadow price for distance* and other push–pull variables. Their base model was:

$$M_{ij} = P_i P_j (R_i + E_j)/d_{ij}$$

where:

$M_{ij} =$ magnitude of movement from place i to place j (of r places)
$d_{ij} =$ the distance between places i and j measured in discrete units (e.g., kilometers of road, money units, travel time, etc.)
$R =$ the push variables
$E =$ the pull variables
$R_i =$ the aggregate value of all push variables from place i
$E_j =$ the aggregate value of all pull variables to place j

Dorigo and Tobler (1983) applied the model, in its various modifications, to census migration estimates between the nine regions of the United States for the period 1965–1970. The fit of the model to the inflow and outflow migration data was about the same as is usual for gravity models ($R^2 > 80\%$). The resulting push, pull, turnover, and attractivity values in a geographic context showed the Pacific Region to be the most "repulsive," with high housing costs and metropolitan air pollution being possible reasons. Conversely, the same area was shown to be the most "enticing," perhaps because of the attractive scenery, climate, and life-styles afforded. Dorigo and Tobler did consider the outcomes as contradictory, because through summation their model turns these two into a larger turnover, and also it assigns a high net activity to the region. In the same way, one may compare those regions that were losing population, such as the mid-Atlantic. This was an area that had a strong drawing power, but it was overrun by the *push* effect. It was possible to examine each region in this manner.

Models such as these attempt to quantify the value of place attributes. This illustrates the functionalist structure embedded in the gravity model and the *push–pull* approaches to migration studies, whereby people respond to the stimulus of some idealized external attributes. Geographers seeking to evaluate the importance of various forms of urban structure found this approach appealing (e.g., Harrison & Sarre, 1976; Carp, 1975; Vance, 1976). But such approaches tend to exhibit the problem of "intellectual fallacy" discussed in Chapter 1, Section 1.6.

12.4.6 Attractors and the Attracted: An Alternative Approach

To understand the forces that are driving differential levels and rates of population growth and decline across the cities and regions of a nation, Maher and Stimson (1994) have proposed that this involves identifying: (1) the *attractors*, or what are the particular characteristics of regions that attract population; and (2) the *attracted*, or those individuals and households that are making decisions about changing location.

Net movement in migration is an artificial construct that summarizes the result of a complex set of flows both inward to and outward from any region. Areas of high in-movement tend also to be areas of high out-movement. Population growth through migration only occurs where the movement in is substantially greater than the movement out. So it is the ability of an area to both attract and hold a population that is important. *Attractors* can be divided into a number of categories, any combination of which may result in population growth.

Natural attractors are those features of the physical environment that are desirable but not necessarily indispensable accompaniments of living. Differences in physiographic regions also are important in that physical environments are often sought after for their amenity as well as their recreation potential. Topographic structures (particularly the attraction of elevated areas) can operate as an attractor at a local scale, as can the presence of bodies of water (coast, lakes, rivers, etc.). Other environmental amenity attractors include special features such as unique flora and fauna, or a sense of remoteness from dense settlement.

Such physical attractors on their own have a growing but still somewhat limited appeal. Man-made attractions provide important adjuncts to natural attractors, and they relate to *economic* and *sociocultural* attractions of areas. Differential economic opportunity is one of the most potent forces that attracts population from one location to another. This operates largely at an aggregate scale, prompting moves from areas of lower opportunity or lesser rewards to those with greater prospects. Thus a good deal of international migration, as well as interregional movement, where destinations are chosen because of the prospect of a new or better job for oneself or one's children, is driven by economic factors. Although there are a number of constraints on labor mobility, it is reasonable to attribute a good deal of large-scale movement to this factor. There is more to economic attractors than solely jobs. Services need to be accessible and of good quality. Infrastructure must be in place to enable interaction and exchange to occur. Thus roads, electricity and telephones are basic infrastructural requirements. Others such as water supply, drainage, sewerage disposal, and so on, may also act as attractors, but they are not as significant as are elements in the first category.

The *sociocultural milieu* includes social networks, social cohesion, availability of affordable housing, and the existence of necessary social infrastructure. The existence of social networks is particularly important for long-distance migration of which immigration is one aspect. Migration flows are closely likened with information availability—information that flows back from a destination to others at the origin who are potential movers is an important influence on choice. These social networks often provide the cultural and linguistic support necessary when incoming migrants are coming from different cultural environments. Information networks are important, and the presence of relatives or friends are a means of cutting down the possibility of social isolation.

A further element that can be categorized as an attractor of population is *accessibility*, which ties activities together. Most activities are interdependent. Residential location for some is premised on access to workplace, for others to schools or shops and for others again, to necessary services. Other things being equal, accessible locations tend to be more desirable than those with poor accessibility. However accessibility is not just provided by proximity but by mobility. As long as a minimum level of accessibility is available, it is frequently traded off against other attributes.

A final element of the sociocultural milieu can be identified generally as representing *life-style*. This may reflect a desire for freedom from routine, a search for a relaxed atmosphere, or the desire for a degree of informality. It also may be reflected in a demand for certain cultural or recreational facilities.

Each location can be seen to have a mix of each of these *attractors*, and they may appeal to various groups quite differently. It has been argued also that certain attractors have a differential effect on movers at different scales. The notion of attractors can work in two separate ways: (1) they can bring people into the region; (2) they can act to retain the existing population. It is the imbalance of in-and out-movers that, in the end, results in population growth.

Population change, through differential migration flows, is the result of a complex interplay between the nature of the region (*attractors*) and the nature of the population seeking to capitalize on those attractions. Those groups of people who respond to opportunities, activities, or attributes of particular regions have been termed the *attracted*, and a basic distinction can be made among those making locational decisions. One group can be characterized as *initiators*; the other as *reactors*.

Initiators are those who are actively seeking to change their personal circumstances. They are motivated more by what they perceive as an opportunity to improve their circumstances, (whether social, economic, or environmental), rather than responding to events that have altered their existing circumstances. The decisions could relate to:

1. Economic circumstances (for example, a change in job, a reorientation of career, or a search for different business circumstances).
2. Housing (for example, the desire for a new or better dwelling unit).
3. Social contacts (for example, seeking to establish or retain contact with family, friends, or social peers).

Some initiators can be defined as the *footloose*; and for these, there are few constraints on location. Thus they are free to seek a location that emphasizes life-style, amenity, or social contacts. This group includes recent retirees who, particularly when reinforced with the sale of an existing house, have the capacity to choose from a wide range of possible locations on climatic, life-style, or other amenity factors.

In reality, often these factors combine to motivate a household both to move and to choose where to move to. It is impossible to empirically classify the initiators from other types of movers, but the categorization has utility in terms of the condition under which such movement will occur and thus gives some appreciation of where population growth will occur.

A second category of the *attracted* are those moving to areas of strong population growth. Maher and Stimson (1994) identified as *reactors* those responding to events or processes that either prompt or necessitate a move. Whereas *initiators* are moving primarily because of the attractions of the destination and the perceived opportunity for the improvement of circumstances, *reactors* are moving because of events at the origin—that is, *push* factors. However, the basis of the decision on the destination area will relate to prospects for compensating for the push mechanism. The essential difference between initiators and reactors is the *reason* they are moving, rather than where they are moving to. The significance of the difference between them, however, may have implications for the long-term redistribution of population. A *pull* process may be more permanent than a *push* factor. The range of reasons to prompt reactors to move is similar to those for initiators. For example, economic motivation is important to both groups. However, for reactors it can be hypothesized that the decision to move is a reaction to an event such as a loss of previous employment, a decline in earnings, compulsory job transfer, or retirement. Depending on the economic capacity of the household, a decision is made to select a new location in the same manner as the initiator. Some housing-related reasons also can be identified as reactive. Events such as household formation or dissolution, and increasing or decreasing household size, are events associated with the need for housing and may precipitate a decision to move in a reactive sense. These are frequently related to *life course events,* such as leaving home, marriage, childbearing, or separation or loss of a partner. Although less common, some movement may be reactions to environmental change—and especially factors such as perceived deterioration of neighborhood amenity through increased traffic, social change, and/or change in the physical character of surrounding dwellings.

Although reactors are facing *push* pressures, they do not always have to move. If they do, it is assumed that they are able to then make choices about where they relocate.

A final group of *attracted* can be identified as those who have little or no choice about movement (their move is forced), and they also have little or no choice about where to move. These can be classified as *involuntary* or *forced* movers—involuntary both in terms of not having choice about moving and with little choice as to where to move. This group comprises those with little economic power, those not in the labor force, or those in the lower socioeconomic status areas of work. Many are dependent on pensions or benefits; and for many, their housing tenure is insecure. There are still *attractors* for this group, but the attractions frequently are either the existence of low housing costs, sought by family groups in poor economic circumstances, or warm climates and beaches, which appeal to the young footloose unemployed, as is the case for sun belt locations.

According to Maher and Stimson (1994), regardless of the categories of the attracted, the attractors remain relatively consistent:

1. The availability of jobs or business opportunities.
2. The existence of appropriate housing.
3. A climatic and/or physical/locational situation with high amenity and lifestyles.

12.5 Limitations of the Gravity Model Approaches

The continuing use of *gravity models* and a focus on *push–pull* factors in migration studies generally has been influenced by two important considerations. The first, and the more important, is the availability of increasingly inexpensive and efficient computer facilities. The second is the method by which macro level migration modelers utilize the assumptions of their models.

It was because of computers that "an explosion of migration research" has been undertaken at the macro level (Greenwood, Mueser, Plane, & Schlottmann, 1992). Computer facilities have made secondary data analysis possible at a relatively low cost. These include large statistical data bases, such as the census, and much of the proliferation of migration research has been focused on the manipulation of these statistical data bases. This has produced an enhanced explanation of variations in the flow of migration both over space and in time.

I. R. Gordon (1992:123) has suggested that the macro level gravity approach has been dominated by two approaches using aggregated data:

1. The *unconstrained approach*, where volumes of migrants to and/or from an area are related proportionately with accessibility.
2. The *fully constrained* approach, which analyses the spatial distribution of independently determined volumes of migrants to and/or from a specific area.

Mohlo (1986) classifies these approaches into six distinct categories, all based on computational modeling:

1. The basic *gravity model* approach.
2. A *human capital* approach, derived from Sjaastad (1972), where migrants base their decisions on anticipated future benefits.
3. A *stress response and rational decision-making* approach, where *push* and *pull* factors are optimized (e.g., Dorigo & Tobler, 1983, discussed above in Section 12.4.5).
4. The *random utility*, or *discrete choice* approach, which adds a random utility aspect to the human capital approach by integrating discrete choice

functions because of the nondivisibility of typical behavioral assumptions in the rationalizing decision maker (e.g., Maier & Weiss, 1991).

5. A *search*, or *multistream* approach, which attempts to incorporate the inherent differences in motivation for migration. In recognizing that there are many levels of source information regarding migratory opportunities (Armhein, 1985) as well as principal reasons for wanting to migrate, statistical data need to be disaggregated to reflect these motivational variables (e.g., I. R. Gordon, 1992).

6. An *adjustment lag*, or *temporal aspects* approach, which seeks to define time factors, such as response lags to information and the cyclical nature of migration, and which recognizes aggregated and/or disaggregated migration flows (e.g., Mohlo, 1986; Gordon & Vickerman, 1982).

12.5.1 Return and Onward-Moving Migrants

Any modeling approach is restricted by the paradigm to which it belongs. Because of the reliance on secondary analysis of census data, providing analysis of net migration in a cross-sectional framework for a specified time period, computational models cannot address the interactive flows of return and onward-moving migrants (e.g., Morrison & DaVanzo, 1986). The models fail to consider those factors that may drive the migration decision-making process (Cadwallader, 1989). For example, Fielding (1993) looked beneath the net migration flow into the greater London region in southeast England to reveal a substantial but generally counterbalancing migratory flow into and out of the region. This arises because aggregate migration models attempt to project migration flows over time by adducing a trend parameter (from the comparison of two or more comparable data sets), which may be exponential. However, where growth patterns are dependent on nonrenewable resources, exponential trends provide an incomplete representation of an overall trend curve of population growth, and they fail to incorporate negative feedback mechanisms that might counterbalance and/or limit the exponential growth trend.

12.5.2 Predictive Capacity versus Understanding of Process

The wide use of macro level computational models using disaggregated statistical data has been in part a response to the need to understand and predict aggregated migrant flows between the regions within nations, sometimes in the context of being part of regional economic adjustment programs. But a difficulty in the models is that all migrants are assumed to make the same rational choice, particularly in responding to employment, social and amenity conditions at origins and destinations. Alternatively, disaggregated behavioral models investigating the spatial choices of migrants take account of the nonrational foibles of human decision making and challenge the implicit structure of the aggregate models. Although these computational models may be useful for predicting aggregate migrant behaviors, the social relevance of such outcomes in policy determinations is likely to be at best partial. Thus there is a dilemma for migration researchers in choosing between methodologies that will provide good aggregate predictive capabilities or those that will furnish improved explanations of the reasons why people move.

12.5.3   The Mobility Transition Hypothesis: Problems with the Assumptions of Models

A further important consideration is that macro-level migration studies generally have been affected with what Zelinsky (1971) has termed the *mobility transition hypothesis*, which largely has determined the manner in which assumptions have been drawn for many modeling techniques. His hypothesis was that there are definite patterned regularities in the growth of personal mobility through space–time, and these regularities comprise an essential component of the modernization process. Zelinsky's concerns were threefold:

1. The assumption of *patterned regularities* induces an erroneous link between the notion of flow and stock.
2. The application of teleology to modernization, which encourages the problem of *intellectual fallacy*.
3. The overall idea that *modernization encourages mobility.*

As Zelinsky's hypothesis underpins much of the macro-level models of migration, the basis of his three concerns necessitates further consideration.

The assumption within Zelinsky's hypothesis of *patterned regularities* suggests a link between the notion of *flow* and the mathematical question of *stock*. Except for a few nations, such as the United Kingdom, where the General Household Survey asks questions on the intentions of households to migrate, the long-distance residential relocational *stock* is assumed in migration models to be the total resident population in the spatial area under consideration. This is incorrect. In the United Kingdom it has been shown that many people only make one move, that being out of the parental household, and often to a nearby location (R. M. Smith, 1979). Similarly, people may wish to migrate, but because of financial and/or family responsibility reasons, they may be so constrained that they are never able to materialize their wish to relocate. For example, Hughes and McCormack (1985, 1987) found a positive correlation between levels of regional unemployment and intentions to migrate, but they found contradictory evidence in the actual flow of out-migration. Similarly, Fielding's (1993:142–145) work points to a differentiation in the incidence of migration along class lines, with increased levels of migration as one moves up the socioeconomic indicator scale. Also it is only a relatively small proportion of people who undertake long-distance migration, and to assume that the moving population in a region being studied comprises all the residents may not be methodologically sound. In a similar vein, other researchers have noted that the statistics on migrants might not be representative of the population from which they are drawn (I. R. Gordon, 1992; Rogers, 1992). Thus it may be plausible to postulate that potential migrants have characteristics that are different from the population as a whole.

The second aspect noted in Zelinsky's hypothesis was that *modernization* is coupled to the notion of *teleology*, and that this inevitably materializes as a problem of *intellectual fallacy*. The linking of modernization and indicators of social benefit represent the functionalist problem of *ecological fallacy*, where causation is asserted for a number of variables that might only be spatially associated. By ascribing causal relationships, many migration researchers have inferred a *cause–effect* relationship between area-based spatial variables via correlation and regression models, whereas it might be valid to infer only that there was a high degree of *spatial association* between component variables.

The assumption in many migration studies that people rationalize regional wage differentials may be true for localized regions, such as may occur in a large metropolis, but even then the association is weak (I. R. Gordon, 1985), with many workers preferring to commute rather than move when they change jobs (Evers, 1989). For the migration decisions of people in affluent nations where they relocate over long distances, a variety of studies dispel the assumption of migration in response to wage differentials between regions (Gabriel, Shack-Marquez, & Wascher, 1993; Ballard & Clark, 1981). For example, elderly migrants in the United States have indicated that employment served as a locational binding agent (McHugh, 1984; Oldakowski & Roseman, 1986), and the unemployed have indicated that resources are usually insufficient to enable relocation, and/or the perceived benefits are too small to warrant the cost (Bogue, 1977; Coupe & Morgan, 1981; Chalmers & Greenwood, 1985; Hughes & McCormack, 1985, 1987).

Other ideal-type theories used in migration research pose different problems, particularly when connected to *ceteris paribus* limitations, and they might lead to socially inappropriate conclusions being drawn. It is not the modeling approaches per se that are called into question, but rather the manner in which assumptions are drawn. So although Zelinsky's hypothesis has been used to link spatial structure and the behavior of migrants in computational gravity models, there is no evidence to prove that people undertake migration according to these "ideal-type" assumptions.

The third criticism is the idea that *modernization* encourages mobility. This is an assertion that has not been substantiated. Warnes (1992:49–50), for example, illustrates the paucity of this proposition and has argued cogently that the modernization process of industrialized nations has caused a reduction in migration. This was noted much earlier by Dewey (1960). Warnes (1992) argued that migration in affluent nations may be less of a phenomenon associated with economic circumstances than one related to reasons of a personal kind, such as status and/or life-style factors.

The gravity approach has achieved a prominent role in migration research, especially in the study of aggregate flows and their attribution to a variety of possible push or pull factors affecting flows between regions or cities. As an indicator of population movements, macro level approaches using aggregate statistical data do have an important role to play. But as Greenwood et al. (1991:252) suggest, "aggregation is perhaps the number one bogey of migration research," as such studies cannot be used to predict future events. Because of this problem, disaggregated behavioral approaches to the study of migration are important.

12.6 Selectivity Differentials and Motivation in Migration: Towards Micro Models of Migration

The *macro* modeling approaches discussed above represent a significant portion of the literature on migration research. The focus has been on describing, and at times seeking to predict, aggregate population flows using census data variables. Sometimes they are differentiated by gender, age, occupation, and income. Economic indices relating to wage differentials, levels of unemployment, and rates of job generation or loss at origins and destinations mainly have been used as push–pull factors to explain flows. Although migration differentials between regions may point to potentially important locational effects, which need to be considered in testing hypotheses of migration decision making, these approaches have provided limited explanation or understanding of the motivational basis of the decision by individuals and households to migrate.

It is well known that migration is a very *selective* process. Young adults are the most mobile members of the population, and propensities to move decline rapidly after age 25 before increasing slightly (particularly for female individuals) after age 65 years. Decisions to move are closely related to progression through the life/family cycle (Rowland, 1982). Ten life-cycle events have been identified as giving rise to residential adjustment or relocation (see Table 12.2). Of these two—retirement and death of a spouse—are related to elderly groups of the population.

An important early study by Beshers and Nishiura (1961) argued that occupation differentials in migration represent territorial, social, and cultural constraints that discourage migration among persons who have a local versus a territorially extended mode of orientation. In this context, occupation differentials have been regarded as being attributable largely to the modes of orientation toward goal attainment. Work by Brown and Belcher (1966) hypothesized that migration selectivity could be understood on the basis of cosmopolitan versus local social roles. Ritchey (1976) argued later that career mobility and individual career choice decision making, underlie socioeconomic differentials in migration.

In general, there has not been a clearly defined theoretical link between mi-

TABLE 12.2
Ten Life-Cycle Events Related to Residential
Adjustment or Relocation

1. Completion of secondary education.
2. Completion of tertiary education.
3. Completion of occupational training.
4. Marriage.
5. Birth of the first child.
6. Birth of the last child.
7. First child reaches secondary-school age.
8. Last child leaves home.
9. Retirement.
10. Death of a spouse.

Source: Rowland, 1982

gration differentials, selectivity, and motivations for migration. Even aggregate life-cycle and cohort approaches have given emphasis to a deterministic view in which migration is seen as a response to social and environmental pressures. Similarly, human ecologists (for example Hawley, 1968) have viewed migration as a dynamic process of change in areas of the city through which an equilibrium is maintained. The causes for movement are seen in the physical and social environment without reference to individual motives and values. However, researchers have recognized income-based differentials as being a major explanatory factor in migration. It is evident also that the various economic theories of migration have failed to provide other than aggregate explanations, focusing mainly on labor force migration within a pull–push framework, in which the place of migration is seen as a mechanism for the effective allocation of the labor force of an economy, and the role of migration is seen within an overall context of economic change at places of origin and destination.

12.6.1 Mobility and Stage in the Family/Life Cycle

The sociologist Rossi (1955) provides an alternative approach for migration studies by concentrating on housing relocation over short distances. He connects *family life-cycle stages* to *housing preference* as an additional variable in studying mobility, concluding that people relocated as a response to the stimulus of altered needs generated by household life-cycle changes. Here *life cycle* refers to the explicit material needs of families as they move through the various stages of the reproductive cycle.

This approach generated considerable research interest, especially as it relates to the urban housing industry (Laslett, 1989; Hogan, 1987; Ermisch, 1983). Some geographers have continued with *life cycle* as a migration variable, but as Harper (1991:25) points out, the "use of the life-cycle indicator alone as a predictor of mobility is clearly inadequate." If life cycle were to be influential, one would not need to take into account the impact of life-style and/or cultural variables, such as income, status, ethnicity, and location (Fielding, 1989; Lewis, 1982; Bonnar, 1979). Even so, as variables in the migration decision-making process, life-cycle factors still may provide a partial explanation for some migrant behavior (Grundy, 1992; Lewis, 1982; Newton, 1978). However there remain limitations in the concept "life cycle" because it is rigid in how it connects idealized functional ages to particular social activity. As an alternative, the term *life course* (after Levinson, 1978) avoids age and structural stereotypes, and it is regarded by some as being less deterministic (Warnes, 1992:177).

12.6.2 Motivation Underlying Migration

Where, then, do we stand with respect to an understanding of the motivation underlying migration?

DeJong and Fawcett (1979) suggest that there are two salient points concerning motivations for migration:

1. Motivations for migration have roots or counterparts in environmental or structural factors that help explain aggregate mobility patterns.
2. The key to an integration is the identification of linkage processes through which macro stimuli are translated into relevant considerations for individual decision making.

They identified a number of *major motive categories*—economic; social mobility/social status; residential satisfaction; motive to maintain community-based social and economic ties; family and friend affiliation; attaining personal preferences—all of which may be regarded as having positive direction or negative effects on the decision to move. The strength of motive to move was estimated, and it was shown that these differ between developed and developing nations, and within developed nations these vary between long-distance and short-distance moves. Furthermore, potential migrant groups may be identified that are most affected by a specific motive category. In this way selectivity and differentials in migration may be better understood.

DeJong and Fawcett (1979) drew several conclusions from the schema presented in Table 12.3:

1. Major motives for migration are the maximization of actual or expected economic returns, social mobility and social status attainment, residential satisfaction, affiliations with family and friends, and attainment of personal preferences.
2. The major motive for not moving is the desire to maintain community-based social and economic ties.

TABLE 12.3

Summary of Major Motive Categories for Migration

| | | Estimated Strength of Motive | | |
| | | | Developed Nations | |
Major Motive for Migration	Direction of Relationship with Decision to Move	Developing Nations	Long-Distance Moves	Short-Distance Moves
Economic motive	Positive	Strong	Strong	Weak
Social mobility/social status	Positive	Moderate	Moderate to weak	Weak
Residential satisfaction	Positive (with level of dissatisfaction at area of origin)	Weak	Moderate	Strong
Desire to maintain community-based social and economic ties	Negative	Strong	Moderate	Moderate
Family and friend affiliation	Positive	Strong	Moderate	Moderate to weak
Attaining personal preferences	Positive	Weak	Weak, but increasing	Moderate to weak

Source: After De Jong and Fawcett, 1979:37.

3. There is a clear difference between developing and developed countries. In developing countries, micro level migration is not sufficiently detailed to give an analysis of motives for move by distance categories.

4. The dominant motive for rural to urban migration in developing countries is actual or expected economic returns. Also important are the facilitating influence of family and friends at urban areas of destination. In contrast, migration is constrained by the motive of maintaining community-based economic ties.

5. In developed countries, short-distance mobility is strongly motivated by housing and neighborhood residential satisfaction, with the economic motive being relatively unimportant, as job change may not be involved. In contrast, long-distance migration is strongly related to the economic motive. The maintenance for community-based social and economic ties appears to be relatively unimportant for both short- and long-distance moves.

6. A general proposition is that individuals will choose an area that will permit the maximization of their goals based on a hierarchy of economic values and minimize sociocultural and language adjustment problems. If economic goals are assumed to be dominant, then it will follow that the choice of where to move would maximize the likelihood of income or job-related values. Further, the nonlocal choice of destination is based largely on where family or friends are located, particularly in developing countries. Thus family and friends are important sources of information. The roles of prior experiences and of formal and informal information also figure in the development of awareness space.

7. A set of factors relates to structural facilitators and constraints, such as the housing stock, distance, transport flows, subsidy schemes, and so on, which may modify the feasibility of some alternative location choices.

8. A further set of salient factors exist, such as forced moves brought on by wars, political pressures, housing evictions, urban renewal projects, occupational transfers, or the impact of environmental catastrophes, such as drought, flood and famine.

12.6.3 Migration as a Decision Process

Consideration of selectivity differentials in migration and the investigation of the motivational basis for migration increasingly became a focus for the development of *micro* models in which the decision-making process of individuals and households was to be regarded as having paramount importance in developing a fuller explanation of migration. In this context, we are concerned with the *decision choice process*. This involves a consideration of both why people *move* and why people do *not move*.

It needs to be emphasized that there are considerable difficulties in investigating this complex question of decision making and choice in migration. An obvious problem is that of *ex post facto* rationalization that dominates responses to questions in surveys concerning why people moved or why people did not move. Today it is well recognized by researchers that the motives adduced by migrants for moves in the past may hide, rather than reveal, underlying causes of movement. The memory of people becomes blurred with the progression of time. Important objectives and dramatic events can stand out in peoples' memories and influence their recall, leading to distortions in the reporting of the cumulative effects of hopes, aspirations, and fears, which probably are the real causes leading to a decision to leave one place and move to another. It is difficult to overcome these errors and biases in data generated through survey research methods, which are the typical modes used in micro level decision and choice process studies of migration.

It has been suggested by some writers that the micro-level approach to migration studies is of lesser importance than the macro-approaches discussed earlier in this chapter. This view is usually put in the context that macro level studies are more useful to inform policies and planning, since they deal with the broad processes that public policies seek to influence as they tend to focus on the volumes and directions of migration flows. However, it may be argued just as convincingly that an understanding of micro level decision processes in migration may provide improved guidance for public policies that are intended to influence population distribution. As DeJong and Fawcett (1979) argue:

> To put it another way, traditional migration studies tell more about places than they do about people, with corresponding implications for applications of the findings. Migration streams to large cities can of course be deflected by making lucrative employment and attractive amenities available elsewhere, but what else can be done in a shorter term perspective and at a lower, more feasible cost? Studies that illuminate the process of migration decision making we would argue, will suggest alternative means by which such decisions can be influenced through public policies and programs. Knowledge about the range and importance of relevant motivations can be used to advantage in programs to exert direct influence on migration through educational persuasion, and can also be of value for the design of policies that would change the structure of incentives and disincentives for migration. (p. 45)

12.6.4   A Value Expectancy Model of Migration

When migration is considered within a decision choice framework, it may be analyzed as a general behavioral or psychological process. One particular approach is the *value expectancy model* based on the work of the psychologist Crawford (1973:54), who wrote that: "despite differences in terminology, the expectancy value behavior theorists all propose that the strength of the tendency to act in a certain way depends on the *expectancy* that the act will be followed by a given consequence (or goal) and the *value* of that consequence (or goal) to the individual."

The value expectancy model assumed that people usually will behave in a forward-looking, positive, way making choices that they believe will maximize their well-being. It sought to specify personally valued goals that might be met by moving (or staying) and to assess perceived linkage in terms of expectancy between migration behavior and the outcome. This is a *cognitive* model, as it deals with mental events, and it is cast essentially in a cost–benefit framework. Thus, migration is an instrument of behavior, and decision making is based on a cognitive assessment that involves subjective anticipation and weighting of the factors involved in achieving specified goals.

The basic components of the value expectancy models are *goals* (values and objectives) and *expectancies* (subjective probabilities). Pairs of value expectancy components may be assumed to have multiplicative rather than additive relationships. We may express the value expectancy model of migration as follows:

$$M = V_i E_i$$

where:

$V =$ the value of the outcome, that is, migration
$E =$ the expectancy that migration will have to the desired outcome
$M =$ the strength of the motivation for migration, that is, the sum of the VE products

The multiplicative assumption means if either the importance of a particular value is low, or the expectancy concerning it is weak, then that component will contribute little to total motivation. In using this approach to analyze migration, it is necessary first to identify the goals or the values that people hold that are likely to be associated with spatial mobility. DeJong and Fawcett (1979) have identified the seven broad values or goals listed in Table 12.4. These are wealth, status, comfort, stimulation, autonomy, affiliation, and morality. It is possible to link each category to a list of potential indicators of the value. In constructing a questionnaire, they suggest that it is possible for a respondent to measure the personal importance of each. The categories are described in the following terms:

1. *Wealth* includes a wide range of factors related to individual economic reward.
2. *Status* encompasses a number of factors connected with social standing or prestige.
3. *Comfort* means a goal to achieve better living or working conditions.
4. *Stimulation* means exposure to pleasurable activities, in contrast to relief from an unpleasant situation.
5. *Autonomy* has many dimensions, but generally referring to personal freedom and the ability to live one's own life.
6. *Affiliation* refers to the value of being with other persons, in connection with or as a result of migration.
7. *Morality* is related to values and belief systems (such as religious beliefs systems) that proscribe good and bad ways of living.

Testing the values of respondents by how they rate the selected indicators listed in Table 12.4 makes it possible to compute value expectancy peer scores. The total score is an index of behavioral intention. DeJong and Fawcett (1979) stated that in the migration context there can be several scores for each individual; one for the place of current residence, and one for each of several alternative destinations.

The highest score would represent a propensity to move or to stay. The migra-

TABLE 12.4
Values and Goals Related to Migration, With Selected Indicators and Parallel Aspects of Personal Motives and Socio-Structural Characteristics

Expected Values/ Goals	Selected Indicators	Personal Motives		Social-Structural Characteristics	
		Attraction	Avoidance	Pull	Push
Wealth	Having a high income; stable income Having economic security in old age Being able to afford basic needs; some luxuries Having access to welfare payments and other economic benefits	To increase income and wealth, economic security	To escape from (relative) poverty, insecurity; economy expanding	Economic opportunities, availability of high-paying jobs,	Shortage of jobs, low-paying jobs, economically depressed area
Status	Having a prestigious job Being looked up to in the community Obtaining a good education Having power and influence	To improve social status	To free self from low-status situation, seek new role	Social mobility by achievement, opportunities for change	Social status by ascription, lack of flexibility
Comfort	Having an "easy" job Living in a pleasant community Having ample leisure time Having comfortable housing	To have easier, more comfortable life	To avoid harsh environment, physical labor, long work hours	Urban amenities, white-collar and service jobs with regular hours and wages	Poor rural setting, labor-intensive agricultural production
Stimulation	Having fun and excitement Doing new things Being able to meet a variety of people	To seek stimulation fun, variety	To avoid boredom	Public entertainment, diversity of activities "bright city lights"	Uniformity and simplicity of village life, no night life
Autonomy	Being economically independent Being free to say and do what you want Having privacy Being on your own	To have freedom of choice, personal autonomy	To be relieved of traditional role constraints, free from influence of others	Impersonal social structure, urban anonymity	Tight social controls, close surveillance by family/community
Affiliation	Living near family, friends Being part of a group/community Having a lot of friends Being with spouse/prospective spouse	To be with friends, family or spouse, join particular community	To avoid loneliness, social isolation	Availability of certain persons or groups	Nonavailability of certain persons or groups
Morality	Leading a virtuous life Being able to practice religion Exposing children to good influences Living in a community with a favorable moral climate				

Source: After De Jong & Fawcett, 1979.

tion intention score is expected to be predictive of future mobility behavior. If migration behavior is seen to be directed toward the goal of improving or maintaining the quality of life, it is possible to ascertain the expected values of migration, their strength, salience, degree of certainty about outcomes; values for oneself and for others; and the short- and long-term perspectives of individuals. If intervening factors, such as barriers and facilitators, are taken into account, it should be possible to achieve the stated aim of the model. In essence, the value expectancy approach provides a method for measuring at a subjective level many of the factors that are likely to enter into the decision to migrate. There is an instrument for measuring behavior and decision-making based on what DeJong and Fawcett have called a "cognitive calculus," which involves anticipatory weighting of the elements involved in attaining certain goals.

DeJong and Fawcett (1979) proposed a value expectancy model of migration decision making in the form shown in Figure 12.5. It is worth noting, as shown in Table 12.4, that for each of the expected values (except morality), personal motives for migration have counterparts in social structural factors, although the correspondence is not exact. For the first six values, four distinctions are shown: attraction and avoidance in relation to personal motives; and pull and push in relation to social structural characteristics. Individual and household demographic characteristics, societal and cultural norms, personal traits, opportunity structure, and information all operate on the expectancy of attaining values; whereas individual and household demographic characteristics, societal and cultural norms, personal traits, and opportunity structure differentials between areas, operate to effect values or the goals of migration. The values and goals of migration and the expectancy of attaining values interlink to form migration behavioral intentions, which can lead to either a decision to stay or an *in-situ* adjustment.

In summarizing their model, DeJong and Fawcett (1979) have this to say:

> Migration behavior is thus hypothesized to be the result of (1) the strength of the value expectancy derived intentions to move, (2) the indirect influences of background individual and aggregate factors, and (3) the potential modifying effects of often unanticipated constraints and facilitators which may intervene between intentions and actual behavior. These constraints and facilitators may include change in family structure (for example, marriage, divorce, separation, death, increased or decreased family size), financial costs of moving, changes in health, distance, and changes in the location of support anticipated from family and friends. We suggest that these factors may be unanticipated and that they may not have been salient considerations in the original migration intentions, as derived from value expectancy based cost benefit calculations of potential migrants. Forced mobility may also create incongruities between migration intention and behavior. (pp. 59–60)

There are considerable similarities between this value expectancy approach and a number of models developed by behavioral geographers to study the residential location decision process in an intraurban mobility context that are discussed in Chapter 13, Section 13.4.

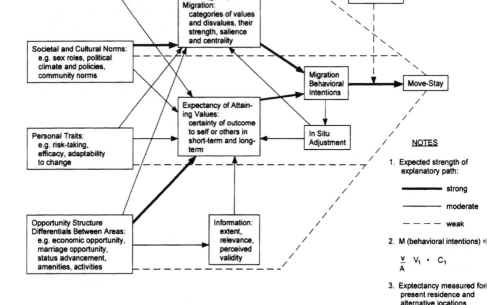

Figure 12.5. A value expectancy model of migration decision making. *Source:* De Jong & Fawcett, 1979.

12.7 Habitas as a Cultural Process in Migration

For a long time anthropologists have represented the social role of culture and its influence on spatial behavior that affords order, direction, and guidance in solving problems. People's lives and their decisions and choices are strongly influenced through *cultural mores*. This perspective is interesting in the context of migration studies as it raises questions about the degree to which individuals and households process information. In addition, cultural mores may be seen to influence strongly those factors proposed in the value expectancy approach discussed above.

One may hypothesize that migration decision makers receive cultural signals that motivate a desire to migrate. Instead of seeking to find structural variables, such as unemployment and relative income advantage, as precursors for migration, alternatively one might seek to find the issues that motivate the individual to regard migration as a plausible option. This difference is subtle, but it is important. For example, this may be seen as important if one notes also that, contrary to what is inferred in much of the literature, in lower-socioeconomic groups a significant number of people are content to stay in their locality, even in the face of employment adversity (Hodge, 1985).

Before developing this discussion further, some demonstration of what is meant by the term *culture* is needed.

Fielding (1992) and Halfacree and Boyle (1993) both have utilized Bourdieu's (1984) notion of *habitas* to illustrate the cultural process that precipitates migration. Here *culture* refers to all the shared ideas, beliefs, values, and knowledge that constitute the societal basis of social action and physical expression; whereas *habitas* is the realm of the individual's own conscious and subconscious perception of what culture means to him or her. Explaining this in another way, accepting that each person needs reference groups in order both to be identified and to be accepted, as well to evaluate and regulate opinions and actions, then *habitas* refers to that inner repository of cultural knowing that each person holds and uses in order to socialize.

Fielding (1992:202–205) suggests that a person's *habitas* is materialized in the way that person uses places. Thus, there are a number of spatial and cultural features that can bring about a propensity for an individual or a household to migrate. People might perceive the *culture* of the place they live as becoming unsuitable, this being due to one or more of the following:

1. A relational disintegration with reference groups (such as ceasing work).
2. A change in social motivation (such as a desire to pursue a different set of social activities).
3. Being marginalized by reference groups (this could be brought about by a major conflict, or a changing community structure because of inmigration)
4. A change in the physical structure of the place (such as a new freeway causing dislocations and barriers, or a place becoming gentrified).

Similarly people may perceive some other location as providing a preferable *reference group,* such as when people migrate to follow family, or when they join a social group, such as a fundamentalist church that persuades members to adopt different social goals. People may aspire to a social standing that, in part, necessitates migration, such as might occur among young, upwardly mobile individuals.

In a study of retirement migration, Longino (1992) identified *moorings* as being important factors influencing the decision to relocate, particularly over longer distances. As well as being social items, moorings embrace also items of a cultural and spatial nature. As Table 12.5 indicates, moorings are all those items through which a person gains access to social and psychological well-being. While each of the items nominated could be utilized as push and pull factors, it is the reference values that people ascribe to the items that contextually separate the two concepts. In addition, moorings are influenced by cultural signals. For example, the personal motivation to remain in a location with respect to any push or pull factor is assumed as small. However, there may be cultural signals that an individual perceives as important and that constitutes an item as a mooring. Thus a mooring is an item that motivates a person to want to remain located in an area, or it is sufficiently unimportant not to bind the person to a particular location.

The difference in this approach compared to the functionalist macro models of migration discussed earlier in this chapter is that although people might migrate

TABLE 12.5
Typical Moorings

Life-Course Issues
Household structures
Household income
Educational opportunities
Caregiving responsibilities
Cultural Issues
Household wealth
Employment structure
Social networks
Cultural affiliations
Ethnicity
Spatial Issues
Climate
Access to social contacts
Access to cultural icons
Proximity of regions of recreational interest

Source: Longino, 1992.

to optimize their experiential circumstances, instead of making the decision according to some functional structure or rationalization of evidence, people do so in response to signals perceived as being transmitted from a variety of cultural symbols. The way that people "read" and even "respond to" those signals may not necessarily be rational.

This cultural-based perspective has considerable methodological appeal in that the actor undertaking migration is not considered to be directed positively by forces outside his or her control. Nor is he or she guided toward rational action by a subconscious. Rather he or she is a "free agent capable of controlling their [sic] own destiny" (Atkinson, Atkinson, Smith, & Hilgard, 1987:433). But being capable of controlling one's own destiny does not necessitate rationality of purpose. A cultural approach to migration research reflects the propensity to place the onus of responsibility back on the person. This accords with the views of Lyotard (1984, 1986), Derrida (1982, 1988) and others, who suggest that rather than testing the functionalist view that bestows an institutional narrative on people to undertake action according to ordered rules, it is preferable to regard each person as making his or her own narrative from a world that lays intensely competitive forms of consumerist imagery over cultural symbols of status, power, and prestige. This is a perspective proposed in recent writing by postmodernists, such as Best and Kellner (1991) and S. K. White (1991). It provides an interesting new perspective in migration studies.

13

Residential Mobility and Location Decisions

13.1 A Framework for Analysis

Much of the micro scale, disaggregated behavioral analysis of migration has been conducted at the intraurban scale. The focus has been on investigating the patterns of intraurban mobility and the reasons underlying the residential relocation decision process.

Residential location decisions of individuals and households at the intraurban scale take place within the constraints of time and various enabling factors. This framework incorporates behavioral processes related to the satisfaction of perceived dwelling needs, the locational preferences of households, and their knowledge of the subset of the housing market in the urban setting in which the decision is made. The mosaic of residential areas in a city that are differentiated by their social space characteristics and by housing submarkets reflect the outcome over time of the aggregate residential decisions of households. Urban, social, and behavioral geographers have proposed various theoretical explanations of the social structure of cities, have developed methods to classify residential areas, and have developed models of the residential location decision process. A framework for conducting such an analysis is proposed by Stimson (1978) in Figure 13.1.

In Chapter 4, Section 4.5, we discussed how the *mosaic of residential areas*, within which residential location choices are made, may be studied in terms of their *social space* characteristics using *factorial ecology* and *residential typology classification* methods based on small area census data to give a territorial expression of urban space. This is indicated in the left-hand panel of Figure 13.1. Individual and household residential location decisions are made in response to stress, as people seek to satisfy their housing needs and to meet their individual locational aspirations. Search behavior is involved, as is choice between alternative locational and housing opportunities. This is indicated in the right-hand panel in Figure 13.1. These spatial actions take place within the action spaces of individuals and households and are related to the cognitive decision process of the individual. Choice of residential location is made with respect to the residential space preferences of people as indicated in the middle panel of Figure 13.1.

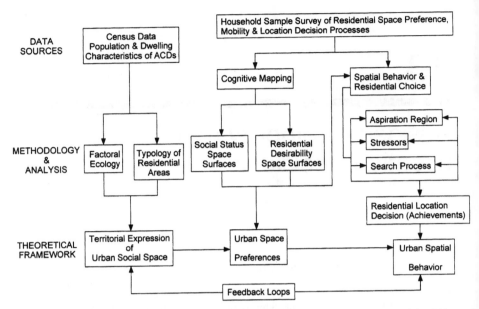

Figure 13.1. A framework for analyzing the residential typology of cities and the residential location decision process. *Source:* Golledge & Stimson, 1987:264.

In this chapter we focus on the behavior of individuals and households in their choice of residential location within the context of intraurban mobility.

Early models of household behavior in the choice of residential location fall into two basic approaches:

1. The first assumes that residential location choice may be described in terms of a *tradeoff* between *transport costs* and the *price of housing*. If we know the shape of the rent gradients and the spatial patterns of access to transportation in the city, then a household is seen to choose a location by equating the savings on housing costs with increments to transport costs. This approach included the use of normative microeconomic models (Alonso, 1964; Muth, 1969; Wingo, 1961). Such approaches were limiting in that they did not include neighborhood effects, and they assumed constant income and continuous rent and transport gradients. Casetti and Papageorgiou (1971) attempted to extend this trade-off approach by incorporating some of these variables.

2. The second approach involves *micro-behavioral* modeling, arguing that accessibility is not the primary determinant of residential location choice, but that amenity and environmental, socioeconomic, psychological, and time factors operate together to produce a multivariate explanation of decision making. This behavioral approach emphasizes the process by which location decisions are made.

But before turning to an analysis of the evolution of disaggregated behavioral

approaches in the study of the decision and choice processes of households within cities, it is necessary to look at the typical aggregate patterns of intraurban
mobility.

13.2 Mobility within Cities

It is important to realize that in aggregate terms *intraurban mobility* constitutes the
vast majority of moves by individuals and households within advanced western nations. Typically such moves involve much less disruption to patterns of living than
in the case for long-distance moves, but often they do involve significant changes in
the activity spaces of households through changes in the relationships between residential location and household activities, and there may be substantial changes in
the location of activities such as work, school, shopping, and so on for members of
a household as a result of a move. Thus Roseman (1971) referred to intraurban
moves as resulting in *partial displacement* as opposed to total displacement in long-
distance relocation.

Within cities it is typical for residential relocation to be predictable and related to significant *life-cycle events* (as discussed in Chapter 12, Section 12.6), such as
children leaving home, marriage, separation and divorce, retirement, and so on. In
contrast to long-distance migration, mobility within cities is more related to adjustments to housing consumption in this life-cycle context. Residential mobility may
be seen as "one outcome is an opportunity structure whereby households express
choices of where and how they live" (Maher & Saunders, 1994:3). It is linked closely to the housing system in cities, reflecting the interrelationship between producers
and consumers of housing, along with a variety of mediating institutions and individuals that include government, financiers, agents, and the legal system (Murie,
Niner, & Watson, 1976).

While many residential relocations within cities are related to life-cycle
events, they also incorporate socioeconomic status-related factors, including housing as a status symbol of achievement, as an investment vehicle, and as a form of
stored wealth.

Kingsley and Turner (1993) have demonstrated in the United States that the
nature of housing supply in a city is a major determinant of mobility patterns of
households but that housing supply also exerts a large influence on preferred housing options of individuals. Maher et al. (1992) have shown in Australia that housing
choice revolves very much around affordability, accessibility, and "appropriateness," and these are influenced by density, infrastructure costs, and environmental
issues.

In addition, *housing tenure* is an important issue. Although in nations like the
United States, Canada, Australia, and some Western European countries, home
ownership is the dominant form of housing tenure, private rental housing tenure is
the preferred choice for many households, and this is often related to stage in the
life cycle. For some low-income and socially disadvantaged groups, private rental or
public housing is the only feasible form of tenure.

13.2.1   Patterns of Intraurban Population Change and Mobility

It is common to analyze intraurban residential mobility patterns and trends and to relate them to population change in cities, using small-area data obtained from the census. In advanced Western economies, it is common to find that the dominant trend in population and mobility within metropolitan regions is one of suburbanization (as discussed in Chapter 4, Section 4.2). Particularly in metropolitan areas in the United States, population decline in the inner cities has been marked. This is related to both the suburbanization of jobs and the concentration of ethnic groups in the inner city, which increasingly has become the domain of low-income, unemployed, and homeless people.

In a relatively few instances, urban geographers have been able to obtain data from longitudinal studies of mobility of households in cities, both from census data and from sample survey research studies. For example, Maher and Saunders (1994) and Maher et al. (1992) have calculated intraurban moves between small areas over 20 years in the Australian city of Melbourne. The data cover four intercensal periods, with moves recorded as flows between origins and destinations in a 56 × 56 matrix of small areas, generating 3,136 cells. The data identified households that had moved their place of residence over the previous 5 years for each census, and patterns of movement were analyzed over four concentric zones for metropolitan Melbourne. Both net and gross flows were analyzed. The level of mobility for the small census areas was analyzed using the net migration ratio, which measures the impact of the net flows in relation to the population at the beginning of each census period. Table 13.1 shows the relative concentration of different sources of movement in Melbourne over the 20-year period 1971 to 1991. It is evident that the rate of movement within the metropolitan area is dominant, comprising about two-thirds of all moves. Figure 13.2 shows how intraurban mobility is most important in the redistribution of the population within the metropolitan area, with the inner zone experiencing substantial but declining loss of population, as did the middle suburbs. The growth in population in more recent years over this 20-year period was in the outer zones and in particular in the metropolitan fringe areas. During this time the total population of the metropolitan area increased by 20%. Figure 13.3 shows the pattern of net intraurban mobility in Melbourne over this 20-year period, and the

TABLE 13.1
Population Turnover by Source of Move: Melbourne, 1971–1976 to 1986–1991 (Percentage of Those Aged 5 or More)

	1971–76	1976–81	1981–86	1986–91
Did not move	56.4	59.8	56.5	58.2
Intrauban	30.7	28.9	30.1	26,3
Other*	5.8	5.4	5.9	5.6
Overseas	5.2	4.6	4.9	5.5
Not stated	2.0	1.3	2.6	4.4

*Other move includes intrastate and interstate.
Source: Maher & Saunders, 1994:6.

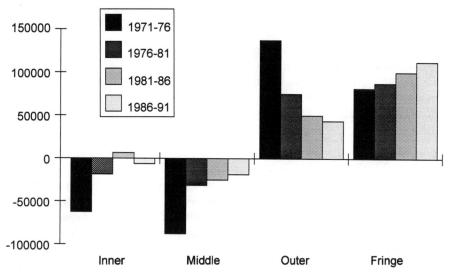

Figure 13.2. Melbourne: population change by zone, 1971–1991. *Source:* Maher & Saunders, 1994:7.

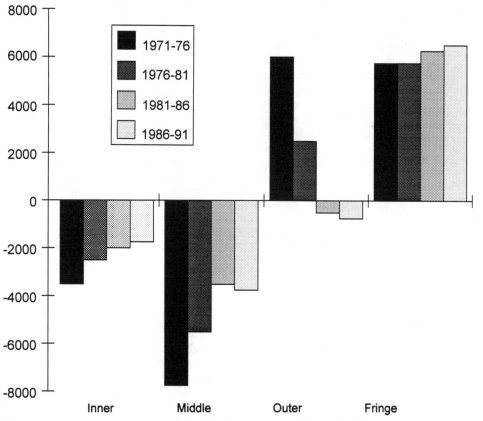

Figure 13.3. Melbourne: net intraurban mobility by zones, 1971–1991. *Source:* Maher & Saunders, 1994:8.

pattern reinforces that evident in Figure 13.2. Although net intraurban mobility does not account for all the overall population change in Melbourne (births, deaths, and in- and outmigration to areas beyond the metropolitan areas are influences), what is evident is that "the effects of relocation decisions made within the urban area are dominant in changing the distribution of the population" (Maher & Saunders, 1994:8).

13.2.2 Life Cycle of Suburbs

An important concept related to the dynamics of population change and mobility within cities is the notion of *life cycle of residential areas,* in a manner similar to that of households. It is apparent that a complex developmental sequence occurs over time that affects both household structure and size, plus the population of suburbs and the use made of housing. Thus, for any suburb within a metropolitan area, there will be a considerable change over time in the characteristics of its population and households. As Maher and Saunders (1994) state:

> A typical sequence may well be in the initial development plan, when population growth is very rapid because households are moving to an area previously undeveloped, followed by a stabilization phase, where the number of dwellings has reached a maximum, the number of households is also stable. Once younger members of the household reach adulthood however, they begin leaving their family home, attracted by independence, work and education. Older members of the household generally remain in the family dwelling, and thus there are few opportunities for replacement of the population moving out. What residential mobility does occur at this stage tends to maintain the characteristics of the population rather than change it dramatically. At later stages however, after the original households have moved on (either spatially or spiritually), more rapid neighborhood change may occur. The extent to which this happens is a function both of factors internal to the neighborhood and external factors that determine the nature of locational choice. These are more a function of the composition of mobility streams than of actual numbers of movers in or out of an area. However, significant changes in neighborhood social composition may well have an impact on subsequent rates of movement. (p. 12)

Over time within a city these dynamics may result in a variety of outcomes, including gentrification of inner suburbs, redistribution in the population of inner and middle suburbs, rapid growth of outer and fringe suburbs, and stability of population in many middle and outer suburbs.

13.3 From Aggregate to Individual Choice Models

In discussing the evolution of behavioral models of the residential location choice process in cities, Smith, Clark, Huff, and Shapiro (1979:3–5) identified the following broad approaches. The first two are empirical, and the next three are theoretical in nature:

1. *Spatial aspects of the residential relocation process at an aggregate level,* which looked at population flows at an aggregate level between census tracts and

neighborhoods. They include: Brown and Longbrake's (1970) study of the degree to which people move between similar socioeconomic districts of the city; Simmon's (1974) study of the flow of people between census tracts; and Clark's (1970) study of income and house value constraints on how people choose neighborhoods in the city. Such studies give a general indication of search procedures used in residential decision making.

2. *Household spatial behavior at the micro level*, which were concerned with *search behavior*. They include studies of the search behavior of households, using questionnaires administered after the search and relocation process was completed to analyze the spatial pattern of search (Hempel, 1969; Barrett, 1973), or information gathering about housing vacancies (Baresi, 1968; Palm, 1976a).

3. *Markov modeling of the structure of intraurban migration*, which are the comprehensive group of modeling strategies used to specify the structure of mobility rates and analyze longitudinal characteristics of migration. The focus was on when migration was likely to occur rather than on where it would go and was based largely on life-cycle determinants. Studies included investigations of social mobility (McGinnis, 1968) and the mobility for heterogeneous populations (Ginnsberg, 1971, 1973). But these models were not concerned with the decision-making process of individual households.

4. *Migration in relation to the housing stock or neighborhood characteristics*, which are aggregate models of neighborhood change, such as studies by Quigley (1976) on housing characteristics in small areas, the movement of people through these housing stocks, and the impacts on communities and housing characteristics.

5. *Issues of the decision process in residential relocation*, epitomized by the Brown and Moore (1970) model in which the process is separated into two stages, namely the decision to move and the actual relocation process. They also refined Wolpert's concepts of action space, search behavior, and place utility.
To those need to be added:

6. *Decision-making and search models*, which seek to predict the mobility pattern within a city by incorporating spatial and temporal aspects of residential choice in a probability framework (Smith et al., 1979) and which suggested that the probability of moving will be influenced by both cumulative resistance (inertia) and dissatisfaction (stress), with the probability of moving being a function of the resultant of these two conflicting forces (Huff & Clark, 1978).

13.4   Residential Aspirations, Preferences, and Achievement

Research on *residential aspirations* is fragmented, but results seem to suggest that social variables in the urban environment are important. For example, Beshers (1962) noted social rank or status as important; Bell (1958) referred to life-style factors; stage in the life and family cycle were emphasized by Pickvance (1973); and Selye (1956) highlighted value orientation. Figure 13.4 illustrates the factors that seem to be relevant in a *causal model of residential mobility*. It has been a diffi-

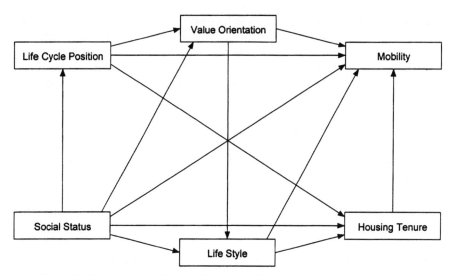

Figure 13.4. A causal model of residential mobility. *Source:* Pickvance, 1973:281.

cult task to quantify attributes that measure preferences for housing types and residential locations. As Brown and Moore (1970) indicate:

> The urban population is differentiated on social, economic, and locational dimensions according to different sets of environmental needs. This differentiation is also directly applicable to the definition of systematic differences between the aspiration regions of intended migrants. In applying such differentiation to the search and evaluation phase, however, a comparable vector describing vacancy characteristics is needed, together with an understanding of the way in which the subjective aspiration levels of the intended migrant relate to objective measures of vacancy characteristics. Systematic distortions in migrant perceptions may be identified by establishing functional relationships between migrant aspirations and vacancy characteristics needed for each of the basic dimensions of the aspiration region or the variables basic to the migrant's evaluation procedures. (p. 10)

Despite considerable empirical investigation, relatively little progress has been made toward the development of a comprehensive explanatory model of residential mobility.

While residential mobility is seen to be triggered by discontent with the present residential location or dwelling, it is, as Boyce (1969:2) notes, highly voluntary. Researchers tended to look for push and pull factors. Ermuth (1974) asked these questions:

> Are social factors of the environment more important than physical or aesthetic factors? Do households in different subareas of the metropolitan area view housing quality and the social life of a community differently? How important are spatial factors, such as accessibility considerations and relative location of the subareas within the

whole urban area? Also, to what extent do households trade off and substitute one factor for another? (p. 3–4)

13.4.1 Why Families Move: Rossi's Classical Study

The classical work on residential mobility was Rossi's book, *Why Families Move* (1955). He investigated the high rate of residential mobility in the United States because it was one of the most important forces underlying change in urban areas. He posed questions on the mobility of different types of families and on the motivations underlying residential shifts. It was an early noteworthy major study using the *social survey*, employing modern social research methods in the study of residential mobility and choice. The focus was on residential relocation, and the study provided important insights into and empirical evidence on housing preferences in terms of the characteristics of the dwelling type and its immediate social and economic environment. Rossi compiled a *mobility potential index* and a *complaints index*, and he proposed an *accounting schema* for moving decisions shown in Table 13.2. It was shown how the position of the household in the family life cycle and its members' attitudes toward the home and residential neighborhood were important predictors of the household's current desires for moving. Rossi (1955:9) states that his findings: "Indicate the major function of mobility to be the process by which families adjust their housing to the housing needs that are generated by shifts in family composition that accompany life cycle changes."

13.4.2 Preference and Desirability Studies

Other important works followed, but not until the 1960s. Various approaches were adopted. For example, Wilson (1972) studied *livability factors* in cities. Hoinville

TABLE 13.2
Rossi's Accounting Schema for Moving Decisions

Stage I	Stage II	Stage III
The decision to move	The search for a new place	The choice among alternatives
REASONS FOR LEAVING		
A. Decision forced by outside circumstances (2/5 households). Reasons: job, change in marital status, home destruction, etc.	A. Channels of information a) Formal: papers, real estate agents b) Informal: friends, on actual and prospective vacancies	A. Only 1 opportunity offered. Reasons presumed to be identified with specifications (1/2 households); clinching factors
B. Decision made voluntarily (3/5 households) because of dissatisfaction with old place; space complaints generally	B. Specifications (features desired in new home). Most families looking for particular kinds of dwellings, i.e., particular size and features.	B. Several opportunities offered. Choice made because of comparative attractions of alternatives; generally took lower cost place

Source: Rossi, 1955:174.

(1971) attempted to derive measures of value of *residential amenity loss or gain* arising from planning decisions in a trade-off situation. The Highway Research Board in the United States (HRB, 1969) provided data for 43 cities across the country on national housing and environmental *preferences*. Michelson (1966) looked at the *value orientation* of people and the nature and extent of *social interactions*. A further study by G. L. Peterson (1967) employed color photographs of neighborhoods to simulate visual appearance, hypothesizing *residential desirability* is a *multidimensional* phenomenon that could be simplified by an orthogonal model of preferences. He found that the most significant dimensions were general physical quality and cultural conditions.

A typical study incorporating more specific locational variables in the analysis of *residential environmental preferences* and choice was that by Menchik (1972). He employed a regression model to look at first- and second-order preference choice for individuals and groups using sets of variables relating to the beauty of the natural and built environments, accessibility, characteristics of house and lot, and familiarity attributes of the residential environment. He suggested that preferences represented trade-offs among these attributes and that comparable measures could be derived for people's present residences that reflect their actual choices. He found that preferences do express themselves to some extent through market choice and that different people prefer different residential characteristics.

In a detailed study of *residential satisfaction* and *environmental preferences* in the American city of Des Moines, Iowa, Ermuth (1974) set out to test the premise that the satisfaction of a household with the housing it chose and currently occupied could be used to make statements about housing preferences. He found that:

> The ex post facto housing choice behavior of the household therefore is conceptualized as being based on a revealed subjective preference scaling of all possible housing and residential environment opportunities. Given a particular housing choice of a household, these preferences are assumed to be a function of his perception and evaluation of relevant urban and environmental characteristics. (p. 15).

People react to environmental stimuli on the basis of their internalized organization of events (their cognitive map). A variety of judgmental phenomena can be related to how cognitive maps are used. Preference judgments, therefore, can be meaningfully represented as transformations of the respondent's cognitive maps.

The nature of *residential space preferences* in the Australian city of Adelaide was studied by Stimson (1978), using a questionnaire survey administered by interviewers face-to-face to a sample of 727 individuals at 130 clusters of residential locations across the metropolitan area. Respondents were asked to rate on a 5-point scale the level of residential desirability of 100 suburbs scattered throughout the metropolitan area. Figure 13.5 shows how the most highly desirable suburbs are located mainly to the east of the city, to the south, along the Adelaide hills, along the coast, and in some inner-city, old high-status areas. In contrast, the old industrial suburbs in the inner city, particularly in the west, the northwestern and southwestern sectors, and the northern and northeastern fringe areas, had the lowest-ranking residential space preferences.

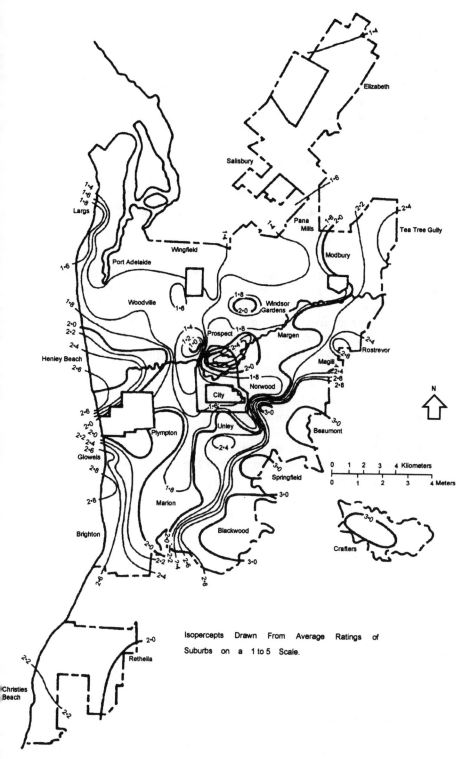

Figure 13.5. Residential space preference surface of residents of Adelaide. *Source:* Stimson, 1978.

To ascertain the factors Adelaide residents perceived to be important in selecting a residential location and dwelling, respondents were asked whether or not they thought 20 items in a list of factors hypothesized as affecting residential location choice were considered desirable or undesirable. The degree to which the factors were important or unimportant was also gauged. The 20 factors were grouped into five broad sets of attributes relating to:

- Location-specific attributes related to action space and proximity.
- Location-specific attributes related to the location of an area in the urban system and its residential environment.
- Location-specific attributes related to provision of public services.
- Socioeconomic status attributes.
- Dwelling-specific attributes and site characteristics.

Respondents were asked also to indicate whether or not each of the 20 attributes within these five broad sets appeared negatively (i.e., it was undesirable), positively (i.e., it was desirable), or not at all (i.e., it was indifferent) in their aspiration regions. The results are given in Figure 13.6.

From this study of Adelaide residents, Stimson (1978, 1982) concluded that residential aspirations constitute a particularly complex phenomenon, and that the most important considerations appeared to be:

- Proximity to convenient shopping and transport.
- The newness of a residential area.
- Access to the CBD.
- Prestige and status of a residential area.
- Convenience and facilities for children.The price of an area and the provision of sewerage.

Thus, proximity to elements in household activity spaces, community facilities, socioeconomic status, life/family cycle stage, and environmental factors are shown to be important attributes in the aspiration region of Adelaide residents. This was similar to findings from empirical studies conducted in other cities. Also, subgroups of the population, differentiated on the basis of factors related to socioeconomic status, stage in life/family cycle, and social-area type of present residential location, displayed differences in their assessments of the direction and degree of importance of the 20 specific attributes. Furthermore, it was demonstrated that there were differences in the occurrence of many of these attributes and the aspirations of people differentiated on the basis of the socioeconomic quintile group in which their present residences were located and in the search field that was covered during the last residential location decision process.

It would seem that residential preferences are influenced strongly by factors such as socioeconomic status and stage in the life/family cycle. It appears also that people are constrained by economic considerations such as income and likelihood of gaining access to housing finance, plus individual taste and preferences, when

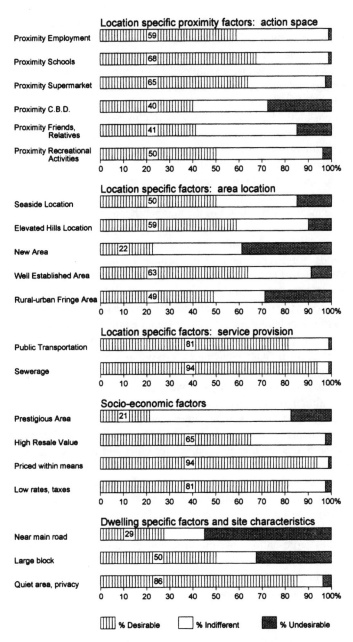

Figure 13.6. Ratings of attributes in the residential aspirations of Adelaide residents. *Source:* Stimson, 1978.

forming their aspiration regions and in assessing the residential desirability of specific locations in a city.

In a somewhat different study, Cadwallader (1979) sought to investigate the *evaluative dimensions* used by a sample of 148 residents in Madison, Wisconsin, to rate eight neighborhoods (containing about 10 city blocks of single-family dwelling units) on 11 semantic differential scales. The hypothesis was that households rate neighborhoods on the basis of three evaluative dimensions:

1. The impersonal environment, comprised mainly of physical attributes of the neighborhood.
2. The interpersonal environment, comprised mainly of the social attributes of the neighborhood.
3. The locational attributes of the neighborhood.

Data derived from a 148 × 11 data matrix for each of the eight Madison neighborhoods were examined using principal components analysis. Results showed that in almost every case the percentage of total variance explained by these three components was fairly low, suggesting that the 11 scales were not easily collapsed into underlying dimensions, and the factor structures associated with each of the eight Madison neighborhoods were far from identical. Further analysis was conducted to test for perceived similarity among the neighborhoods, measured by a similarity rating in which respondents were asked to identify for each neighborhood those other neighborhoods they regarded as being most similar. It was shown that neighborhoods perceived to be similar were cognized in terms of similar dimensions. Thus, although evaluative dimensions were not the same for all neighborhoods, people tended to use similar dimensions when evaluating similar structures of these components for each of the eight neighborhoods. It was found that for four of the neighborhoods the highest loadings for quiet and privacy were on the same component, and for four of the eight neighborhoods the highest loadings for quiet and spaciousness were on the same component. Elementary linkage analysis was used to uncover the major groupings among the variables. The two major groupings to emerge were:

1. Scales representing spaciousness, housing quality, distinctiveness, quiet, and privacy.
2. Scales representing neighborhood reputation, yard upkeep, safety, type of people, and park facilities.

The findings of these and other studies conducted in the 1970s investigating *residential space preferences* and *aspirations* identified recurrent themes. The selection of a new residential location and the choice of dwelling involved considered attributes that were surrogates for accessibility factors with respect to:

1. Locations of functions that are important in household activity spaces.
2. Physical environmental and aesthetic qualities of the neighborhood.
3. Community services and facilities.

4. Social environment factors such as prestige, socioeconomic status, and ethnicity.
5. Stage in the life/family cycle as it relates to housing needs.
6. Familiarity with the residential area and individual site and dwelling characteristics.

Newton (1978) summarizes the aspirations and locational choice of households by advancing these four propositions:

1. When changing residence in a city a mover household may choose, or be forced, to live among households with similar social and demographic characteristics to its own.
2. When changing residence within a city a mover household will select an area which contains housing and amenities with the necessary attributes to satisfy its residential needs.
3. When changing residence within a city a mover household will relocate in that part of the city which meets its accessibility requirements to nodes with which there is frequent, routinized, contact.
4. When changing residence within a city a mover household will be constrained in its residential choice by: (i) the availability and/or (ii) the cost, of various housing bundles located at different points within the city. (pp. 33–34)

13.4.3   Stressors and the Level of Residential Satisfaction

Residential satisfaction is considered to have been achieved when there is *congruence* between (or greater overlap of) residential *achievement* vis-à-vis *aspiration* (Brown & Moore, 1970). When aspirations and achievements are out of congruence, a *stressor* is said to exist, which triggers *residential search*. This is an overt act reflecting a decision to relocate, be it tentative or otherwise. In Wolpert's words (1966:93), a *stressor* is: "any influence, whether it arises from the internal environment or the external environment which interferes with the satisfaction of basic needs or which disrupts or threatens to disrupt the stable equilibrium."

Stressors may be *internal*, such as change in family size, or *external*, such as expiration of a lease on a dwelling. Furthermore, they may be *dwelling specific*, such as the need for an extra bedroom, or *location specific*, such as a change in job leading to an increased length of journey to work. A stressor may act upon either the *achievement* level or *aspirations*, that is, it may alter what the decision maker already has (i.e., achievement), or it can alter aspiration.

Once it appears, a stressor will alter considerably the relationship between achievement and aspiration. The degree to which a particular stressor upsets the relationship will vary between decision makers according to factors such as socioeconomic status, stage in the life/family cycle, experience, and personality. Also, each individual will have a threshold of aspiration level that adjusts on the basis of these factors, thus giving rise to different levels of tolerance to stress. Rossi (1955) indicated that urban structure itself (manifest through things such as commercial neighborhood blight, industrial sites, and a change of racial and ethnic composition of a neighborhood) is an important factor that may influence an individual to seek a new residential location. Rossi (1955), Butler et al. (1969), and Clark (1970) demonstrate how the evolution of households through the life and family cycle may give rise to stress. Lansing and Hendricks (1967) point to the effects of changes in transport technology on accessibility between neighborhoods as a factor acting as a stressor. Finally, upward socioeconomic mobility involving a change in income and/or job level, can lead to change in aspirations that give rise to stressors. It is likely also that overt pressures of advertising may have a similar effect.

Much has been written on the nature of stress in city life, and there is a distinct possibility that some urban social dysfunction and pathologies are spatially correlated with stress indices. Urban stresses have been used to encompass factors such as noise, pollution, traffic density, crowding, high-rise living, and the general complexity of scale of urbanization. Stress may be seen to manifest itself in the form of detrimental effects to physical and mental health, crime, fear, breakdown of community interaction and cohesiveness, and alienation. However, it is difficult to furnish proof of cause and effect relationships. In general there are few detailed empirical data available on the nature of stressors as they give rise to residential dissatisfaction and play a role in the residential relocation decision process in cities.

Two studies illustrate the general nature of stressors.

Clark and Cadwallader (1973) looked at five specific *stress-producing factors* that were surrogates for the types of stressor sets that have been discussed and identified by various researchers. Five stressors were identified in relation to stress processes and stress response, as given in Table 13.3 in a study of Santa Monica residents, Los Angeles. The five groupings of stressor factors are interesting, as they relate to the characteristics of the dwelling and life/family-cycle stage requirements, social space affinity characteristics of the neighborhood, proximity factors, and environmental factors. If the stress level is high for a household, then there would be a high degree of desired, if not actual, movement to adjust to the stress level. In the Santa Monica study, Clark and Cadwallader found a significant correlation between the desire to move and locational stress. The greater the household level of stress, the greater the potential mobility exhibited by the household. They claimed that:

> There is a need to specifically introduce elements of the urban spatial structure, such as the nature of the local neighborhood, and to analyze these elements with reference

TABLE 13.3
Stressors, Stress and Stress Reactions in Los Angeles

	Processes	
Stressor	Stress measurement	Stress response
1. Size–facilities of dwelling	Derived from stress mode,	a. Desire to move
2. Access to work	using individual household	b. Actual movement
3. Access to friends	attitude scale on stressors.	c. Modification of dwelling
4. Kind of people in neighborhood	N.B. subjectively	d. Public action, e.g.,
5. Air pollution (smog)	evaluated by households	petition for more services

Source: Clark & Cadwallader, 1973:35.

to their stress-producing potential. It is possible that this kind of approach will more directly allow the development of a model in which mobility can be fitted into the context of neighborhood and community change.

Furthermore, the researchers found that:

The decision to move can be viewed as being a function both of the household's present level of satisfaction and of the level of satisfaction it believes may be attained elsewhere. The differences between these levels can be viewed as a measure of "stress" created by the present residential location. (Clark & Cadwallader, 1973:30–31).

In the study of Adelaide by Stimson (1978, 1982) referred to previously, it was found that general levels of *satisfaction/dissatisfaction* of residents with their residential achievements were such that 69% were satisfied, 24% were neither satisfied nor dissatisfied, and only 7% were dissatisfied with their current residence. Thus it would seem that only about one-third of residents suffered from or were highly vulnerable to a stressor. When the sample was broken down into population subgroups, significant differences emerged in level of satisfaction/dissatisfaction. For example, multiple-head households tended to have high levels of dissatisfaction. The number of years people had been resident at their present location tended to influence level of residential satisfaction, the latter increasing with increased period of residence. In general, migrant groups tended to be less satisfied with their residences than the Australian-born population. Owner-occupiers had considerably greater levels of residential satisfaction than did both private rental tenants and public housing tenants, but all groups still had relatively high levels of satisfaction. The higher-status occupation groups and the higher-income groups had greater levels of satisfaction with residential attainment than did the lower-status occupation and lower-income groups, although the differences were not statistically significant. There were large and highly significant differences among the various age groups with level of satisfaction tending to increase up to about 40 years of age, after which it declined somewhat before increasing again for the older age groups.

It would be expected also that the degree of satisfaction with residential location would relate to the spatial and temporal aspects of the search process. In the Adelaide study, for recent movers there was no statistical difference in levels of satisfaction with respect to the number of months spent in the search process and the mean intensity of the search, but there was a statistically significant difference in the level of satisfaction with respect to the mean number of suburbs reached. This tended to peak at around the six suburb level.

Where dissatisfaction with the residential location was evident, and where there had been residential relocation, a stressor had been at work. The results for the Adelaide survey are given in Table 13.4. For example, the most important single reason to emerge was the desire for people to move into their own home after being a renter (25.5%). A change in the location of employment or an increase in the difficulty of access to place of work were important stressors (15.1%). Marriage was a further important factor (13.5%). Dwelling-specific factors, particularly relating to the dwelling being too small, were important stressors (10.2%). Expiration of lease,

TABLE 13.4
Stressors Causing Decision to Relocate in Adelaide

Stressor	Number	%
1. Desire to own permanent residence	155	25.5
2. Change in location of employment	92	15.1
3. Marriage	82	13.5
4. Place of residence only temporary while looking for permanent dwelling	62	10.2
5. Desire for more room or larger house; desire for less room or smaller house	58	9.5
6. Change in size of family	37	6.1
7. Rental lease expired; evicted	28	4.6
8. Rent/rates too high	27	4.4
9. Left parental home	26	4.3
10. Residence too old	15	2.5

Source: Stimson, 1978.

or eviction from rental accommodation, and the cost of rent or mortgage, were relatively insignificant stressors.

The implications of these and similar studies conducted by researchers in numerous cities have important implications for planning in that it may be possible to identify areas of high stress within a city that produce a high degree of desired mobility. Such identification could help show how stress may be overcome to avoid problems such as urban decay or decline and to better design new suburbs in cities. Overall, a high importance was attached to home ownership, home–work relationships, dwelling-specific factors influenced by changes in the status and size of the family and household, and the expiration of rental lease arrangements. These factors readily manifest themselves as stressors, triggering the residential relocation process.

13.5   A Residential Location Decision Process Model

Behavioral theories and models of the residential location decision process have emphasized the existence of multiple goals among subgroups of the population seeking a residential location, the existence of imperfect information, and the use of nonoptimal decision logic. And they have recognized the existence of constraints (economic and institutional) within which choice is made. The decision maker has been viewed as having a goal of *satisfaction*, and behavior has been seen as inherently rational. As Wolpert (1965:161) states: "Man is limited to a finite ability to perceive, calculate and predict and to an otherwise imperfect knowledge of the environment, [but] he still differentiates between alternative courses of action according to their relative utility or expected utility."

The decision maker realizes the limitations of his or her knowledge, and alternatives that minimize uncertainty will be preferred, so that there is a tendency to postpone decisions and rely on feedback information. This incorporates what Simon (1957) referred to as "intentionally rational behavior."

The decision maker is viewed as being satisfied when there is some measure of concordance between what we have termed achievements and aspirations. *Aspirations* are effective (i.e., they do not include wild dreams) and they relate only to what the decision maker perceives as being possible (i.e., feasible). Thus, if a decision maker pursues a goal of satisfaction in an intendedly rational manner, it is valid to assume that the *state of equilibrium* between achievement level and aspiration level is predictive of future behavior. Brown and Moore (1970:1) operationalized the concept of *place utility*, which essentially measures an individual's level of satisfaction or dissatisfaction with respect to a given residential location.

Figure 13.7 is a model encompassing the above postulates; it incorporates the phases that Brown and Moore describe as the *intraurban migration process.*

Initially the decision maker is at a satisfactory location because the current dwelling/location combination tolerably meets needs, that is, there is congruence between achievement level and aspirations. This satisfactory state turns to one of dissatisfaction through the appearance of a *stressor* as discussed earlier.

If a stressor leads to perception of *dissatisfaction*, then the first decision response will have been made, and this may take one of three forms:

1. To *modify aspirations* and adjust *in situ* leading to a decision to stay.
2. To *modify achievement in situ* leading to a decision to stay.
3. To *start looking for a new location* because of a feeling that only by moving will the decision maker restore *equilibrium* between *achievement* and *aspirations.*

If alternative (3) is followed, a tentative decision to *relocate* is taken; dwelling and location-specific requirements (i.e., aspirations) are formulated; and the *search process* begins. This involves *information gathering* during which time the *achievement region* is conceptualized. Brown and Moore (1970:6) refer to a number of crucial elements that may be identified in a search situation, namely:

1. Information available to the searcher that will be biased because of selectivity.
2. Information possessed initially by the searcher that will foster more awareness of space in some locations than others.
3. The manner in which a searcher utilizes information, which determines the way aspirations are built up, perception of the degree of congruence between achievement and aspirations, and will affect the probability of success in finding a suitable location.
4. Time is a stress factor encouraging search behavior and exists as a variable in the learning process.

Following the gathering of information, there is a *review of aspirations* based on what has been *learned*. At this stage, the decision maker has one of three choices:

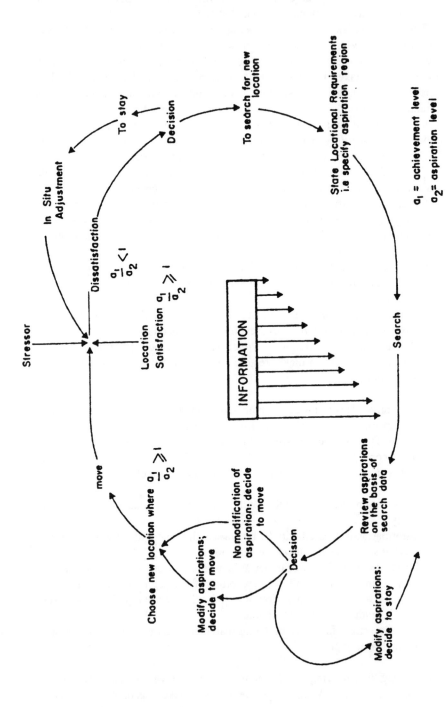

Figure 13.7. The residential location decision process: a behavioral model. *Source:* Brown & Moore, 1970.

a_1 = achievement level
a_2 = aspiration level

- To stay put.
- To modify his aspirations but decide to move.
- To move without modifying his aspirations.

A *mover* will be distinguished from a *stayer* by the congruence or a greater overlap between *achievements* and *aspirations*, whereas a stayer needs to review his or her *aspirations* to bring them into *congruence* with his or her *achievements*. Those who decide to move choose a new location on the basis of their *search* data and with reference to their *aspirations*. The move is made, and the cycle is completed, that is, *achievements* are back on *congruence* with *aspirations*.

It is evident that this approach to studying the residential migration decision process identified three important phenomena:

- The *aspirations* of people, including consumer preferences for housing and urban locational environments and achievement levels.
- The nature of *stressors*.
- *Search* behavior.

In this model, intraurban residential relocation is seen as being dependent upon the ways in which the decision maker uses environmental information and how he or she reacts to stressors in searching space in choosing between locational alternatives that have place utility. Thus the process is linked in an important way to the cognitive mapping process and the cognitive maps of individuals. As Brown and Moore (1970:1) explain: "If the place utility of the present residential site diverges sufficiently from his immediate needs, the individual will consider seeking a new location. The resulting search for and evaluation of dwelling opportunities takes place within the confines of the intended migrant's action space." In this context *place utility* is a measure of the overall attractiveness or unattractiveness of the location, relative to alternative locations, as perceived by the decision maker.

13.6 The Nature of Residential Search and Choice

For over three decades now much of the behavioral research on residential relocation has focused on the search and choice process. Clark (1993:228–229) has proposed a general model that structures *search* and *choice* in residential location as a series of dynamic processes, as set out in Figure 13.8. The residential relocation process is divided into:

1. A decision to move, that is, the appearance of stressors that reflect the equilibrium between aspirations and achievements.
2. A decision to view alternatives, that is, to search.
3. The actual relocation behavior, that is, the choice of outcome.

Because of the complexity of the search and choice process, modeling and empirical investigations have tended to focus on one of these three components of

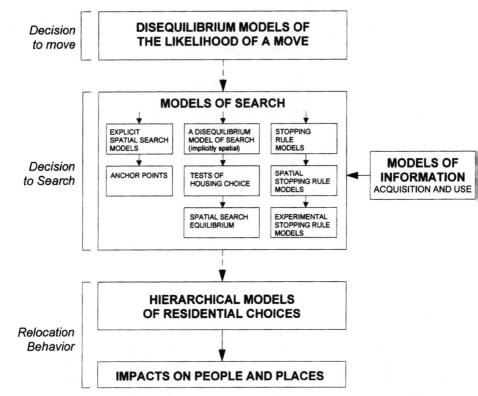

Figure 13.8. Modeling the dynamics of search and choice. *Source:* Clark, 1993:229.

this process, so that it is better to think of the decision to move and the choice itself as the structure with the search process imbedded within overall mobility decision making.

13.6.1   Spatial Models of Search

Much of the early work on residential search and choice had a root either in economics or in psychology.

The *economic* tradition focused on models about the trade-off between accessibility to work and distance. Also, economic models addressed the trade-off between continuing search and incurring the associated costs and stopping search. Work by Clark and Flowerdew (1982) emphasized the influence of information and the way in which uncertainty and the flow of information influenced decisions to continue searching. Phipps and Laverty (1983) emphasized the nature of optimism in search strategies within utility-maximization frameworks; and Ionnides (1979) sought to frame housing search in this context.

The *psychology* literature was more concerned with the choice among alternatives, emphasizing how alternatives are evaluated rather than how they are collected and identified, with the emphasis being on individual judgments (Slovic, Fischoff, & Lichtenstein, 1977).

Behavioral geographers have built on these approaches. Building on the economic approach, they developed both work on the use of stopping rules (Flowerdew, 1978) and disequilibrium approaches to residential mobility in general. Smith et al. (1979) and Clark and Smith (1979), and Smith and Clark (1980) proposed a decision-making and search model outlined in Figure 13.9. In these approaches, *location* is simply one attribute of the alternatives searched, with the expected utility attached to the search for a dwelling in a particular neighborhood being based on the household's prior estimates of the housing characteristics in the area and the expected costs of search in that area. *Search* is seen as a *utility-equilibrium process* emphasizing that an individual's decision to search is a function of the difference between the expected utility of further search and the utility of the best vacancy found to date (Smith & Clark, 1980). It is proposed that households compute the *locational stress* of neighborhoods in a city and search in the neighborhood where the likelihood of success is greatest. A vacancy with greater utility than the current house (or vacancy) becomes the best alternative, and search continues until stress is driven to nonpositive values in all neighborhoods. Clark and Smith tested this model in a section of Los Angeles.

Building on the early work of Adams (1969) on *sectoral bias* in search behavior, and on work by Brown and Holmes (1971) that showed how search emerged and then became more organized as a result of information feedback, J. O. Huff (1986) developed two explicitly spatial models focusing on the nature of selection of areas and in relationship to search constraints. This is demonstrated in Figure 13.10.

1. The *area-based search model* is the outcome of considering the search process as one in which households make decisions to visit a particular vacancy conditional on the area containing that vacancy. Thus areas with large numbers of vacancies have a greater chance of being included in the search space. But the prior search experience of households also can influence the likelihood of an area being visited. Selection of an area is followed by more intensive activity in that area with high probabilities of existing vacancies near the last vacancy visited. The model was tested in an area of Los Angeles by J. O. Huff (1986:217), who concluded that it "provided strong behavioral evidence of the existence of geographically defined submarkets that limit the domain of search for individual households."

2. The *anchor point search model* similarly assumes that the observed search pattern reflects the underlying distribution of vacancies; but in this case the *bias* in the spatial search pattern is related to locational preferences of a household and specifically to the way a household concentrates search in a small area around a focal point. *Search intensity* declines around that point. The empirical basis of this model suggests that a subset of households undertake a locationally hesitant search strategy by concentrating effort near the mean center of the search pattern. This anchor point model matched actual behaviors for over 50% of the Los Angeles sample.

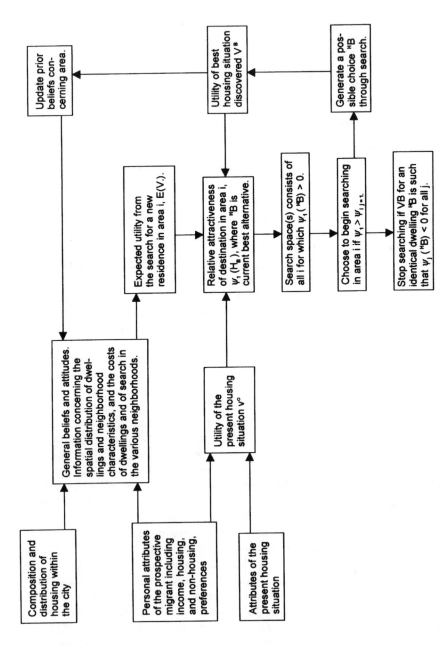

Figure 13.9. Schematic diagram of a decision-making and search model. *Source:* Smith et al., 1979:15.

Area based search

Anchor point based search

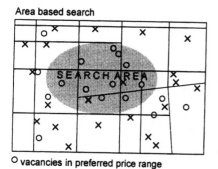

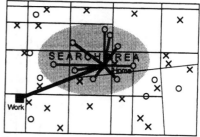

○ vacancies in preferred price range

○ vacancies in preferred price range

Figure 13.10. The spatial context of search. *Source:* Clark, 1993.

These area-based and anchor point models provide substantial evidence to suggest that there are strong spatial regularities in residential search, and they demonstrate that there is a basis on which to build operational models of residential search in order to predict where households will search and where they will be likely to move.

Importantly, these models of spatial search recognize the variations that exist in urban *housing markets*, which are well known to be differentiated spatially both by tenure and by price. They underpin the creation of residential environments, and their existence is reflected in the search and choice process of individual households (Clark, 1993:305). It is evident that households do not search in a vacuum but that they are able to segment the housing market by these important variables which then enter the search and choice process. Bourne (1981) suggests that, at the neighborhood level, the distinguishing feature of the housing market is that of limited spatial entity, which is difficult to define.

Empirical studies have demonstrated how low-income and minority households in the United States have more limited spatial search patterns than do higher-income households. Deurloo, Clark, and Dieleman (1990) have shown in the Netherlands how households relocate to residential environments very similar to those in which they originate. A study of recent movers in Adelaide, Australia, by Stimson (1982) demonstrates the clear directional bias and the generally short distance of residential relocations, with renters in particular tending to move within the same or to a nearby neighborhood, as shown in Figure 13.11.

13.6.2 The Length of Search

The *time dimension* obviously is an important component of the residential search and choice process because it affects:

- The length of search.
- How many alternatives does a household view.
- How long do households search before making a decision.
- What rules do households use to make the decision.

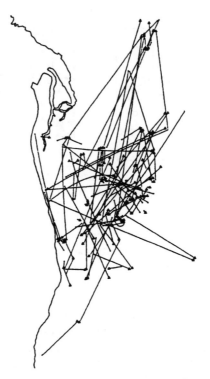

Figure 13.11. Spatial patterns of recent movers in Adelaide. *Source:* Stimson, 1978.

It is established that housing search is a relatively short process. This is demonstrated in numerous empirical studies in a variety of cities and states, including Toronto (Barrett, 1973), Connecticut (Hempel, 1969), Los Angeles (Clark, 1982), and Adelaide in Australia (Stimson, 1982).

The length of search and the number of units searched are closely related. An index of *search intensity*—measuring the number of houses visited, neighborhoods searched—and measures of the *distance* of search and the *length* of search all confirm the limited time and the narrow spatial context of search in the residential relocation process. For example, Figure 13.12 illustrates the intense nature of search in Los Angles identified by J. O. Huff (1986).

Both economic- and psychological-based hypotheses have been proposed to explain the relatively short duration of the search process. For example, Smith et al. (1979) suggests that *satisficing* behavior by households means that they are *risk averse* in selecting alternative houses, stopping search when they have found a housing unit that satisfies certain minimal conditions. Also, the *cost of search* is a constraint, and this is particularly important with respect to renters who tend to conduct very short searches. J. O. Huff (1984) proposes that, given the known costs of search and the expected utility or net gain from search, there is a minimum net gain such that any vacancy found that is greater than or equal to it will be selected, even though further search might uncover a better alternative. If the gains are smaller and

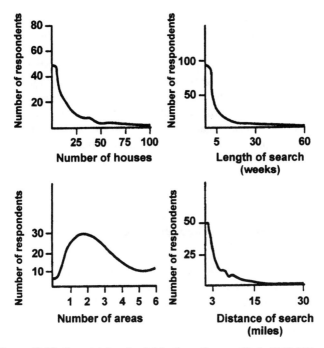

Figure 13.12. Search intensity distributions. *Source:* Clark, 1993:307.

if the distribution of the variance of the net gains over all vacancies or areas in the domain of the search is low, then the search period will be short. It is well established that the size of a household's choice set is likely to be a fraction of the amount of variance existing in the population of alternatives (Michelson, 1979; Meyer, 1980). With respect to low-income households, their search time is constrained highly by both income, which restricts the parts of the housing market that are feasible for individuals or households to search, and by the transport costs involved in searching, which restrains the time spent (Cronin, 1982).

13.6.3   Information in Search and Choice

Researchers have shown how the availability, amount, and quality of information influence directly the nature of the choice process. Thus residential relocation choice will be influenced by the level of detailed information and relative house prices of alternative visits at alternative locations and the use made of *information gatekeepers* or *intermediaries*, such as real estate agents, newspapers, relatives, or friends, and other sources. Long ago Rossi (1955) showed that lower-income households relied more heavily on personalized information than did higher-income households. Palm (1976a,1976b) has demonstrated that real estate agents provide spatially biased information to searchers. MacLennon and Wood (1982) have shown that information sources changed over time. Personal knowledge and use of friends or relatives declines as the time of search increases, while the role of newspapers increases with time. And in the later stages of search, the role of aural information becomes replaced by specific data on housing availability.

Clark and Smith (1979) used simulation methods in a study in Los Angeles to demonstrate that searchers are sensitive to the cost of information, with their search efficiency being significantly less with high-cost information. Good real estate agents speed up the search process—not an unexpected finding. They found also that whereas both owners and real estate agents provide price information, in general owners provide much more exact locational information. There was some evidence to show that information provision changes in stable and price-escalating markets.

13.7 Relating Residential Relocation Behavior to the Urban Structure

Much of the work of behavioral geographers on residential relocation has focused more on *why* people move and less on the impacts of mobility on the structure of urban areas. As Clark (1993) notes, there is a need to know much more about the complex interconnections between the urban structure and human behavior:

> It was no longer enough to show that people often moved short distances in neighborhoods that they were familiar with, that they moved to bring their housing consumption nearer to equilibrium consumption, that younger households moved more often or that tenure was important (that renters moved more often and owners less often). It was important to relate mobility to the built environment and to ask questions about the interaction of search, choice, and impacts both on the changing neighborhood of cities and, by extension, on the population composition of neighborhoods. (p. 300)

The result has been a bipolar approach to our understanding of residential relocation behavior. Questions about why people move have become replaced either by studies of the sociopsychological process of decision making or studies of the geographic outcomes of those decisions.

In Chapters 8 and 9 we saw how the study of *activity patterns* provides a means of investigating the relationship between environment and behavior. The same may be said about the study of search patterns and household residential choices as providing a link between the built environment, neighborhoods of the city, and the people who live in them. As Clark (1993) says,

> It also allows us to raise important questions about how housing opportunities are known and used, and on the spatial extent of peoples' experiences of the built environment. Even more important, the search and relocation behaviors raise questions about the extent to which choices are exercised freely or are constrained by household factors and factors of the market. (p. 301)

Increasingly researchers have adopted the view that the actions of households are grounded in the structure of the *housing market*. Individual households are constrained by the availability of housing by tenure, price, and location, and there is considerable evidence to show that the same processes are at work in both con- trolled and uncontrolled markets (Dieleman, Deurloo, & Clark, 1989). Although there is considerable debate about the issue of choice and constraint in housing mar- kets, the market is a major factor in choice in addition to individual household char- acteristics and circumstances. Clark (1993) has suggests that

> A better understanding of locational preferences and how those preferences are formed and how they influence choices in the housing market are an important part of understanding how the urban structure will evolve. Current interest in urban restruc- turing, in poly-centric urban form and the spatial organization of a pluralistic urban form will be better served by a theoretically informed understanding of residential search and choice. (p. 301)

Research on residential search and choice may be fitted into a developing theme that has been proposed as a way to view the processes of moving, location, and housing choice. During the last decade, there has been an increasing concern to develop a broader conceptual structure within which the notions of housing choice and residential mobility might be embedded. This has led to a focus on the *housing career* and the *life course* as approaches to understanding the complex interrelated activities of search and choice.

13.7.1 Housing Careers

Housing careers have been proposed by Kendig (1984) as an organizing principle to examine the interaction of housing choices and household family composition, link- ing housing tenure decisions and family life cycle. Kendig suggests that housing ca- reers are the means for linking mobility and the life cycle. This builds on earlier work by Michelson (1979), which suggested that households pass through a series of dwellings that increasingly meet their long-term housing aspirations. The paths of individual households through the housing stock are influenced by broader soci- etal changes—such as increasing incidence of divorce, remarriage, and *de facto* household arrangements, including multiple-family households—as well as life transitions and local housing markets.

13.7.2 Life Course

Life course provides a broader context in which housing careers may be situated. Life course is the concept for viewing the progress of an individual, a household, or groups through their lives (Mayer & Tuma, 1990). The progression through life is regarded as a sequence of events, which include a wide range of occurrences related to education, family formation, career decisions, retirement, and so on. Housing needs and choices interrelate to decisions about family and career. The objective of life-course analysis is to look at individual life events and the pattern of life trajec- tories in the context of social processes that generate these events and trajectories (Clark, 1993:312).

The concept is related to the notion of *migration careers*, which required data derived either from residential histories of households or from a longitudinal framework of data collection. This involves complex and usually expensive survey instruments of data collection, and as a result studies have tended to be small scale in nature focusing on specific groups in the population or on specific geographical areas. However, the more recent availability of census unit record sample data sets permits a more comprehensive analysis of migration and, in particular, an investigation of those segments of the population that are most prone to migrate.

The future of research on the residential relocation decision and choice process is likely to address more explicitly the complexities of the decision-making process in which spatial choice is occurring, both in the city and between and among regions. This will address the "processes whereby households link work and residence, and the processes of mobility, migration, and commuting are used to effect that linkage" (Clark, 1993:312). These are crucial issues in urban planning. Policies that seek to address the relationships among housing consumption, types of dwellings, and the locational preferences of different types of households need to take account of the behavioral basis of the individual decision-making processes that result in the aggregate patterns of mobility within cities.

14

Geography and Special Populations

14.1 Geography and "The Other"

"The Other" are the outcasts of a community. A community consists of a population with a shared moral, political, and social environment, and with membership typically based on geography, race, ethnicity, or language. As Lakshmanan (1994) notes,

> The community is thus a unifying inclusive idea. The members of a community are actors in its history and entitled to be viewed as community [members] as individuals with minds, wills and voices, creating collective self-knowledge. Individuals who do not share the attributes of community membership are excluded from full participation in the community's activities and form "The Other." (p. 1)

Thus, "The Other" is denied the opportunity of fully sharing in community experiences and participating in their activities. They are assumed to be not able or willing to fully understand the thoughts and emotions of the communities from which they are excluded.

Perhaps the most prominent example of the creation of groups of "The Other" can be found in the condition known as the "nation state." In such a case, nonnationals are defined as "The Other." For much of the history of European settlement in North America, for example, those of Caucasian stock, despite their varied nation state of origin, were loosely regarded as community, whereas those of Native American, African American, Asian American heritage, or from other national groups, have typically been regarded as "The Other." Along with those of different national origins, those who differ in religious persuasion (such as Jews or Muslims), and those of different or mixed ethnicities (such as indigenous people, Chicanos, and Japanese or Chinese Americans), at various times have inherited the position of "The Other." In today's developed economies, groups such as women, the disabled, gays and lesbians, or those with terminal diseases such as AIDS, represent the most visible groups of "The Other." But barely scratch the surface, and groups of homeless, the poor, ethnic or religious minorities, and different cultural groups also fall into this category. All constitute "The Other" and suffer the consequent disadvantages of discrimination, disenfranchisement, and disassociation from the majority community.

Things are, however, changing. In the United States, over "the last four decades, the Civil Rights Revolution and a variety of ethnic pride movements demanded greater rights and greater acknowledgment of the contribution of groups literally identified as 'The Other.' Across America, minorities proclaimed their pride and demanded their rights" (Lakshmanan, 1994: 4).

Although present for the greater part of human history, "The Other" have attained their greatest visibility in the latter part of this century. Who are "The Other"? "The Other" are groups who are discriminated against, who face significant physical, psychological, and societal barriers that produce disability. They are the disenfranchised, the maltreated, and the ignored members of society. They are the ethnic, religious, nationalistic, and cultural minorities who are often denied equal opportunity and equal access to the advantages and benefits of the society in which they reside. At various times, these groups have forced society to recognize the disadvantages under which they must live and have obtained concessions in both the legal and moral domains designed to help offset the barriers and problems they face. In the past two decades, Western Marxists have tried to lay the blame for the discriminatory treatment of these groups at the feet of capitalist societies. But "The Other" exist in all societies; capitalist and noncapitalist, Western and non-Western, developed and less-developed, democratic or dictatorial, left-wing or right-wing societies. As one might expect, the nature of differentiation does vary among societies, economies, political structures, ideologies, and environmental domains.

In this chapter we examine some of the problems of those who are sometimes called "The Other," and whom we will call "Special Populations." The specific populations with which we will deal include those suffering movement disability (wheelchair-bound or mobility-constrained individuals), the blind or vision impaired, and people who are mentally challenged or retarded. In Chapter 15, we examine two other special populations, women and the elderly.

14.2 Disability

14.2.1 Definitions of the Disabled

The number of countries passing laws similar to the Americans with Disability Act (1990) is increasing. Still there is considerable difficulty in obtaining data on disability due to a lack of uniform terminology, inconsistent reporting, and a lack of institutional screening of reported statistics. There are few national registries specifically devoted to collecting statistics on disabilities. To help overcome this fundamental problem, the World Health Organization has adopted the following standard definitions:

1. *Impairment*: A lack of function of a bodily part.
2. *Disability*: A physical limitation in performing a specific task in a given society.

3. *Handicap*: A limitation of activity or social function as a result of impairment, which in turn results in some social or psychological disadvantage.
4. *Disadvantaged*: A state of being in which it is difficult to perform the accepted and expected activities typically undertaken in a society because of discrimination, differentiation, lack of equal opportunity, or simply because the social system does not facilitate the constrained behaviors of disadvantaged groups.

All too often "disability" is equated just with the problem of physical mobility. This is but one area of disability, which generally includes:

1. Lack of physical mobility (e.g., requiring assistive tools such as wheelchairs, crutches, walkers, etc.).
2. Vision impairment and blindness.
3. Speech impairment (or language deficiency).
4. Hearing impairment or deafness.
5. Haptic impairment (e.g., touch insensitivity, crippling by diseases such as arthritis).
6. Cognitive impairment and brain damage (e.g., difficulty in handling spatial problems, short-term memory problems, or being dominated by anxiety or stress).
7. Reading impairment (e.g., dyslexia).
8. Phobias (acrophobia, claustrophobia, and so on).
9. Mental challenges (e.g., mild, moderate, and severe retardation).
10. Suffering from debilitating diseases (e.g., cancer, heart diseases).

Members of all these groups are both implicitly and explicitly discriminated against with reference to mobility, movement, employment, sociocultural behavior, education, and information flows.

14.2.2 Distribution of Discriminatory Practices

Disenfranchisement occurs in all societies and economies regardless of their ideological persuasion. This is inevitable considering a basic dictum of most societies that "the good of the many should outweigh the good of the few." Thus doors have been developed with round handles that require turning. This simple fact discriminates against the person with crippling arthritis or the paraplegic or others with similar disabilities. Recognizing that this is so has not prevented us from continuing to build doors with round doorknobs. But some steps to counteract difficulties caused by this simple unintendedly disabling architectural device have been made; most fire doors are equipped with a bar latch that, once depressed, can give access to an emergency exit. Similarly, mass transit facilities have been developed with little regard to the needs of the disabled; but in some countries movements are underway to remedy this problem. For example, on buses special seating near doors is often reserved. Doors open automatically rather than having to be pushed; transformed buses and paratransit facilites provide opportunities for wheelchair riders to use these service; and in light rail or suburban rail services, gaps between cars have been altered so that they no longer can be perceived as entrances by the vision-impaired or blind individuals.

There are, however, many other discriminatory and disenfranchising activities that are not quite as obvious as these. For many years, the blind traveler was unable to take a guide dog on transportation systems or into shops or restaurants. In today's information superhighways, the software is tied to point and click procedures and to a Windows© [Graphical User Interface (GUI)] operating format that currently denies participation by the blind and vision-impaired population and erects barriers as well for those who have haptic disability. Wheelchair users often are located in peripheral places at public gatherings because of fire/safety regulations ("they block the aisles"). Recreational or leisure-time activities such as exploration of natural environments provide little opportunity for the physically disabled or the blind and vision-impaired population. Many deinstitutionalized disabled people have been relegated to the lowest levels of the job market and given little chance for improvement of their economic or social status because of barrier-filled workplace environments; and these barriers are both environmental and societal.

In many countries, disabled people are educated in special classes. Schoolmates, with whom they only interact at recreational and lunch breaks (and often not even then), often deride them and label them as freaks. Although some mainstreaming experiments are being undertaken, this effectively takes the disabled student out of the assistive environment and into harsh reality where he or she faces double jeopardy.

Tables 14.1 and 14.2 present examples of physical and human interaction barriers that are faced by different disabled groups. These summaries indicate that the wheelchair-bound and the blind or vision-impaired population are the two groups that perhaps face the greatest quantity and variety of barriers. We explore these issues further, beginning with the wheelchair-bound group.

14.3 Wheelchair Populations

14.3.1 Environmental Design for Wheelchair Users

Goldsmith (1983) examines the problem of environmental design for the disabled comparing ideological and cultural influences on design in the United States and the United Kingdom and discussing their moral and ethical basis. He argues that the formulation of an environmental design ideology only could have come from the United States because it embodies a *self-help* view on enablement, equal treatment, and idealism. In comparison, the United Kingdom is seen to be dominated by *social welfare* cultural values of compensation, special treatment, and pragmatism.

Environmental design issues represent a therapeutic attack on the environment at a *macro* rather than a *micro* level. They focus on a physical remedy rather than a social one. In 1961 in the United States, *The American Standards Specifications for Developing Barrier-free Environments* bill was passed, thus making buildings and facilities accessible to and usable by the physically disabled. Revised in

TABLE 14.1
Physical Barriers for Disabled Persons

Visually Impaired	Age	Wheelchair	Otherwise Physically Disabled	Phobic
Construction/repair	Stairs	Construction repair	Construction/repair	Stairs
Weather	Door handles	Surface textures	Weather	Lack of rails
Lack of railings	Door latches	Layout	Unprotected natural	Elevator
Ramp availability	Door closures	Gradient	environments	Distance
Elevators	Ramp availability	Design features/utilities	User access	
Distance	Distance	Unprotected natural	Stairs	
Door location	Elevators	features	Lack of rails	
Door handles	Lack of rails	User accessibility	Ramp availability/	
Door latches	Gradient	Weather	curbs	
Door closures	Unprotected natural	Stairs	Elevator	
Nonstandard fixtures	hazards	Lack of rails	Distance	
Traffic hazards	Weather	Ramp availability/curbs	Doors	
Travel access		Elevator	Gradient	
Surface textures		Distance		
Overhead obstructions		Doors—width, handles,		
Lack of barriers		location, closures		
Lack of cues				
Gradient				

Source: Golledge, in preparation.

1980, these standards assume two moral axioms: the goal of *normalization* and *nondiscrimination;* and the value of *independence.* These two assumptions are difficult to implement simultaneously, for sometimes they are in conflict. Within the United Kingdom, enabling and compensatory actions to favor disabled individuals (particularly wheelchair users) included the invalid tricycle (discontinued 1976), the provision of unisex public toilets, curb cuts, and sign plates. The invalid tricycle originally was introduced to help the ambulatory activities of severely disabled people, particularly war veterans. In the 1970s it was criticized as being antisocial and stigmatizing, and was withdrawn by 1976. The unisex public toilet—a device seen to be needed where wives or other helpers of the opposite sex could help disabled husbands (or vice versa)—was encouraged during the 1970s and became a feature of the urban landscape. Strategically located signposts pinpointed the location of the special facilities for disabled people. Signage, however, was not always clear and unambiguous. Curbcuts, while seen as wheelchair enabling, in terms of increasing mobility, were often not developed with respect to a coordinated plan, thus it was sometimes difficult to develop a route by which curbcuts provided easy access to the street. Nor were curbcuts standardized in terms of location, slope, or width. Problems associated with these shortcomings are obvious.

Goldsmith (1983) argues that this was in contrast to the design criteria laid down in the *enabling legislation* in the United States. Goldsmith further evaluates the relative worth of the self-help (United States) and social welfare (United Kingdom) models for guiding environmental design modifications. He argues that the former indicates the amenability to regulatory controls and regulations, whereas the

TABLE 14.2
Human Interaction Barriers for Disabled Persons

Visually Impaired	Hearing Impaired	Age	Learning Disabilities	Retardation	Wheelchair	Otherwise Physically Disabled	Brain Damage
Signs	Travel access	Traffic hazards	Traffic hazards	Traffic hazards	Traffic/transit	Design/utilities	Traffic hazards
Layout	Auditory cues	Signs	Signs	Signs	Signs/symbols/icons	Traffic hazard	Signs
Unprotected natural hazards (bodies of water, cliffs, etc.)		Layout		Traffic access	Travel	Travel	Travel
						Layout	Lack of cues

Source: Golledge, in preparation

TABLE 14.3
Comparison of Self-Help and Social Welfare Ethics

Self-help	Social Welfare
1. Enablement	Compensation
2. Independence	Welfare services
3. Self-reliance	Mutual supportiveness
4. Equal treatment (normalization)	Special treatment/discrimination
5. Idealism	Pragmatism
6. Indoctrinality	Gradualism and incrementalism

Source: Goldsmith, 1983:200.

latter shows a preference for education and exportation. The conclusion was that the self-help culture has a more practical approach than the social welfare culture. The comparisons are indicated in Table 14.3.

14.3.2   Enabling Legislation

Relevant legislation is detailed in Table 14.4. Examples include the United Kingdom's *Chronically Sick and Disabled Persons Act* of 1970, the *Americans With Disability Act* of 1990 (United States), the *Building Standards for the Handicapped Act* of 1965 (Canada), and in Australia the *Design for Access by Handicapped Persons Act* of 1968. Note that the *American Standards Act* of 1961 significantly impacted all these consequent legislations.

Despite their societal context, all of these acts emphasize that the number one enemy of disabled people is inaccessibility and that this impinges directly on independence. The basic argument is that rehabilitation techniques have limited potential for success because eventually they have to project client's into societally determined systems of education, recreation, socialization, and employment. All of these are replete with architectural or other design-based barriers.

TABLE 14.4
Disability Legislative Acts

1. American Standard Specifications for Making Buildings and Facilities Accessible to and used by the Physically Handicapped—U.S. Congress, 1961 (Revised 1980).
2. Canadian Building Standards for the Handicapped, 1965.
3. Access for the Disabled to Buildings, United Kingdom, 1967.
4. Design for Access for Handicapped Persons, Australia, 1968.
5. Regulations for Access for the Disabled to Buildings, Sweden, 1969.
6. Architectural and Transportation Barriers Compliance Board, United States, 1981.
7. The Chronically Sick and Disabled Persons Act, United Kingdom, 1970.
8. The Disabled Persons Act, 1981.
9. Amendment to the Urban Mass Transportation Act, United States, 1970.
10. Americans with Disability Act, 1990.

14.3.3 Barrier-Free Environments

The ethic of the barrier-free movement manifested itself in the *American Standards Act* of 1961 and focused on the *macro* environment. Traditional emphasis had been on the micro environment including medical aid, welfare support, derived employment, special education, assistive housing and transportation, and assistive technological equipment. This effort was concentrated on individual disabled people. Individual profiles of needs were first evaluated with respect to developing a self-help program and environment in the home, the workplace, and in transport between them. The profiles were developed by special rehabilitation advisory groups. After review by rehabilitation councils (generally state or federal appointed), some portion of the recommended aids was allocated to the individual and training sessions set up to help them become familiar with assistive technology. The transfer of interest from the individual (or *micro*) environment to the physical design of the *macro* environment was a major change of direction. It changed the intervention process from one focusing on personal and social interventions to a process that emphasized physical environmental design changes. The consequent development of macro features included automatically opening doors, telecommunication devices for the deaf (e.g., TV programs closed captioned for the hearing-impaired viewer), and electronic obstacle avoiders for the blind.

All were emphasized with the aim of helping a disabled person to function independently in the everyday environment. Examples of design changes within buildings and in the external built environment include: ensuring that elevator controls can be reached by those in wheelchairs; establishing signs at eye level for the wheelchair-bound person, lowering the height at which water fountains operated or window locks could be found; reducing the height of work surfaces within the home (e.g., in the kitchen) and in the office (e.g., computer workstations); providing lowered washbasins in public toilets and washrooms; providing wider doors; changing doorknobs to bar levers to facilitate opening; providing wider parking areas for disabled people; providing wheelchair ramps as an alternative to steps or stairs to enter buildings or to move between floors where elevators were not available.

14.3.4 Cross-Disability Problems

The emphasis of much of the above-mentioned legislation and environmental design interventions has been dominated by the needs of the wheelchair brigade. Many of the mandated environmental changes do not help those with other ambulatory disabilities (e.g., those with crutches, walkers, walking sticks, etc.; those with vision-impairment or blindness, hearing impairment or deafness; or those with disabilities of coordination).

For the person dependent on a wheelchair, there are still frequent encounters with environmental problems such as steps instead of ramps, narrow doorways, and small elevator cars. These are difficulties not usually encountered by people with other disabilities, and the architectural efforts to intervene in the environment to alleviate problems for the wheelchair user have done little to help relieve the problems of others. In fact, curbcuts can be a major hazard for the blind and visually impaired if they are cut gradually so that the blind traveler cannot sense their change in slope. The result is an increased likelihood of walking into a roadway and exposing oneself unnecessarily to dangers associated with the stream of passing traffic, whether it is vehicular, bicycle, or another form. Some efforts are being made to compensate for this problem by installing tactile tiles at the beginning of curbcuts or cutting grooves so that there is a change in surface texture that can be picked up by a blind or vision-impaired traveler using a cane. It should be noticed that when comparing a wheelchair-bound person to those with other disabilities, such an individual is least-like the population at large in terms of environmental needs. In other words, those with sight, touch, hearing, or developmental disabilities can operate within an environment in a manner very similar to that of their nondisabled peers. The wheelchair user, however, constantly faces many barriers that are not seen as such by other disabled groups. There appear to be differences in the needs of the ambulant and chair-bound disabled populations, and these differences can produce conflicting design modifications.

In the United States, the legislative base for barrier-free transportation is the 1970 amendment to the *Urban Mass Transportation Act*. This states it is "the national policy that elderly and handicapped persons have the same right as other persons to utilize mass transportation facilities and services" (Cannon & Rainbow, 1981, (as quoted in Goldsmith, 1983:208). This nondiscriminatory principle was incorporated into the Washington Metro and the Bay Area Rapid Transit (BART) system in the San Francisco area. The United Kingdom has not had the same doctrinal stance with respect to the design of mass transportation systems. In Australia in 1994, however, supporters of those with disabilities managed to hold up the distribution of 200 new buses to state transportation agencies because they failed to incorporate facilities for wheelchair people.

14.3.5 Transportation

Considerable advances have been made in the private transportation sector. Previously nondriving disabled people with feet or leg impairments can now get custommade vehicles in which all of the necessary controls can be accessed by the hands. This is a definite advantage for some nonambulatory people, restoring to them the freedom and independence of travel in the macro environment that is partially denied them because of movement limitations.

14.3.6 *The Americans With Disabilities Act* (1990)

The *Americans With Disabilities Act* (ADA) is a federal antidiscrimination and equal opportunity act brought into law in January, 1990. This legislation prohibits discrimination against people with disabilities in employment, public service, public accommodations, education, and telecommunications. It states emphatically that no qualified person with a disability can be denied participation in a job. It clearly is not an affirmative action law but, rather, an antidiscrimination and equal opportunity law. A reoccurring theme is one of accessibility, whether it be in the macro environment or accessibility to programs such as are provided in educational services. ADA offers the following definitions:

1. A *physical impairment* includes any physiological condition, cosmetic disfiguration, or anatomical loss or damage: it includes, but is not limited to, heart disease, HIV, cancer, diabetes, hearing, motor or vision impairment, paraplegia, and others.
2. A *mental impairment* includes cognitive (e.g., learning disabilities) and psychological disabilities.
3. The *length and severity* of incapacity and its long-term impact may all be used as criteria to determine disability status.

As well as evidence of impairment, there must be a record of such an impairment.

ADA (*Public Law* 101-336) is the most sweeping civil rights legislation since the signing of the *Civil Rights Act* of 1964. Signed into law on July 26, 1990 by President George Bush, its formal purpose is to eliminate discrimination against people with disabilities.

14.3.7   Assistive Technology

For those who are ambulatory disabled and wheelchair confined, there are a number of tools and devices that have been developed to help assist movement and other life-sustaining activities. These include the following: wheelchairs (manual or motorized); motorized tricycles (custom-built cars); crutches; canes; walkers; prosthetics (limbs or joints); flatbed rollers; paratransit; modified or refitted vehicles; reachers and grabbers for obtaining things from shelves; elevators and escalators; stationary and moving ramps; moving sidewalks; curbcuts; remote controls; telecommuting; and so on.

The major aim of these *assistive devices* is to allow the disabled wheelchair-bound person to participate in as many everyday activities as possible including employment, other economic and social activities, shopping, recreation, socializing, education, communication, and so on. Some of these are oriented toward *macro* environmental modifications (e.g., moving sidewalks and curbcuts), whereas others are oriented more to the *micro* or individual level (e.g., customized vehicles, prosthesis, and manual or motorized wheelchairs).

None of these devices alone provides the opportunity for full reintegration of the disabled person into the everyday functioning of society. However, they all contribute to the improvement in the quality of life, to enabling disabled individuals to perform activities and participate in functions otherwise denied to them, and provide compensatory alternatives for activities that would otherwise require the use of limbs (Table 14.5). There appears no doubt that a combination of effort at both the macro and micro level is needed to provide the confidence and independence of action for this and other populations currently denied them because of their disability.

TABLE 14.5
Samples of Assistive Devices for Disabled People

Vision	Physical/ambulatory (Continued)
Long cane	Paratransit
Guide dog	Elevator/escalator
Human assistant	Modified/refitted vehicles
Laser cane	Environmental modification
Mowat Sensor	Stationary/moving ramps
Sonic guide	Moving sidewalks
Nottingham Obstacle Sensor	Curb cuts
Opticon (OCR)	Synthesized speech
Kurzweil Reading Machine	Voice operated computers
Voice synthesizers	
NOMAD	Touch
Tactile maps/arrays	Digitized/synthesized speech
PGS	Tape recorders
Braille	Video
Voice operated computers	Operating lights/sound
Talking signs	Remotes
Auditory beacons	
Electronic strips	Speech
Motion detectors	Sign language
Pressure detectors	Computer communication
Bar code readers	Written aids
Beacons	Artificial larynx
Vision enhancing devices	Prerecorded speech
Infrared detectors	Nonspeech sounds
	Music
Hearing	
Hearing aids	Reading
Sign language	Scanning devices
Close captioning	Digitized/synthesized speech
TDD	Kurzweil Reader
Flasher warning systems	Opticon
Hearing enhancing devices (parabolic mikes)	Talking signs/maps
	Video/TV/radio (mass communication)
Physical/ambulatory	Icons/earcons
Wheelchair (manual/motorized)	People readers
Crutches	Glasses/contacts
Canes	Inverting lenses
Walkers	
Prosthetics (limbs/joints)	

14.4 Blindness and Vision Impairment

14.4.1 Spatial Abilities With and Without Vision

Byrne and Salter (1983) have compared congenitally totally blind subjects with a group of sighted subjects matched for age and sex with respect to their ability to estimate distances and directions among familiar locations in an urban neighborhood. They found that a sense of direction was greatest for the blind subjects when the home was designated as the origin point; when other points were designated as the origin, increasing error crept into directional estimates. Both groups, however, appeared to perform well on distance estimation.

Herman, Chatman, et al. (1983) examined the spatial ability of sighted, blindfolded sighted, and congenitally blind subjects in terms of walking through an unfamiliar large-scale space in which target locations were occluded. Individuals were taken to each target location and asked the position of others. The results showed that the groups that had past visual experience did better in terms of acquiring spatial knowledge.

Many people have never observed a blind person navigating through an area with only a long cane as a wayfinding aid. They are, therefore, unfamiliar with the types of obstacles, physical and otherwise, the traveler must face and overcome on a daily basis. At the beginning of this chapter, we discussed some of the physical obstacles and societal constraints that act as barriers to the blind. Often these are easily circumnavigated by a sighted person, and it is outside the realm of that person's experience to consider that they may constitute a barrier. Awareness of barriers is often reduced because of the lack of experience with disabled people in one's usual behavior setting. But why should this be so?

14.4.2   The Spatial Distribution of Blindness

According to the World Organization Program for Prevention of Blindness, 27 to 35 million persons have binocular vision of 0.05 or below, and 41 to 52 million have vision of 0.1 or below. Given a world population of approximately 6 billion, somewhere between 15 and 20% are said to suffer a visual disability. Thus, vision impairment or blindness must be considered a significant disability at any scale. In addition, more than 90% of the world's blind population lives in developing countries.

The rate of blindness increases greatly as age increases (Table 14.6). The most frequent reason for blindness are cataracts. Insufficient numbers and uneven spatial distributions of eye care services in developing countries have resulted in an inability to deal with the cataract problem. In West Africa and Central America alone, it is estimated that 85 million people are at risk and face potential blindness.

There are approximately 1.8 billion children below 15 years of age in the world of whom 1.5 million are blind (Table 14.7). A major cause of this impairment is vitamin A deficiency. Again, this problem is particularly prominent in developing countries. On the whole, blindness is more prone to occur in premature or low-birth-weight babies than in others.

In developed countries, diabetic retinopathy is the most important cause of adult blindness. Other significant causes are glaucoma and macular degeneration. Nakajima (1991) assumes that there are about 30 million blind people in the world; 17 million cases are due to cataract, 6 million due to trachoma, 3 million to glaucoma, 0.5 million to xerophtalmia, and 3.5 million to other causes.

There is a direct positive relationship between size of population in an area and blindness (Figure 14.1). In the United States, for example, state statistics show a high linear relationship, except for a few outliers (such as Florida) where the proportion of elderly retired persons is much larger than it is in other states. Table 14.8

TABLE 14.6
Prevalence of Severe Vision Impairment

(a) Age Distribution, United States, 1989*			
Age Range	Population	Impairment per 1000	Number Severe Vision Impairment
0–24	90,428,000	0.528	47,746
25–34	43,835,000	1.23	53,917
35–44	36,503,000	1.68	61,352
45–54	25,897,000	4.8	119,505
55–64	21,593,000	7.8	168,425
65–74	18,182,000	47.0	854,554
75–84	9,761,000	99.0	966,339
85 +	3,042,000	250.0	760,500

(b) Age Statistics from Great Britain†		
	Number (thousands)	Proportion of age group (%)
16–19	76	2.1
20–29	264	3.1
30–39	342	4.4
40–49	453	7.0
50–59	793	13.3
60–69	1,334	24.0
70–79	1,687	40.8
80 +	1,254	71.4

*Source: Unpublished Data, Department of Commerce, Bureau of Census. Prevalence based on National Center for Health Statistics, Supplement on Aging, 1984.
†Source: OPCS survey of disabled adults.

TABLE 14.7
Estimates of Blindness in Children (All Causes) by Region

Region	Persons aged 0–15 years (millions)	Estimated Cases of Blindness (n)	Prevalence (n/1,000)
North America, USSR, Europe, Japan, Oceania	240	72,000	0.3
Asia	1,200	1,080,000	0.9
Africa	240	264,000	1.1
Latin America	130	78,000	0.6
Total	1,810	1,494,000	0.8

Global estimates of blindness in children, 1989. From World Health Organization: Prevention of Childhood Blindness 1991. The report of the symposium on childhood blindness, held in London, June, 1990. Source: Nakajima, 1991.

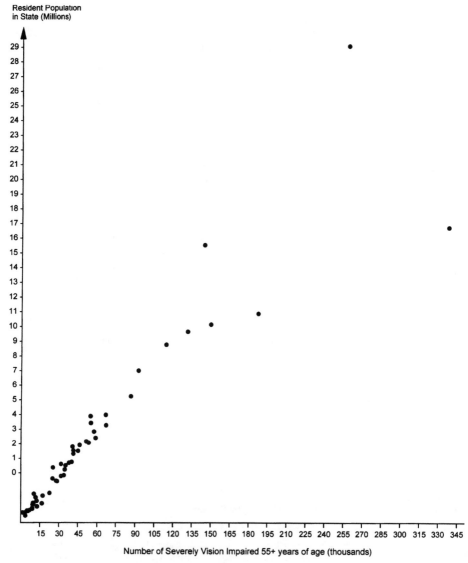

Figure 14.1. Scatterplot of association between population size and severe vision impairment 55+ years, by state, United States, 1989.

shows the distribution of people with a work disability by states in the United States. Note the somewhat higher proportion in most Southern states. Blindness does not appear to affect one sex more so than another across all age groups (Adamsons & Taylor, 1990).

Institutions such as the World Health Organization (WHO) of the United Nations and the World Blind Union try to collect world-wide information on blindness, its causes, its distribution across age groups, its appearance in the sexes, and its ap-

TABLE 14.8
People With a Work Disability, United States, by State

State	Number	%	State	Number	%
Alabama	256,907	10.6	Alaska	14,798	5.4
Arizona	155,114	9.1	Arkansas	175,668	12.7
California	1,279,189	8.2	Colorado	139,413	7.2
Connecticut	131,563	6.5	Delaware	30,785	7.9
District of Columbia	43,259	9.9	Florida	594,133	9.9
Georgia	360,534	10.4	Hawaii	38,181	5.9
Idaho	50,303	8.7	Illinois	529,724	7.3
Indiana	277,282	8.0	Iowa	130,044	7.2
Kansas	112,640	7.6	Kentucky	260,983	11.4
Louisiana	250,218	9.6	Maine	68,258	9.7
Maryland	223,903	8.0	Massachusetts	270,537	7.3
Michigan	548,782	9.3	Minnesota	180,577	7.0
Mississippi	117,434	11.8	Missouri	280,333	9.1
Montana	40,515	8.1	Nebraska	67,958	7.0
Nevada	42,509	7.8	New Hampshire	44,362	7.5
New Jersey	327,094	6.9	New Mexico	67,013	8.2
New York	865,589	7.7	North Carolina	371,231	9.7
North Dakota	26,955	6.7	Ohio	606,718	8.8
Oklahoma	203,213	10.8	Oregon	166,103	9.9
Pennsylvania	642,761	8.5	Rhode Island	52,445	8.6
South Carolina	196,202	9.8	South Dakota	31,585	7.6
Tennessee	303,421	10.4	Texas	688,618	7.6
Utah	64,862	7.5	Vermont	27,712	8.5
Virginia	298,695	8.4	Washington	235,684	8.8
West Virginia	149,815	12.3	Wisconsin	199,407	6.8
Wyoming	18,522	6.1			

Source: U.S. Census and Disabled Adults Survey: 1980 Census.

pearance by country and region. But it is widely accepted that it is impossible to say how many blind or vision-impaired people there are in the world today. Reasoned estimates rather than exact counts are the prime source of information, although most of these estimates are extremely conservative. One reason for widely varying estimates lies in the accepted definitions of blindness. Conservative criteria would argue that 3/60 vision or less in the better eye represents a reasonable definition of blindness. Given such criteria, WHO estimates over 30 million blind persons in 1984 (Watson, 1990:245). In using simple extrapolations, it is estimated that there will be 40 million blind people by the year 2000. It is estimated further that approximately 9 out of every 10 blind persons live in a developing country and that as many as 7 out of every 10 of those persons had a blindness that could have been prevented or cured by treatment.

Within the United States, statistics provided by the Braille Institute (personal communication, 1994) show that there are 13 million blind, vision-impaired, or low-vision people; that 90% of legally blind people have some residual vision; that 70% of severely vision-impaired persons are age 65 years or older; and that 50% of severely vision-impaired individuals can be classified as legally blind. Severe vision impairment is defined as being unable to read an ordinary newspaper even with glasses or other corrective lenses, and legal blindness is defined as having a vision acuity of 20/200 or less in the better eye with corrective lenses, or a central vision acuity of more than 20/200 if the peripheral field is restricted to a diameter of 20° or less.

A conservative estimate thus indicates that 2.5 people out of every 1,000 in the United States are legally blind and that 7.2 people out of every 1,000 are severely vision impaired. By rough extrapolation, this provides a conservative estimate of approximately 2.5 to 3 million legally blind or severely vision-impaired people in the United States. Blindness is third only to cancer and AIDS as the biggest health fear expressed by the general public.

Given the relatively significant number of blind and vision-impaired people throughout the world and their concentration in developing economies, it is obvious that an ethical problem of some magnitude exists. In developed economies, considerable financial, personal, and societal resources are expended to deal with blind and vision-impaired populations. These amounts are increasing as populations age and are more subject to infirmaty or other disabling problems. If we are at all concerned with the quality of life of such populations, we must be aware of the community impact on the state of individual and group welfare and well-being of these partly disenfranchised and discriminated groups. As spatially aware professionals, geographers have an expertise that can be used to examine problems and processes of activity and interaction that take place between disabled populations and their physical, built, and societal environments.

14.5   Design of Environments for Those Without Vision: A Case Study

We have seen how in many societies there is evidence of efforts to compensate for different types of disability (e.g., the modification of the design of the physical environment for those who are wheelchair bound). Much less attention has been paid to pursuing environmental modifications for information-dispensing programs that would increase the mobility and activity scheduling of blind or vision-impaired people.

Being blind and alone outdoors is an intimidating experience for the blind traveler. In several Danish towns, specific projects have been initiated. In Fredericia there is a guide line from the railroad station to the town center and other selected locations. For the sake of visually impaired people, the guide lines vary in color, have tactile indication of the route, and indicate by size or change in surface a change of direction, bus stops, and pedestrian crossings.

In this town, both in pedestrian streets and at many traffic lights, the road surfaces have been changed, thus enabling blind people to move about more safely. Furthermore, a relief "You Are Here" map has been designed for blind people. These are but a few of the design changes instituted in this town to assist blind and vision-impaired persons. Other actions taken are described in readily available technical reports. These contain the following messages.

1. To leave the familiar and secure walls of one's home is difficult and confusing for many blind people and dangerous to all of them.

2. Our cities, the traffic, stations, bus stops, and public buildings are not designed to allow blind citizens to move about easily and safely.
3. By implementing relatively simple measures in an urban environment, blind people can move about far more safely and with less difficulty. These measures include:
   a. Guide lines can be used with different floor, pavement, and ground surfaces to indicate the way to central city facilities. A design criterion is that the surface must not be an inconvenience to other people.
   b. The walking area around the guide lines must be obstacle-free—that is, no advertising signs, sales carts, benches, rubbish bins, parked baby buggies, or bicycles, and so on must obstruct the area.
   c. Loudspeaker announcements at stations, in trains and buses both at normal and unexpected stops are mandated to provide spatial updating; safety crossing at traffic lights is provided by auditory signals.

In several train stations, a different surface marks the platform edge at a distance of one to two meters. Conscious efforts are being made to make the entrance hall of Copenhagen central railroad station accessible to blind people by laying a new flagged surface.

An example of how buildings and their surroundings can be designed so that they can be used by all kinds of people can now be seen in Fredricia at the Fuglsang Centre, the new center of the Danish Association of the Blind. Here conditions appropriate for disabled people have been integrated into its basic architecture both indoors and out. Besides full accessibility for physically disabled people, the Fuglsang Centre has many fine points, which make it easy and enjoyable for blind people to move around in the house. For example, the floor along the corridor walls is dark blue and raised in some areas, and in the middle of the corridor it is gray and smooth. Every change of direction is indicated by a turret skylight, which gives a bright light compared to that of the corridors; the acoustics change, and the floor surface changes appearance at intersections or decision points.

Most floors are covered with beechwood parquet, the vibrations of which can be felt by deaf people when somebody traverses them. They act as an effective soundingboard for music and speech. All radiators and the like are located in niches so that unexpected obstacles are avoided. Outside, paths with different surfaces and outdoor areas with scented plants make it easier for blind persons to find their way about.

14.6 Spatial Abilities of Vision-Impaired People

In this section we examine some of the spatial abilities and competencies of vision-impaired people. To do this we consider three scenarios:

1. Research on cognitive maps in the context of wayfinding by those without vision.
2. Geographic contributions to the development of technical assistive devices for the blind.

3. Developments in the field of mass transportation that are explicitly oriented toward improving urban mobility of the blind.

14.6.1 Wayfinding by Those Without Vision

Vision is the spatial sense par excellence. Without vision, most spatial information has to be conveyed either through spoken words or by auditory localization or touch. Whereas vision is wide ranging, and a simple turn of the head can expose an extraordinarily rich environment in one's field of view, for the vision impaired, this simple type of overview is impossible. From the general ambient noise in the environment, specific sounds can be identified once a perceptual threshold has been reached. These sounds can be classified (e.g., bird calls, train whistles, church bells, traffic noise, speech), and to some extent, they can be localized by the head-turning process. Accurate auditory perception, however, is usually restricted to relatively close environments. How then do blind people obtain more general information about their environment, particularly about places that are distant and spatially distributed? Verbal explanations are, where possible, enhanced by direct familiarization and experience—that is, by walking through the environment. This provides proprioceptive information as one learns how much time, effort, and coordinated movement are required to traverse a segment of space. Thus, experiencing an environment while traveling through it is a critical way that those without vision collect environmental information. But verbal description, auditory localization, and touch often are inadequate or inefficient ways of imparting information about complex environments. Sometimes when supplemented by preprocessing or on-route aids (such as tactual maps), wayfinding behavior can improve. But in general, as for those with vision, the principal navigational and wayfinding aids consist of the declarative knowledge structure stored in long-term memory, the procedural rules designed to integrate bits of that knowledge in response to a need level, and the spatial abilities necessary to ensure that a cognitive map includes facsimile information about the objective environment in which travel must take place.

As with many other forms of behavior, we often are able to perform an act without being able to articulate or completely understand how we perform that act. Thus, even though wayfinding and navigation are assumed to be complex spatial tasks that sometimes defy description, the majority of people from young infants to the aged can find their way through complex environments.

A blind or vision-impaired traveler in a given environment must be able to remember and recall information about origins and destinations that are outside of the immediate area, must be able to undertake sequential learning of path segments and turn angles, must be able to associate environmental information with key anchorpoints (e.g., choice points where turns must be made), must be able to take shortcuts, and must be able to update his or her location on a regular basis. To do this, blind persons require not only information about *environmental cues*, but also information about the *characteristics of the route* they are traveling, along with some general *layout information*. Like those with vision, frequently they are involved in pretravel route planning and path selection, and as part of their general behavior, they undertake choice activities including selection of travel mode and anticipation and identification of obstacles or barriers.

Unlike those with vision, who can use assistive devices such as maps or simply visually identify key landmarks while traveling through an environment, the blind or vision-impaired person must develop specific *wayfinding skills*. These skills are often supplemented by using assistive devices called *orientation* and *mobility aids*. These include an *obstacle avoider* (e.g., a white cane, a sonic obstacle sensor, a guide dog, or a sighted companion). Blind and vision-impaired persons have difficulty in wayfinding when:

- They have little prior experience with the environment.
- They have difficulty recalling instructions about the paths they have to follow.
- They are faced with design irregularities in the environment (e.g., gradually curving streets or walkways and irregularly spaced intersections).
- They have other sensory deficiencies that inhibit processing of needed information.
- They lack prior training in wayfinding, orientation, and mobility.
- When strategies developed in one environment have to be transferred to a different environment.

A blind traveler needs two different types of information to assist with wayfinding and to provide necessary safety:

1. First, there is information about the *proximal environment* that is represented in one's cognitive map. This provides spatial information about local cues and identifies obstacles. Information may be required about the sequence of driveways passed on a given route, the number of telephone poles or doors bypassed, the location of building entrances, the identification of curbs and street intersections, and local effects such as noises, smells, wind direction, sun angle, and so on (Welsh & Blasch, 1980; Wiedel, 1983; Dodds, 1988; Tatham & Dodds, 1988).

2. Second, *information* is needed about *larger scale geographic space*, and this may include knowledge of the location of near and more distant buildings, information about changing terrain, and the definition of and differentiation among path segments. This type of information generally is built up via exploratory search or repetitive travel behavior in a local environment. The information is coded and stored as part of the traveler's cognitive map. And whereas the person with sight can simply scan an environment and identify critical cues by unique colors or shapes or dominance of visible form or specific function, the blind traveler must either learn about these second hand or not at all. With auditory perception becoming the major substitute for the long-distance information processing otherwise provided by vision, interpretation of natural sounds, their auditory localization, the estimation of distance and direction between and among simultaneously or sequentially occurring sounds, and the ability to unpack the information contained in verbal descriptions are all major tools of the blind traveler. There is every evidence that these are less

effective than vision in terms of providing precision, varied memory load, or cognitive demands.

In one of the most comprehensive studies of wayfinding practices in blind and vision-impaired persons, Passini, Dupré, and Langois (1986) cataloged an elaborate set of activities involved in wayfinding. First they assembled profiles of the members of groups of congenitally and adventitiously blind subjects, including: subjective evaluations of their own mobility; their own assessment of what constituted difficult settings; the variety of technical or human aids used in their wayfinding activities; the information they used during actual wayfinding behavior; their history of mobility and orientation instruction; their evaluation of the magnitude of difficulty associated with different physical obstacles and dangers likely to be encountered during wayfinding; and their cognitive mapping ability. Then they examined also the abilities of subjects to represent the structure and feature content of large-scale environments, their evaluation of situations that produced spatial disorientation, their attitude toward wayfinding tasks, and the nature of the environmental cues or the physical structure of pathways that contributed most to successful wayfinding and safety. The initial observation was that all persons, regardless of the group to which they belonged, were able to develop a wayfinding proficiency. This was defined as the ability to move unaided throughout a specific environment. As partial proof, Passini et al. (1986) identified a particular client who was a totally blind piano tuner who reached his clients daily using public transportation for the larger distances and a Braille map to lead him to specific final destinations.

In the analysis of survey responses, Passini et al. (1986) found the most difficult places experienced by the visually handicapped included the following: shopping complexes; department stores; hotel lobbies; train and bus stations; airports; parking lots; open spaces and parkland areas; and other places that were either too crowded, too exotic, or that lacked distinctive auditory reference points to assist with wayfinding. Indoors the subjects indicated that important information-identifying reference points were often masked by sound-absorbing materials such as carpets or acoustic tiles, or disguised by a high level of mixed background noises existed.

Another common difficulty was the inability of blind and vision-impaired persons to understand the layout of the environments through which they had to find their way. This appears particularly true when layouts are ambiguous, such as where specific patterns are repeated multiple times over a given area of space (e.g., repetitively landscaped housing units on blocks, segments of stores arranged exactly the same way, and so on). However, the repetitive example of a rectangular street system with uniform size blocks was often given as an example of a simple but highly commended organizing principle.

Passini et al. (1986) found also that individuals with some residual vision relied heavily on that residual vision during locomotion. All of the subjects used auditory cues as reference points when they were distinct and memorable enough as both location and directional indicators (e.g., the direction of traffic on streets). In wayfinding, touch is usually interpreted through a long cane that can assist in following a route along sidewalks, identifying curbs and intersections, warning of obstacles, relating information on surface texture, and decreasing the probability of bumping into other travelers. Smell, on the other hand, often is regarded only to be of value where significant smells can be identified. Examples include areas of strong pollution (as in some industrial areas), or specific locational indicators such as the smell of the ocean or other water body, the smell of food, or the smell of animals. Some individuals also use the sensing of information by the skin, such as when identifying the direction of wind blowing on the face or arm.

14.6.2 Theories of Difference

Andrews (1983) has summarized relevant theory relating to the spatial abilities of blind and vision-impaired people into three categories: difference theory, inefficiency theory, and deficiency theory.

1. *Difference theory* emerged from research in the 1920s and beyond (e.g., von Senden, 1932) which suggested that congenitally blind people who had no prior visual experience could not possibly develop an understanding of space or spatial concepts. There has been abundant evidence that this theory can be rejected.

2. *Deficiency theory* argues that one may expect to observe differences in the levels of task performance between sighted and nonsighted subjects working on spatial problems because the levels of understanding of configurational, integrational, and layout features have to be reasoned or inferred by those without vision.

3. *Inefficiency theory*, on the other hand, suggests that those without vision have the same range of abilities as are found in those with vision, but they will perform spatially related tasks in an inefficient manner.

Of the three theories, deficiency and inefficiency theories appear to be the most favored at this time.

Evidence against difference theory typically shows that congenitally and adventitiously blind individuals (i.e., those blind from birth, and those blinded later in life) can perform the same set of tasks in complex spatial layouts as can those with vision. Passini et al. (1990), for example, compared the performance of blindfolded sighted, congenitally blind, and adventitiously blind subjects on a number of wayfinding tasks under controlled laboratory circumstances. They constructed a labyrinth, and members of the three groups were asked to perform wayfinding, retracing, and directional pointing tasks. They found that congenitally blind people tended to perform better than the other groups. At a micro scale, Hollins and Kelley (1988) used table-top tasks in which subjects learned locations of objects via touch, and then after having been taken to the other side of a table from which the original learning took place, they were asked to reconstruct the locational patterns. The results indicated that both blind groups were capable of solving the changing perspectives problem and performing spatial updating as well as learning locations and layouts. When attempting pointing tasks after changing perspective, however, blind subjects tended to do worse than the blindfolded sighted group.

The cues used by blind travelers often are quite different from those used by sighted ones. Golledge (1992) has shown no overlap between sets of cues identified by blind and sighted travelers following the same route across a western United

States university campus (Table 14.9). Features used by those with sight often included things such as building heights, unique shapes or colors, architectural distinctiveness, or other visual attributes. These are missing completely from the cue list of blind travelers where obstacles and local reference points dominate, such as fences, stairwells, changes in surface texture, sounds, smells, and warning devices, such as curbs and signals at road crossings.

Let us turn now to an example of how cognitive maps might be constructed and used by blind or vision-impaired individuals.

14.6.3 Cognitive Maps of the Blind: A Case Study

The concept of a cognitive map was developed in Chapter 7. It is now used commonly in many disciplines. One of the first investigations of cognitive maps of blind people was undertaken by Casey (1978). Using both blind and sighted high school

TABLE 14.9

Blind versus Sighted Route Descriptions: Lists of Cues for Blind and Sighted Travelers for the Route from the Administration Building to South Hall (University of California, Santa Barbara Campus)

Blind Traveler	Sighted Traveler
Administration Building to bicycle path.	Exit through pillars outside Cheadle Hall.
Exit Administration Building from right door.	Turn right before kiosk.
Smooth texture at exit for a few feet.	Walk past courtyard and planter boxes on right.
Pass to concrete with crack.	Enter North Hall underpass to courtyard.
Find central crack in walk (five cracks total).	Computer Center on right
Make right turn.	Kerr Hall on right
Continue on track.	Turn right at entrance road between Kerr Hall
Five planters with trees on left	and Arbor Snack Shop.
Buildings (intermittent) on right	Proceed along sidewalk near Kerr Hall to end
Grass, bushes (intermittent) on right	of block.
Pass two grass areas on left, divided by paths.	Cross street to entrance of South Hall.
Locate end of path.	
End of grass on left	
Change in walk texture	
Sounds of bicycles at front	
Sounds of people in mall coffee shop at front	
Curve to South Hall.	
Cross to fence at bicycle path.	
Track fence to opening (lamp at end)	
Caution: cross bicycle path to left	
Turn right.	
Continue on track (driveway on right with curb).	
Sounds of bicycles on left	
Building echoes on left	
Sounds of occasional traffic on right	
Caution: cross driveway (dip and smooth texture).	
Follow left edge of grass to corner.	
Left turn.	
Enter South Hall (left or right)	

Source: Golledge, 1991a.

students, Casey had them construct a tactile map of their campus. He found that although some blind students made quite well-organized and accurate maps, in general the maps of the blind were not well organized or integrated when compared to those made by blindfolded or partially sighted subjects. To collect relevant information, Casey provided subjects with a 24-inch by 16-inch cloth-covered game board and 22 wooden models. Each model represented a specific building on the campus, and was labeled in print and Braille. Velcro attached to the bottom of each model allowed it to be firmly located on the game board. Flexible Velcro strips were provided to symbolize roads or pathways. Independent judges evaluated the resulting configurations on a line scale representing degree of accuracy.

The results of this experiment showed that maps made by the partially sighted were significantly superior in organization and accuracy than those constructed by the congenitally blind. The former group included significantly more features on their maps (16 out of 22 as opposed to 12 out of 22 for the blind). They also completed the exercise in a shortened time, but the difference was not statistically significant. Congenitally blind subjects had a tendency to linearalize curved paths, to segment or chunk information rather than integrate it, and to represent material along well-traveled routes much better and more completely and more accurately than they represented off-route information. Overall, layout features in these external representations of internal knowledge structures were inferior for the congenitally blind individuals. Figure 14.2 shows these differences.

In a later study, Lockman, Rieser, and Pick (1981) used nonmetric multidimensional scaling to uncover the latent spatial structure and subjective interpoint distances among three landmarks from an environment that was well known to blind and blindfolded sighted persons alike. The authors used the triadic comparison procedures in which subjects were asked to determine which two out of any three places were closest together or furthest apart. The responses were input into a nonmetric MDS, and a two-dimensional configuration of the places was obtained. The configurations were then rotated and stretched or shrunk to a point of maximum coincidence, and a two-dimensional correlation was developed. Correlations achieved by the blindfolded sighted groups were superior to those obtained from the blind group.

14.6.4   Spatial Orientation

Spatial orientation refers to a person's ability to relate personal location to environmental frames of reference. These frames of reference might be local and relational as with respect to landmarks or street systems, or they might be related to a global and widely accepted frame of reference such as traditional geographic latitude and longitude coordinate systems and the cardinal compass directions. Thus, spatial orientation is seen to be an important component of the larger process of cognitive mapping.

Rieser, Guth, and Hill (1982) suggest that the major components of spatial orientation include:

1. Knowledge of spatial layout of destinations and landmarks along the way.
2. The ability to keep track of where they are and in which direction they are heading.

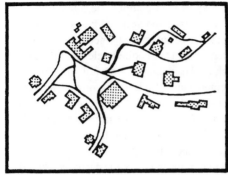

(a) The standard map of the school campus.

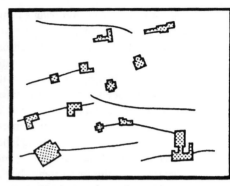

(b) An illustration of a tactile map by a congenitally blind subject.

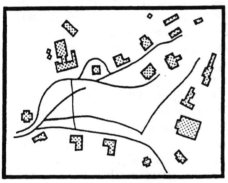

(c) An illustration of a tactile map made by a blindfolded partially sighted subject.

Figure 14.2. Blind versus sighted cognitive maps. (a) The standard map of the school campus. (b) An illustration of a tactile map made by a congenitally blind subject. (c) An illustration of a tactile map made by a blindfolded partially sighted subject. *Source:* Casey, 1978:299.

3. Comprehension of the organizing structural principles embedded in a given environment.

Without the first, travel can take place only by using search and exploration strategies. Without the second, spatiotemporal components of movement become disorganized. Without the third, needless effort is expended in trying to memorize every component that can be encoded as a simple pattern (e.g., a grid-like street system).

Possible ways to learn layout as a means of facilitating wayfinding include:

1. Gridlike search.
2. A sequential memorization of segments and turn angles.
3. Understanding the shape of routes so that, at least in theory, shortcuts could be inferred.
4. Organizing a geographic map-like representation so that route patterns are integrated in a common frame of reference.

5. Updating one's position via path integration so that knowledge of location with respect to origin is always current.

The frame of reference is particularly important in orientation, and it is most important for comprehending or giving verbal descriptions needed to follow specific paths. If a frame of reference can be established, orientation also can be defined independent of the facing direction of the potential traveler or speaker, which in turn allows independent orientation to the correct heading or the estimation of direction to selected cues. Currently, perhaps the only way the blind traveler can get access to this type of information is from a tactual map or from verbal descriptions. Verbal descriptions of feature layout in the vicinity of a path are often fuzzy because of the spatial prepositions used in natural language.

A significant problem, however, is that tactual maps of local environments are rarely available. Those tactual maps that are available are generally used in teaching situations in classrooms and are not freely and inexpensively available to the traveling public. Although, as mentioned earlier, some attempts are being made to locate auditory and tactual maps at critical locations (such as railway stations, airports, or bus stops), this form of information processing that is widely available to those with vision in the form of differently scaled maps is rarely available to the blind traveler. However, if such tactual maps could be made available, they would be most useful both in planning and undertaking a trip—more so than the auditory, tactile, or vibratory devices that are used in obstacle avoidance or cue identification by travelers today (Golledge, 1991a).

14.7 Wayfinding Strategies

14.7.1 Simple Heuristics

Strategies used for wayfinding include cognitive mapping, dead-reckoning or path integration, sequential learning of linear path segments and turn angles, or controlled branch and bound search procedures. Passini emphasizes that to undertake successful travel, "the blind person requires much effort, much attention and concentration and, given the great number of reference points used, a remarkable memorization capacity. Learning new routes and experiencing new settings often leads to tension, anxiety and feelings of insecurity" (Passini, 1986:906).

Perhaps the simplest design for a *path-following heuristic* is to segment a route by focusing on critical choice points and remembering the number of segments and the turn angles that occur between them. This emphasizes route-type learning which indeed is the most common form of spatial learning used by blind and vision-impaired travelers. Involved in such a representation is an egocentric frame of reference, the definition of segment sequence and segment length, and a feeling for the integration of space and time required to follow a preselected path between a given origin and destination. This form of wayfinding strategy lends itself to improvement with repetition. It does not help develop a capacity to choose alternative routes, to take shortcuts when available, to develop successful search strategies when well-known routes are blocked by unexpected obstacles such as construction, or to assist in the substitution of one destination for another in real time as travel is being undertaken.

Using cognitive maps, Dodds et al. (1982) studied congenitally and adventitiously blind 11-year-old children to examine questions of the development and structure of their mental maps. Tasks included pointing, map drawing, and spatial reasoning concerning the characteristics of two simple routes. Repeated trials were undertaken, although it was shown that each child could travel either route successfully after only one trial. Dodds, et al. concluded that four of the congenitally blind children were dominated by egocentric or self-referent spatial coding strategies and evinced an almost complete lack of understanding of the general spatial structure of the environment through which they traveled. On the other hand, all the adventitiously blind subjects used external frames of reference similar to what would be used by sighted subjects viewing conventional two-dimensional maps.

Dodds et al. (1982) suggest that an examination of mental maps of the blind is of both practical and theoretical interest. They designed a number of experiments to develop a means of making reliable and valid inferences about the structure of spatial representation of blind people by using maps drawn by the subjects themselves. By comparing maps of congenitally blind with those of the adventitiously blind (who have had some previous vision), they evaluated the role previous visual experience plays in the encoding and decoding of serially gathered spatial information.

In their task, the subjects learned two routes consisting of a walk around two separate blocks. After the first familiarization trial, subjects carried a board containing a pointer and a scale marked in 5° intervals; before and after each turn on the route, they set the pointer toward home and target goal. For each trial, each subject presented 10 pairs of orientation responses and 1 drawn map. Pointing validation was undertaken by placing the pointer scale at selected locations on the hand-drawn map such that home and target goal directions could be estimated. This process, called *construct validation*, involved using such orientations as predictors for the pointer responses to be obtained during the next trial. Significant correlations were obtained among the sighted students, thus implying that a valid means of inferring mental representation had been developed. A reliability test showed that drawings produced under blindfolded conditions showed highly significant correlations between subjective and objective orientations. Having indicated (using blindfolded sighted subjects in the pilot test) that the experimental design appeared to be valid and reliable, Dodds et al. then repeated the experiment using congenitally and adventitiously blind 11-year-old children. In this context, the task was set up as a game involving learning how to get to an imaginary friend's house (the target goal) from a given origin. The subjects were required to pay attention to where they were going because they would have to take the experimenter on subsequent trips. The experimenter initially accompanied the children on the trip and pointed out turns on the way. Between trials, subjects were driven back to the home base by a different route to preclude the learning of relevant information via reverse travel. There were seven subjects in all with four completing route 1 first, and three completing route 2 first. Of these subjects, one congenitally blind child was dropped from the study because, although she could learn the route, she appeared to have little or no sense of direction. Using the measure of absolute pointing error in degrees (i.e., difference between subjective and objective directions) Dodds et al. (1982) showed that adventitiously blind children perform better than did congenitally blind ones and that all the children were better oriented at some route points than at others. After the first trial, only one congenitally blind subject could produce a route drawing consistent with sighted conventions.

14.7.2 Spatial Updating

Golledge, Klatzky, and Loomis (1995) suggest that *spatial updating* appears to be an important component of all wayfinding capabilities. In other words, for successful travel, it is not sufficient just to know where places stand in relation to each other, but one must also know where the traveler stands in relation to other places in the environment. These authors suggest that spatial updating can be undertaken in a variety of ways, such as:

- Recognizing that one is near a known landmark whose location is well known.
- Reconstructing previous movements in order to mentally retrace a path and infer one's current location.
- Perceptually relating locomotion to environmental knowledge.

They suggest further that practical things a good traveler might use to assist in wayfinding by helping update position include:

- Street or avenue numbering systems to determine distance of travel.
- Use of known street systems as a frame of reference to allow estimation of orientation and direction.
- Recognition of distinctive patterns in the environment such as street systems, or sidewalk/building or sidewalk/curb/grass combinations in residential neighborhoods.
- Environmental cues, such as wind direction or sun angle.
- Updating by automatic relation of personal locomotion to environment knowledge.

For those without vision, orientation requires the development of strategies for locating objects in unfamiliar environments and spatially updating a variety of self-to-object relations while traveling. To date, no definitive study exists of the strategies that are used by persons with little or no vision with respect to how they may navigate or follow paths and consequently learn novel environments by experience. Although some systematic spatial search procedures such as grid-line following and perimeter-to-center exploration are described in orientation and mobility texts, there appears to be no detailed simple list of strategies that potentially can be used in wayfinding problems. Golledge, Klatzky, et al. (1995) suggest that some contributions to such a list might be:

1. Independent systematic exploration. Here strategy depends on the purpose for which environmental exploration will be undertaken.
2. A clearly defined origin (e.g., significant landmark or choice point needs to be selected).
3. An exploratory strategy aimed at defining the nature of the functional structure embedded in the local environment (e.g., the regularity of street systems and size of blocks, etc.) needs to be implemented.
4. During initial and consequent exploration, critical landmark cues (whether they be single objects, intersections, linear segments, small neighborhoods, or other geographic component) need to be defined as anchorpoints for the cognitive representation of the environment.
5. A cognitive representation that embeds in it location, layout, and functional characteristics should be developed.
6. Path selection criteria (e.g., minimize effort, minimize time, maximize aesthetics, minimize obstacles, minimize turns, etc.) should be chosen.

If these steps are followed, then a destination should be clearly imaged and defined, and it should also be possible to select a path that leads to the destination.

By accessing preexisting information and past experiences, exploration or search strategies that appear to have achieved previous success in the type of environment now faced should then be activated and the wayfinding process implemented. If success is achieved (i.e., the objective is reached), information can then be transferred from working memory to a long-term memory storage for access at other times or in other similar situations.

Hill et al. (1993) have identified the following search patterns for locating objects:

- Haptic (hands and arms) exploration of one's immediate surroundings (including environments such as table tops).
- Cane or other hand-held objects designed to extend one's reach.
- Independent movement such as unassisted or guided walking to explore larger spaces.
- Establishing object-to-object relations using anchorpoints and polar vectors, grid-line routinized searches, or perimeter to perimeter searches.

Other strategies that can be used to solve the wayfinding problem include step counting and fractional angle turning to provide input to the mental trigonometry used to discover the final turn angle and distance to be traveled. Alternatively, one may try to construct a mental image of the spatial relations among the points including distances and turned angles, thereby using mental geometry to help solve the task.

Rieser et al. (1982) conducted two experiments to investigate the spatial updating process. In the first, 18 adults of normal or above-normal intelligence ranging from 18 to 38 years and 12 blind or vision-impaired participants were used as subjects. In one task, subjects were guided from an origin to each of a set of objects in a room then were taken to a new origin and asked to point at the objects (the locomotion task condition). This was designed to access the extent to which subjects were able to perceive where they walked in relation to the locational arrangement of targets in the experimental setting (i.e., the use of a perceptual strategy to update position). In the second task, subjects were asked to *image* standing at a specific location and then to point to where each of the other objects would be with respect to that imaged location (the imagination task condition). This assessed the extent to which participants were able to figure out new directions and use a computational strategy to update position. Blindfolded sighted, blind, and vision-impaired subjects all reported the imagination task as being much more difficult. Here, however, Rieser et al. (1985) reported that the early-blinded subjects found both locomotion and imagination tasks difficult, whereas the blindfolded sighted and late-blinded adults found the locomotion task comparatively simple. Further experiments were then undertaken with the locomotion and imagination tasks being repeated with and without sound cues. No significant changes in actual and perceived performance were noted with and without sound cues. However, the results did suggest that early-blinded people were deficient at perceptual updating or traveling from place to place.

In a third experiment, rather than locomoting, pointing to targets was used in both conditions. The results of this study apparently indicated a difference in the spatial abilities of early versus late-blinded adults. It was suggested that early-blinded people may have to rely more on slower cognitive strategies than faster perceptual strategies for updating purposes.

14.7.3 Wayfinding via the Process of Path Integration

Path integration is a wayfinding strategy in which the traveler senses self-velocity (heading and speed) and integrates to obtain position within some coordinate system and bounding frame of reference (Mittelstaedt & Mittelstaedt, 1982). Although this does not rely explicitly on external positional information (such as optical, acoustical, chemical, or radio signals), some signals may be involved in the sensing of self-velocity. A considerable literature on animal, bird, and other nonhuman species migration and wayfinding activities has shown they have an uncanny ability to perform tasks that lend themselves to path-integration solutions (Mittelstaedt & Mittelstaedt, 1982; von Saint Paul, 1982; Müller & Wehner, 1988; Fujita et al., 1993).

Research on adult and children wayfinding, particularly where sight is lacking, provides some indication that a form of path integration processing may be ongoing. Klatzky et al. (1990) used pathway completion, a task in which the subject follows an outbound multisegment pathway and attempts to return directly to a home base, to examine wayfinding without vision. In that research, it was assumed that subjects used some sort of metric representation in performing the given task. This assumption was motivated by the systematic dependence of subjects' pathway completion responses upon quantitative manipulations of the configuration of the outbound paths (Loomis, Klatzky, Golledge, et al., 1993). One extreme was a minimal metric representation that indicated only relative location of the origin in terms of the turn and distance that would be needed for the observer to return home (Fujita, Loomis, Klatzky, & Golledge, 1990; Müller & Wehner, 1988). These parameters constituted the "homing vector," which is updated with each step of the observer (Figure 14.3). Such a representation is history-free, for there is no record of the path the observer took to arrive at the current location. At the other extreme might be what has been called "survey representation" of the task environment. Within some common coordinate systems is represented the origin of locomotion and the observer's path, including current location. Other locations could be represented as well, thus allowing the observer to plan direct paths to any of them. Fujita et al. (1993) examined pathway completion by assuming the operation of four internal processes. These are:

1. Perception of each of the pathway segments and turns by sensing and integrating the translational and rotational velocities respectively.
2. Formation of an internal representation of the outbound paths.
3. Computation of the direct path back toward the origin.
4. Execution of the completed path.

As observed, completion responses could have resulted from any of the component processes. Fujita et al. (1993) argued that evidence suggested perceptual and output processes alone (i.e., (1) and (4) above) cannot entirely account for errors in pathway completion. Previous work found substantial differences between the level of performance on tasks that merely require replication of a linear segment or turn

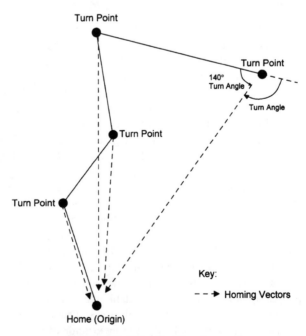

Figure 14.3. Example of homing vector.

(which essentially involves only the perceptual and output components) and on those tasks that require more complex inferences about the multiple linear segments of navigated pathways. Further, correlation and factor analyses suggested that individuals' performances on these two types of tasks are relatively independent (Klatzky et al., 1995).

With respect to the second component of pathway completion activity, an important issue is whether the navigated pathway is represented in some sense or whether subjects simply maintain the history-free homing vector that allows them to return to the origin. Evidence against the history-free hypothesis comes from the time to initiate pathway completion, which Fujita et al. (1993) found increased with the complexity of the previously traversed route. If subjects had been continuously computing homing vector and discarding information about the path, then the latency to start homeward should have been independent of pathway complexity. In light of this result, it was concluded that subjects probably develop and use representations of their outbound paths.

From their pathway representation, travelers can be assumed to compute a homeward trajectory. This computational process might be equivalent to cognitive trigonometry in that it introduces no systematic source of error. Note that measurement of an image (analogous to perceptually accessing angles and leg lengths when a path is viewed on a map) is not distinguishable from an abstract trigonometric computation. Alternatively, subjects might use weak heuristics with systematic biases in order to solve their problem. The Fujita et al. (1993) model assumed the first of these alternatives (i.e., that the encoded values are fed into a computational process, which, unlike a heuristic, introduces no further systematic error). Description of the encoded representation is sufficient, then, to characterize subject performance. The Fujita et al. model provides a mechanism for computing the internal representation that corresponds to a simple multisegment path. Thus, it allows the computation of a hypothetical encoding function that relates actual values of segment lengths and turns to internalized encoded values. Errors in pathway completion are produced by these encoding functions with a high degree of accuracy. Tests of the model found that subjects encoded both turns and extents with considerable regression toward the mean of the values they had experienced. The group encoding functions for distance and turn were linear with nonzero intercepts and slopes substantially less than one. This indicated that subjects overestimated small values and underestimated large ones—a result that appears frequently throughout the spatial cognition literature. It might be hypothesized that much of this tendency reflects imprecision in proprioceptive sensing and would be reduced if not eliminated if subjects had visual information about self-motion, even if they did not have vision of the pathway as a whole.

A wide range of locomotive tasks verify the presence of abilities and differences in the time taken to perform locomotion tasks. For example, congenitally blind individuals have been found to be disadvantaged in locomotion-based knowledge of familiar environments when this knowledge is assessed by completing tasks such as pointing to known landmarks, estimating distances, or creating maps (Dodds et al., 1982; Casey, 1978; Rieser, Hill, Taylor, Bradfield, & Rosen, 1992; Rieser, Lockman, & Pick, 1980). For example, Rieser, Guth, and Hill (1986) trained sighted and blind subjects to know positions of several objects relative to a single origin and then asked them to point to the objects from another position after real or imagined locomotion to that new position. Their conclusion was that congenitally blind subjects did not update their positions during locomotion. This was confirmed by high errors in both the imagined and real locomotion conditions. In a replication study, however, Loomis, Klatzky, Golledge, et al. (1993) found considerable individual variation in task performance with some of the congenitally blind performing extremely well in both real and imagined locomotion conditions. Using different configurations or points located on a standard-sized circle, they found no consistent trend to emphasize either angular or distance error in any of the congenital, adventitious, or blindfolded sighted groups and were not able to differentiate significantly among these groups in performance of triangle completion tasks (Figure 14.4).

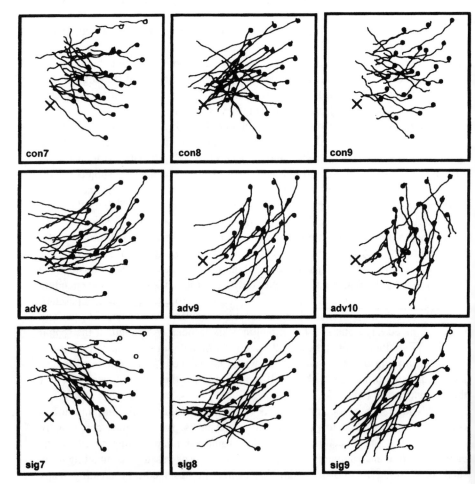

Figure 14.4. Examples of triangle completion task for congenital blind, adventitiously blind, and blindfolded sighted.

14.7.4   Wayfinding Using Assistive Devices

Wayfinding refers to a person's cognitive and behavioral abilities to determine the path between a specified origin and destination and to successfully negotiate the path. For successful travel, information must be obtained about reference points used to identify start, end, and current location, cues that signal when changes of direction are required, recognition of the appropriate turn angle that links consecutive segments, and a mechanism for determining where one is at any particular point in time with respect to the origin (path integration). When traveling through an environment, a blind traveler needs two types of information:

1. Information about the proximal environment represented in one's cognitive map used to identify local cues and obstacles.
2. Information about the larger-scale geographic space through which travel takes place (i.e., of the layout of near and distant landmarks, changing terrain, and differentiation among the path segments).

Both these sets of information usually are developed by experiencing travel and search in terms of exploratory behavior. Information sensed is coded and stored as part of a cognitive map. When the blind person is undertaking additional travel in the same environment or over the same route, the cognitive map information is recalled from memory and helps in the wayfinding task.

When travel takes place frequently along essentially the same route, *travel habits* emerge. This is so for populations without vision equally as well as for populations with vision. One way to improve the efficiency and accuracy of travel behavior, is to preprocess information in the planning stage prior to travel. This type of activity is often undertaken by rehearsing verbal descriptions of the route.

Recently, however, a new tool developed by geographers (Parkes & Dear, 1990) provides a rich preprocessing opportunity.

The device called NOMAD is a combined auditory and tactile aid. It consists of a touch-sensitive pad in a solid frame in which is mounted a notepad computer and a voice synthesizer. The touch-sensitive surface is geocoded using a 1-cm grid. Information can be entered into the computer via a keyboard after touching a particular location. Brief or long descriptions of relevant material can be attached to specific locations, paths, or areas. Maps, graphics, charts, photographs, or other imagery can be copied onto chemically treated capsule paper. When exposed to heat, the capsules expand in proportion to the darkness of the lines or features copied on the paper. This produces the tactile image with a vertical rise of up to 2 mm. Information is obtained by exploring a shape of the feature being examined (e.g., exploring the layout of a given environment or interior of a building), and as the tactile shapes are explored, messages attached to specific locations are auditorialized. Thus, the individual gets both a tactile and verbal description of what he or she is sensing.

In terms of wayfinding, NOMAD is extremely useful as a preprocessing device. Prior to travel one simply explores the total environment through which travel will take place, and obtains a memory refresh of the layout or configuration of specific cues. The path to be followed can also be previewed and the number of segments and the turn angles between segments memorized. Off-route auditory information can also be encoded with respect to both distance and direction from key landmarks or choice points along the way. The device, therefore, consists of a process of making available to those without vision an equivalent of a cartographic map. Since most tactile maps have in the past been large, complex, expensive to produce, and used more for display in classroom situations, NOMAD represents a major change in direction. It is designed to be a travel aid as well as to enlarge the spatial experiences of users. For example, one can explore the outlines of the major continents of the world—a practice that is rarely, if ever, available to blind people. Local neighborhood maps, interiors of buildings or shopping centers or airports, and other small- and large-scale environments can be previewed and stored as part of one's cognitive map. Thus, independent travel is potentially increased as one is able to access stored information along with specific route information. Now that the device has been developed, it remains for geographers to begin production of a series of tactual maps equivalent in scope and variety to those normally available to sighted populations.

14.8 The Use of Bus Transit by the Blind

An increasing amount of attention has been given to the destination choice and transportation mode selection processes by disabled individuals. In the United States, there are over a million people registered as being legally blind. Of these, approximately 110,000 use a white cane as a mobility aid, and 10,000 use a guide dog. This leaves almost 1 million legally blind people who have to rely on public transportation or transportation provided by friends and families for their locomotive activities.

Approximately 50% of the legally blind population in the United States do not use mass transportation. This means that approximately 500,000 blind people distributed over the country are essentially confined to their homes or must rely on friends, neighbors, or family to take them out of their local environment. Obviously this raises questions concerning the features or characteristics of mass transit that produce such widespread lack of acceptance. Is it the design of the mode? The timing in which the service is offered? The limited routes that are followed? Questions of safety or fear? Difficulties in deciding where one is on-route? Or is it simply the way that drivers or controllers and other passengers treat disabled people?

In this section we examine this particular problem by focusing on what vision-impaired and blind disabled people perceive to be the sources of frustration faced as the result of their nondriving status. We examine also how this affects lifestyles and attempt to get some idea of the physical, psychological, emotional, and societal barriers that face this disabled group when they attempt to pursue activities in the large-scale complex environment.

Much has been written about the American infatuation with the automobile. Cars are perceived as symbols of social and economic standing, they provide independence and freedom of movement, are constrained only by economic well-being, and are the single most dominant form of transportation in the country. In 1990, there were approximately 168 million drivers in the United States (67% of the 250 million children and adults in the country drive a car!). According to a recent World Almanac, 10 million blind persons are under age 19, and more than 13 million are over age 70. Although the majority of drivers are of working age, many of the perceived problems of driving are focused on those in the teenage or elderly subgroups.

Those who are blind or vision impaired, as well as many who have low vision, are unable to acquire a driver's license. What is their alternative for travel in complex environments such as large cities?

Approximately 425 cities across the United States provide some form of mass transportation or specialized transit systems. These are said to be suitable for use by disabled people. However, blindness and vision impairment erect both physical and social obstacles that diminish the possible use of mass transportation by this group. What are some of these features that inhibit use of publicly provided systems?

1. *Negative stereotypes:* Many blind or vision-impaired persons do not want to have their presence emphasized. They often try to maintain a sense of anonymity. Corn and Sacks (1994) surveyed 110 adults in 24 states to provide information on the nature of these frustrations and problems. They found that in attempts to maintain some sense of anonymity and not to stick out in a crowd, 40% of their participants reported being hit by a car.

2. *The social importance of driving:* Independence of movement is one thing that is consistently lacking across both vision-impaired and blind travelers. It becomes very important for adolescents that they not be singled out as being unable to participate in the social habit of driving. At the other end of the scale, lack of access to a vehicle may severely inhibit the ability of an older person to use his or her environment for those minimal support functions necessary for living. In the Corn and Sacks study, 71% of respondents rated themselves as independent travelers, whereas 19% stated they needed assistance, particularly in unfamiliar areas. Ten percent required sighted guides for all trips outside their own homes. When asked what mode of transportation was used frequently for a range of activities, responses such as mass transportation, walking, and soliciting assistance from a family member or friend were cited most frequently. Rides with neighbors was cited least frequently (see Table 14.10). Mass transportation was used most frequently for traveling to and from work and school-related activities. Visits to department stores for comparative shopping purposes were also dominated by mass transport use. It was rarely used, however, for convenience shopping (e.g., groceries), pleasure, and recreational activities, for going to church, or for visiting friends or relatives. Professional needs (doctors or dental appointments), or visits to financial institutions, often are undertaken using taxis. Other personal errands were undertaken in private vehicles using hired drivers. Curiously, the majority (67%) of subjects in the Corn and Sacks (1994) study perceived the costs of aided transportation were lower than the costs of transportation for people who drive! A critical geographic factor that emerged was that many vision-impaired and blind people felt that they were constrained to live in certain areas in order to have access to mass transportation or to be able to pay for different types of transportation services. Often this produces double jeopardy

TABLE 14.10
Level of Frustration Experienced from Obstacles to Access to Transportation,
by Visual Status

Obstacle	Mean Scores for Functionally Blind Respondents	Mean Scores for Low-Vision Respondents	r Value
Preplanning means of transportation	2.66	3.34	.24*
Relying on others' time	3.16	3.97	.28**
Accepting rides that are offered	1.50	2.33	.39***
Carrying object throughout the day	2.66	3.55	.31**
Explaining to others why I can't drive	1.50	3.07	.54***

Source: Corn & Sacks, 1994:60. *$p \leq .01$; **$p \leq .001$; ***$p \leq .0001$.

where poorer or older persons suffering a disability have to live in some of the central city areas, paying higher rents and/or taxes, so that they can get the greater access to public transportation facilities.

3. *Impact on life-style:* Many blind and disabled people feel considerable amounts of frustration with respect to the use of mass transportation. Some of these frustrations extend to the paratransit or individualized transportation services available to them. The obstacles rated highest in producing the greatest frustration were waiting for rides that were late, or being unable to arrange for a ride on an important occasion. Other important frustrations resulted from having to walk in dangerous areas in order to access public transportation. Yet other frustrations were produced by having to rely on the time schedules of others for rides. There was also some concern about accepting rides that were offered because of a feeling that such offers were made out of pity. Others felt a reluctance to accept rides that were offered because of the inconvenience they would put the well-intentioned driver to. And concerns were more general in nature including the fact that blind or vision-impaired people often have to explain to others why they can't make certain appointments or visit certain places simply because they can't drive.

Other information was collected by Corn and Sacks by asking respondents to indicate on a five-point rating scale the extent to which people such as spouses, parents, friends, physicians, and other health care providers, and clergy generally understood the emotional and the logistical impact of being a nondriver. The respondents rated neighbors, the general public, physicians and health care providers, and clergy as having little understanding about the emotional or logistical impact of nondriving. Spouses, significant others, parents, and roommates tended to best understand these impacts (see Table 14.11).

There was unanimous agreement that life-style was severely impacted by not being able to drive. Most said they were strongly or moderately affected by not being able to drive in terms of the extent of social activity, opportunities for dating, and opportunities for employment. Isolation was the common result of nondriving. Apart from work, nondriving impacted most on recreation and leisure activities, independent living, independence of traveling, personal security, choice of location

TABLE 14.11
Rank Ordering, by Means, of Understanding by Others of the Emotional Impact of Nondriving on Adults with Visual Disabilities*

Other persons	Male		Female		Blind		Low Vision		Total	
	Mean	Rank	Mean	Rank	Mean	Rank	Mean	Rank	Mean	Rank
Husband, wife, or significant other	3.96	2	4.06	2	4.28	2	3.70	1	4.01	1
Roommate	4.00	1	4.22	1	4.29	1	3.60	2	4.00	2
My parents	3.35	5	3.73	3	3.60	4	3.50	3	3.55	3
My children	3.04	9	3.60	5	3.50	5	2.85	8	3.16	7
Close relatives	2.86	10	3.02	9	3.16	9	2.72	10	2.94	10
Close friends	3.31	6	3.61	4	3.61	3	3.23	6	3.43	4
Neighbors	2.22	12	2.10	12	2.11	11	2.22	11	2.16	13
People with whom I work	3.12	8	3.21	8	3.28	8	2.77	9	3.01	9
My employer	3.41	4	3.58	6	3.50	5	3.26	5	3.37	6
People with whom I go to school	3.13	7	2.86	10	2.11	11	3.16	7	3.14	8
The public at large	1.88	14	1.55	14	1.71	14	1.62	14	1.66	14
Professionals in the field of blindness and visual impairment	3.49	3	3.36	7	3.36	7	3.41	4	3.38	5
Doctors and other health service providers	2.22	12	1.92	13	1.89	13	2.18	12	2.04	12
Churches or synagogues	2.29	11	2.13	11	2.31	9	2.09	13	2.20	11
Total means	2.91		2.94		2.99		2.79			

*From 1 = no frustration to 5 = high amount of frustration. *Source:* Corn & Sacks, 1994:62.

for housing, and out-of-home socializing. The limitations produced by an inability to have free mobility in an urban context also produced a wide variety of fears with respect to the interaction with an environment.

In this section we have shown how a disability relating to lack of vision can severely impact the emotional, activity, economic, and social lives of disabled persons. Most vision-impaired or blind people are of necessity nondrivers. Few, however, accept the fact that available mass transportation makes a reasonable substitute for the terminal elasticity and independence of travel provided by the privately owned automobile. Problems in scheduling, fear and safety, congestion, and long waiting times between opportunities combine to negatively impact the likelihood of public transportation being used as the primary traveling mode.

14.9 The Mentally Challenged

14.9.1 Attitudes Toward Retardation

For many years mentally retarded people were regarded as outcasts, and often they were confined to institutions where they were brutalized personally and kept remote from ordinary everyday living. Little discrimination was made between the mentally retarded and the insane. Both were confined, often chained or straitjacketed, and subject to chemical, electronic, or other medical treatments often without their knowledge or permission.

Since the 1970s, programs of deinstitutionalization in the United States have emerged, the rapidity with which they have been implemented often depending on the economic situation of the institutional support system (e.g., local, state, or federal organization). Often accounting decisions to cut budgets were accompanied by orders to cut down the size of inmate populations. In the case of retarded people, many who were borderline (mildly or moderately retarded) were released in care of family, friends, group homes or halfway houses, or simply thrown on the mercy of society. Some were placed in private nursing homes, but many have been turned out of their protective environment and have been expected to operate and survive within the context of everyday urban environments. Social workers and others concerned with these populations expressed dissatisfaction with such release policies, for although some of the released individuals consistently show a capacity to perform basic personal, vocational, and social skills needed to survive in a complex environment such as a city, most have not been taught how to comprehend the spatial or functional structure of the environments in which they have been released.

Groups concerned with the constitutional rights of disabled people generally, and with the retarded subpopulation in particular, stress the right of such people to remain, grow, and participate within the community. They have called for the development and provision of community services oriented toward these ends. As a result of the lobbying of concerned groups in the United States, federal and state legislation appears to have adopted the principle of least restrictive alternatives, which in turn mandates the establishment of a continuum of services according to the needs of persons who are mentally challenged—particularly those who are mentally retarded throughout their life-spans.

14.9.2   Spatial Competence

One consequence of this was that questions were raised as to the relative competence of retarded individuals to live and negotiate within their communities—their neighborhoods, their municipalities, their cities, and the national systems in which they belong. Particular questions were raised concerning the singularity of their cognitive levels and the degree to which these correlate with their abilities to accomplish things such as ability to understand the environments in which they live. Realizing that many members of this subgroup *can* live ordinary lives in mainstream society, suggestions were made to change the descriptive term for this subgroup to "the mentally challenged."

Despite the furor and concern produced by deinstitutionalization, the fact is that the vast majority of those who fall within the general category of mental retardation *have* resided in the community even at the height of institutional commitments. Yet little is known on issues such as how aware they are of environmental cues that help them navigate through their environment, how well they can differentiate between the external form (or facades) of different urban functions, how well they can use community services in times of need, or how well they can use the complex interlocking forms of public transportation to assist them in travel.

One of the few studies of the spatial competence of such populations was that undertaken by Golledge, Parnicky, and Rayner (1980) between 1977 and 1980 at Ohio State University and the University of California at Santa Barbara. The aims of this project were:

1. To discover what aspects of complex, large-scale external environments such as cities retarded individuals are aware of.
2. To determine how these aspects are used in day to day interactions in the city.
3. To discover the level of awareness that moderately and mildly retarded people have of spatial features and spatial relations (e.g., location, spatial sequencing, proximity, separation, direction, distance) that occur in the environments in which they live.
4. To discover the type and range of environmental cues used in the encounters with these environments.
5. To examine the completeness and/or the complexity of their cognitive representations or mental maps of such environments.
6. To examine the nature of distortions and disturbances that exist in these cognitive representations and to compare them with the cognitive representations of the same environments by nonretarded subgroups.
7. To discover the rates at which retarded people learned to travel in and use the spatial environments.
8. To examine the concepts that can be built into a spatial environmental competence scale.
9. To examine the concepts that can be built into a spatial locomotorability scale.

The specific terms that are essential to understanding the material developed for this study include the following:

1. *Levels of awareness:* These refer to the ability of subjects to identify, locate, or use individual or environmental cues and to integrate cues in a structured fashion such that they can be objectively represented in a maplike form.

2. *Cognitive representation* (cognitive map): As defined in Chapter 7, this is the incomplete, schematized, and augmented representation of the relative locations and attributes of phenomena found in their everyday spatial environments.

3. *Location cues:* These are the names by which specific locations or places are known or identified.

4. *Environmental cues:* These refer to stimuli that act as identifying criteria for locational information about elements of the environment. These are commonly identified, recognized, and used by individuals operating in specific environments. Some of the names attached to these cues vary from place to place (e.g., Von's Supermarket as distinct from Lucky's Supermarket). It is recognized that classes of cues (such as road signs, landmarks, degrees of congestion) exist in all environ-

ments and can be made *place specific* by allocation of a unique name (e.g., corner of Fifth and High versus corner of High and Jones).

5. *Retarded population:* This is taken to refer to significant subaverage intellectual functioning existing concurrently with deficits in adaptive behavior and manifest during the developmental period of life. Of this general population, we concerned ourselves only with those deemed capable of living and acting independently within an urban environment (usually described in professional and public literature as the "mildly" and "moderately" retarded).

14.9.3 Environmental Characteristics Known to Retarded People: Frames of Reference

Adequate frames of reference are required for simple tasks such as orientation, direction in both the abstract spatial or geometric sense and in terms of the language used in common speech. For example, developing a path between any two nodes and being able to retrace it, requires understanding of the concept of direction and the more abstract cognitive notion of response symmetry. The lack of an ability to develop a reference system makes it difficult to understand direction and even relative distance. There appears to be no evidence that retarded people are incapable of developing high levels of spatial competence and, in doing this, developing an understanding for different frames of reference. Most commonly used frames of reference are egocentric; that is, relating things in the environment to oneself in a polar coordinate or vector fashion. Local relational systems appear to be the next most commonly used—examples include understanding the basic regularity in an underlying street system. Abstract geographic coordinate locations and global compass directions appear to be less easily understood.

14.9.4 Spatial Concepts: A Case Study of Retarded Groups

Here we stress not so much the specific nature of the frame of reference by retarded individuals but, instead, emphasize that many spatial concepts such as proximity, dispersion, location, clustering, orientation, similarity, and so on, require some type of frame of reference for their complete understanding. As one develops an internal representation or cognitive map of an external environment, it is possible that different frames of reference are used for different sets of cues, features, or events. As with nonretarded population groups, it appears that spatial information in cognitive maps is hierarchically organized, focused on specific anchorpoints, and capable of reasonably quick access. In the nonretarded population case, it is assumed using the anchorpoint theory that the hierarchically organized information will eventually take on formal metric properties and that spatial relations involving complex operations such as rotation, translation, transformation, and perspective viewing will be performed at an abstract level (implying that a survey-type representation of environmental features has been attained). However, the investigators could not produce clear evidence of the ability of retarded subgroups to adequately respond in this level of abstract thinking. There is no doubt that these individuals are able to produce a declarative knowledge base by recognizing distinct features and attributing names and significance weights to them. They are able also to link path segments together to make up complex multisegment routes between specific origins and destinations and to travel successfully in both a forward and reverse direction. However, tasks such as integrating information from separately learned routes, or two-dimensional mapping of the location patterns of cues, or even directional and distance judgments between sets of cues were all difficult tasks for the retarded subgroups.

The retarded groups were able to locate features correctly, sequence them correctly, identify features that were on and off routes that they had learned, to segment and connect paths into routes, and to perform forward and retrace operations along familiar routes (see Figure 14.5). Examples of subjects' abilities to perform sequencing tasks can be seen in Table 14.12. Attempts to construct two-dimensional

SESSION I	SESSION II
1. Naming places in Columbus 2. Identifying slides of 20 places in Columbus 3. Preliminary map board responses 4. Plotting pictures of places in neighborhood area on map board 5. Naming what one sees in city, suburban, country areas 6. Identifying slides of places in city, suburban, country areas	1. Personal characteristics 2. Educational data 3. Occupational data 4. Income and expenses 5. Family information 6. Recorded test scores 7. Residential information 8. Mobility information 9. Knowledge of community resources 10. Familiarity with city at large
SESSION III	SESSION IV

1. Wide Range Achievement Test
 	
   a. Spelling
   b. Reading
   c. Arithmetic
 	
2. AAMD Adaptive Behavior Scale
 	
   a. Social Skills
   b. Deviant Behaviors
	1. Places On/Off Routes
	
	   a. To Grocery Store
	   b. To Workshop
	   c. To Shopping Mall
	   d. To Downtown Department Store
	   e. To Neighborhood Movie
	
	2. Sequence Places Along Routes
	
	   a. To Grocery Store
	   b. To Workshop
	   c. To Shopping Mall
	   d. To Downtown Department Store
	   e. To Neighborhood Movie

SESSION V	CONTRAST GROUP

1. Bicycle or Walk to and from:
 	
   - Neighborhood Food Shop
   - Neighborhood Movie
 	
2. Go in Auto to and from:
 	
   - Workshop
   - Shopping Mall
 	
3. Go in Bus or Auto to and from:
 	
   - Downtown Department Store
	1. Naming Places in Columbus
	2. Identifying 20 Slides of Columbus
	3. Preliminary Map Board Responses
	4. Plotting Pictures of Places in Neighborhood on Map Board
	5. Naming What One Sees in City, Suburban and Country Areas
	6. Identifying Slides of Places in City, Suburban and Country Areas
	7. Sequencing Places Along Routes
	8. Personal Background Information

Figure 14.5. List of tasks for retarded subjects and contrast group, Columbus study.

TABLE 14.12

Pearson and Spearman Correlation between Cognitive Route Sequencing Data and Reality—Columbus Contrast Group

	Route Destination									
	Big Bear		ARC West		Gold Circ		Lazarus		Univ Flick	
Subj ID	r_p	r_s	r_p	r_s	r_p	r_s	r_p	r_s	r_p	r_s
102	.99	1.00*	.99	1.00*	.99	.94*	.93	.94*	.99	1.00*
103	.90	.94*	.99	1.00*	.98	1.00*	.99	1.00*	.99	1.00*
104	.97	1.00*	.96	.94*	.95	1.00*	.99	1.00*	.99	1.00*
105	.96	.94*	.60	.60	.67	.83*	.93	.94*	.97	1.00*
106	.90	.94*	.99	1.00	.96	1.00*	.66	.60	.87	.83*
107	.96	1.00*	.63	.66	.94	1.00*	.94	.94*	.97	.94*
109	.57	.43	.23	.03	.88	.94*	.37	.26	.43	.25
110	.98	1.00*	.92	.89*	.96	1.00*	.99	1.00*	.97	.97*
112	.92	.94*	.92	.94*	.96	1.00*	.99	1.0*	.98	.98*
113	.99	1.00*	.50	.66	.97	.94*	.83	.77*	.94	.94*
114	.98	1.00*	.43	.37	.94	1.00*	.99	1.00*	.99	1.00*
115	.98	.95*	.98	1.00*			.90	.94*	.97	.97*
116	.98	1.00*	.67	.77*	.94	.94*	.95	.95*	.99	1.00*
117	.99	1.00*	.99	1.00*	.99	1.00*	.95	.94*	.99	1.00*
118	.96	1.00*	.65	.66	.90	.94*	.77	.77*	.97	.94*
119	.95	.94*	.94	.94*	.51	.49	.81	.83*	.86	.89*
120	.97	1.00*	.52	.60	.96	1.00*	.787	.71	.88	.84*
121	.95	1.00*	.98	1.00*	.94	.94*	.99	1.00*	.99	1.00*
122	.96	1.00*	.42	.49	.89	.94*	.68	.66	.98	.94*
123	.99	1.000*	.90	.94*	.99	1.00*	.99	1.00*	.94	1.00*
124	.99	1.00*	.908	1.00*	.93	.77*	.92	.83*	.98	1.00*
125	.98	1.00*	.99	1.00*	.95	.94*	.99	1.00*	.97	.94*
126	.99	1.00*	.96	1.00*	.98	1.00*	.88	.77*	.96	.94*
127	.98	1.00*	.96	1.00*	.96	1.00*	.87	.83*	.98	.94*
128	.97	.94*	.98	1.00*	.98	1.00*	.96	.94*	.99	1.00*
129	.99	1.00*	.93	.94*	.95	1.00*	.54	.48	.92	.89*
130	.99	1.00*	.61	.66	.91	.94*	.98	1.00*	.91	.89*
131	.99	1.00*	.79	.83*	.94	.94*	.63	.60	.96	.94*
132	.75	.83*	.82	.83*	.71	.83*	.81	.83*	.87	.77*

*Spearman rank correlation coefficients statistically significant at .10 level.

representations of environments by applying nonmetric multidimensional scaling techniques to paired proximity judgments fail to produce configurational structures that markedly differ from random arrangement of points. Table 14.13 shows the degree of correlation between subjective and objective (i.e., two-dimensional Euclidian) representations of selected environments for both populations of retarded individuals and a control population of nonretarded people.

Two areas were chosen in which to pursue the problem stated earlier. One was located between a major Midwestern university and the downtown area of a city in the United States. This area provided a range of differential community features including shops, restaurants, movie theaters, schools, hospitals, parks, residential areas, institutions, and many other urban functions. The sheltered workshops where

TABLE 14.13
Bidimensional Correlation between Map Board Configuration and Reality

Subject ID	Bidimensional Correlation Coefficient
Santa Barbara Experimental Group	
901	0.18
902	0.71
903	0.92
904	0.76
905	0.71
906	0.29
Santa Barbara Contrast Group	
701	0.91
702	0.58
703	0.96
704	0.93
705	0.91
706	0.71
707	0.99
708	0.65
709	0.98
710	0.79
711	0.99
712	0.99
713	0.91
714	0.95

most of the individuals who participated in the study worked or trained were located just outside the boundaries of the study area, which had to be traversed in order to get to the workplace. A second area was a segment of a small town in southern California that had less functional diversity and had more suburban and rural features. Subjects involved in the experiment in the main urban area included 15 retarded and 32 contrast group members, and in the western suburban area, 9 retarded and 20 contrast group members.

The tasks involved were split into a number of sessions as indicated in Figure 14.5. In the first session, activities such as naming and identifying places from slides, locating sequences of cues along routes to work, and so on, were collected. A second session collected personal characteristics from individuals, from supervisors of group homes, and from the state agency that was the primary designated care unit for the retarded subgroups. Access was made available to recorded test scores, mobility information, and city familiarity estimates. A third session involved giving a Wide-Range Achievement Test (WRAT) and the AAMD Adaptive Behavior Scale Test to subjects. In the fourth session, the staff from group homes were consulted regarding places where subjects worked, shopped, and went for recreation and entertainment, and estimates were made of their ability to travel independently from the group homes. In this session, on and off route recognition tasks were also undertaken. Examples of the cues used in session I are given in Table 14.14, and cues used in the on- and off-route experiment are given in Table 14.15.

The final session involved meeting with the contrast group of 32 nonretarded subjects. These were primarily socioeconomically disadvantaged individuals located in the same neighborhood as were the group homes for the retarded subjects. It was hypothesized that, because of the socioeconomic disadvantages of this group,

TABLE 14.14
Locations of Environmental Cues, Session I

Slides of Columbus Sites
1. State Fairgrounds, Between E 11th and E 17th West of I-71
2. Riverside Hospital, 3535 Olentangy River Rd.
3. State Capitol, Between Broad and State Sts. East of High St.
4. Lazarus Dept. Store, W. Town St. on High St.
5. OSU Stadium, 410 W. Woodruff
6. University City Cinema, 2943 Olentangy River Rd.
7. Northland Shopping Mall, 1649 Morse Rd. At Karl Rd.
8. Lincoln Tower (OSU), 1800 Cannon Dr.
9. Drake Union (OSU), 1849 Cannon Dr.
10. Eastland Shopping Mall, 2677 S. Hamilton Rd.
11. Veterans Memorial Bldg., 300 W. Broad St.
12. Greyhound Bus Depot, 111 E. Town St.
13. Columbus Park of Roses (Whetstone Pk.), 3923 N. High St.
14. Gold Circle, 3360 Olentangy River Rd.
15. Battelle Institute, 505 King Ave. (at Cannon Dr.)
16. Port Columbus Airport, Northeast of Columbus on Janes Rd.
17. Ohio Union (OSU), 1739 N. High St.
18. Ohio Village, 17th Avenue and I-71
19. Graceland Shopping Mall, 30 Graceland (N. High Street)
20. Long's Book Store, 1836 N. High St.
Plotting Structures in the Neighborhood
1. Neil Avenue Mennonite Church, Corner of Neil and 6th Aves.
2. Bank Ohio, 1221 N. High St.
3. Battelle Institute, 505 King Ave. (at Cannon Dr.)
4. Ohio Union (OSU), 1739 N. High St.
5. Playground NW corner, 5th Ave. And Highland
6. Bowling Alley, 1547 N. High St.
7. Bus Stop, King and Perry
8. Big Bear Store, 777 Neil Ave.
9. Franklin Cnty. Cncl. For Retarded Citizens, 777 Neil Ave.
10. Neighborhood Store, 246 W. 5th Ave.
11. Thurber Towers, 645 Neil Ave.
12. Nisonger Center, 1580 Cannon Dr. (OSU)
13. Lincoln Towers (OSU), 1800 Cannon Dr.
14. Goodale Park, between Goodale and Buttles
15. Residence, 1085 Neil Ave. (and 3rd Ave.); SW Corner
16. Residence, 1276 Neil Ave. (and 5th Ave.); NE Corner
17. OSU Football Stadium, 410 W. Woodruff
18. Doctors Hospital, 1087 Dennison Ave.
19. Lawsons Grocery Store, Northeast Corner, 5th Ave. And Highland

TABLE 14.15
Session IV: Cues Used in On–Off Routes and Sequencing: Identifying Places On–Off Routes

On Route	Off Route
Destination: Big Bear (Lane Ave.) 386 W. Lane Ave.	
1. King's Way Fellowship Hall King/Neil intersection (SW corner)	1. Speech & Hearing Center of Col. And Central Ohio 1515 Indianola Ave.
2. Holiday Inn 328 W. Lane Ave.	2. Food Stamp Issuance Center 168 N. High St.
3. St. John's Arena	3. OSU 10th/Neil N/W Corner
4. Battelle Institute 505 King Ave	4. Kroger University City Shopping Center on Olentangy River Rd.—N. Of Dodridge intersection (W. Side)
5. Lincoln Tower 1800 Cannon Drive	5. Scioto Downs 6000 S. High St.
Destination: ArCraft West—1000 Kinnear Rd.	
1. King's Way Fellowship Hall (address above)	1. Center of Science & Industry (COSI) 280 E. Broad St.
2. Truck Depot Kenney/King Ave. (NW corner)	2. Country Cue-cornfield
3. Apartments, Kenney Rd. 1 block S of Kinnear/Kenney	3. Residence on E First Ave. between High/Indianola (S side)
4. OSU from King Ave. facing north	4. OSU Lane/High (SW corner)
5. Intersection King/Olentangy River Road	5. Intersection Dodridge/Olentangy

they would be less likely to have exaggerated knowledge of the total urban environment and more likely to have a more detailed neighborhood knowledge that would provide a basis for comparison. The contrast group replicated all the tasks of the retarded groups.

The results of these studies showed that there were no substantial differences between the retarded subjects and the contrast group in terms of the ability to locate and identify environmental cues, to place them in sequences with appropriate intervening distances indicated between them, to travel routes, to identify cues that were on and off routes (i.e., potential indicators of being lost), and to recognize and use neighborhood features and community services. Differences did appear when more abstract reasoning tasks were required, particularly where two- or three-dimensional inferences had to be made. This tended to indicate that, whereas retarded persons were quite competent in a one-dimensional and topological domain, more ordered metric domains, such as two- and three-dimensional Euclidean environments, proved difficult to understand or to operate within. Attempts at obtaining interpoint distance estimates (even when only on a monotomic level), when analyzed by a nonmetric MDS, produced almost random configurations of points. Standard grids, distorted to represent the configurational structures obtained from retarded subgroups, were markedly distorted compared to those obtained from the contrast group members (see e.g., Figure 14.6). Nevertheless, this geographic experiment

Objective locations of cues   Poor resident's mental map

Middle income resident's locational map   Retarded resident's locational map

Figure 14.6. Examples of displaced cues.

showed that, with training and experience, retarded people who are about to be de-institutionalized can be provided with simple training either on or off the institutional grounds to help them to develop skills of environmental cue and feature recognition, cue and feature sequencing, functional specialization of features commonly found in the environment, and independent navigation and wayfinding skills for familiar and well-traveled routes. These latter skills could be developed in the context of having to transfer between different bus lines and of recognizing priming cues when approaching decision points or destination points. In this case, the geographer is able to use disciplinary skills and knowledge to suggest ways of improving the spatial competence of retarded persons.

15

Gendering and the Elderly

15.1 The Functional Significance of Two Major Divisions of Society

We saw in Chapter 3 how two aspects of societal change that are assuming increasing importance are the changing nature of male and female roles and the aging of the population.

It is ironic that for so long social scientists gave scant attention to the issue of gender, and that the specific role of *women,* who constitute half the population, was largely neglected apart from a simple differentiation of the population into male and female members. This was similar to differentiation made among age groups, occupations, and industry categories when discussing phenomena such as the social structure of the city and the differences in the spatial behaviors of various groups in the population. Although a large body of social theory developed around the themes of class and race as constructs in creating social and economic differentiation in spatial and societal systems, it has been only in the last couple of decades that issues of *sex* and *gendering* have been given the serious consideration they both require and deserve in the study of social and economic processes of change and the behavior of people in space. The focus has shifted to the issue of gender in the context of changing functional roles and relationships in society, particularly with respect to the role of the family and household structure, changing work–home relationships, and the role of women in the locational and other decision processes in households. Questions of gender in segregation, alienation, security, and access to appropriate services provision in urban and nonurban areas have been addressed not only by feminist scholars, but also as fundamental societal issues in the development of new theories on society, space, and behavior.

The issues of *age* and *aging* also has assumed increasing importance in the social services. Though age had been treated as an important explanatory variable in spatial epidemiology, health care behavior, marketing, and recreation during the last two to three decades, the sheer weight of numbers of the elderly, and in particular of the *old* aged, has resulted in aging and the elderly becoming topics of serious concern in public policy formulation for social and economic support programs of governments. This is especially true in advanced western societies, due to a whole range of factors related to reduced levels of mortality and morbidity through improved medical technologies and life-style. The "graying" of society has major implications for government expenditure in a whole range of areas as the levels of dependency and demands for aged-related services increase. In addition, the increasing proportion of retirees, who are either asset rich or superannuated or both, has created a major reduction in marketing and the production of goods and services that are oriented to the needs of the elderly in fields as diverse as housing, recreation and tourism, health, consumer durables, and transport. The elderly are now a major market for manufactured products and services, as well as being a major policy concern for governments that provide income support and social, welfare and health services for them.

In this concluding chapter, we address selected issues relating to gender and the role of women in society, and to elderly components of the population. We do this in the context of the social structure of the city, human spatial behavior, and locational decision making.

15.2   Women and Gender Issues

Feminist scholars have pointed to sexist biases in human geography that have important implications for theory, research design, data collection and analysis; and they have proposed how these issues may be addressed and how new approaches might be incorporated into the mainstream of research into economic, social, and behavioral aspects of human endeavor (Monk & Hanson, 1982). Here we refer to a number of advances that have been made by researchers in human geography who have sought to redress these deficiencies, and we refer to a number of fields in social and behavioral geographic research where new insights have developed into the roles of women in society, space, and behavior.

15.2.1   An Epistemological Basis for the Lack of Attention
to Gender Issues

It is interesting to speculate why it is that, for the most part, human geography researchers have assiduously avoided questions that embrace half the human race (Monk & Hanson, 1982:12). It has been suggested that this is because knowledge is a social creation, and that until recently too few women were involved in research. Geographers were more concerned with studying spatial dimensions of social class, for example, than of social roles, such as sex roles. Also, the logical positivist tradition, with its separation of facts from values, meant that most geographic research was little concerned with leading social change (King, 1976).

Significantly, alternative paradigms have done relatively little to incorporate a feminist perspective. For example, Marxists may have championed social change, but few have explored explicitly the effects of capitalism on women. Phenomenologists have provided a more humanistic geography, but they have produced few insights into the lives of women (Monk & Hanson, 1982:12). Also, although much research in human geography has sought to provide a more rational basis for informed decision making, it is evident that most planning—and probably most planners—

gives little if any explicit attention to the role and needs of women per se. This is quite evident in fields such as transportation planning.

Feminist social scientists have been questioning what have been, and probably what largely remain, the definitions of concepts such as *status, class, work, labor force*, and *power*, which tend to reflect male spheres and attitudes.

15.2.2 Gender Bias in Geographic Research

Monk and Hanson (1982) provide a list of problems relating to research content and method in human geography that reflect gender bias. These problems include:

1. *Inadequate specification of research problems*, as evidenced in much of the research on suburbanization, which neglects a focus on the role of women.
2. *The gender bind theory*, where the omission of gender as an explanatory variable can impoverish explanation, as evidenced in much of the work on travel and trip behavior.
3. *The assumption of traditional gender roles*, where the focus is on social or biological roles, as evidenced in social area analysis and factorial ecology studies of cities.
4. *The avoidance of research myths that directly address women's lives*, where in many cultural and urban geographic studies, the specific role of women and their explicit needs are neglected, as they have been in much of the work on health care, retailing, and transportation.
5. *The selection of data variables*, where often it is assumed that global variables, such as occupation, vehicle ownership, or old age, are adequate measures, whereas they rarely provide an adequate indication of the sex-related differentials that are inherent in such constraints.
6. *The selection of respondents in surveys*, where often the head of household is chosen as the person from whom or about whom data are collected or where randomized selection of individuals may mark gender differences in attitudes and behavior within and between members of the household.
7. *Inadequate secondary data sources*, where, for the sake of convenience, data on women are omitted, as in most migration studies.

In policy research in particular, it is crucial that gender issues are addressed explicitly. If they are not, then grossly simplistic and inappropriate policy formulation and program design can result. A continuing problem is evident in the field of urban transportation planning, where models and policies almost invariably are based on the journey to work for the full-time worker as a single-purpose trip.

15.3 Rethinking Urban Residential Structure from a Feminist Perspective

A central concern of social geography has been the study of the residential structure of the city, and in the 1950s, 1960s, and 1970s, there were literally hundreds of studies conducted using social area analysis and factorial ecology to establish the dimensions of urban social structure and the spatial patterning of those dimensions in metropolitan areas. Subsequently, the investigation turned to seeking explanations of that urban residential structure and its implications. Since the 1970s there have been sweeping changes to the demographic and social structure of households, fundamentally changing the structure and pattern of residential areas.

15.3.1 A Critique of Social Area and Factorial Ecology Analysis

Pratt and Hanson (1988) provide a critique of the theoretical weaknesses of the social area and factorial ecology-type studies in view of these changes. This critique indicates that their dependence on secondary data available from the census, and their tendencies to assume homogeneity with respect to categories of occupations and households undifferentiated by sex, mask the real differences that exist. There exist high levels of sex segregation in the labor force (Reskin & Hartmann, 1986), with men and women holding very different positions. In addition, it is likely that the geography of social class is also different for men and women, and if this is so, then "the social geography of the city is a great deal more complex than the standard models of residential structure would lead us to believe, and causes tracts that may be homogeneous with respect to male social class to not be homogenous when both men's and women's social classes are considered" (Pratt & Hanson, 1988:15–16).

Women were not, of course, ignored completely in social area analysis and factorial ecology methodologies, with the dimensions "familism" or "urbanization" explicitly recognizing fertility, female participation in the labor force, and single-family dwelling units. But these reflected gender stereotype roles of the time.

Social theorists in the 1970s and 1980s also assumed residential areas to be homogeneous with respect to class. For example, Scott (1985:488) wrote: "in the modern American city, a particularly sharp and deeply entrenched line of disjunction can be drawn between white-collar occupations and blue-collar occupations," and such a residential structure was considered to both mirror and reproduce class relations in capitalist societies, with occupationally differentiated neighborhoods assisting the smooth social reproduction of human attributes, attitudes, and behavioral patterns essential for labor markets to function. The notions were that social reproduction facilitates the intergenerational transfer of skills and that the class composition of neighborhoods determines the shape of intraurban labor markets. While advocates of this theory acknowledge that ethnicity and other "life-style" factors additionally influence residential segregation, nevertheless the view is restrictive. In addition, "it is assumed that women who work can be classified in terms of their husbands' occupational standing, and the fact that the conditions of women's labor force participation are radically different from men's is ignored. As is typical of class-based theories, the issue of gender is ignored" (Pratt & Hanson, 1988:18).

There are some factorial ecology studies that have not relied exclusively on male occupations and income data as components of the variable set for analysis and identification of social dimensions. For example, Patterson's factorial ecology of the Vancouver region (cited in Ley, 1983) analyzed 192 variables, including many that measured characteristics of labor force participation of women, and two of seven factors identified were clearly related to the economic status construct, one appearing to measure male labor force participation and the other female participation. The factor loadings of variables across the two factors clearly differentiated between employment categories of industry sectors that had marked sex differences in concentration of workers. The researcher labeled the dominant factor "socioeconomic status" and concluded that the split of the construct into two factors was an artifact of the use of small-area data. However, as Pratt and Hanson (1988:17) point out, "an alternative explanation is that the spatial distribution of female clerical and retail workers differs from that of workers in traditional male-dominated jobs."

The neglect of gender also leads to questions about the reproduction argument in factorial ecology studies, especially in view of contemporary demographic trends that no longer support the stereotype of the nuclear family maintained by a male breadwinner. With the increasing entry and staying of women in the labor force, lower-income workers in a dual-salary household can achieve a household income level that permits them to afford a home in the same residential area as a household supported by a higher-income male worker. Thus "variability in the employment patterns of women may loosen the link between individual incomes, occupational class, and residential segregation" (Pratt & Hanson, 1988:21).

15.3.2 The Interaction between Gender and Class in Residential Segregation

There is a lack of a theoretical base for residential segregation that reflects the interaction between gender and class. For example, the segregation of women into low-paying manufacturing or service jobs means that families headed by women are extremely vulnerable to poverty, with well over one-third of United States female-headed households being poor compared with only about 6 to 7% of two-parent households. Thus, if single-parent, female households are spatially concentrated in the city, then there is a new basis for income-related residential segregation.

In a study of Worcester, Massachusetts, Pratt and Hanson (1988) use small-area census data to measure class and income by sex. They use 1980 Public Use Micro-Data One Percent Sample (PUMS) for the Worcester small metropolitan statical area (SMSA), and two special runs from the census journey-to-work file for the Worcester SMSA. The study is limited by the lack of PUMS data at a finer scale than simply the "central city" and "rest of the SMSA," thus restricting the comparison of male and female class and occupational patterns to city versus suburb rather than at the census tract level. Pratt and Hanson (1988) investigated five propositions:

1. *Female labor-force participation exhibits relatively little spatial segregation.* This is supported by mapping the spatial distribution of the proportion of women over 16 years employed in the labor force, and it is demonstrated that women's class standing did not vary across space as much as did that of men.

2. *Men's income and occupational class are higher than women's.* Analyses show that this is the case and that men and women are differently distributed over the class categories.

3. *Male income and occupational class measures show more spatial variation than do female income and class measures.* Analysis of contingency tables arraying wives' against husbands' class and status supports the contention, with husbands' incomes being higher than those of their wives, and with there being considerable interhousehold heterogeneity with respect to husbands' and wives' occupational classes.

4. *Women's jobs are very different from men's with considerable heterogeneity in social status within households.* A breakdown of jobs into industry sectors confirmed this expectation, and confirms the heterogeneous nature of neighborhoods with respect to male and female occupational class; this holds for an analysis within census tracts.

5. *Female-headed households constitute a new gender-related basis for segregation.* There is a clear concentration of female-headed households within the inner city (at least 25% of all households and as high as 67% in one census tract), reflecting, among other things, the pattern of distribution of public housing.

15.4   Home–Work and Work–Home Relationships

The home–work relationship has been studied by numerous feminist researchers during the last decade or so to investigate the interdependencies between *production* (the workplace) and *reproduction* (home) as part of a rethinking of both residential and workplace locations and how households make locational decisions that conceptualize them as one rather than as separate spheres (Hanson & Pratt, 1988a).

Hanson and Pratt conducted a survey of 600 households in Worcester, Massachusetts, in 1987 using in-depth personal interviews with both men and women of working age. "The main purpose was to probe the ways in which individuals and households decided where to live and where to work" (Hanson & Pratt, 1988:305a). They looked specifically at three issues:

- The effects of home on work.
- The effects of work on home.
- The interdependence between home and work in the local context of the study area.

From this and other studies, it is possible to identify a number of important new perspectives as a result of taking a deliberate focus on gender issues per se.

15.4.1.   Effects of Home Location on Workplace Selection

Taking the perspective that "the work decision—including whether to work full-time, part-time or not at all; where to work; and the type of work to pursue—is a social process" (Hanson & Pratt, 1988a:306), it is shown that the work decision is less automatic for women than it is for men, and that the home environment is important in the work decision both with respect to potential employment opportunities and as a source of support services such as child care. The Worcester study shows that many households did not choose their residential location through a search process that took account of work location, and that this is particularly the case for women.

It is argued by feminist researchers that "the residential suburbanization process in North America effectively removed middle class women from the labor force in the early Post World War II era. Trapped in a residential environment, often without consistent access to a car, and constrained to work close to home by familial responsibilities, suburban women were effectively cut off from opportunities of working outside of the home" (Hanson & Pratt, 1988:307). This "spatial entrapment" of middle-class, white housewives has "spawned a spatial realignment on the part of certain types of employers" (Hanson & Pratt, 1988a:307), as evidenced by the proliferation of diffusion to suburban locations of many office activities that are dominated by female labor, such as back-office data-processing functions. In addition, research in Vancouver by MacKenzie (1980) shows how suburban women also create their own employment locally in activities such as child care, craft manufacture, and small business retailing. The importance of local social networks in the home–work relationship for women was noted also in the Worcester study (Hanson & Pratt, 1988a:308). These findings indicate that the concept "home" needs to be extended to include at least the surrounding neighborhood and intrahousehold interactions, and to encompass as well the potential jobs, social networks, and services that are important in the work decisions of women. For example, Nieva (1985) argues that because women bear the main responsibilities for maintaining the household and for child raising, their employment decisions need to be considered in the context of the division of power and labor within the household. The Worcester study by Hanson and Pratt (1988a:310) shows how there are interdependencies among the work, child care, and home maintenance responsibilities of women.

15.4.2 Effects of Workplace on Locational Choice of Home

For a long time it has been demonstrated that higher-status occupation managerial groups are more mobile, particularly in making interurban moves, than lower-status groups. The Worcester study (Hanson & Pratt, 1988a:310) identifies that only 38% of male household heads in the management or professional categories have lived more than 90% of their lives in the Worcester area, compared with 64 to 66% for skilled and nonskilled manual male workers. However, only 77% of women managers or professionals have selected their home location before their job, compared to 95 to 100% of women in the skilled manual and nonskilled occupations. Thus there are major sex-related biases in the influence of work on home location as well as occupational status or class influences. In addition, the presence of two workers in the household "both increases and narrows residential location options. On the one hand, the increased income extends the range of homes and neighborhoods that are affordable. On the other hand, the residential location must accommodate two job locations" (Hanson & Pratt, 1988a:312).

A major if somewhat dated study by Madden (1980) analyzing the Panel Survey of Income Dynamics data, shows that the location decisions of two-worker households are not different from one-worker households once variables for household size, income, age, and race were controlled. But there are significant differences in the location decisions of single as against married individuals, with the former living closer to work and in smaller housing units. There also are significant differences between households with children as against those without children, with larger households trading off housing quality for larger housing units.

Other studies show that two-income households are less likely to move than one-worker households but that often, women chose jobs that are portable in order to fit more readily into their spouse's job-related residential mobility (Markham, 1987).

15.4.3. Social Relations and Status

There is some evidence to suggest that participation by women in neighborhood and community activities declines with their entry to the labor force (Pincus, 1986). Longitudinal studies demonstrate that paid employment brings a change in the types of organizations that women belong to and in which they participate and that there is a shift to functional professional activities (Edwards, Edwards, & Watts, 1984; Schoenburg, 1980). Hanson and Pratt (1988a:313) found in their Worcester study that 48% of women employed in paid work reported having no friends compared to 35% of women not working for a wage. It has been shown also that two-worker households tend to develop work-based more than neighborhood-based social interactions, which has implications for the meaning of the neighborhood concept.

A further important issue that has arisen since the late 1970s is the impact of escalating house prices in many cities with the implications for affordability for households being linked to the necessity of maintaining a two-income pattern of work activity. In a study in Vancouver, Pratt (1986) interviewed a sample of single-earner households who were renting and found that they were prepared to forego the preferred tenure of home ownership to continue to rent in order to maintain their single-earner household status. In addition to housing markets, local urban economies vary in their mix of jobs and industry sector diversity, thus affecting employment opportunities and residential location and housing-choice decisions. Abrahamson and Sigelman (1987) have examined the effects of occupational structure, metropolitan size, population change, unemployment rate, proportion of families with children, percentage of women in the labor force, and median female education on the occupational sex segregation in cities.

A final interesting issue arises as a result of the emergence in some city regions—such as London, New York, and Tokyo—of major international financial and producer services centers and the rapid expansion of employment in these jobs that require long hours of work and their impacts on residential decisions. Leyshon, Thrift, and Daniels (1987) show how in London this led to a substantial demand for upscale housing in inner London, and work by Ley (1986) in Vancouver shows that employment growth in the downtown in producer services spawned a significant gentrification process.

Home and work are interwoven in complex ways, and as Hanson and Pratt (1988a) suggest:

> There is a need to think of how the home–work link functions for a variety of diverse household types and in a variety of local contexts. A related point is the need to make explanations scale specific; one group for whom the workplace does have priority tends to be middle-class males who are making career-related inter-urban moves. It is at the regional (or national) scale, then—not the intraurban—that the original assumption of workplace preceding residence rings truest. (pp. 318–319)

15.4.4 Commuting and Work: Gender, Occupation, and Race Differences

There now exists a substantial body of research to show that, in advanced western industrialized societies, women work significantly closer to home and have shorter commuting times than men (Hanson & Johnston, 1985; Howe & O'Connor, 1982). However, it appears that the difference between men and women is such that for women it is proportionately less in time than in distance (McLafferty & Preston, 1991:2).

Research indicates that explanatory factors may relate to the responsibilities of women for child care and patriarchal relatives within the home (Madden & White, 1980). However, there is evidence to show that the number of children and their ages and the existence or nonexistence of children does not vary the length of women's work trips. Rather, it is more likely that the differences relate to the different class positions of women in the labor force, with women being more likely to hold low-wage and part-time jobs that provide little payoff for commuting long distances (Bergmann, 1986; Madden, 1981; Rutherford & Wekerle, 1988). And at the intraurban level, the spatial distribution of women's jobs is such that female-dominated occupations are uneven and differ greatly from that of male jobs (Blumen & Kellerman, 1990; Hanson & Pratt, 1986). For example, some firms, especially those concentrating on back-office and clerical functions, intentionally choose suburban locations that are close to a captive female labor pool (Nelson, 1986), whereas firms employing large numbers of women may cluster for other reasons, such as agglomeration or lower land transport costs. Also, in community-service-sector activities such as hospitals, and in the retail and hospitality sectors, there tends to be a high degree of spatial concentration at specific central city and suburban nodal locations. But there is evidence also of dispersed patterns for some occupations.

McLafferty and Preston (1991:4) have noted how many studies: "imply that the spatial structure of opportunities affects women's commuting times and distances, just as it affects all types of spatial interaction. But further research is needed to tease out the relationships between the location of employment opportunities and women's commuting patterns."

It is significant that different racial and ethnic groups among women also display differences in these relationships, reflecting, for example, the well-documented differences in occupational status of white woman and Hispanics, whites and blacks, and blacks and Hispanics in United States cities. These differences also reflect the considerable differences in residential spatial segregation among these groups.

There are differences too in the household characteristics between white and minority women, with the latter being much more likely to be single parents, facing enormous responsibilities for child care and financial support (Simms & Malveaux, 1986). The effects of single parenthood on commuting are complex but little understood, and it may be that as a rule wage earners who are single female parents may be forced to endure longer work trips to find decent-paying jobs in cities (Stolz, 1985).

A major study conducted by the Port Authority of New York and New Jersey (1986) across the 24 counties that constitute the New York Consolidated Metropolitan Statistical Area (CMSA) in 1980, clearly shows significant differences between male and female commuting times, further differentiated by race and occupation, with the latter having shorter commutes. Table 15.1 gives the results. White people spend less time traveling to work than black people; managerial workers have the longest commute times among the occupational groups, particularly for the advanced producer services and distributive worker occupations; and overall, income and race produce the largest impacts on commuting time. More disaggregated analysis using analysis of variance shows that the commuting times for black women are consistently similar to those for black men. In five of six industry categories they commute slightly longer, reflecting the high concentration of black female employment in domestic jobs involving long commutes to high-status suburbs. Hispanic women displayed similar patterns. These gender effects diminish when specific industries are analyzed to account for race and income. It is possible that the differences are exacerbated by the predilection of many employers to locate near white female labor.

TABLE 15.1
Mean Commuting Time by Occupation, Gender, and Race In New York and New Jersey

Factor		Commuting time (min)		
		Mean	Standard Deviation	n
Gender	Male	33.9	24,3	69,998
	Female	28.3	22.5	60,675
		$F = 1941.0*$		
Race	White	29.9	23.5	102,182
	Black	37.2	23.7	18,117
	Hispanic	34.9	23.0	10,110
		$F = 849.7*$		
Occupation	Blue collar	30.2	22.4	18,956
	Pink collar	30.1	23.3	59,007
	White collar	30.3	25.3	25,208
	Managerial	35.7	25.3	27,502
		$F = 459.0*$		

*Significantly different from zero (at $F = 0.05$)
Source: McLafferty & Preston, 1991:8.

15.5   Do Spatial Abilities Vary between the Sexes?

Self and Golledge (1994) and Self, Gopal, Golledge, and Fenstermaker (1992) suggest that spatial ability is one of the few aspects of cognitive functioning in which evidence has been produced that male and female subjects can perform differently on some spatial tasks. But the general literature appears to be inconclusive.

15.5.1   Sex Differences in Spatial Aptitude

For every spatial aptitude task on which men perform at a superior level, other spatial aptitude tasks on which women's performance is superior can be cited. And indeed, on some tasks simply by changing the measure of aptitude or ability (e.g., from response times to accuracy), the dominance of one sex over the other alternates.

Although instances of male or female dominance on some spatial tasks can be cited, there appears to be no overwhelming evidence at this stage of the consistent dominance by one sex. Increasing evidence, however, is pointing to the fact that there are some tasks (e.g., geometric rotation and visualization tasks) on which there is a tendency for men to perform better.

But there is an increasing indication that women pay more attention to the details of an environment and that they are better able to externally represent such environments in verbal or pictorial form. There is also some preliminary (and somewhat inconclusive) evidence that women perform spatial tasks such as navigation and sketching better if they are provided with sets of dominant landmarks to anchor their representations. Men, on the other hand, are said to do better than women at navigational or sketching tasks if only frames of reference are provided. However, geographers Gilmartin and Patton (1984) and Gilmartin (1986) found no significant differences between the sexes in terms of their development of map-use skills.

15.5.2   Sex Differences in Spatial Behavior

While it is uncertain that there are differences in spatial abilities between male and female individuals, there is some evidence of different behavior patterns.

Some recent work by Golledge, Bell, and Dougherty (1994) does indicate that women increase the accuracy of their responses more than do men when given multiple trials at solving a geographic problem. This may indicate that the abilities are equivalent but that societal and cultural gendering practices (e.g., in terms of role expectations, aspirations, or familial influence on behaviors and activities during the formative years) may have resulted in poorer preparation of female individuals to handle spatial problems. The fact that they respond so well to teaching, however, such that afterward their performance cannot be statistically differentiated from that of their male peers, is strong evidence of an equivalence of abilities in the spatial domain.

Thus, Self and Golledge (1996) suggest that even if differences are found between the spatial abilities of male and female individuals, these can often be mitigated by training, reinforcement, and repeated trials. In other words, any recorded lower performance by female subjects is not necessarily due to the lack of ability or a sex-related difference but, rather, to a lack of opportunity to develop spatial skills.

15.5.3 Some Experiments

Newcombe, Bandura, and Taylor (1983) defined a list of about 80 activities that a panel of judges deemed to be spatial (Figure 15.1). These activities were sex-typed (i.e., labeled predominantly male activities, predominantly female activities, or neutral). A significant proportion (40) were classed as dominantly male activities, and 21 were classed as dominantly female activities.

Over a decade later, Self and Golledge (1994) repeated this experiment and indicated a substantial paradigmatic shift in people's evaluations of the degree to which types of spatial activities were perceived to be dominated by one sex or the other. They found that only eight of the activities were classified as dominantly male activities, only two were dominantly female activities, and an overwhelming proportion were classified as *neutral* (Self, Golledge, & Montello, 1995). Table 15.2 lists the results.

The production and reading of maps is an essential spatial task critical to the discipline of geography. Gilmartin (1986) has examined the ability of male and female subjects to construct mental images from verbal descriptions and male–female differences in the ability to recall geographic information. She suggested that although some evidence was found concerning male dominance in imagery and visualization tasks, there was little evidence of a similar type of dominance when she examined a variety of other map-reading and map interpretation tasks.

15.5.4 Discrimination and Equal Opportunity Issues

As the 20th century draws to a close, there are strong societal currents that address problems of discrimination and equal opportunity. To ensure this process is successful, there is an ongoing need for more information at the individual, group, and societal level concerning the abilities, beliefs, and activities of traditionally disenfranchised groups.

The extreme paradox is the simple fact that women represent more than 50% of the world's population, but in many circumstances, they are still classified as "The Other." Current activity has been concentrated on encouraging women to enter traditionally male-dominated professions—particularly in scientific, technical, and managerial fields.

In this surge, geography can play a part. Often seen as a "soft" science, geography is an ideal waystation and training ground for women. Here they are exposed to the spatial skills that often appear in other "harder" sciences. But these skills are often developed in a nonhostile environment that does not require the same comprehensive background (e.g., in mathematics) that typifies requirements in other physical and life sciences. This is not to say that geography does not need or use formal mathematics; it does in virtually every geographic domain—human, physical, and technical. But background skills are often taught *in situ* and in an applied context.

For those (both men and women) who see formal math training as a threaten-

Sex-typed activities are activities generally viewed as more typical either for males or females. From your current point of view, please sex-type these spatial activities listed below. Classifications to be used are: masculine (m), feminine (f) or neutral (n). If anything listed does not, in your opinion, appear to involve spatial ability, please circle it after you classify it. In addition, indicate your level of participation in each activity by marking (0) for those activities you have done no more than once, and (1) for those you have done more than once.

Sex- Your
type part.

____	____	touch football
____	____	tackle football
____	____	bowling
____	____	field hockey
____	____	figure skating
____	____	water ballet
____	____	pingpong
____	____	ice hockey
____	____	horseshoes
____	____	hunting
____	____	navigate in car
____	____	glass blowing
____	____	carpentry
____	____	touch typing
____	____	photography (adjusting focus)
____	____	juggling
____	____	mechanical drawing
____	____	marching band
____	____	layout for newspaper, yearbook
____	____	knitting
____	____	target shooting
____	____	tailoring
____	____	building model planes
____	____	building train or racecar sets
____	____	building go-carts
____	____	painting (three dimensional)
____	____	drawing (three dimensional)
____	____	gymnastics
____	____	ballet (pirouettes)
____	____	ballet (choreography)
____	____	softball
____	____	advanced tennis
____	____	soccer
____	____	volleyball
____	____	darts
____	____	squash
____	____	sculpting
____	____	leatherwork (with seams)
____	____	arranging furniture
____	____	interior decorating
____	____	car repair
____	____	plumbing
____	____	jumping horses

Sex- Your
type part.

____	____	archery
____	____	golf
____	____	rock climbing
____	____	canoeing (shooting rapids)
____	____	dodgeball
____	____	baton twirling (toss in air)
____	____	baton twirling (more than 1 baton)
____	____	pottery (wheel)
____	____	frisbee
____	____	basketball
____	____	baseball
____	____	beginning racquetball
____	____	advanced racquetball
____	____	jewelry (mount stones)
____	____	fencing
____	____	high jumping
____	____	disco dancing (with falls)
____	____	tap dance (own routine)
____	____	make/fix stereos, radios
____	____	sketch clothes designs
____	____	sketch house plans
____	____	sketch auto designs
____	____	quilting
____	____	shooting pool
____	____	pole vaulting
____	____	ski jumping
____	____	ski slalom
____	____	skateboarding
____	____	embroidery (no pattern)
____	____	foosball
____	____	air hockey
____	____	electrical circuitry
____	____	sledding (around obstacles)
____	____	surfing
____	____	video games
____	____	bike riding
____	____	roller skating
____	____	checkers/chess/board games (Monopoly, Life, etc.)
____	____	using compass
____	____	weaving (design own wrap)
____	____	diving

Figure 15.1. Sex-typing of spatial activities. *Source:* Self & Golledge, 1994.

TABLE 15.2
Pilot Results from the Sex Typing Experiment

75% or more of subjects rated these activities:

Masculine	Feminine	Neutral
Tackle football	Water ballet	Softball
Baseball	Knitting	Advanced tennis
Ice Hockey		Volleyball
Hunting		Jumping horses
Building model planes		Diving
Building train or racecar sets		Drawing
Make/fix radios, stereos		Painting
Sketch auto designs		Sculpting
		Photography
		Navigate In car
		Layout for newspaper/yearbook
		Marching band

Source: Data collected by Carole Self, 1989

ing barrier, such a practice provides a halfway house or gateway to more formal science and technical expertise.

Geography, along with its traditional training practices, takes on a new role in the future. By stressing that women can develop similar types of spatial skills to those commonly attributed to men and by developing educational programs that allow the development of socially and culturally retarded (or gendered) abilities, geography can play a role in removing discrimination and differentiation based on sex, and, in so doing, lead the way into the future rather than trail behind other disciplines as they make their move.

15.6 The Elderly

Another increasingly important component of the population is elderly people who represent one of the most rapidly increasing groups in modern society, since as age-specific mortality rates decline, age expectancy increases. And the aged tend to include more women than men, especially with increasing age.

15.6.1 Diversity in Aging

The elderly population includes persons of diverse economic, social, ethnic, and psychological characteristics (Golant, 1986). People age differently, and elderly people differ widely in their experience with the incidence and severity of debilitating functional impairments or chronic disease associated with biological aging. They also differ in how they cope. It is necessary to conceive old age as being two substages of human development in which there is:

- The *old–old* population, roughly aged over 75 or 80 years.
- The *young–old* population, roughly aged 60 to 74 or 79 years.

Retirement is not always a choice of aging persons, and once retired, they exhibit considerable differences in how they spend their time. While an increasing number of retirees are superannuated and are relatively secure financially, the majority are dependent on either the aged pension or on low incomes that have deteriorated in the low interest-rate period of the 1990s. The incidence of poverty among the aged remains high. Increasing longevity of the aged is extending periods of good health and making it less appropriate to view old age as a unitary period of life; but at the same time, there is a massive increase in the impact on the public health budget as aging of the population progresses.

The unprecedented growth of the population aged 65 years and over in the United States, and in particular the rapidly increasing size of the old-old population (persons over 75 years of age), and the further increase in their numbers that is assured because of the inevitable aging of larger cohorts of younger populations and the promise of longer life expectancy, have placed gerontological issues at the forefront of societal concerns and have produced a fertile field for researchers. Golant, Rowles, and Meyer (1988) have identified three major issues that have emerged that are relevant to the elderly population and their housing and living needs:

1. It has become apparent that elderly people might occupy, utilize, and experience environments in ways distinctively linked to the aging process, and it appears likely that older people may respond or cope differently with environmental stresses than younger people. This implies that the locations and living environments of older people might be modified or manipulated to relieve age-related stress.
2. The social problems, injustices, and inequalities that afflict old people are linked to their location.
3. Interest has developed in the distinctive social, political, economic, and aesthetic features of human settlements that contain large elderly population concentrations, and in the impact of such concentrations of older people on the quality of life of younger residents and on surrounding communities. (pp. 451–452)

Here we review some of the experiences that relate to three broad aspects of aging and the housing needs and characteristics of the elderly, namely:

1. The patterns, antecedents, and consequences of elderly people's residential location choices and residential mobility/migration behavior.
2. The utilization (activity patterns), meanings (perceptions and interpretations), and impact of elderly people's residential environments.
3. The characteristics, options, and trends in retirement community housing provision for the elderly, and the implications for planning for housing environments for the elderly.

15.6.2 Mobility Patterns of the Elderly

Research in both the United States (Golant et al., 1988) and Australia (Hugo, 1984; Hugo, Rudd, & Downie, 1984) has shown that residential location and migration patterns of the elderly since the 1970s tend to exhibit the following characteristics:

1. There are significant concentrations in the inner cities, in rural and economically depressed nonmetropolitan areas, in regions experiencing out-migration of the younger populations, and in numerous retirement-oriented and/or amenity-rich regions.
2. There is a tendency toward suburbanization, movement into nonmetropolitan, retirement-oriented enclaves, an increasing dispersal to an increasing diversity of locations outside traditional retirement regions, and the growth of elderly population concentrations as a result of planned retirement housing projects in all locations.
3. The vast majority of the aging population retire *in situ*.

Specific social and economic characteristics of places that are *destinations for in-migration* of elderly people and characteristics of places *losing* elderly people are shown in Table 15.3.

Research in a number of different situations has identified specific individual and household characteristics of elderly people that account for both their local and long-distance mobility and their migration patterns. These include:

- Income, gender, employment status, marital status, ethnic status, and social status.
- Family ties, especially to children.
- Place of birth.
- Amount of previous mobility experience.
- Need for assistance because of poor/deteriorating functional health.
- The need to reduce living costs.

TABLE 15.3
Characteristics of Places Gaining and Losing Elderly Populations

Attracting Factors

- residential environments distinguished for their retirement
- amenities (e.g., oceans, lakes, golf courses),
- retirement housing accommodation and retirement communities
- high socioeconomic status
- 'sun belt' climates
- lower housing rents
- suburban settings
- the existence of large concentrations of retired people

Diffusing Factors

- older urban cores
- rural, economically depressed areas
- isolated areas

Source: After Golant, Rowles & Meyer, 1988

- Previous residential ties.
- The location of former friends.

Studies in the United States (Golant, 1984a; Wiseman, 1987) found that those most likely to move are, on the one hand, the older homeowners who feel that caring for their houses is too expensive, takes too much time, or is too tiring, and on the other hand, renters who feel that their buildings are not kept clean and in good repair. Fear of crime, poor weather conditions, difficulties with neighbors, and dissatisfaction with neighborhood facilities were also factors triggering relocation.

However, on a national basis in the United States, Bohland and Rowles (1987) found that the growth of the elderly population was becoming more locationally widespread, that older people are moving to more dispersed locations, and that they are no longer restricted to traditional retirement and/or amenity locations of the Sun Belt. A consequence is that fewer counties are experiencing significant regional increases in their percentage of elderly people, with increasing trends toward aging-in-place and dying-in-place being in part responsible.

The *inertia of aging-in-place* is strong and involves the high level of overall immobility among elderly populations. It highlights social and psychological factors that tie old people to their homes, the status quo, a familiar neighborhood, friends and family, and it is also often associated with low income, poor health, and perceptions that more suitable housing is not available.

An interesting aspect of the migration of the elderly in the United States has been the rise of planned communities for older people, such as Sun City, Arizona (Gober, 1985) with distinctive physical, organization, and population age-segregated characteristics reflecting a new cultural and social value.

Special-purpose elderly housing is also on the rise (Golant, 1987), and these units are proliferating in number and options that incorporate elements that have traditionally been outside the realm of conventional housing, such as commercial food services, paid custodial care, formal health-care provision, and family caregiving, all located within the same housing space.

15.7   Retirement and Residential Relocation

15.7.1   *Push–Pull* Factors and Segmentation

The movement of retired persons from where they had lived while working prior to retirement to where they relocate during retirement has been studied both in the context of *push–pull* factors and *segmentation* among the cohort (Loomis, Sorce, & Tyler, 1989). *Push–pull* factors are identified in Table 15.4.

Segmentation is also evident in retirement relocation. Warner (1983) has identified three groups of the elderly population, each of which have different motives for changing their housing, namely:

1. The *recreation and amenity market,* where the prime motivations are primarily to family and friends who may have located in the area.

TABLE 15.4
Push-Pull Factors in the Residential Relocation of Retirees

Push Factors

- an inability to care for one's home
- loneliness and isolation
- poor condition of present dwelling
- fear and crime
- insecurity
- death of a family member or close friend
- a change in health, finances or social activity or an anticipated change
- being a tenant

Pull Factors

- the attractive elements of a new home
- companionship
- safety and security
- social activities
- cost
- pressures of family or friends
- locational attributes of the disadvantaged
- availability and quality of health care

Source: After Loomis, Sorce & Tyler, 1989

2. The *metropolitan convenience market*, where the primary motivation is a desire to maintain home ownership status but with decreased maintenance responsibilities—especially among single elderly women and couples in somewhat poor health.

3. The *supported independence market*, where the main participants are older people who seek on-site medical assistance and stability.

Some studies have clearly differentiated between the motives of elderly people who move to a continuing care community and those who are attracted to other retirement housing options. Hunt, Feldt, Maraus, Patalan, and Vakalo (1983) found that *continuing care* residents are attracted by the medical and financial security offered, whereas the choice of *new towns* or *retirement subdivisions* is because of the desire for beginning or maintaining an active life of leisure as the main source of appeal, hence the attraction of sun belt, lakes, the seaside, and golf-course location.

In a survey conducted in New York State concerning the conditions of healthy, middle-income people aged 60 years and over whose income qualified them for retirement community housing, Loomis et al. (1989:32–33) found that the level of interest in this form of housing alternative was related to both psychographic and demographic variables such that:

1. Extroverted people, who consider themselves to be quite security minded and physically active and fairly self-reliant, expressed a higher degree of interest in the proposed retirement community than did people who were highly family oriented, not at all self-reliant, nor very extroverted.

2. Single, widowed, and divorced retirees expressed more interest in the proposed project than did married people.
3. Renters expressed more interest in the project than home owners.
4. People who had only one relative in the local area indicated a higher level of interest in the project than was expressed by retirees with two or more family members in the area.
5. Women were more likely than men to move as a result of a physical inability to care for their present home.
6. Men were more likely than married people to consider moving because of illness.
7. People who lived with only one person cited widowhood as a reason to move more often than living with two or more people.

15.7.2 Environment and Well-Being

Golant et al. (1988) identify three major issues that are relevant to the elderly population and their housing and living needs.

1. Elderly people are occupying, utilizing, and experiencing environments in ways distinctly linked to the aging process, and they may also respond to or cope with environmental stresses differently than younger components of the population, with the implication that the locations and living environments of older people might be modified or manipulated to relieve age-related stress.
2. Social problems, injustices, and inequalities that afflict older people are linked to their locations and environments, and it may be necessary to develop age-specific policies and solutions focusing on changes in location or modification of environment.
3. There is increasing interest in the distinctive social, political, economic, and aesthetic features of human settlements that contain large elderly population concentrations and in the impact of such concentrations for the quality of life of younger residents and on surrounding communities.

Research conducted mainly in the United States suggests that the way elderly people perceive and interpret their residential environment and its impact on their lives is an important factor influencing their well-being and generating their propensities to relocate. A variety of attributes specific to the home, the neighborhood, the community and the region may impinge either positively or negatively on levels of well-being, in addition to factors affecting the physical and psychological health of the elderly person or couple. It has been suggested that the relationships between the elderly and their environments should be viewed within a theoretical framework, and that the concept of "person–environment congruence or fit" is a useful means in doing so (Marans, Hunt, & Vakalo, 1984; Kahana, 1982). The notion is that people vary in the types and relative strengths of their needs, while environments vary in their capacity to satisfy those needs. "A good fit, or congruence, occurs when the environment is able to satisfy individual needs. Conversely, when the environment is unable to satisfy those needs, there is a nonfit, or incongruence, between the person and the environment" (Marans et al., 1984:60).

An alternative approach is that suggested by French, Cobb, and Rodgers (1974), who suggest that there are two types of "fit" between the individual and the environment. One is represented by the degree to which the environment provides resources or supplies to meet individual needs. The second kind of "fit" considers the capabilities and resources of the individual in meeting the demands and requirements of the environment. A "misfit," or incongruence, of either kind is seen as threatening to the individual's well-being.

According to Marans et al. (1984), the environment of the elderly operates at two levels:

• The retirement community itself (*internal* environment).
• The wider community in which the retirement community is located (*external* environment).

The resources and services of these internal and external environments include public and private facilities and programs that range from health care and shopping to parks and transport. Social support systems (consisting of family, friends, and organizations) and opportunities for productive activity (employment and volunteer work) are also resources. The abilities and resources of the elderly can be viewed as a set of competencies (biological health, cognitive skills, etc.) and factors such as financial status and the like. The physical characteristics of the retirement community, the availability of services, rules, and regulations, and social norms represent further factors that impinge on the interaction of the individual resident with the retirement community environment. Additionally, links and exchanges occur between the elderly and their interior and exterior environments, and reciprocal relationships exist between retirement communities and their surroundings. For example, local government revenues can provide facilities and in turn are increased by the rate-revenue provided, and retirement communities can provide employment opportunities in the local economy.

Marans, Feldt, Pastalan, Hunt, and Vakalo (1983) have proposed a conceptual model showing the relationships among environment, related activities, health, and individual well-being that, in the context of retirement communities, bring together these factors that affect congruence or fit (see Figure 15.2).

Studies conducted among elderly urban populations in the United States by Golant et al. (1988) indicate that the factors that seem to be important in generating higher levels of satisfaction of the elderly with their residential locations and settings are:

• Accessibility to a pharmacy and doctor.
• Participation in group activities.
• Proximity and access to shopping centers.
• Security, safety, and friendliness of the neighborhood.
• The addition of energy conservation measures.

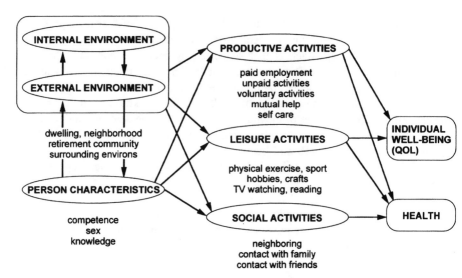

Figure 15.2. Conceptual model showing relationships among environment, selected activities, health, and individual well-being. *Source:* Marans, 1983.

It has been found that the activity patterns of older people provide important insights into developing an understanding of how the attributes of the housing and wider environment satisfy their daily needs and goals. Golant (1984b) has proposed that "the smaller the space–time locus of activity behavior, the more likely that people's environmental experiences derive from memories and fantasies of earlier mental transactions."

It is difficult to generalize about the frequency, regularity, purpose, and locational context of the activities of older people, as the way they utilize their environment reflects a range of factors, such as: age, sex, employment status, marital status, ethnicity, race, religion; health; personality; life stage; and economic status (Golant et al., 1988:456). The crucial issue is the degree to which these activity patterns result from individual preference or are involuntarily imposed on older people because of impaired behavioral or mental functioning and constraints, such as lack of access to a car, the absence of public transport, difficulties of night-time activity participation, fear of crime and safety, and lost social opportunities resulting from the death of a spouse or friend and the movement away of a friend.

It is important that issues such as these be carefully considered and researched to ensure that the location, design, operation, and management of a specific retirement community or complex incorporate those attributes, services, and functions that are appropriate for enhancing the achievement of maximum congruence or fit with the aspirations, perceptions, and needs of the particular client group(s) to which a development project is marketed.

15.8 Retirement Housing

15.8.1 A Typology of Retirement Housing

For the retirees who choose to change their place of residence, there are two important factors to consider: the situation of a *dwelling* and the choice of a *location* (setting and community) where that dwelling is located. Among the relatively small but still significant, and at times increasing, population of older people who relocate their residence, a number are choosing to live in some form of retirement community or complex. Retirement housing has become an increasingly important "niche" component of the housing market. Since the 1950s in the United States, a seemingly bewildering array of what are called retirement communities have developed, and since the late 1960s other countries such as Spain, Portugal, Costa Rica, and Australia have also seen the development of multipurpose retirement housing.

A wide range of definitions and classifications constitute what has been loosely referred to as retirement communities. Marans et al. (1984) refer to the different meanings the term assumed among researchers in the United States:

> Some consider retirement communities to be relatively self-contained small entities, spatially separated from large population centers. Others have used the concept to refer to any communal living arrangement composed entirely or primarily of retired people. Whether or not a community is located in an isolated setting, the common ingredient of the plan examined is the population, which for the most part is not employed or actively working in a regular occupation. (pp. 61–62)

They point out how retirement communities have been defined as:

1. Housing that involves congregate, segregate, institutional living (Webber & Osterbind, 1961).
2. A small community relatively independent, segregated, and noninstitutional, where population is mostly older people, separated more or less completely from their regular or career occupations in joining nonpaid employment (Webber & Osterbind, 1961).
3. Planned, low-density developments of permanent buildings designed to house-active adults over the age of 50 and equipped to provide extensive services and leisure activities (Barker, 1966).
4. Planned, low-density, age-restricted developments, constructed by private capital, and offering extensive recreational services and relatively low-cost housing for purchase (Heintz, 1976).
5. One form of planned housing for the elderly that is privately developed and self-contained, imposing an age limitation on their residents and containing purchased homes and shopping, medical, and active leisure services (Lawton, 1980).
6. Any subsidized or nonsubsidized living environment to which most of the residents have moved since they retired (Longino, 1980).

A typology of retirement communities was suggested by Marans et al. (1984) that incorporates four attributes: scale; population characteristics; level of services; and sponsorship. Five types of retirement communities and complexes were identified: retirement new towns; retirement villages; retirement subdivisions; retirement residences; and continuing-care retirement centers.

15.8.2   Attributes Sought by the Elderly in Retirement Housing

While the decision-making process for those retired and elderly people who move into some form of retirement community or specific-purpose, aged housing community is complex, it would seem that the major attractions are the availability of health care, the expectation of being able to live there for the rest of their lives, safety and security, and the specific locational attributes of the setting.

Studies in the United State have shown that those among the retired and elderly who are most likely to move into a retirement community or complex are:

- Married couples or widowed women.
- Aged in their mid-70s plus.
- Those with postschool education.
- Those from a professional or own-business background.
- Those moving directly from a single-family home that they owned.
- Those who lived alone or with a spouse before moving.
- People who feel in control of their lives.
- People who are still able to drive a car or have access to a car.

Interestingly, about two-fifths of those making such moves are in good health, while a further two-fifths are suffering from some disability or health problem that limits their activities (Parr, Green, & Behncke, 1989).

Surveys of residents of retirement communities in the United States have found that they have quite specific concerns about the community or complex in which they plan to live (Table 15.5). Also there are a series of *push* and *pull factors* that appear to be important in the decisions to relocate to a retirement community

TABLE 15.5
Major Concerns of the Elderly in Choosing a Retirement Community

Effective handing of maintenance
A caring and concern for staff
Clarity of policies and rules
Opportunities for residents to influence management decisions
An effective functioning resident's council
Sufficient privacy
Satisfaction with social interactions
Feeling free to have family and friends visit
Being kept informed about the management and operation of the community/complex
Access to appropriate health care
Overall value for money (not just cost)

(see Table 15.4). The decision-making process for those who move into a retirement community complex focuses on the availability of health care and the expectation of being able to live there for the rest of their lives along with safety and security as major considerations. But in general,

> residents are well educated, healthy individuals generally unaccustomed to group living and used to being in control of their lives. They are also future oriented. They are planners. Most don't need personal assistance or extensive health care now but want to be sure it is available in the future. (Parr et al., 1989:26)

It is clearly evident from the now considerable volume of studies that the decision of retirees to move and the degree of interest in a particular retirement community are related to *life-style*. As Loomis et al. (1989) note:

> As life expectancy of Americans increases, researchers will notice wider gaps in the lifestyles of people over age 60. Life long personality traits, abilities and interests, coupled with increasing health and financial concerns and family mobility patterns may result in people in these ages being more different than alike. America's cultural evolution of sales at any price is shifting to a marketing approach to sales and service delivery. No retirement community will appeal to everyone,—the market is too diverse to be satisfied by a single product. The most successful projects will be those that best understand to whom they appeal and why and which integrate these features into meaningful need-based programs. (pp. 33–34)

15.8.3 Policy and Planning Issues

Although it could be expected that, given the choice, the large majority of older people prefer to remain in retirement in the homes they have lived in during the years up to retirement, nonetheless, there is likely to be an increasing demand for the retirement community option. There are a number of factors to suggest that this is a continuing and increasing trend, including:

1. An increase in the number of retirees with sizable accumulated assets, including indexed retirement income schemes and equity in property through home ownership and second-dwelling investments.
2. Increasing awareness of the specific health care needs of older people and how these can be met in the context of specifically designed new living arrangements provided within retirement communities.
3. Increasing concern among the elderly over security and safety available in a protected and age-segmented environment.
4. The attraction of lifestyle-enhancing facilities, and especially leisure facilities, provided by retirement communities.

Local authorities tend to exercise most planning and development approval and land use zoning controls, and often they have tended to develop and implement town planning schemes and instruments that are characterized by residential zones and development practices that severely limit the locations where retirement community housing developments can occur, especially for those projects that incorporate institutional care facilities. It is common for retirement housing to be precluded from Residential Type A Zones and for there to be stringent "by consent" approval conditions.

Often retirement community developments are restricted by a local authority to future urban areas and to business and commercial zones, particularly where institutional care components are a significant part of a project. In addition, parking requirements often are unrealistically large, and plot ratio and density requirements often are too rigid.

There are, however, considerable variations in both the attitude of local authorities to retirement community housing developments and in the flexibility of both planning schemes and planning approval practices that can facilitate or constrain the development of innovative approaches to the provision of specific-purpose retirement housing developments.

There is potential merit in the widespread adoption of what are known as "performance-based criteria for assessment" in the planning and approval process at the local authority level in dealing with retirement community housing development projects.

What are some of the important emerging future issues and policy implications that can be drawn from the trends in mobility of the elderly population and the apparent increasing demand that has emerged for specialized housing and retirement communities for the aged? Some of these are canvassed below.

It has been suggested by Golant (1987:538–539) that it is important to: "consider how the personal characteristics of a place's in-migrants (for example their lifestyle, health status, values and personal resources) differ from those of its current elderly residents if one is going to speculate on the impact that a new population will have on a community's social and political fabric."

The United States experience clearly identifies how the migration streams of elderly people reveal a strong preference for the lower density and less congested living of the metropolitan suburbs, and as a result, the central cities are experiencing a decline in their older populations. It is apparent that the demand for specialized housing and communities for elderly people is increasing.

From a policy perspective, the question is being asked whether these new waves of suburban and nonmetropolitan elderly inmigrants will exacerbate the growing gap between the service demands of elderly people and the resources of these recipient cities to meet them? Investigations of the impacts of increasing concentrations of the elderly are becoming a major focus, especially in sun belt retirement states and in the newly emerging regional retirement areas such as western North Carolina and the lakeshore areas of northern Michigan (Golant et al., 1988:460). Issues for which answers are required include:

1. What is the impact upon local economics of the in-migration of young–old, bringing their retirement income spending potential with them?
2. What are the costs to cities when, as this population ages, it develops increasing needs for medical and social services?

3. How are local economics affected by the season arrival and departure of "snowbirds" (i.e., people who leave the "Snow Belt" for the Sun Belt) (Happel, Hogan, & Sullivan, 1983)?

4. Do elderly populations jeopardize service provision for the young by creating a powerful constituency able to vote down school bonds or, through their political influence, effectively veto community programs that are not of direct benefit to its members?

5. To what extent does the presence of large numbers of elderly people in the community result in the development of a distinctive sociocultural incline persuaded by norms of behavior and value systems that imbue the place with a gerontological personality (Golant et al., 1988:460)?

6. Does spatial segregation of the elderly, particularly when it is ineffectionalized and incorporated in zoning, increase the potential for intergenerational tensions?

7. To what extent is society obligated to maintain those sections of its elderly population who choose to live or remain in geographically remote but familiar, preferred, or segregated environments regardless of the cost of providing services that are needed?

15.8.4 Financing and Sponsorship of Retirement Communities

With respect to financing housing projects for the elderly, the most significant barriers that have emerged in the United States by the late 1980s were:

1. Increasing construction costs in general.

2. The decrease in FHA low-interest mortgage loans and rent subsidies for low-income elderly tenants making it more difficult to develop housing at a cost that is affordable for the low- and middle-income elderly.

3. Uncertainty about the continued expansion of the market.

4. Increasing financial failure of some newly constructed retirement communities leading to increasing lender apprehension about the financial feasibility of specialized projects that have a single market and single use.

5. The potential erosion of tax exempt bond financing, which had emerged as the most frequently used type of financing for housing for the elderly.

In response to these difficulties facing the developers or financers of specialized housing for the elderly, it has been suggested by Early (1989) that it is crucial for the reputation and future of the industry that existing and planned housing projects for the elderly have sufficient cash flow and liquid reserves to meet debt service payments and deliver the amenities and services promised to residents. New financing options are needed in response to the changes pending in tax laws, and these could include the promotion of:

- Taxable bonds.
- Conventional mortgages.
- Government National Mortgage Association (GNMA) taxable securities.

- Linking the above with federal and private mortgage insurance, syndication, and other types of credit enhancements.

It is necessary also for new programs to be developed and/or for existing programs to be expanded to where low-income elderly people can gain affordable access to specialized housing.

An interesting concept that has emerged in the later 1980s is *ecogenic housing* for the elderly that seeks to provide opportunities for companionship and socialization, safety and protection from crime, enhanced productivity in terms of cooperative effects regarding activities of daily living, privacy, and autonomy, and provision of health services all within a family life-style at an affordable cost (Pastalan, 1989). The idea is that at least two economically unrelated people share common areas including kitchen and living areas, but both have their own private sleeping, social, and sanitation areas. Each person is independent and constitutes a separate household. It has particular relevance for providing a noninstitutionalized family setting, and living at a price that is moderate and affordable for low-income elderly people.

Sponsorship of retirement communities is a further important factor. Tyler, Loomis, and Sorce (1989) note how an SRI Gallop and Hospitals Poll conducted in 1986 revealed that a religious sponsor significantly increased the attractiveness of a project as did a nonprofit sponsor, but a nonreligious, nonprofit sponsor was the most attractive of all. He replies that sponsored projects seem to be in danger of losing part of their market attractiveness, especially as they are seen to be somewhat lacking in innovative management and service provision, and certainly the nonprofit mode of operation is seen increasingly as desirable.

The experience and stability of a developer and/or builder are also important. Increasingly, "the experienced sponsor or developer (of specialized housing for the elderly) with a good relationship with a financial institution will have competitive advantage in the market place" (Early, 1989:64).

The development of better policies for providing housing for the elderly thus involves complex issues that relate to efficiency, social justice, preferred life-styles and meeting the health and social needs of the elderly. As Golant (1989:104) concludes: "The challenge for the future will be how to best package these components (i.e., shelter, social services, long-term care, and the family support network) in ways that are most cost effective, flexible, and sensitive and suitably addresses the wide ranging needs of the elderly consumer."

15.8.5 Segregated Communities

While the vast majority of the elderly retire *in situ*, an increasing proportion are relocating to areas of high amenity, typically in sun belt locations or to special-purpose retirement housing communities that are proliferating in all geographic locations and are providing an increasingly attractive living, life-style, and supportive care environment for the elderly.

Certainly the "graying" of society is creating a significant and powerful new political force in many nations, and in particular in local areas where there are high concentrations of the elderly. "Gray power" is now a powerful factor for government policies and for private-sector enterprises producing and marketing specialized products for the aged.

There are concerns about the social implications of the development of segregated residential communities, both on a large scale, such as the Sun City retirement city in Arizona, with a population in excess of 50,000, and on a small scale within normal urban settings, through the development of gated communities. However, the latter are not provided exclusively for the elderly.

Retirement communities have been shown to have advantages and disadvantages depending on whose perspective is emphasized. As suggested by Golant et al. (1988:453): "On the one hand, these accommodations represent very satisfying environments to older people who select them. But on the other hand, they worried some local populations, who feared that these residents would make excessive demands on their community's social and medical services."

Finally, it is as well to remind ourselves, as Golant (1986) has done, that:

> the patterns observed among the present elderly population and the solutions recommended to address their needs (including their demand for and attitudes towards retirement communities, their facilities, design, management and financing) . . . may be inaccurate and unsuitable a decade from now because of the different composition and attitudes and behaviors of future cohorts (generations) of (the) elderly. (p. 438)

References

Abbott, C. (1981). *The New Urban America*. Chapel Hill: University of North Carolina Press.

Abelson, R. P., & Levy, A. (1985). Decision making and decision theory. In G. Lindsey & E. Aronson (Eds.), *Handbook of Social Psychology* (Volume 2, pp. 231–309). New York: Random House.

Abrahamson, M., & Sigelman, L. (1987). Occupational sex segregation in metropolitan areas. *American Sociological Review, 52,* 588–597.

Acredolo, L. P. (1976). Frames of reference used by children for orientation in unfamiliar spaces. In G. T. Moore & R. G. Golledge (Eds.), *Environmental Knowing* (pp. 165–172). Stroudsburg, PA: Dowden, Hutchinson and Ross.

Adams, J. S. (1969). Directional bias in intra–urban migration. *Economic Geography, 45,* 302–323.

Adamsons, I., & Taylor, H. R. (1990). Major causes of world blindness: Their treatment and prevention. *Current Opinion in Ophthalmology, 1,* 635–642.

Adler, T. J., & Ben–Akiva, M. E. (1979). A theoretical and empirical model of trip chaining behavior. *Transportation Research B, 13,* 243–257.

AHURI. (1994). *The Australian Urban System: Trends and Prospects.* Prepared for the Urban and Regional Development Review. Melbourne: Australian Housing and Urban Research Institute, in conjunction with Spiller Gibbins & Swan Pty Ltd.

Aitken, S. C. (1994). *Putting Children in Their Place.* Washington, DC: Association of American Geographers

Aitken, S. C., Cutter, S. L., Foote, K. E., & Sell, J. L. (1989). Environmental perception and behavioral geography. In G. L. Gaile & C. J. Wilmott (Eds.), *Geography in America* (pp. 218–238). Columbus: Merrill Publishing Company.

Aitken, S. C., & Prosser, R. (1990). Residents' spatial knowledge of neighborhood continuity and form. *Geographical Analysis, 22*(4), 301–325.

Ajzen, I. (1985). From intentions to actions: A theory of planned behaviour. In J. Kuhl & J. Beckman (Eds.), *Action Control: From Cognition to Behaviour* (pp. 11–39). Berlin: Springer–Verlag.

Alonso, W. (1960). A theory of the urban land market. *Papers and Proceedings of the Regional Science Association, 6,* 149–158.

Alonso, W. (1964). The historic and the structural theories of urban form: Their implications for urban renewal. *Land Economics, 49,* 227–231.

Amedeo, D. (1993). Emotions in person–environment–behavior episodes. In T. Gärling & R. G. Golledge (Eds.), *Behavior and Environment* (pp. 83–116). Amsterdam: Elsevier Science Publishers.

Amedeo, D. & Golledge, R. (1975/1986). *An Introduction to Scientific Reasoning in Geography.* Melbourne, FL: Krieger Publishing.

Amedeo, D. & York, R. A. (1984). Grouping and effective responses to environments: Indications of emotional norm influence in person and environment relations. *In EDRA 15, Proceedings of the*

Environmental Design and Research Association Conference (pp. 122–130). Washington, DC, EDRA..

American National Standards Institute. (1961). *American national standards specifications for making buildings and facilities accessible to and usable by physically handicapped* (A117.1–1961). New York: The Institute.

American National Standards Institute. (1980). *American national standards specifications for making buildings and facilities accessible to and usable by physically handicapped* (A117–1–1980). New York: The Institute.

Anderson, D. A. (1982). *Predicting aggregate choices from individual judgment equations: The case of job choice after graduation.* Unpublished paper presented to the Meetings of the NSW Statistical Society, Sydney, June.

Anderson, J. (1971). Space–time budgets and activity studies in urban geography and planning. *Environment and Planning, 3*, 353–368.

Anderson, N. H. (1974) Information integration theory: A survey. In D. Krantz, R. Atkinson, & R. Luce (Eds.), *Contemporary Developments in Mathematical Psychology* (pp. 236–301). San Francisco: Academic Press.

Anderson, T. (1955). Intermetropolitan migration: A comparison of the hypotheses of Zipf and Stouffer. *American Sociological Review, 20*, 287–291.

Andrews, S. K. (1983) Spatial cognition through tactual maps. In J. Wiedel (Ed.), *Proceedings of the 1st International Symposium on Maps and Graphics for the Visually Handicapped* (pp. 30–40). Washington, DC: Association of American Geographers.

Appleton, J. (1975a). Landscape evaluation: The theoretical vacuum. *Transactions of the Institute of British Geographers, 66*, 120–123.

Appleton, J. (1975b). *The Experience of Landscape.* New York: John Wiley & Sons.

Appleyard, D. (1969). Why buildings are known. *Environment and Behavior, 1*, 131–156.

Appleyard, D. (1970). Styles and methods of structuring a city. *Environment and Behavior, 2*, 100–117.

Armhein, C. (1985). Interregional labour migration and information flows. *Environment and Planning A, 17*, 1111–1126.

Atkinson, R. L., Atkinson, R. C., Smith, E. E., & Hilgard, E. R. (1987). *Introduction to Psychology.* San Diego: Harcourt Brace Jovanovich.

Axhausen, K. W., & Gärling, T. (1992). Activity–based approaches to travel analysis: Conceptual frameworks, models, and research problems. *Transport Reviews, 12*(4), 323–341.

Axhausen, K. W., (1991, February). Eurotopp: Towards a dynamic and activity–based modeling framework. In *Proceedings of DRIVE Conference, Advanced Telematics in Road Transport.* Amsterdam: Elsevier Scientific Company.

Baird, J. C., Wagner, M. & Noma, E. (1982). Impossible cognitive spaces. *Geographical Analysis, 14*, 204–216.

Baldassare, M. (1992). Suburban communities. *Annual Review of Sociology, 18*, 475–494.

Ballard, K. P., & Clark, G. L. (1981). The short-run dynamics of inter–state migration: A space-time economic adjustment model of in-migration to fast growing states. *Regional Studies, 15*(3), 213–228.

Barber, P. O., & Lederman, S. J. (1988). Encoding direction in manipulatory space and the role of visual experience. *Journal of Visual Impairment and Blindness, 82*(3), 99–106.

Baresi, C. (1968). The role of the real estate agent in residential location. *Sociological Focus, 1*, 59–71.

Barker, M., (1966). *California Retirement Communities.* The Center for Real Estate and Urban Economics, Institute of Urban and Regional Development, University of California.

Barrett, F. (1973). *Residential Search Behavior.* (Geographical Monograph No. 1). Toronto: York University.

Beale, C. L. (1975). *The Revival of Population Growth in Nonmetropolitan America.* Washington: USDA, Economic Research Service.

Beck, R. (1967). Spatial meaning, and the properties of the environment. In D. Lowenthal (Ed.), *Environmental perception and behavior* (pp. 18–29). Chicago: University of Chicago, Department of Geography, Research Paper No. 109.

Beck, R. J., & Wood, D. (1976a). Cognitive transformation of information from urban geographic fields to mental maps. Environment and Behavior, 8, 199–238.

Becker, R. J., & Wood, D. (1976b). Comparative development analysis of individual aggregate cognitive maps of London. InG. T. Moore & R. G. Golledge (Eds.), *Environmental Knowing: Theories, Research and Methods* (pp. 173–184). Stroudsburg, PA: Dowden, Huchinson and Ross.

Becker, H. S. (1965). A theory of the allocation of time. *Economic Journal, 75,* 493–517.

Becker, H. S. (1970). Life history and the scientific mosaic. In H. S. Becker (Ed.), *Sociological Work: Methods and Substance* (pp. 63–73). Chicago: Aldine.

Bednarz, M. J. (Ed.). (1977). *Barrier-free Environments.* Stroudsburg, PA: Hutchinson and Ross.

Belcher, D. (1973). *Giving Psychology Away.* San Francisco: Canfield.

Bell, W. (1958). Social choices, life styles and suburban residence. In W. M. Dobriner (Ed.), *The Suburban Community* (pp. 225–247). New York: Putnam.

Ben-Akiva, M. E., & Lerman, S. R. (1985). *Discrete Choice Analysis.* Cambridge: MIT Press.

Bergmann, B. (1986). *The Economic Emergence of Women.* New York: Basic Books.

Berry, B. J. L. (1973). *Growth Centres in the American Urban System.* Cambridge, MA: Ballinger.

Berry, B. J. L. (1981). *Comparative Urbanization.* New York: St. Martin's Press.

Berry, B. J. L., Barnum, H., & Tennant, R. (1962). Retail location and consumer behavior. *Papers of the Regional Science Association, 9,* 65–106

Beshers, J. (1962). *Urban Social Structure.* New York: Free Press.

Beshers, J., & Nishiura, E. N. (1961). A theory of migration differentials. *Social Forces, 39,* 214–218.

Bessant, J., & Haywood, B. (1988). Islands, archipelagos and continents: Progress on the road to computer-integrated manufacturing. *Research Policy, 17,* 349–362.

Best, S., & Kellner, D. (1991). *Postmodern Theory: Critical Interrogations.* London: Macmillan.

Blades, M. (1990). The reliability of data collected from sketch maps. *Journal of Environmental Psychology, 10*(4), 327–340.

Blades, M. (1991). Wayfinding theory and research: The need for a new approach. In D. M. Mark & A. V. Frank (Eds.), *Cognitive and Linguistic Aspects of Geographic Space* (pp. 137–165). Dordrecht: Kluwer Academic Press.

Blades, M., & Spencer, C. (1994). The development of children's ability to use spatial representations. In H. W. Reese (Ed.), *Advances in Child Development.* San Diego: Academic Press.

Blakely, E. J. (1989). *Planning Local Economic Development.* Sage Library of Social Research, Vol. 168. Newbury Park: Sage Publications.

Blakely, E. J. (1992). Shaping the American Dream: Land Use Choices for America's Future. Mimeographed paper, Berkeley: University of California.

Blakely, E. J., & Bowman, K. (1986). *Taking Local Development Initiatives.* Canberra: Australian Institute of Urban Studies, Publication No. 129.

Blakely, E. J., & Stimson, R. J. (1992). *New Cities in the Pacific Rim,* Monograph No. 43. Berkeley: Institute for Urban and Regional Development, University of California.

Blaut, J. (1982). Nationalism as an autonomous force. *Science & Society, 46*(1), 1–23.

Blaut, J. & Stea, D. (1969). *Place Learning.* Worcester, MA: Graduate School of Geography, Clark University, Place Perception Report No. 4.

Blommestein, N., Nijkamp, P., & van Veenendal, W. (1980). Shopping perceptions and preferences: The multidimensional attractiveness of consumer and entrepreneureal attitudes. *Economic Geography, 56,* 155–174.

Blumen, O., & Kellerman, A. (1990). Gender differences in commuting distance, residence, and employment location: Metropolitan Haifa, 1972 and 1983. *The Professional Geographer, 42,* 54–71.

Bogue, D. J. (1969). *Principles of Demography.* New York: Wiley.

Bogue, D. J. (1977). A migrant's eye-view of the costs and benefits of migration to a metropolis. In A. A. Brown & E. Neuberger (Eds.), *Internal Migration, a Comparative Perspective* (pp. 167–182). New York: Academic Press.

Bohland, J. R., & Rowles, G. D. (1987). *Changes in the distribution of the elderly United States population.* Final Report to the National Science Foundation, No. SES-8218690.

Bolton, R. (1989). *An economic interpretation of a "sense of place."* Research Paper Series, Williamstown, MA: Department of Economics, Williams College.

Bonnar, D. (1979). Migration in the south-east of England: An analysis of the inter-relationships of housing, socio-economic status and housing demand. *Regional Studies, 13,* 345–359.

Borchert, J. (1972). American metropolitan evolution. *Annals, Association of American Geographers, 62,* 87–98.

Bourdieu, P. (1984). *Distinction: A Social Critique of the Judgement of Taste.* London: Routledge & Kegan Paul.

Bourne, L. (1981). *The Geography of Housing.* London: Arnold.

Boyce, R. R. (1969). Residential mobility and its implication for urban spatial change. *Proceedings of Association of American Geographers, 1,* 22–26.

Brewster, S., & Blades, M. (1989). Which way to go? Children's ability to give directions in the environment and from maps. *Journal of Environment Education and Information, 8*(3), 141–156.

Briggs, R. (1969). *Scaling of preferences for spatial location: An example using shopping centers.* MA Thesis, Columbus, OH: Ohio State University.

Briggs, R. (1972). *Cognitive distance in urban space.* Unpublished PhD Dissertation, Columbus, OH: Ohio State University.

Briggs, R. (1973). Urban cognitive distance. In R. Downs & D. Stea (Eds.), *Image and environment: Cognitive mapping and spatial behavior* (pp. 361–388). Chicago: Aldine.

Briggs, R. (1976). Methodologies for the measurement of cognitive distance. In G. T. Moore & R. G. Golledge (Eds.), *Environmental Knowing* (pp. 325–334). Stroudsburg, PA: Dowden, Hutchinson and Ross.

British Standards Institution. (1979). *Code of practice for access for the disabled to buildings* (BS5310: 1979). London: The Institution.

Brög, W., & Erl, E. (1981). Application of a model of individual behavior (situational approach) to explain household activity patterns in an urban area to forecast behavioral changes. Paper presented to the International Conference on Travel Demand Analysis: Activity Based and Other New Approaches. Oxford, July.

Brown, L. A., & Belcher, J. (1966). Residential mobility of physicians in Georgia. *Rural Sociology, 31,* 439–457.

Brown, L. A., & Holmes, J. (1971). Search behavior in an intraurban migration context: A spatial perspective. *Environment and Planning, 3,* 307–326.

Brown, L. A., & Longbrake, D. G. (1970). Migration flows in intra-urban space: Place utility considerations. *Annals of the Association of American Geographers, 60,* 368–384.

Brown, L. A., & Moore, E. A. (1970). The intra-urban migration process: A perspective. *Geografiska Annaler, 52B,* 1–13.

Bryant, K. J. (1982). Personality correlates of sense of direction and geographical orientation. *Journal of Personality and Social Psychology, 43,* 1318–1324.

Bryson, J. M., & Einsweiler, R. C. (Eds.). (1988). *Strategic Planning: Threats and Opportunities for Planners.* Washington, DC: Planners Press, American Planning Association.

Bryson, J. M., & Roering, W. D. (1988). Applying private sector strategic planning in the public sector. In J. M. Bryson & R. C. Einsweiler (Eds.), *Strategic Planning: Threats and Opportunities for Planners.* Washington, DC: Planners Press, American Planning Association.

Bullock, N., Dickens, P., Shapcott, M., & Steadman, P. (1974). Time budgets and models of urban activity patterns. *Social Trends, 5,* 45–63.

Bunch, D. S., Bradley, M., Golob, T. G., & Kitamura, R. (1993). Demand for clean-fuel vehicles in California—A discrete-choice stated preference pilot project. *Transportation Research A: Policy and Practice, 27,* 237–253.

Burgess, E. (1925). The growth of the city. In R. Park, E. Burgess, & R. McKenzie (Eds.), *The City.* Chicago: University of Chicago Press.

Burnett, P. (1973). The dimensions of alternatives in spatial choice processes. *Geographical Analysis, 5,* 181–204.

Burnett, P. (1976). Behavioral geography and the philosophy of mind. In R. Golledge & G. Rushton (Eds.), *Spatial Choice and Spatial Behavior* (pp. 23–50).Columbus, OH: Ohio State University Press.

Burnett, P., & Hanson, S. (1979). Rationale for an alternative mathematical approach to movement as complex human behavior, *Transportation Research Record, 723,* 11–24.

Burnett, P., & Thrift, N. J. (1979). New approaches to travel behavior. In D. Hensher & P. Stopher (Eds.). *Behavioral Travel Demand Modelling* (pp. 116–136). London: Croom Helm.

Burnley, I. H., & Murphy, P. R. (1993). Socio-economic structure of Sydney's peri-metropolitan region. *Journal of the Australian Population Association, 10,* 127–144.

Burroughs, W., & Sadalla, E. (1979). Asymmetries in distance cognition. *Geographical Analysis, 11,* 414–421.

Burton, I., & Kates, R. W. (1964). The perception of natural hazards in resource management. *Natural Resources Journal, 3,* 412–441.

Burton, I., Kates, R., & Snead, R. (1969). The human ecology of coastal flood hazard in megalopolis. Research Paper No. 115. Chicago: Department of Geography, University of Chicago.

Butler, E. W., Chapin, F. S., Hemmens, G. C., Kaiser, E. J., Stegman, M. A., & Weiss, S. F. (1969). *Moving Behavior and Residential Choice: A National Survey.* National Cooperative Research Program, Highway Research Board, Washington, DC, Report No. 81.

Butler, R. (1980). Concept of a tourist area cycle of evaluation: Implications for management of resources. *Canadian Geographer, 24*(1), 5–12.

Buttimer, A. (1969). Social space in interdisciplinary perspective. *Geographical Review, 59,* 417–426.

Buttimer, A. (1972). Social space and the planning of residential areas. *Environment and Behavior, 4,* 279–318.

Byrne, R. W. & Salter, E. (1983). Distances and directions in cognitive maps of the blind. *Canadian Journal of Psychology, 37,* 293–299.

Cadawallader, M. T. (1973a). A methodological examination of cognitive distance. In W. F. E. Preiser (Ed.), *Environmental Design Research Association 4 (EDRA-4),* vol. 2 (pp. 193–199). Stroudsburg, PA: Dowden, Hutchinson and Ross.

Cadwallader, M. T. (1973b). *An Analysis of Cognitive Distance and Its Role in Consumer Spatial Behavior.* Unpublished Ph.D. dissertation. Department of Geography, University of California at Los Angeles.

Cadwallader, M. T. (1975). A behavioral model of consumer spatial decision making. *Economic Geography, 51,* 339–349.

Cadwallader, M. T. (1976). Cognitive distance in intra-urban space. In G. T. Moore & R. G. Golledge (Eds.), *Environmental Knowing: Theories, Research and Methods* (pp. 316–324). Stroudsburg, PA: Dowden, Hutchinson & Ross.

Cadwallader, M. T. (1979). Problems in cognitive distance: Implications for cognitive mapping. *Environment and Behavior, 11,* 559–576.

Cadwallader, M. T. (1981). Towards a cognitive gravity model: The case of consumer spatial behavior. *Regional Studies, 15,* 275–284.

Cadwallader, M. T. (1989). A conceptual framework for analysing migration behaviour in the developed world, *Progress in Human Geography, 13,* 494–511.

Campbell, F., & Krieger, R. (1981). Illusion and disallusion: The ideology of community in an industrial town. In M. Bowman (Ed.), *Beyond the City* (pp. 71–95). Melbourne: Longman Cheshire.

Campbell, J. (1970). A stochastic model of human movement. In *Discussion Paper no. 13.* Department of Geography, University of Iowa.

Cannell, C. F., Miller, P. V., & Oksenberg, L. (1981). Research and interviewing techniques. In S. Leinhardt (Ed.), *Sociological Methodology* (pp. 389–437). San Francisco: Jossey–Bass.

Cannon, D., & Rainbow, F. (1981). *Legislative background on transportation access.* National Center for Barrier Free Environment Access (Information Bulletin). Washington, DC: The Center.

Canter, D. V. (1970). *Architectural Psychology.* London: RIBA.

Canter, D. V., & Canter, S. (1971). "Close together in Tokyo." *Design and Environment, 2*(2), 60–63.

Canter, D. V., & Larkin, P. (1993). The environmental range of serial rapists. *Journal of Environmental Psychology, 13*(1), 63–70.

Carp, F. M. (1975). Life-style and location within the city. *The Gerontologist, 75,* 27–33.

Carpman, J. R., Grant, M. A., & Simmons, D. A. (1985). Hospital design and wayfinding: A video simulation study. *Environment and Behavior, 17*(3), 296–314.

Carr, S., & Schissler, D. (1969). The city as a trip: Perceptual selection and memory in the view from the road. *Environment and Behavior, 1,* 7–35.

Carroll, J. D., & Chang, J. J. (1970). Analysis of individual differences in multidimensional scaling via an N-way generalization of "Eckart–Young" decomposition. *Psychometrika, 35,* 283–319.

Casetti, E., & Papageorgiou, Y. Y. (1971). A spatial equilibrium model of urban structure. *Canadian Geographer, 15,* 30–37.

Casey, S. M. (1978). Cognitive mapping by the blind. *Journal of Vision Impairment and Blindness, 72*(8): 297–301

Castells, M. (1989). *The Information City: Information Technology, Economic Restructuring and Urban and Regional Process.* Oxford: Basil Blackwell.

Cervero, R. (1989). *America's Suburban Centres: The Land Use Transportation Link.* Boston: Unwin-Hyman.

Chalmers, J. A., & Greenwood, M. J. (1985). The regional labor market adjustment process: Determinants of changes in rates of labor force participation, unemployment and migration. *Annals of Regional Science, 19*(1), 1–17.

Champion, T. (1989). *Counterurbanisation.* London: Edward Arnold.

Chapin, F. S. (1965). *Urban Landuse Planning* (2nd ed.). Urbana, IL: University of Illinois Press.

Chapin, F. S. (1968). Activity systems and urban structures: A working schema. *Journal of American Institute Planners, 34,* 11–18.

Chapin, F. S. (1974). *Human Activity Patterns in the City: What Do People Do in Time and Space?* New York: John Wiley.

Chapin, F. S. (1978). Human time allocation in the city. In T. Carlstein, D. N. Parkes, & N. J. Thrift (Eds.), *Timing Space and Spacing Time II: Human Activity and Time Geography* (pp. 13–26). London: Edward Arnold.

Chapin, F. S., & Brail, R. K. (1969). Human activity systems in the metropolitan United States. *Environment and Behavior, 1,* 107–130.

Chatfield, C., Ehrenberg, A. S. C., & Goodhardt, G. J. (1966). Progress on a simplified model of stationary purchasing behavior. *Journal of the Royal Statistical Society A, 129,* 317–367.

Cheshire, P. (1989). Explaining the recent performance of the European Community's major urban regions. *Urban Studies, 27,* 311–333.

Chinitz, B. (1991). A framework for speculating about future urban growth patterns in the US. *Urban Studies, 28*(6), 939–959.

Chombart de Lauwe, P. H. (1952). Introductory statement. In P. H. Chombart de Lauwe (Ed.), *Paris et l'Agglomeration Parisienne: L'Espace Social dans une Grande Ville.* Paris: PUF.

Chombart de Lauwe, P. H. (1960). L'evolution des besoins et la conception dynamiques de la famille. *Revue Fran‚oise de Sociologie, 1,* 403–425.

Chombart de Lauwe, P. H. (1965). *Essais de Sociologie, 1952–1964.* Paris: Universitaires de France.

Christaller, W. (1933). *Die Zentralen Orte in Suddeutschland* (trans. C. W. Baskin, 1963, as *Central places in Southern Germany*). Englewood Cliffs, NJ: Prentice Hall.

Clark, W. A. V. (1969). Information flows and intra-urban migration: An empirical analysis. *Annals of the Association of American Geographers, 1,* 38–42.

Clark, W. A. V. (1970). Measurement and explanation in intra-urban mobility. *Tijdschrift voor Economische en Sociale Geografie, 61,* 49–57.

Clark, W. A. V. (1981). Residential mobility and behavioral geography: Parallelism or independence? In K. R. Cox & R. G. Golledge (Eds.), *Behavioral Problems in Geography Revisited* (pp. 182–205). New York: Methuen.

Clark, W. A. V. (1982). Recent research on migration and mobility: A review and interpretation. *Progress and Planning, 18,* 1–56.

Clark, W. A. V. (1986). *Human migration.* Newbury Park, CA: Sage Publications.

Clark, W. A. V. (1993). Search and choice in urban housing markets. In T. Gärling & R. G. Golledge (Eds.), *Behavior and Environment: Psychological and Geographical Approaches* (pp. 298–316). Amsterdam: Elsevier/North Holland.

Clark, W. A. V., & Cadwallader, M. (1973). Locational stress and residential mobility. *Environment and Behavior, 5,* 29–41.

Clark, W. A. V., & Flowerdew, R. (1982). A review of search models and their application to search in the housing market. In W. A. V. Clark (Ed.), *Modelling Housing Market Search* (pp. 4–29). London: Croom Helm.

Clark, W. A. V., & Rushton, G. (1970). Models of intra-urban consumer behavior and their applications for central place theory. *Economic Geography, 46,* 486–497.

Clark, W. A. V., & Smith, T. R. (1979). Modeling information use in a spatial context. *Annals of the Association of American Geographer, 69* (4), 575–588.

Cloke, P., Philo, C. & Sadler, D. (1991). *Approaching Human Geography: An Introduction to Contemporary Theoretical Debates.* London: Paul Chapman.

Cohen, S. S., & Zysman, J. (1987). *Manufacturing Matters: The Myth of the Post-Industrial Economy.* New York: Basic Books.

Converse, P. D. (1949). New laws of retail gravitation. *Journal of Marketing, 14*(3), 379–384.

Converse, P. E. (1972). Country differences in the use of time. In A. Szalai, P. E. Converse, P. Feldheim, & E. K. Scheuch (Eds.), *The Use of Time: Daily Activities of Urban and Suburban Populations in Twelve Countries* (pp. 145–177). The Hague: Mouton.

Corn, A. L., & Sacks, C. Z. (1994). The impact of non-driving on adults with visual impairments. *Journal of Vision Impairment and Blindness, January/February, 53*–68.

Coshall, J. T. (1985a). The form of micro-spatial consumer cognition and its congruence with search behaviour. *Tijdschrift voor Economische en Sociale Geografie, 76*(5), 345–355.

Coshall, J. T. (1985b). Urban consumer's cognitions of distance. *Geografiska Annaler B: Human Geography, 67,* 107–119.

Costanzo, C. M., Hubert, L., & Golledge R. G. (1984, April). *Directional bias in criminal behavior patterns.* Paper presented at the Association of American Geographers, Washington, DC.

Couclelis, H. (1991). A linguistic theory of spatial cognition. Unpublished paper, Department of Geography, University of California–Santa Barbara, Santa Barbara, CA.

Couclelis, H., & Golledge, R. G. (1983). Analytical research, positivism, and behavioral geography. *Annals of the Association of American Geographers, 73,* 331–339.

Couclelis, H., Golledge, R. G., Gale, N. D., & Tobler, W. R. (1987). Exploring the anchor-point hypothesis of spatial cognition. *Journal of Environmental Psychology, 7,* 99–122.

Coupe, R. T., & Morgan, B. S. (1981). Towards a fuller understanding of residential mobility: A case study in Northampton, England. *Environment and Planning A, 13,* 201–215.

Cox, K. R. (1973). *Conflict, Power, and Politics in the City: A Geographic View.* New York: McGraw–Hill.

Cox, K. R., & Golledge, R. G. (Eds.). (1969). *Behavioral Problems in Geography: A Symposium.* Evanston, IL: Northwestern University Press.

Crawford, T. (1973). Beliefs about birth control: A consistency theory analysis. *Representative Research in Social Psychology, 4,* 53–65.

Cronin, R. (1982). Racial differences in the search for housing. In W. A. V. Clark (Ed.), *Modeling Housing Market Search* (pp. 81–105). London: Croom Helm.

Cullen, I. G. (1976). Human geography, regional science, and the study of individual behavior. *Environment and Planning A, 8,* 397–409.

Cullen, I. G. (1978). The treatment of time in the explanation of spatial behavior. In T. Carlstein, D. N. Parkes, & N. J. Thrift (Eds.), *Timing Space and Spacing Time II: Human Activity and Time Geography* (pp. 27–38). London: Edward Arnold.

Cullen, I. G., & Godson, V. (1975). Urban networks: The structure of activity patterns. In D. Diamond & J. B. MacLaughlin (Eds.), *Progress in Planning: Part I* (pp. 1–96). Oxford: Pergamon.

Cullen, I. G., & Philps, E. (1975). *Diary Techniques and the Problems of Urban Life.* Washington, DC: Social Science Research Council.

Cunningham, C. J. (1984). Recurring natural fire hazards: Case study of the Blue Mountains, New South Wales, Australia. *Applied Geography, 4,* 5–27.

Cunningham, C. J., Jones, M., & Taylor, N. (1994). The child-friendly neighbourhood: Some questions and tentative answers from Australian research. *International Play Journal, 2,* 79–95.

Cunningham, C., Kelly, S., Greenway, J., & Jones, S. (1993). *Attrition of community knowledge of natural hazards: A case study of the Blue Mountains, New South Wales, Australia.* Unpublished paper, Department of Geography, University of New England, New South Wales, Australia.

Curry, L. (1966). Seasonal programming and Bayesian assessment of atmospheric resources. In W. R. D. Sewell (Ed.), *Human Dimensions of Weather Modification,* Research Paper No. 105 (pp. 127–138). Chicago: Department of Geography, University of Chicago.

Daly, M. T., & Stimson, R. J. (1992). Sydney: Australia's gateway and financial capital. In E. J. Blakely & R. J. Stimson (Eds.), *New Cities in the Pacific Rim,* Monograph No 43. Berkeley: Institute of Urban and Regional Development, University of California.

Damm, D., & Lerman, S. R. (1981). A theory of activity scheduling behavior. *Environment and Planning A, 13,* 703–718.

Daniels, P. W., O'Connor, K. B., & Hutton, T. A. (1992). The planning response to urban services growth: An international comparison. *Growth and Change, 22,* 3–26.

DaSilva, J. A. (1985). Scales for perceived egocentric distance in a large open field: Comparison of three psychophysical methods. *American Journal of Psychology, 98*(1), 119–144.

Datel, R. E., & Dingemans, D. J. (1980). Historic preservation and urban change. *Urban Geography, 1,* 229–253.

Datel, R. E., & Dingemans, D. J. (1984). Environmental perception, historic preservation, and sense of place. In T. F. Saarinen, D. Seamon, & J. L. Sell (Eds.), *Environmental perception and behavior: An inventory and prospect.* University of Chicago, Department of Geography Research Paper No. 209 (pp. 131–144). Chicago: University of Chicago Press.

Day, R. A. (1976). Urban distance cognition: Review and contribution. *Australian Geographer, 13,*

Dear, M. (1987). *Community solutions to the homelessness problem.* Paper presented at the Association of American Geographers Annual Meeting, Portland, OR.

Dear, M. (1991). Gaining community acceptance. In C. Poster & E. Longergan (Eds.), *Resolving Locational Conflicts, Not-in-My-Backyard Part 1: Human Service Facilities and Neighbourhoods.* Tucson, AZ: Roy P. Drachman Institute for Land and Regional Development Studies, University of Arizona, and Information and Referral Services Inc.

De Jong, G. F., & Fawcett, J. T. (1979). *Motivations and migration: An assessment and value–expectancy research model.* Paper presented at the workshop in micro level approaches to migration decisions, 10th Summer seminar in population. University of Hawaii, Honolulu.

Densham, P., & Rushton, G. (1988). Decision support systems for locational planning. In R. G. Golledge & H. Timmermans (Eds.), *Behavioural Modelling in Geography and Planning* (pp. 56–90). London: Croom Helm.

Denzin, N. K. (1978). *The Research Act: A Theoretical Introduction to Sociological Methods.* New York: McGraw-Hill.

Derrida, J. (1982). Signature, event, context. In A. Bass (Trans.), *Margins of Philosophy* (pp. 307–330). Brighton, Sussex: Harvester Press.

Derrida, J. (1988). Afterword: Toward an ethic of discussion. In G. Graff (Ed.), *Limited Inc.* (pp.111–142). Evanston, IL: Northwestern University Press.

Deurloo, M. C., Clark, W. A. V., & Dieleman, F. M. (1990). Choice of residential environment in the Randstad. *Urban Studies, 27,* 335–351.

Devlin, A. S. (1973). Some factors in enhancing knowledge of a natural area. In W. Preiser (Ed.), *Environmental Design Research,* vol. 2 (pp. 200–207). Stroudsburg, PA: Dowden, Hutchinson and Ross.

Dewey, R. (1960). The rural–urban continuum: Real but relatively unimportant. *American Journal of Sociology, 66,* 60–66.

Dewey, R. (1990). The rural–urban continuum: Real but relatively unimportant. *American Journal of Sociology, 66,* 60–66.

Dicken, P. (1992). *Global Shift: The Internationalization of Economic Activity* (2nd ed.). New York: Guilford Press.

Dieleman, F. M., Deurloo, M. C. & Clark, W. A. V. (1989). A comparative view of housing choices in controlled and uncontrolled markets. *Urban Studies, 26,* 457–468.

Dobson, R., Dunbar, F., Smith, C. J., Reibstein, D., & Lovelock, C. (1978). Structural models for the analysis of traveller attitude-behavior relationships. *Transportation, 7,* 351–364.

Dodds, A. G. (1988). *Mobility Training for Visually Handicapped People.* London: Croom Helm.

Dodds, A. G., Howarth, C. I., & Carter, D. (1982). The mental maps of the blind: The role of previous visual experience. *Journal of Visual Impairment and Blindness, 76*(1), 5–12.

Donaghu, M. T., & Barff, R. (1990). Nike just did it: International subcontracting and flexibility in athletic footwear production. *Regional Studies, 24,* 537–552.

Dorigo, G., & Tobler, W. R. (1983). Push–pull migration laws. *Annals of the Association of American Geographers, 73,* 1–18.

Douglas, J. D. (1979). *Investigative Social Research: Individual and Team Field Research.* A Sage University Paper. Newbury Park, CA: Sage Publications.

Downs, A. (1985). Living with advanced telecommunications. *Society, 23*(1), 26–34.

Downs, R. M. (1970a). The cognitive structure of an urban shopping center. *Environment and Behavior, 2,* 13–39.

Downs, R. M. (1970b). Geographic space perception: Past approaches and future prospects. In C. Board, R. J. Chorley, P. Haggett, & D. R. Stoddart (Eds.), *Progress in Geography* (Vol. 2, pp. 65–108). London: Edward Arnold.

Downs, R. M. (1981a). Maps and metaphors. *The Professional Geographer, 33*(3), 287–293.

Downs, R. M. (1981b). Maps and mappings as metaphors for spatial representations. In L. Liben, A. Patterson, & N. Newcombe (Eds.), *Spatial Representation and Behavior across Life Span: Theory and Applications* (pp. 143–166). New York: Academic Press.

Downs, R. M., & Stea, D. (1973). Cognitive maps and spatial behavior: Process and products. In R. Downs & D. Stea (Eds.), *Image and Environment* (pp. 8–26). London: Edward Arnold.

Dunning, J. H. (1980). Towards an eclectic theory of international production: Some empirical tests. *Journal of International Business Studies, 11,* 9–31.

Durkheim, E. (1893). *De la Division du Travail Social.* Paris: Felix Alcan.

Earle, C., Dilsaver, L., Ward, D., Hugill, P. J., Mitchell, R. D., Harris, C., Ostergren, R. C., Lemon, J., Hartnett, S., Goheen, P., Wyckoff, W., Galloway, J. H., Grim, R., Trimble, S. W., Cooke, R. U., Botts, H., Kay, J., & Colten, C. E. (1989). Historical geography. In R. L. Gaile & C. J. Willmott (Eds.), *Geography in America* (pp. 156–191). Columbus, OH: Merrill.

Early, M. C. K. (1989). Housing for the elderly: Financing options. *Journal of Housing for the Elderly 5*(1), *Special Issue, Lifestyles and Housing of Older Adults: The Florida Experience,* 51–65.

Edwards, P. K., Edwards, J. N., & Watts, A. D. (1984). Women, work and social participation. *Journal of Voluntary Action Research, 13,* 7–22.

Edwards, W. (1954). The theory of decision making. *Psychological Bulletin, 51,* 380–417.

Egenhofer, M. J. (1991). Deficiencies of SQL as a GIS query language. In D. Marks & A. Frank (Eds.), *Cognitive Linguistic Aspects of Geographic Space* (pp. 477–491). Dordrecht: Kluwer Academic Publishers.

Eliot, J., & McFarlane-Smith (1983). Different dimensions of spatial ability. In *Studies in Science Education.* Oxford: NFER-Nelson Publishing Company.

Elson, D. (1988). Transnational corporations in the new international division of labour: A critique of 'cheap labour' hypotheses. *Manchester Papers on Development, IV,* 352–376.

Ericson, K. A., & Oliver, W. (1988). Methodology for laboratory research on thinking: Task selection, collection of observation and data analysis. In R. J. Sternberg & E. E. Smith (Eds.), *The Psychology of Human Thought* (pp. 392–428). Cambridge: Cambridge University Press.

Ericson, K. A., & Simon, H. A. (1984). *Protocol Analysis: Verbal Reports as Data.* Cambridge: MIT Press.

Ermisch, J. (1983). *The Political Economy of Demographic Change.* London: Heinemann.

Ermuth, F. (1974). *Residential Satisfaction and Urban Environmental References.* Toronto: Geographical Monographs, York University.

Evans, P. (1976). *Motivation.* London: Methuen.

Evers, G. H. M. (1989). Simultaneous models for migration and commuting: Macro- and microeconomic approaches. In J. Dijk, H. Van Folmer, H. W. Herzog, & A. M. Schlottmann (Eds.), *Migration and Labour Market Adjustment* (pp. 177–198). Dordrecht: Kluwer.

Festinger, L. (1994). *Conflict, Decision and Dissonance,* Stanford: Stanford University Press.

Fielding, A. J. (1989). Inter-regional migration and social change: A study of South East England based upon data from the longitudinal study. *Transactions of the Institute of British Geographers, 14,* 24–36.

Fielding, A. J. (1992). Migration and culture. In T. Champion & A. T. Fielding (Eds.), *Migration Processes and Patterns, Vol. 1: Research Progress and Prospects* (pp. 201–212). London: Belhaven Press.

Fielding, A. J. (1993). Migration and the metropolis: An empirical and theoretical analysis of inter-regional migration to and from South East England. *Progress in Planning, 39,* 71–116.

Fincher, R. (1991). *Immigration, Urban Infrastructure and the Environment.* Canberra: Bureau of Immigration Research, AGPS.

Fincher, R. (1994). Appendix 5, Social Structure. In *The Australian Urban System, Trends and*

Prospects. Melbourne, Victoria. Report for the Urban and Regional Development Review, The Australian Housing and Urban Research Institute.

Fishbein, M., & Ajzen, I. (1975). *Belief, Attitude, Intention and Behavior: An Introduction to Theory and Research.* Reading, MA: Addison-Wesley.

Fishman, R. (1990). America's new city: Megalopolis unbound. *The Wilson Quarterly, 14*(1), 24–45.

Flood, J. (1992). Internal migration in Australia: Who gains, who loses. *Urban Futures, Special Issue, 5,* 44–53.

Flowerdew, R. (1978). Search strategies and stopping rules in residential mobility. *Transactions of the Institute of British Geographers, 1,* 47–57.

Ford, L. R. (1979). Urban preservation in the geography of the city in the USA. *Progress in Human Geography, 3,* 211–238.

Fotheringham, A. S. (1983). A new set of spatial interaction models: The theory of competing destinations. *Environment and Planning A, 15*(1), 15–36.

Fotheringham, A. S. (1985). Spatial competition and agglomeration in urban modelling. *Environment and Planning A, 17,* 213–230.

Frank, A. U. (1992). Qualitative spatial reasoning about distances and directions in geographic space. *Journal of Visual Languages and Computing, 3,* 343–371.

Freeman, C. (1987). The challenge of new technologies, in OECD. In *Interdependence and Cooperation in Tomorrow's World.* (pp. 123–156). Paris: OECD.

Freksa, C. (1992). Using orientation information for qualitative spatial reasoning. In A. U. Frank, I. Campari, & U. Formentini (Eds.), *Theories and Methods of Spatio-Temporal Reasoning in Geographic Space* (pp. 162–178). Berlin: Springer-Verlag.

French, J. R. P., Jr., Cobb, S., & Rodgers, W. L. A. (1974). A model of person–environment fit. In G. V. Coelho, D. A. Hamburg, & J. E. Adams (Eds.), *Coping and Adaption* (pp. 316–333). New York: Basic Books.

Frey, W. H. (1988). The re-emergence of core regions growth: A return to the metropolis? *International Regional Science Review, 11*(3), 261–267.

Frey, W. H. (1990). Metropolitan America: Beyond the transition. *Population Bulletin, 45,* 1–49.

Frey, W. H., & Spear, A. (1988). *Regional and Metropolitan Growth and Decline in the United States. Census Monography.* New York: Russell Sage Foundation.

Frey, W. H., & Spear, A. (1992). The revival of metropolitan population growth in the United States: An assessment of findings from the 1990 census. *Population and Development Review, 18*(1), 129–147.

Frey, W. H. (1993). People in places: Demographic trends in urban America. In J. Sommer & D. A. Hicks (Eds.), *Rediscovering Urban America: Perspectives on the 1980s* (pp. 3-1–3-106). Washington, DC: US Department of Housing and Urban Development, Office of Policy Development and Research.

Frey, W. H., & Speare, A. (1991). *US Metropolitan Area Growth 1960–1990: Census Trends and Explanation. Research Report No. 91-212.* Ann Arbor: University of Michigan Population Studies Centre.

Friedmann, J. (1986). The world city hypothesis. *Development and Change, 17,* 69–83.

Fröbel, F., Heinrichs, J., & Kreye, O. (1980). *The New International Division of Labour.* Cambridge: Cambridge University Press.

Fuguitt, G. V. (1985). The nonmetropolitan population turnaround. *Annual Review of Sociology, 11,* 259–280.

Fujita, N., Klatzky, R. L., Loomis, J. M., & Golledge, R. G. (1993). The encoding-error model of pathway completion without vision. *Geographic Analysis, 25*(4), 295–314.

Fujita, N., Loomis, J. M., Klatzky, R. L., & Golledge, R. G. (1990). A minimal representation for dead-reckoning navigation: Updating the homing vector. *Geographical Analysis, 22*(4), 326–335.

Gabriel, S. A., Shack–Marques, J., & Wascher, W. L. (1993). Does migration arbitrage regional labour market differentials?. *Regional Science and Urban Economics, 23*(2), 211–233.

Gale, N. D., Doherty, S., Pellegrino, J. W., & Golledge, R. G. (1985). Toward reassembling the image. *Children's Environment Quarterly, 2*(3), 10–18.

Gale, N. D., Golledge, R. G., Pellegrino, J., & Doherty, S. (1990). The acquisition and integration of neighborhood route knowledge in an unfamiliar neighborhood. *Journal of Environmental Psychology, 10,* 3–26.

Gärling, T. (1989). The role of cognitive maps in spatial decisions. *Journal of Environmental Psychology, 9*(4), 269–278.

Gärling, T., Böök, A., & Lindberg, E. (1979). The acquisition and use of an internal representation of the spatial layout of the environment during locomotion. *Man–Environment Systems, 9,* 200–208.

Gärling, T., Böök, A., & Lindberg, E. (1985). Adults' spatial memory representations of the spatial properties of their everyday physical environment. In R. Cohen (Ed.), *The Development of Spatial Cognition* (pp. 141–184). Hillsdale, NJ: Lawrence Erlbaum.

Gärling, T., Brännäs, K., Garvill, J., Golledge, R. G., Gopal, S., Holm, E., & Lindberg, E. (1989). *Household activity scheduling.* Paper presented at the Fifth World Conference on Transportation Research, Yokohama.

Gärling, T., & Golledge, R. G. (1989). Environmental perception and cognition. In E. H. Zube & G. T. Moore (Eds.), *Advances in Environment, Behavior, and Design* (Vol. 2, pp. 203–236). New York: Plenum Press.

Gärling, T., & Golledge, R. G. (Eds.). (1993). *Behavior and Environment: Psychological and Geographical Approaches.* Amsterdam: Elsevier/North-Holland.

Gärling, T., Kwan, M-P., & Golledge, R. G. (1994). Computational-process modeling of household activity scheduling. *Transportation Research B.*

Gärling, T., Säisä, J., Böök, A., & Lindberg, E. (1986). The spatiotemporal sequencing of everyday activities in the large-scale environment. *Journal of Environmental Psychology, 6,* 261–280.

Garreau, J. (1991). *Edge City: Life on the New Frontier.* New York: Doubleday.

Gibson, J. J. (1966). *The Senses Considered as Perceptual Systems.* Boston: Houghton-Mifflin.

Gilmartin, P. P. (1986). Maps, mental imagery and gender in the recall of geographical information. *The American Cartographer, 13*(4), 335–344.

Gilmartin, P. P., & Patton, J. C. (1984). Comparing the sexes on spatial abilities: Map-use skills. *Annals of the Association of American Geographers, 74*(4), 605–619.

Ginnsberg, R. B. (1971). Semi-markov processes and mobility. *Journal of Mathematical Sociology, 1,* 233–262.

Ginnsberg, R. B. (1973). Stochastic models of residential and geographic mobility for heterogeneous populations. *Environment and Planning, 5,* 113–124.

Gober, P. (1985). The retirement community as a geographical phenomenon: The case of Sun City, Arizona. *Journal of Geography, 84,* 189–198.

Godelier, M. (Ed.). (1982). *Les Sciences de l'Homme et de la Société en France (Social Sciences in France).* Paris: Le Documentation Fran,ais.

Golant, S. M. (1982). Individual differences underlying the dwelling satisfaction of the elderly. *Journal of Social Issues, 38*(3), 121–133.

Golant, S. M. (1984a). The effects of residential and activity behaviours on old people's environmental experiences. In I. Altman, J. Wohlwill, & M. P. Lawton (Eds.), *The Elderly and the Environment* (pp. 239–278). New York: Plenum Press.

Golant, S. M. (1984b). *A Place to Grow Old: The Meaning of Environment in Old Age.* New York: Columbia University Press.

Golant, S. M. (1986). The suitability of old people's residential environments: Insights from the geographical literature. *Urban Geography, 7,* 437–447.

Golant, S. M. (1987). Residential moves by elderly persons in U.S. central cities, suburbs and rural areas. *Journal of Gerontology, 42,* 534–539.

Golant, S. M. (1989). The residential moves, housing locations, and travel behaviour of older people: Enquiries by geographers. *Urban Geography, 10,* 100–108.

Golant, S. M., & Burton, I. (1969). The meaning of a hazard-application of the semantic differential. Natural Hazard Working Paper No. 7. Boulder, CO: University of Colorado.

Golant, S. M., Rowles, G. D., & Meyer, J. W. (1988). Aging and the aged. In G. L. Gaile & C. J. Willmott (Eds.), *Geography in America* (pp. 451–466). Columbus, OH: Merrill Publishing Co.

Gold, J. R. (1980). *An Introduction to Behavioral Geography.* New York: Oxford University Press.

Goldsmith, S. (1976). *Designing for the Disabled* (3rd ed.). London: RIBA Publications.

Goldsmith, S. (1983). The ideology of designing for the disabled. In D. Amedeo (Ed.), *Proceedings of the 14th International Conference of the Environmental Design Research Association (EDRA 14)* (pp. 198–214). Lincoln, NE: EDRA.

Golledge, R. G. (1965). *A probabilistic model of market behavior: With a reference to the spatial aspects of hog marketing in eastern Iowa*. PhD Dissertation, Department of Geography, University of Iowa.

Golledge, R. G. (1967). Conceptualizing the market decision process. *Journal of Regional Science, 7*, 239–258.

Golledge, R. G. (1970b). Some equilibrium models of consumer behavior. *Economic Geography, 46*, 417–424.

Golledge, R. G. (1975). *On reality and reality*. Paper presented at AAG, Milwaukee, WI.

Golledge, R. G. (1976a). *Cognitive Configurations of a City* (Vol. 1). Columbus, OH: The Ohio State University Research Foundation.

Golledge, R. G. (1976b). *Cognitive Configurations of a City* (Vol. 2). Columbus, OH: The Ohio State University Research Foundation.

Golledge, R. G. (1976c). Methods and methodological issues in environmental cognition research. In G. T. Moore & R. G. Golledge (Eds.), *Environmental Knowing* (pp. 300–313). Stroudsburg, PA: Dowden, Hutchinson and Ross.

Golledge, R. G. (1978a). Representing, interpreting and using cognized environments. *Proceedings of Regional Science Association, 41*, 169–204.

Golledge, R. G. (1978b). Learning about urban environments. In T. Carlstein, D. N. Parkes, & N. J. Thrift (Eds.), *Timing Space and Spacing Time I: Making Sense of Time* (pp. 76–98). London: Edward Arnold.

Golledge, R. G. (1985). General principles and approaches to economic impact assessment. In R. J. Stimson & R. Sanderson (Eds.), *Assessing the Economic Impacts of Retail Centres: Issues, Methods and Implications for Government Policy*. Publication no. 122 (pp. 98–105). Canberra: Australian Institute of Urban Studies.

Golledge, R. G. (1991a) Tactual strip maps as navigational aids. *Journal of Visual Impairment and Blindness, 85*(7), 296–301.

Golledge, R. G. (1991b). Cognition of physical and built environments. In T. Gärling & G. Evans (Eds.), *Environment, Cognition, and Action: An Integrated Approach* (pp. 35–62). New York: Oxford University Press.

Golledge, R. G. (1992). Place recognition and wayfinding: Making sense of space. *Geoforum, 23*(2), 199–214.

Golledge, R. G. (1993). Geography and the disabled: A survey with special reference to vision impaired and blind populations. *Transactions of the Institute of British Geographers, 18*, 63–85.

Golledge, R. G. (1996). Disability, barriers and discrimination. In L. Chatterjee & Å. Anderson (Eds.), *Contestable Differences: A Global Dilemma*. (In preparation).

Golledge, R. G., Bell, S., & Dougherty, V. (1994). *The cognitive map as an internal GIS*. Paper presented at AAG meeting, San Francisco, California.

Golledge, R. G., & Bell, S. M. (1995). Reasoning and inference in spatial knowledge acquisition: The cognitive map and an internalized geographic information system. Unpublished manuscript, Department of Geography, University of California–Santa Barbara, Santa Barbara, CA.

Golledge, R. G., Briggs, R., & Demko, D. (1969). The configurations of distances in intra-urban space. *Proceedings of the Association of American Geographers, 1*, 60–65.

Golledge, R. G., & Brown, L. A. (1967). Search, learning and the market decision process. *Geografiska Annaler B, 49*, 116–124.

Golledge, R. G., Dougherty, V., & Bell, S. (1995). Acquiring spatial knowledge: Survey versus route-based knowledge in unfamiliar environments. *Annals of the American Association of Geographers, 85*(1), 134–158.

Golledge, R. G., Halperin, W. C., Gale, N., & Hubert, L. (1983). Exploring entrepreneurial cognitions of retail environments. *Economic Geography, 59*, 3–15.

Golledge, R. G., & Hubert, L. (1981). Matrix reorganization and dynamic programming: Applications to paired comparisons and unidimensional seriation. *Psychometrika, 46*, 429–441.

Golledge, R. G., & Hubert, L. (1982). Some comments on non-Euclidean mental maps. *Environment and Planning, 14,* 107–118.

Golledge, R. G., Hubert, L. J., & Costanzo, C. M. (1982, April). *Combinatorial data analysis procedures for evaluation of alternate models.* Paper presented at the 13th Annual Modeling and Simulation Conference, Pittsburgh, PA.

Golledge, G. R., Klatzky, R. L., & Loomis, J. M. (1995). Cognitive mapping and wayfinding by adults without vision. In J. Portugali (Ed.), *The Construction of Cognitive Maps* (pp. 215–246). Dordrecht, Kluwer Academic Publishers.

Golledge, R. G., Parnicky, J. J., & Rayner, J. N. (1979a). An experimental design for assessing the spatial competence of mildly retarded populations. In R. G. Golledge, J. J. Parnicky, & J. N. Rayner (Eds.), *The Spatial Competence of Selected Populations,* vol. 1 (pp. 105–119). The Ohio State University Research Foundation Project 761142/711173—Annual Report, July. NSF Grant #SOC77-26977-A02.

Golledge, R. G., Parnicky, J. J., & Rayner, J. N. (1979b). An experimental design for assessing the spatial competence of mildly retarded populations. *Soc. Sci. & Med., 13D,* 291–295.

Golledge, R. G., Parnicky, J. J., & Rayner, J. N. (Eds.). (1980). *The spatial competence of selected populations.* The Ohio State University Research Foundation Project 761142/711173 Final Report, November. NSF Grant SOC77-26977-A02.

Golledge, R. G., Rayner, J. N., & Rivizzigno, V. L. (1975). *The recovery of cognitive information about a city.* Paper read at the North American Classification Society and Psychometric Society Meetings, Iowa City.

Golledge, R. G., Rayner, J. N., & Rivizzigno, V. L. (1982). Comparing objective and cognitive representations of environmental cues. In R. G. Golledge & J. N. Rayner (Eds.), *Proximity and preference: Problems in the multidimensional analysis of large data sets* (pp. 233–266). Minneapolis, MN: University of Minnesota Press.

Golledge, R. G., Richardson, G. D., Rayner, J. N., & Parnicky, J. J. (1983). The spatial competence of selected mentally retarded populations. In H. L. Pick & L. P. Acredolo (Eds.), *Spatial orientation and spatial representation* (pp. 77–100). New York: Plenum Press.

Golledge, R. G., Rivizzigno, V. L., & Spector, A. (1976). Learning about a city: Analysis by multidimensional scaling. In R. G. Golledge & G. Rushton (Eds.), *Spatial Choice and Spatial Behavior* (pp. 95–116). Columbus, OH: Ohio State University Press.

Golledge, R. G., Ruggles, A. J., Pellegrino, J. W., & Gale, N. D. (1993). Integrating route knowledge in an unfamiliar neighborhood: Along and across route experiments. *Journal of Environmental Psychology, 13*(4), 293–307.

Golledge, R. G., Rushton, G., & Clark, W. A. V. (1966). Some spatial characteristics of Iowa's farm population and their implications for the grouping of central place functions. *Economic Geography, 43,* 261–272.

Golledge, R. G., Smith, T. R., Pellegrino, J. W., Doherty, S., & Marshall, S. P. (1985). A conceptual model and empirical analysis of children's acquisition of spatial knowledge. *Journal of Environmental Psychology, 5,* 125–152.

Golledge, R. G., & Spector, A. (1978). Comprehending the urban environment: Theory and practice. *Geographical Analysis, 10,* 403–426.

Golledge, R. G., & Stimson, R. J. (1987). *Analytical Behavioural Geography.* London: Croom Helm.

Golledge, R. G., & Timmermans, H. (1990). Applications of behavioural research on spatial problems, I: Cognition. *Progress in Human Geography, 14*(1), 57–99.

Golledge, R. G., & Zannaras, G. (1973). Cognitive approaches to the analysis of human spatial behavior. In W. H. Ittelson (Ed.), *Environmental Cognition* (pp. 59–94). New York: Seminar Press.

Goodwin, P. (1992). *Adaptive Responses to Congestion.* Technical Monograph. Oxford: Transportation Studies Unit, Oxford University.

Gopal, S. (1988). *A Computational Model of Spatial Navigation.* Unpublished PhD Dissertation, Department of Geography, University of California, Santa Barbara.

Gopal, S., Klatzky, R., & Smith, T. R. (1989). NAVIGATOR: A psychologically based model of environmental learning through navigation. *Journal of Environmental Psychology, 9,* 309–322.

Gopal, S., & Smith, T. R. (1990). Human wayfinding in an urban environment: Performance analysis using a computational process approach. *Environment and Planning A, 22,* 169–191.

Gordon, I. R. (1985). The cyclical sensitivity of regional employment and unemployment differentials. *Regional Studies, 19*(2), 161–179.

Gordon, I. R. (1992). Modelling approaches to migration and the labour market. In T. Champion & T. Fielding (Eds.), *Migration Processes and Patterns, Volume 1, Research Progress and Prospects* (pp.119–134). London: Belhaven Press.

Gordon, I. R., & Vickerman, R. (1982). Opportunity, preference and constraint: an approach to the analysis of metropolitan migration. *Urban Studies, 19*, 247–261.

Gordon, M. M. (1964). *Assimilation in American Life.* New York: Oxford University Press.

Gordon, P., & Richardson, H. (1993, April). *Sustainable congestion.* Paper submitted to 4th Industrial Workshop on Technological Change and Urban Forms: Productive and sustainable cities. University of California, Berkeley.

Gould, P. (1963) Man against his environment: A game theoretic framework. *Annals of the Association of American Geographers, 53,* 290–297.

Gould, P. (1966). *Questions arising from the description of mental maps: British school leavers.* State University, PA: The Pennsylvania State University.

Gould, P., & White, R. (1974). *Mental Maps.* Harmondsworth, England: Penguin.

Gould, W. T. S., & Prothero, R. M. (1975). Space and time in African population mobility. In L. A. Kosinski & R. M. Prothero (Eds.), *People on the move* (pp. 39–49). London: Methuen.

Graham, E. (1976). What is a mental map? *Area, 8,* 259–262.

Graham, E. (1982). Maps, metaphors, and muddles. *Professional Geographer, 34*(3), 251–260.

Green, N. L. (1990). Histoire comparative des migrations. *Annales ESC, 6* (Nov–Dec), 1335–1350.

Greenwood, M. J., Mueser, P. R., Plane, D. A., & Schlottmann, A. M. (1991). New directions in migration research: Perspectives from some North American regional science disciplines. *Annals of Regional Science, 25,* 237–270.

Grundy, E. (1992). The household dimension in migration research. In T. Champion & T. Fielding (Eds.), *Migration Processes and Patterns, Volume 1, Research Progress and Prospects* (pp. 165–174). London: Belhaven Press.

Gulliver, F. P. (1908). Orientation of maps. *Journal of Geography, 7,* 55–58.

Haber, L. R., Haber, R. N., Penningroth, S., Novak, K., & Radgowski, H. (1993). Comparison of nine methods of indicating the direction to objects: Data from blind adults. *Perception, 22,* 35–47.

Hägerstrand, T. (1952). *The Propagation of Innovation Waves. Lund Studies in Geography B: Human Geography.* Lund: Gleerup.

Hägerstrand, T. (1953/1967). *Innovation Diffusion as a spatial process* (A. Pred, Trans.). Chicago: University of Chicago Press.

Hägerstrand, T. (1957). Migration and area. In D. Hannenberg, T. Hägerstrand, & B. Odeving (Eds.), *Migration in Sweden: A Symposium. Lund Studies in Geography B: Human Geography.* Lund: Gleerup.

Hägerstrand, T. (1970). What about people in regional science? *Papers of the Regional Science Association, 24,* 7–21.

Hägerstrand, T. (1972). Tartorsgrupper som Regionsmahellen, in Regioner att Leva I. In *Rapport fran ERU* (pp. 141–172). Stockholm: Allmanna Forlagert.

Hägerstrand, T. (1975). Space, time, and human conditions. In A. Karlqvist, L. Lundqvist, & F. Snickars (Eds.), *Dynamic Allocation of Urban Space* (pp. 3–12). Farnborough: Saxon House.

Hägerstrand, T. (1976). The time–space trajectory model and its use in the evaluation of systems of transportation. Paper presented at the International Conference on Transportation Research, Vienna.

Haggett, P. (1965). *Locational Analysis in Human Geography.* New York: St. Martin's Press.

Halfacree, K., & Boyle, P. J. (1993). The challenge facing migration research: The case for a biographical approach. *Progress in Human Geography, 17(3),* 333–348.

Hall, P. (1966). *World Cities.* New York: McGraw-Hill.

Hall, P. (1991). *Cities and Regions in a Global Economy. Working Paper No. 550.* Berkeley: Institute for Urban and Regional Development, University of California.

Hall, P., & Markusen, A. (1990). *The Rise of the Gun Belt.* Oxford: Oxford University Press.

Hall, P., & Preston, P. (1988). *The Carrier Wave: New Information Technology and the Geography of Innovation, 1846–2003.* London: Unwin Hyman.

Halperin, W. C. (1985). The analysis of panel data for discrete choices. In P. Nijkamp, H. Leitner, & N. Wrigley (Eds.), *Measuring the Unmeasurable* (pp. 561–586). The Hague: Martinus Nijhoff Publishers.

Hamel, J., Dufow, S. D., & Fortin, D. (1993). *Case Study Methods. Qualitative Research Methods Series, No. 32. A Sage University Paper.* Newbury Park, CA: Sage Publications.

Hanson, R. C. & Simmons, O. G. (1968). The role path: A concept and procedure for studying migration to urban communities. *Human Organization, 27,* 152–158.

Hanson, S. (1978). Measuring the cognitive levels of urban residents. *Geografiska Annaler B, 59,* 67–81.

Hanson, S. (1980). Spatial diversification and multipurpose travel: Implications for choice theory. *Geographical Analysis, 12,* 245–257.

Hanson, S. (1984). Environmental cognition and travel behavior. In D. T. Herbert & R. J. Johnston (Eds.), *Geography and the urban environment: Progress in research and application* (Vol. 6, pp. 95–126). London: John Wiley.

Hanson, S., & Huff, J. (1986). Classification issues in the analysis of complex travel behavior. *Transportation, 13,* 271–293.

Hanson, S., & Huff, J. (1988). Systematic variability in repetitious travel. *Transportation, 15,* 111–135.

Hanson, S., & Johnston, I. (1985). Gender differences in work-trip length: Explanations and implications. *Urban Geography, 6,* 193–219.

Hanson, S., & Pratt, G. (1988a). Reconceptualizing the links between home and work in urban geography. *Economic Geography, 64*(4), 299–321.

Hanson, S., & Pratt, G. (1988b). Spatial dimensions of the gender division of labor in a local labor market. *Urban Geography, 9,* 180–202.

Happel, S. K., Hogan, T. D., & Sullivan, D. (1983). The social and economic impact of Phoenix area winter residents. *Arizona Business, 30,* 3–10.

Harper, S. (1991). People moving to the countryside: case studies of decision-making. In T. Champion & C. Watkins (Eds.), *People in the Countryside: Studies of Social Change in Rural Britain* (pp. 22–37). London: Paul Chapman.

Harris, G. D., & Ullman, E. E. (1945). The nature of cities. *Annals of the American Academy of Political and Social Science, 242,* 7–17.

Harrison, J., & Sarre, P. (1976). Personal construct theory, the repertory grid, and environmental cognition. In G. T. Moore & R. G. Golledge (Eds.), *Environmental Knowing* (pp. 375–385). Stroudsburg, PA: Dowden, Hutchinson and Ross.

Hart, R. A. (1974). The genesis of landscaping: Two years of discovery in a Vermont town. *Landscape Architecture, 65,* 356–363.

Hart, R. A. (1981). Children's spatial representation of the landscape: Lessons and questions from a field study. In L. Leiben, A. Patterson, & N. Newcombe (Eds.), *Spatial Representation and Behavior Across the Lifespan* (pp. 195–232). New York: Academic Press.

Hart, R. A. (1984). The geography of children and children's geographies. In T. Saarinen, D. Seamon, & J. Sell (Eds.), *Environmental Perception and Behavior: An Inventory and Prospect* (pp. 99–129). Chicago: University of Chicago, Department of Geography Research Paper No. 209.

Hart, R. A., & Conn, M. K. (1991). Developmental perspectives on decision making and action in environments. In T. Gärling & G. W. Evans (Eds.), *Environment, Cognition and Action: An Integrated Approach* (pp. 277–294). New York: Plenum Press.

Hart, R. A., & Moore, G. T. (1973). The development of spatial cognition: A review. In R. M. Downs & D. Stea (Eds.), *Image and Environment: Cognitive Mapping and Spatial Behavior* (pp. 246–288). Chicago: Aldine.

Hartshorn, T. A., & Muller, P. O. (1992). The suburban downtown and urban economic development today. In E. S. Mills & J. F. McDonald (Eds.), *Sources of Metropolitan Growth.* New Brunswick, NJ: Centre for Urban Policy Research, the State University of New Jersey.

Harvey, D. (1972). Social justice and spatial systems. *Social Geography, 1,* 87–106.

Harvey, D. (1973). *Social Justice and the City.* Baltimore: Johns Hopkins University Press.

Harvey, D. (1985). *Consciousness and the Urban Experience.* Baltimore: Johns Hopkins University Press.

Harvey, D. (1989). *Conditions of Postmodernity.* Blackwell: Oxford University Press.

Hawley, A. (1968). Ecology: Human ecology. In D. Sills (Ed.), *International Encyclopedia of the Social Sciences* (pp. 328–332). New York: Crowell, Collier and McMillan.

Hayes-Roth, B., & Hayes-Roth, F. (1979). A cognitive model of planning. *Cognitive Science, 3,* 275–310.

Heap, S. H. (1989). *Rationality in Economics.* Oxford: Blackwell.

Heintz, K. M. (1976). *Retirement Communities: For Adults Only,* New Brunswick, NJ: The Center for Urban Policy Research, Rutgers—The State University of New Jersey.

Hemmens, G. C. (1970). Analysis and simulation of urban activity patterns. *Socio-economic Planning Sciences, 4*(1), 53–66.

Hempel, D. M. (1969). Search behavior and information utilization in the home buying process. *Proceedings of the American Marketing Association, 30,* 241–249.

Hempel, D. J. (1970). *A Comparative Study of the Home Buying Process in Two Connecticut Markets.* Storrs, CT: Center for Real Estate and Urban Economic Studies, University of Connecticut.

Henderson, J., & Castells, M. (Eds.). (1987). *Global Restructuring and Territorial Development.* London: Sage.

Hensher, D., & Louviere, J. (1979). Behavioral intentions as predictors of very specific behavior. *Transportation, 8,* 167–182.

Hensher, D., & Stopher, P. (1979). *Behavioral Travel Demand Modelling.* London: Croom Helm.

Herman, J., Chatman, S., & Roth, S. (1983). Cognitive mapping in blind people: Acquisition of spatial relationships in a large scale environment. *Journal of Visual Impairment and Blindness, April,* 161–166.

Herman, J. Herman, T. G., & Chatman, S. P. (1983). Constructing cognitive maps from partial information: A demonstration study with congenitally blind subjects. *Journal of Visual Impairment and Blindness, 77*(5), 195–198.

Hicks, D. A., & Rees, T. (1993). Cities and beyond: a look at the nation's urban economy. In J. Somer & D. A. Hicks (Eds.), *Rediscovering Urban America: Perspectives on the 1980s.* Washington, DC: US Department of Housing and Urban Development, Office of Policy Development Research.

Hill, E. W., Rieser, J. J., Hill, M. M., Hill, M., Halpin, J., & Halpin, R. (1993). How persons with visual impairments explore novel spaces: Strategies of good and poor performers. *Journal of Visual Impairment and Blindness, Oct,* 295–301

Hill, M. H. (1985). Bound to the environment: Towards a phenomenology of sightlessness. In D. Seamon & R. Mugerauer (Eds.), *Dwelling, Place and Environment* (pp. 99–111). Dordrecht: Martinus Nijhoff.

Hirsh, M., Prashker, J. N., & Ben-Akiva, M. E. (1986). Dynamic model of weekly activity pattern. *Transportation Science, 20,* 24–36.

Hirtle, S. C., & Jonides, J. (1985). Evidence of hierarchies in cognitive maps. *Memory and Cognition, 13*(3), 208–217.

Hodge, I. (1985). Employment expectations and the costs of migration. *Journal of Rural Studies, 1,* 45–57.

Hogan, T. D. (1987). Determinants of the seasonal migration of the elderly to sunbelt states. *Research on Aging, 8*(1), 23–37.

Hoinville, G. (1971). Evaluating community preferences. *Environment and Planning, 3,* 33–50.

Holdings, C. S. (1994). Further evidence for the hierarchical representation of spatial information. *Journal of Environmental Psychology, 14*(2), 137–147.

Hollins, M., & Kelley, E. K. (1988). Spatial updating in blind and sighted people. *Perception and Psychophysics, 43,* 380–388.

Hollyfield, R. L., & Foulke, E. (1983). The spatial cognition of blind pedestrians. *Journal of Visual Impairment and Blindness, 5,* 204–209.

Honikman, B. (1976). Personal contruct theory and environmental meaning: Applications to urban design. In G. T. Moore & R. G. Golledge (Eds.), *Environmental Knowing* (pp. 88–98). Stroudsburg, PA: Dowden, Hutchinson and Ross.

Horowitz, J. (1982). Specification tests for probabilistic choice models. *Transportation Research A, 16,* 383–394.

Horowitz, J. (1983). Statistical comparison of non–nested probabilistic discrete choice models. *Transportation Science, 17,* 319–350.

Horton, F. E., & Reynolds, D. R. (1969). An investigation of individual action spaces: A progress report. *Proceedings of the Association of American Geographers, 1,* 70–75.

Horton, F. E., & Reynolds, D. R. (1970). *Action Space Formation: A Behavioral Approach to Predicting Urban Travel Behavior.* Washington, DC: Highway Research Record.

Howe, A., & O'Connor, K. (1982). Travel to work and labor force participation of men and women in an Australian metropolitan area. *The Professional Geographer, 34,* 50–64.

Hoyt, H. (1939). *The Structure and Growth of Residential Neighborhoods in American Cities.* Washington, DC: Federal Housing Administration.

HRB. (1969). *Moving Behavior and Residential Choice: A National Survey.* Report No. 81. Washington, DC: Highway Research Board.

Hubert, L. J., and Schultz, J. (1976). Quadratic assignment as a general data analysis strategy. *British Journal of Mathematical and Statistical Psychology, 23,* 214–226.

Hudson, L. M. (1981). *Image: A theoretical and operational framework.* Unpublished Doctoral Dissertation, School of Earth Sciences, Macquarie University, Australia.

Huff, D. L. *(1962).* Determination of Intra-urban Retail Trade Areas. Real Estate Research Program, University of California, Los Angeles.

Huff, D. L. (1964). Defining and estimating a trade area. *Journal of Marketing, 28,* 34–38.

Huff, J. O. (1984). Distance decay models of residential search. In G. Gaile & C. Wilmott (Eds.), *Spatial Statistics and Models (pp. 345–366).* New York: Reidel.

Huff, J. O. (1986). Geographic regularities in residential search behavior. *Annals of the Association of American Geographers, 76,* 208–227.

Huff, J. O., & Clark, W. A. V. (1978). Cumulative stress and cumulative inertia: A behavioral model of the decision to move. *Environment and Planning A, 10,* 1101–1119.

Hughes, G. & McCormack, B. (1985). Migration intentions in the UK. Which households want to migrate and what succeed? *Economic Journal Conference Supplement, 91*(4), 919–937.

Hughes, G., & McCormack, B. (1987). Housing markets, unemployment and labour market flexibility in the UK. *European Economic Review, 31,* 615–645.

Hugo, G. J. (1984). *The Aging of Ethnic Populations in Australia: An Electronic Social Atlas.* Melbourne: National Research Institute of Gerontology and Geriatric Medicine, University of Melbourne.

Hugo, G. J. (1994). Population Trends. Appendix 1 in *The Australian Urban System: Trends and Prospects.* Report prepared for the Urban and Regional Development Review. Melbourne: Australian Housing and Urban Research Institute.

Hugo, G. J., Rudd, D. M., & Downie, M. C. (1984). Adelaide's aged population: Changing spatial patterns and their policy implications, *Urban Policy and Research, 2,* 17–25.

Humphreys, J. S., & Whitelaw, J. S. (1979). Immigrants in an unfamiliar environment: Locational decision-making under constrained circumstances. *Geografiska Annaler B, 61,* 8–18.

Hunt, M. E., Feldt, A. G., Marans, R. W., Patalan, L. A., & Vakalo, K. L. (1983). Retirement communities: An American original. *Journal of Housing for the Elderly, 1,* 262–275.

Hurwicz, L. (1945) The theory of economic behavior. *American Economic Review, 35,* 909–925.

Ionnides, Y. M. (1979). Market allocation through search: Equilibrium adjustment and price dispersion. *Journal of Economic Theory, 11,* 247–249.

Isard, W. (1956). *Location and Space-Economy,* Cambridge, MA: MIT Press.

Isard, W., & Dacey, M. F. (1962). On the projection of individual behavior in regional analysis. *Journal of Regional Science, 4,* 1–32.

Ittelson, W. H. (1951). The constancies in perceptual theory. *Psychological Review, 58,* 285–294.

Ittelson, W. H. (1960). *Visual Space Perception.* New York: Springer.

Ittelson, W. H. (1973). Environment perception and contgemporary perceputal theory. In W. H. Ittelson (Ed.), *Environment and Cognition* (pp. 1–19). New York: Seminar Press.

Jacob, P. (1989). Presentation [presentation]. In P. Jacob (Ed.), *Epistémologie: L'Age de la Science [Epistemology: The age of science]* (Vol. 2, pp. 9–17). Paris: Odile Jacob.

Jakle, J. A., Brunn, S., & Roseman, C. C. (1976). *Human Spatial Behavior.* North Scituate, MA: Duxbury Press.

Jerome, H. (1926). *Migration and Business Cycles.* New York: National Bureau of Economic Research.

Johnston, R. J. (1972). Activity spaces and residential preferences: Some tests of the hypothesis of sectoral mental maps. *Economic Geography, 48*(2), 199–211.

Johnston, R. J. (1973). Social area change in Melbourne, 1961–66: A sample exploration. *Australian Geographical Studies, 11,* 79–98.

Jones, M. (1972). Urban path-choosing behavior: A study in environmental cues. In W. J. Mitchell (Ed.), *Environmental Design: Research and Practice, Vol. 1* (pp. 11-4-1–11-4-10). Los Angeles: School of Architecture and Urban Planning, University of California.

Jones, P. M. (1979). HATS: A technique for investigating household decisions. *Environment and Planning A, 11,* 59–70.

Jones Lang Wootton. (1990). *Business Space in Perspective: A Review of "High Tech" Development in Australia.* Sydney: JLW Property Research Pty. Ltd.

Joseph, A. E., Keddie, P. D., & Smit, B. (1988). Unraveling the population turnaround in rural Canada. *Canadian Geographer, 32*(1), 17–30.

Kahana, E. (1982). A congruence model of person–environment interaction. In M. P. Lawton, P. G. Windley, & T. O. I. Byerts (Eds.), *Aging and the Environment: Directions and Perspectives* (pp. 97–121). New York: Springer Publishing.

Kahneman, D., Slovic, P., & Tversky, A. (1982). *Judgement under Uncertainty: Heuristics and Biases.* London: University of Cambridge Press.

Kaluger, G., & Kaluger, M. F. (1974). *Human Development: The Span of Life* (2nd ed.). St. Louis, MO: C. V. Mosby.

Kaplan, R. (1976). Way-finding in the natural environment. In G. T. Moore & R. G. Golledge (Eds.), *Environmental Knowing* (pp. 46–57). Stroudsburg, PA: Dowden, Hutchinson and Ross.

Kaplan, R. (1991). Environmental description and prediction: A conceptual analysis. In T. Gärling & G. Evans (Eds.), *Environment, Cognition and Action: An Integrated Approach* (pp. 19–34). New York: Oxford University Press.

Kaplan, S. (1973). Cognitive maps in perception and thought. In R. M. Downs & D. Stea (Eds.), *Image and Environment: Cognitive Mapping and Spatial Behavior* (pp. 63–78). Chicago: Aldine.

Kaplan, S. (1976). Adaption, structure and knowledge. In G. T. Moore & R. G. Golledge (Eds.), *Environmental Knowing* (pp. 32–45). Stroudsburg, PA: Dowden, Hutchinson and Ross.

Kaplan, S., & Kaplan, R. (1982). *Cognition and Environment.* New York: Praeger.

Kasarda, J. (1991). Global air cargo–industrial complexes as development tools. *Economic Development Quarterly, 5:3.*

Kates, R. W. (1966). Stimulus and symbol: The view from the bridge. *Journal of Social Issues, 22,*(4), 21–28.

Kelly, G. A. (1955). *The Psychology of Personal Constructs* (2 vols.). New York: Norton.

Kemeny, J. (1983). *The Great Australian Nightmare.* Melbourne: Georgian House.

Kemeny, J. G., & Thompson, G. L. (1957). Attitudes and game outcomes. In A. W. Tucker & P. Wolfe (Eds.), *Contributions to the Theory of Games, Annals of Mathematical Studies.* Princeton, NJ: Princeton University Press.

Kendig, H. (1984). Housing careers, life cycle, and residential mobility: Implications for the housing market. *Urban Studies, 4,* 271–283.

King, L. J. (1976). Alternatives to a positive economic geography. *Annals of the Association of American Geographers, 66,* 293–308.

King, L. J., & Golledge, R. G. (1969). Bayesian Analysis and Models in Geographic Research. In *Geographical Essays Commemorating the Retirement of Professor L.H. McCarty.* Ames, IA: University of Iowa, Department of Geography, Discussion Paper No. 12, pp. 15–46.

King, L. J., & Golledge, R. G. (1978). *Cities, Space and Behavior.* Englewood Cliffs, NJ: Prentice-Hall.

Kingsley, G., & Turner, M. A., (1993). *Housing Markets and Residential Mobility.* Washington DC: The Urban Institute Press.

Kirasic, K. C. (1991). Spatial cognition and behavior in young and elderly adults: Implications for learning new environments. *Psychology and Aging, 6*(1), 10–18.

Kirasic, K. C., Allen, G. L., & Haggerty, D. (1992). Age-related differences in adults macrospatial cognitive processes. *Experimental Aging Research, 18*(1), 33–39.

Kirk, J., & Miller, M. L. (1986). *Reliability and Validity in Qualitative Research. Qualitative Research Methods Series No. 1. A Sage University Paper.* Newbury Park, CA: Sage Publications.

Kirk, W. (1951). Historical geography and the concept of the behavioral environment. *Indian Geographical Journal, Silver Jubilee Edition,* 152–160.

Kitamura, R. (1984). A model of daily time allocation to discretionary out-of-home activities and trips. *Transportation Research B, 18,* 255–266.

Kitamura, R., Kazuo, N., & Goulias, K. (1990). Trip chain behavior by central city commuters: A causal analysis of time-space constraints. In P. Jones (Ed.), *Developments in Dynamic and Activity-Based Approaches to Travel Analysis* (pp. 145–170). Avebury: Aldershot.

Kitamura, R., & Kermanshah, M. (1984). A sequential model of interdependent activity and destination choice. *Transportation Research Record, 987,* 81–89.

Kitamura, R., Kostyniuk, L. P., & Uyeno, M. J. (1981). Basic properties of time-space paths: Empirical tests, TRB. *Transportation Research Record, N794,* 8–19.

Kitchin, R. M. (1994). Cognitive maps: What are they and why study them? *Journal of Environmental Psychology, 14*(1), 1–19.

Klatzky, R. L., Golledge, R. G., Loomis, J. M., Cicinelli, J. G., & Pellegrino, J. W. (1995). Performance of blind and sighted in spatial tasks. *Journal of Vision Impairment and Blindness, 89*(1), 70–82.

Klatzky, R. L., Loomis, J. M., Golledge, R. G., Cicinelli, J. G., Doherty, S., & Pellegrino, J. W. (1990). Acquisition of route and survey knowledge in the absence of vision. *Journal of Motor Behavior, 22*(1), 19–43

Kleinginna, P. R., Jr., & Kleinginna, A. M. (1981). A categorized list of emotional definitions, with suggestions for a consensual definition. *Motivation and Emotion, 5*(4), 345–379.

Knight, V. R. (1989). City building in a global society. In R. V. Knight & G. Gappert (Eds.), *Cities in a Global Society, Vol. 25, Urban Affairs Annual Review* (pp. 324–334). Newbury Park: Sage Publications.

Knox, P. (1987). *Urban Social Geography,* 2nd ed. Burnt Mill, Essex: Longman.

Kobayashi, K. (1976). An activity model: A demand model for transportation. *Transportation Research A, 10,* 67–79.

Kobayashi, K. (1979). An activity model: A validation. In D. Hensher & P. Stopher (Eds.), *Behavioral Travel Demand Modelling* (pp. 101–115). London: Croom Helm.

Kontuly, T., & Bierens, H. J. (1990). Testing the recession theory as an explanation for the migration turnaround. *Environment and Planning A, 22,* 252–270.

Kontuly, T., & Vogelsang, R. (1988). Explanations for the intensification of counterurbanization in the Federal Republic of Germany. *Professional Geographer, 40*(1), 42–54.

Kotler, P. (1965). Behavioral models for analyzing buyers. *Journal of Marketing, 29,* 37–45.

Kozlowski, L. T., & Bryant, K. J. (1977). Sense of direction, spatial orientation, and cognitive maps. *Journal of Experimental Psychology Human Perception and Performance, 3,* 590–598.

Kruskal, J. B., Young, F. W., & Seery, J. B. (1976). *How to Use KYST-2, a Very Flexible Program to Do Multidimensional Scaling and Unfolding.* Murray Hill, NJ: Bell Labs.

Kuipers, B. (1978). Modeling spatial knowledge. *Cognitive Science, 2,* 129–153.

Kunzmann, K., & Wagener, M. (1991). *The Pattern of Urbanisation in Western Europe 1960–1990.* IRPUD: Universitat Dortmund.

Kutter, E. (1973). A model for individual travel behavior. *Urban Studies, 10,* 238–258.

Ladd, F. (1970). Black youths view their environment: Neighborhood maps. *Environment and Behavior, 2,* 74–99.

Lakoff, G. (1987). *Women, Fire, and Dangerous Things: What Categories Reveal about the Mind.* Chicago: University of Chicago Press.

Lakshmanan, T. R. (1994). The *"other," social friction and culture wars.* Paper presented at the International Workshop on the New Distributional Ethics, Boston University, November 8–9.

Lakshmanan, T. R., & Hansen, W. G. *(1965). A retail market potential model.* Journal of American Institute Planners, 31, 134–143.

Landau, B. (1988). The construction and use of spatial knowledge in blind and sighted children. In J. Stiles-Davis, M. Kritchevsky, & U. Bellugi (Eds.), *Spatial Cognition: Brain Bases and Development.* (pp. 343–371). Hillsdale, NJ: Lawrence Erlbaum Associates

Landau, B., & Jackendoff, R. (1993). "What" and "where" in spatial language and spatial cognition. *Behavioral and Brain Sciences, 16,* 217–265.

Landau, B., Spelke, E., & Gleitman, H. (1984). Spatial knowledge in a young blind child. *Cognition, 16,* 225–260.

Langdale, J. V. (1989). The geography of international business telecommunications: The role of leased networks. *Annals of the Association of American Geographers, 79,* 501–522.

Langdale, J. V. (1991). Telecommunications and international transactions in information services. In S. D. Brunn & T. R. Leinbach (Eds.), *Collapsing Space and Time: Geographic Aspects of Communication and Information* (pp. 193–214). London: Harper Collins Academic.

Lansing, J. B., & Hendricks, G. (1967). *Automobile Ownership and Residential Density.* Ann Arbor: Survey Research Center, University of Michigan.

Laslett, P. (1989). *The Fresh Map of Life: The Emergence of the Third Age.* London: Weidenfeld & Nicolson.

Laurendeau, M., & Pinard, A. (1970). *The Development of the Concept of Space in the Child.* New York: International Universities Press.

Lawton, M. P. (1980). *Environment and Aging,* Monterey, CA: Brooks-Cole.

Lee, E. S. (1969). A theory of migration. *Demography, 3,* 47–57.

Lee, T. R. (1962). Brennan's law of shopping behavior. *Psychological Reports, 11,* 662.

Lee, T. R. (1964). Psychology and living space. *Transactions of the Bartlett Society, 2,* 11–36.

Lee, T. R. (1968). Urban neighborhood as a socio-spatial schema. *Human Relations, 21,* 241–268.

Lee, T. R. (1970). Perceived distance as a function of direction in the city. *Environment and Behavior, 2,* 40–51.

Leiser, D., & Zilberschatz, A. (1989). The TRAVELLER: A computational model of spatial network learning. *Environment and Behavior, 21,* 435–463.

Lenntorp, B. (1976). *Paths in Space–Time Environments: A Time Geographic Study of Movement Possibilities of Individuals. Lund Studies in Geography B: Human Geography.* Lund: Gleerup.

Lenntorp, B. (1978). A time geographic simulation model of individual activity programmes. In T. Carlstein, D. N. Parkes, & N. J. Thrift (Eds.), *Timing Space and Spacing Time II: Human Activity and Time Geography* (pp. 162–180). London: Edward Arnold.

Levinson, D. J. (1978). *The Seasons of a Man's Life* (with C. N. Darrow, E. B. Klien, M. H. Levinson, & B. McKee). New York: Knopf.

Lewin, K. (1935). *A Dynamic Theory of Personality.* New York: McGraw–Hill.

Lewis, G. J. (1982). *Human Migration.* London: Croom Helm.

Lewis, G. J., McDermott, P., & Sherwood, K. B. (1991). The counter-urbanization process: Demographic restructuring and policy response in rural England. *Sociologia Ruralis, 31*(4), 309–320.

Ley, D. (1972). *The black inner city as a frontier outpost: Images and behavior of a North Philadelphia neighborhood.* Unpublished Doctoral Dissertation, Department of Geography, Pennsylvania State University.

Ley, D. (1983). *A Social Geography of the City.* New York: Harper & Row.

Ley, D. (1986). Alternative explanations for inner-city gentrification: A Canadian assessment. *Annals of the Association of American Geographers, 76,* 521–535.

Ley, D., & Samuels, M. (1978). *Humanistic Geography.* Chicago: Maaroufa.

Leyshon, A., Thrift, N., & Daniels, P. (1987). *The Urban and Regional Consequences of the Restructuring of World Financial Markets: The Case of the City of London.* University of Bristol and Portsmouth Polytechnic Working Papers on Producer Services.

Liben, L. S. (1981). Spatial representation and behavior: Multiple perspectives. In L. S. Liben, A. H. Patterson, & N. Newcombe (Eds.), *Spatial Representation and Behavior across the Lifespan.* (pp. 3–36). New York: Academic Press.

Liben, L. S. (1991). Environmental cognition through direct and representational experiences: A lifespan perspective. In T. Gärling & G. W. Evans (Eds.), *Environment, Cognition and Action—An Integrated Approach.* (pp. 245–276). New York: Plenum Press.

Liben, L. S., & Downs, R. M. (1989). Understanding maps as symbols: The development of map concepts in children. *Advances in Child Development and Behavior, 22,* 145–201.

Lieblich, I., & Arbib, M. (1982). Multiple representation of space underlying behavior and associated commentaries. *The Behavior and Brain Sciences, 5*(4), 627–660.

Lindberg, E., & Gärling, T. (1981). Acquisition of locational information about references points during locomotion with and without a concurrent task: Effects of number of reference points. *Scandinavian Journal of Psychology, 22,* 109–115.

Lloyd, R. E. (1982). A look at images. *Annals of the Association of American Geographers, 72*(4), 532–548.

Lloyd, R., & Jennings, D. (1978). Shopping behavior and income: Comparisons in an urban environment. *Economic Geography, 54,* 157–167.

Lockman, J. J., Rieser, J. J., & Pick, H. L. (1981). Assessing blind travelers' knowledge of spatial layout. *Journal of Visual Impairment and Blindness, 7,* 321–326.

Lohman, D. F. (1979). *Spatial Ability: Review and Re-analysis of the Correlational Literature. Aptitude Research Project, Report #8.* Stanford, CA: Stanford University.

Longino, C. F., Jr. (1980). Retirement communities. In F. J. Berghorn & E. D. Schafer (Eds.), *The Dynamics of Aging: Original Essays on the Processes and Experience of Growing Old* (pp. 391–418). Boulder, CO: Westview Press.

Longino, C. F., Jr. (1992). The forest and the trees: Micro-level considerations in the study of geographic mobility in old age. In A. Rogers (Ed.), *Elderly Migration and Population Redistribution: A Comparative Study.* (pp. 23–34). London: Belhaven.

Looman, J. D. (1969). *Consumer spatial behavior: A conceptual model and an empirical case study of supermarket patronage.* Unpublished Masters Thesis, Department of Geography, The Ohio State University.

Loomis, J. M., Da Silva, J. A., Fujita, N., & Fukusima, S. S. (1992). Visual space perception and visually directed action. *Journal of Experimental Psychology: Human Perception and Performance, 18*(4), 906–921.

Loomis, J. M., Klatzky, R. L., Golledge, R. G., Cicinelli, J. G., Pellegrino, J. W., & Fry, P. A. (1993). Non-visual navigation by blind and sighted: Assessment of path integration ability. *Journal of Experimental Psychology, General, 122,* 73–91.

Loomis, J. M., Sorce, P., & Tyler, P. R. (1989). A lifestyle analysis of healthy retirees and their interests in moving to a retirement community. *Journal of Housing for the Elderly, 5(2), Special Issue, The Retirement Community Movement: Contemporary Issues,* 19–35.

Lösch, A. (1954). *The Economics of Location* (W. H. Woglom & W. F. Stolper, Trans.). New Haven: Yale University Press.

Louviere, J. (1974). Predicting the response to real stimulus objects from an abstract evaluation of their attributes: The case of trout streams. *Journal of Applied Psychology, 59,* 572–577.

Louviere, J., & Henley, D. (1977). Information integration theory applied to student apartment selection decisions. *Geographical Analysis, 9,* 130–141.

Louviere, J., & Meyer, R., (1976). *A model for residential impression formation.* Geographical Analysis, 8, 479–486.

Louviere, J. J., & Timmermans, H. (1987). *A review of some recent advances in decompositional preference and choice models.* Paper presented at the Fifth European Colloquium on Quantitative and Theoretical Geography, Bardonecchia, Italy.

Lowenthal, D. (1961). Geography, experience, and imagination: Towards a geographical epistemology. *Annals of the Association of American Geographers, 51,* 241–260.

Lowenthal, D. (Ed.).(1967). *Environmental Perception and Behavior.* Chicago: University of Chicago, Department of Geography, Research Paper No. 109.

Lowenthal, D. (1975). Past time, present place: Landscape and memory. *The Geographical Review, 65,* 1–36.

Lowenthal, D., & Prince, H. C. (1965). English landscape tastes. *Geographical Review, 55,* 186–222.

Lowrey, R. A. (1970). Distance concepts of urban residents. *Environment and Behavior, 2,* 52–73.

Lowrey, R. A. (1973). A method for analyzing distance concepts of urban residents. In R. M. Downs & D. Stea (Eds.), *Image and Environment: Cognitive Mapping and Spatial Behavior* (pp. 338–360). Chicago: Aldine.

Luce, R. D. (1959). *Individual Choice Behavior: A Theoretical Analysis.* New York: John Wiley.

Lynch, K. (1960). *The Image of the City.* Cambridge, MA: MIT Press.

Lynch, K. (1976). Preface. In G. T. Moore & R. G. Golledge (Eds.), *Environmental Knowing*. Strouds-burg, PA: Dowden, Hutchinson and Ross.

Lyotard, J. F. (1984). *The Postmodern Condition: A Report on Knowledge*. Manchester: Manchester University Press.

Lyotard, J. F. (1986). Rules and paradoxes and svelte appendix. *Cultural Critique, 5* (Winter), 209–219.

MacEachren, A. M. (1991). The role of maps in spatial knowledge acquisition. *Cartographic Journal, 28*(2), 152–162.

MacEachren, A. M. (1992). Application of environmental learning theory to spatial knowledge acquisition from maps. *Annals of the Association of American Geographers, 82*(2), 245–274.

MacKay, D. B., & Olshavsky, R. (1975). Cognitive maps of retail location: An investigation of some basic issues. *Journal of Consumer Research, 2,* 197–205.

MacKenzie, S. (1980). *Women and reproduction of labour power in the industrial city: A case study.* Working Paper No. 23, Urban and Regional Studies, University of Sussex.

MacLennan, D., & Wood, G. (1982). Information acquisition: Patterns and strategies. In W. A. V. Clark (Ed.), *Modelling Housing Market Search* (pp. 134–159). London: Croom Helm.

Madden, J. (1980). Consequences of the growth of the two-earner family. Urban land use and the growth in two-earner households. *American Economics Review, 70,* 191–197.

Madden, J. (1981). Why women work closer to home. *Urban Studies, 18,* 181–194.

Madden, J., & White, M. (1980). Spatial implications of increases in the female labor force: A theoretical and empirical synthesis. *Land Economics, 56,* 432–446.

Maher, C. A. (1993). Recent trends in Australian urban development: Locational change and the policy quandary. *Urban Studies, 30,* 797–825.

Maher, C. A., & Saunders, E. (1994, November). *Changing Patterns of Relocation within a Metropolitan Area: Intraurban Mobility in Melbourne 1971–91*. Paper presented to Workshop on Mobility, Melbourne.

Maher, C. A., & Stimson, R. J. (1994). *Regional Population Growth in Australia: Nature, Impacts and Implications*. Bureau of Immigration and Population Research. Canberra: Australian Government Publishing Service.

Maher, C. A., Whitelaw, J., McAllister, A., Francis, R., Palmer, J., Chee, E., & Taylor, P. (1992). *Mobility and Locational Disadvantage within Australian Cities*. Report 2, Social Justice Research Program into Locational Disadvantage, Canberra: AGPS.

Maier, G., & Weiss, P. (1991). The discrete choice approach to migration modelling. In J. Stillwell & P. Congdon (Eds.), *Migration Models: Macro and Micro Approaches* (pp. 17–31). London: Belhaven.

Malecki, E. J. (1980). Dimensions of R and D location in the United States. *Research Policy, 9,* 2–22.

Malecki, E. J. (1981). Public and private sector interrelationships, technological change and regional development. *Papers of the Regional Science Association, 47,* 121–137.

Malecki, E. J. (1985). Research and development and the geography of high technology industries. In J. Rees (Ed.), *Technology, Regions and Policy* (pp. 57–74). Totowa, NJ : Rowman & Littlefield.

Mansky, C., & Lerman, S. (1977). The estimation of choice probabilities from choice based samples. *Econometrica, 45,* 77–88.

Marans, R. W., Feldt, A. G., Pastalan, L. A., Hunt, M .E., & Vakalo, K. L. (1983). Changing properties of retirement communities. In *Housing for a Maturing Population* (pp. 86–111). Washington, Urban Land Institute.

Marans, R. W., Hunt, M. E., & Vakalo, K. L. (1984). Retirement communities. In I. Altman, M. P. Lawton, & J. F. Wohlwill (Eds.), *Elderly People and the Environment* (pp. 57–93). New York: Plenum Press.

Marble, D. F. (1967). A theoretical exploration of individual travel behavior. In W. Garrison & D. f. Marble (Eds.), *Quantitative Geography, Part I (Economic and Cultural Topics)* (pp. 33–53). Evanston, IL: Northwestern University, Department of Geography, Studies in Geography, No. 13.

Marcuse, P. (1996). *Is Australia different? Globalisation and the New Urban Poverty*. Australian Housing and Research Institute (AHIRI), Melbourne, Occasional Paper 3.

Mark, L. S., & Silverman, S. (1971). The effect of concrete operations upon the mental maps of children. Unpublished Seminar Paper, Department of Psychology, Clark University.

Markham, W. (1987). Sex, relocation, and occupational advancement: The "real cruncher" for women. In L. Larwood & B. Gutek (Eds.), *Women and Work: An Annual Review* (Vol. 2) (pp. 207–231). Beverly Hills, CA: Sage.

Maslow, A. H. (1954). *Motivation and Personality*. New York: Harper.

Masters, M. S., & Sanders, B. (1993). Is the gender difference in mental rotation disappearing? *Behavior Genetics, 23*(4), 337–341

Matthews, M. H. (1992). *Making Sense of Place: Children's Characterisation of Place*. Hemel, Hempstead: Harvester Wheatsheaf.

Maurer, R., & Baxter, J. C. (1972). Images of neighborhoods among black-Anglo-and Mexican-American children. *Environment and Behavior, 4*, 351–388.

Mayer, K., & Tuma, N. (1990). *Event History Analysis in Life Course Research*. Madison, WI: University of Wisconsin Press.

McCabe, R. W. (1971). *Shopping Centre Decisions: Evaluation Guides*. Ontario, Canada: Department of Municipal Affairs.

McCabe, R. W. (1974). *Planning Applications of Retail Models*. Ottawa: Ministry of Treasury.

McCalla, G. I., Reid, L., & Schneider, P. K. (1982). Plan creation, plan execution and knowledge execution in a dynamic micro-world. *International Journal of Man–Machine Studies, 16*, 89–112.

McCarty, H. H., & Lindberg, J. B. (1966). *A Preface to Economic Geography*. Englewood Cliffs, NJ: Prentice Hall.

McDermott, P., & Taylor, M. (1976). Attitudes, images and location: The subjective context of decision making in New Zealand manufacturing. *Economic Geography, 52*, 325–347.

McGinnis, R. (1968). A stochastic model of social mobility. *American Sociological Review, 33*, 712–722.

McHale, J. (1969). *The Future of the Future*. New York: George Braziller.

McHugh, K. E. (1984). Explaining migration intentions and destination selection. *Professional Geographer, 26*, 15–25.

McLafferty, S., & Preston, V. (1991). Gender, race, and commuting among service sector workers. *The Professional Geography, 43*(1), 1–15.

McNamara, T. P. (1986). Mental representations of spatial relations. *Cognitive Psychology, 18*, 87–121.

Mehrabian, A., & Russell, J. A. (1974). An Approach to Environmental Psychology. Cambridge, MA: MIT Press.

Meinig, D. W. (Ed.) (1979). *The Interpretation of Ordinary Landscapes: Geographical Essays*. New York: Oxford University Press.

Menchik, M. (1972). Residential environmental preferences and choice: Empirically validating preference measures. *Environment and Planning, 4*, 455–458.

Mentz, H.-J., (1984). *Analyse von Verkehrsverhalten im Haushatskontext. Schriftenreihe des Instituts für Verkehrsplanung und Verkehrswegebau, 11*. Berlin: Technische Universität Berlin.

Merrill, A. A., & Bairol, J. C. (1987). Semantic and spatial factors in environmental memory. *Memory and Cognition, 15*, 101–108

Meyer, R. (1980). A descriptive model of constrained residential search. *Geographical Analysis, 12*, 21–32.

Michelson, W. (1966). An empirical analysis of urban environmental preferences. *Journal of American Institute Planners, 24*, 355–360.

Michelson, W. (1979). *Environmental Choice, Human Behavior and Residential Satisfaction*. New York: Oxford University.

Miles, R. E., & Snow, C. C. (1986). Organisations: New concepts for new forms, *California Management Review, 28*, 62–73.

Milgram, S. (1970). The experience of living in cities. *Science, 167*, 1461–1468.

Milgram, S., & Jodelet, D. (1976). Psychological maps of Paris. In H. Proshansky, L. Rivlin, & W. Ittelson (Eds.), *Environmental Psychology*. Holt, Rinehart and Winston, NY.

Millar, S. (1988). Models of sensory deprivation: The nature/nurture dichotomy and spatial representation in the blind. *International Journal of Behavioral Development, 11*, 69–87.

Miller, H. J. (1991). Modelling accessibility using space–time prism concepts within geographical information systems. *International Journal of Geographical Information Systems, 5*(3), 287–301.

Mitchell, J. K. (1984). Hazard perception studies: Convergent concerns and divergent approaches during the past decade. In T. F. Saarinen, D. Seamon, & J. L. Sell (Eds.), *Environmental Perception and Behavior.* Research Paper #209 (pp. 33–59). Chicago: University of Chicago Press.

Mittelstaedt, H., & Mittelstaedt, M.-L. (1982). Homing by path integration. In J. C. Papi & H. F. Wallraff (Eds.), *Avian navigation.* Berlin: Springer-Verlag (pp. 290–297).

Moar, I., & Carlton, L. R. (1982). Memory for routes. *Quarterly Journal of Experimental Psychology, 34A,* 381–394.

Mohlo, I. I. (1986). Theories of migration: A review. *Scottish Journal of Political Economy, 33*(4), 396–419.

Mollonkopf, J., & Castells, M. (Eds.). (1991). *Dual City: Restructuring New York.* New York: Russell Sage Foundation.

Molotch, H., & Warner, K. (1992, April). *Growth Controls: What They Do and What They Don't.* A New Urban and Regional Hierarchy? Impacts of Modernisation, Restructuring, and the End of Bipolarity. Conference sponsored by International Sociological Association Research Committee, 21, Lewis Centre for Regional Policy Studies UCLA, Comparative Social Analysis Program UCLA, University of California, Los Angeles.

Monk, J., & Hanson, S. (1982). On not excluding half of the human in human geography. *The Professional Geographer, 34*(1), 11–23.

Montello, D. R. (1991). The measurement of cognitive distance: Methods and construct validity. *Journal of Environmental Psychology, 11*(2), 101–122.

Montello, D. R. (in press). A new framework for understanding the acquitiion of spatial knowledge in large-scale environments. In M. Egenhofer & R. Golledge (Eds.), *Spatial and Temporal Reasoning in Geographic Information Systems.* New York: Oxford University Press.

Moore, G. T. (1973). *Developmental Variations Between and Within Individuals in the Cognitive Representation of Large-Scale Environments.* Unpublished M.A. Thesis, Department of Psychology, Clark University.

Moore, G. T., & Golledge, R. G. (1976). Environmental knowing: Concepts and theories. In G. T. Moore & R. G. Golledge (Eds.), *Environmental Knowing* (pp. 3–24). Stroudsburg, PA: Dowden, Hutchinson & Ross.

Morrison, P. A., & DaVanzo, J. (1986). The prism of migration: Dissimilarities between return and onward movers. *Social Science Quarterly, 67*(3), 504–516.

Moser, C., & Kalton, G. (1971). *Survey Methods in Social Investigation.* London: Heineman Educational Books.

Moss, M. J. (1988). Telecommunications shaping the future. In G. Sternleib & J. Hughes (Eds.), *America's New Market Geography: Nation, Region, and Metropolis.* New Brunswick, NJ: Center for Urban Policy Research.

Müller, M., & Wehner, R. (1988). Path integration in desert ants, *Cataglyphis fortis. Proceedings of the National Academy of Sciences of the United States of America, 85,* 5287–5290.

Murdie, R. A. (1969). *Factorial Ecology of Metropolitan Toronto, 1951–61.* Chicago: Department of Geography, University of Chicago.

Murie, A., Niner, P., & Watson, C. (1976). *Housing Policy and Housing Systems.* Birmingham: University of Birmingham Center for Urban and Regional Studies.

Muth, R. F. (1969). *Cities and Housing.* Chicago: Department of Geography, University of Chicago.

Nakajima, A. (1991). Epidemiology of visual impairment and blindness. *Current Opinion in Ophthalmology, 2,* 733–738.

Nasar, J. L. (1979). The evaluative image of a city. In A. D. Seidel & S. Donford (Eds.), *Environmental Design: Research, Theory and Application. Proceedings of the 10th Annual Conference of EDRA* (pp. 38–45). Buffalo, NY: EDRA.

Nasar, J. L. (1982). A model-relating visual attributes in the residential environment to fear of crime. *Journal of Environmental Systems, 11*(3), 247–255.

Nasar, J. L. (1984). Cognition in relation to downtown street scenes: The comparison between Japan and the United States. In D. Duark and D. Campbell (Eds.),*Proceedings of the Environmental Design and Research Association Conference, EDRA 15* Washington, DC: EDRA.

National Housing Strategy. (1992). *National Housing Strategy: Agenda for Action*. Canberra: Australian Government Publishing Service.

Nelson, K. (1986). Female labor supply characteristics and the suburbanization of low-wage office work. In A. Scott & M. Storper (Eds.), *Production, Work and Territory* (pp. 149–171). London: Allen and Unwin.

Newcombe, N., Bandura, M. & Taylor, D. (1983). Sex differences in spatial ability and spatial activities. *Sex Roles, 9,* 377–386.

Newton, P. W. (1978). Modelling locational choice. *New Zealand Geographer, 34,* 31–40.

Nieva, V. (1985). Work and family linkages. *Women and Work, 1,* 162–190.

O'Connor, K. B. (1993). *The Australian Capital City Report, 1993.* Melbourne: Centre for Population and Urban Research, Monash University.

O'Connor, K. B., & Blakely, E. (1990). Suburbia make the city: A new interpretation of city suburbs relationships. *Urban Policy and Research, 7,* 99–105.

O'Connor, K. B., & Edgington, D. (1991). Producer services and metropolitan development in Australia. In P. Daniels (Ed.), *Services and Metropolitan Development: International Perspectives* (pp. 204–225). London: Routledge.

O'Connor, K. B., & Stimson, R. J. (1994). Economic change and the fortunes of Australian cities. *Urban Futures Journal* 4(2)–(3), 1–12.

Ohta, R. J. (1983). Spatial orientation in the elderly: The current status of understanding. In H. L. Pick & L. J. Acredolo (Eds.), *Spatial Orientation: Theory, Research, and Application* (pp. 105–123). New York: Plenum Press.

Ohta, R. J., & Kirasic, K. C. (1983). Learning about environmental learning in the elderly adult. In G. Rowles & R. J. Ohta (Eds.), *Aging and Milieu: Environmental Perspectives on Growing Old* (pp. 83–95). New York: Academic Press.

O'Keefe, J., & Nadel, L. (1978). *The Hippocampus as a Cognitive Map.* Oxford: Clarendon Press.

Oldakowski, R. K., & Roseman, C. C. (1986). The development of migration expectations: changes throughout the lifecourse. *Journal of Gerontology, 41,* 209–205.

Olsson, G. (1965). *Distance and Human Interaction: A Review and Bibliography.* Philadelphia: RSRI.

O'Riordan, T. (1973). Some reflections on environmental attitudes and environmental behavior. *Area, 5,* 17–19.

Orleans, P., & Schmidt, S. (1972). Mapping the city: Environmental cognition of urban residents. In W. J. Mitchel (Ed.), *Environmental Design, Research and Practice. Proceedings of the EDRA 3 Conference* (pp. 1.4.1–1.4.9). Los Angeles: University of California.

Otway, H. J., & Thomas, K. (1982). Reflections on risk perception and policy. *Risk Analysis, 2,* 69–82.

Pacione, M. (1976). A measure of the attraction factor: A possible alternative. *Area, 6,* 279–282.

Pacione, M. (1978). Information and morphology in cognitive maps. *Transactions of the Institute of British Geographers, NS 3,* 548–568.

Pacione, M. (1982). Space preferences, location distances, and the dispersal of civil servants from London. *Environment & Planning A, 14,* 323–333.

Paivio, A. (1969). Mental imagery in associative learning and memory. *Psychological Review, 76,* 241–263.

Palm, R. (1976a). Real estate agents and geographical information. *The Geographical Review, 66,* 266–280.

Palm, R. (1976b). The role of real estate agents as information mediators in two American cities. *Geografiska Annaler B, 51,* 28–41.

Park, R., Burgess, E., & McKenzie, R. (1925). *The City.* Chicago: University of Chicago Press.

Park, R. E. (1924). The concept of social distance. *Journal of Applied Sociology, 8,* 239–344.

Parkes, D., & Dear, R. (1990) NOMAD: *An Interacting Audio-Tactile Graphics Interpreter. Reference Manual, Version 2.0.* Newcastle, NSW, Australia: Institute of Behavioral Science, University of Newcastle.

Parkes, D., & Thrift, N. (1980). *Times, Spaces and Places.* New York: John Wiley.

Parr, J., Green, S., & Behncke, C. (1989). What people want when they move, and what happens after they move: A summary of research in retirement housing. *Journal of Housing for the Elderly, 5*(1), *Special Issue, Lifestyles and Housing of Older Adults: The Florida Experience,* 7–33.

Pas, E. (1982). *A methodology for measuring complex travel-activity patterns.* Paper presented at the annual meeting of the Association of American Geographers, San Antonio.

Pas, E. (1988). Weekly travel-activity behavior. *Transportation Research B, 15,* 89–109.

Pas, E., & Kopplelman, F. (1987). An examination of the determinants of day-to-day variability in individuals' urban travel behavior. *Transportation, 13,* 183–200.

Passini, R. (1984). *Wayfinding in Architecture.* New York: Van Nostrand Rheinhold.

Passini, R. (1986). Visual impairment and mobility: Some research and design considerations. In *Proceedings of the Seventeenth Annual Conference of the Environmental Design Research Association (EDRA 17)* (pp. 225–230). Atlanta and Washington, DC: EDRA.

Passini, R., Dupré, A., & Langlois, C. (1986) Spatial mobility of the visually handicapped active person: A descriptive study. *Journal of Visual Impairment and Blindness, 80*(8), 904–907.

Passini, R., Proulx, G., & Rainville, C. (1990). The spatio-cognitive abilities of the visually impaired population. *Environment and Behavior, 22*(1), 91–118.

Pastalan, L. A. (1989). *The Retirement Community Movement: Contemporary Issues.* New York: The Hayworth Press.

Patton, M. D. (1990). *Qualitative Evaluation and Research Methods* (2nd ed.). Newbury Park, CA: Sage Publications.

Pavlov, I. (1927). *Conditioned Reflexes.* Oxford: Oxford University Press.

Péruch, P., Giraudo, M-D., & Gärling, T. (1989). Distance cognition by taxi drivers and the general public. *Journal of Environmental Psychology, 9,* 233–239.

Peterson, G. L. (1967). A model of preference: Quantitative analysis of the perception of the visual appearance of residential neighborhoods. *Journal of Regional Science, 7,* 19–32.

Peterson, W. (1958). A general topology of migration. *American Sociological Review, 23,* 256–266.

Phipps, A. G., & Clark, W. A. V. (1988). Interactive recovery and validation of households' residential utility functions. In R. G. Golledge & H. J. P. Timmermans (Eds.), *Behavioural Modelling in Geography and Planning* (pp. 245–271). London: Croom Helm.

Phipps, A. G., & Laverty, W. J. (1983). Optimal stopping and residential search behavior. *Geographical Analysis, 15,* 187–204.

Piaget, J., & Inhelder, B. (1956). *The Child's Conception of Space.* London: Routledge and Kegan Paul.

Piaget, J., & Inhelder, B. (1967). *The Child's Conception of Space.* New York: Norton.

Piaget, J., Inhelder, B., & Szeminska, A. (1960). *The Child's Conception of Geometry.* London: Routledge & Kegan Paul; New York: Basic Books.

Pickvance, G. C. (1973). Life-cycle, housing tenure and intra-urban residential mobility: A causal model. *American Sociological Review, 21,* 279–297.

Pigram, J. J. (1993). Human-nature relationships: Leisure environments and natural settings. In T. Gärling & R. G. Golledge (Eds.), *Behavior and Environment: Psychological and Geographicsl Approaches* (pp. 400–426). Amsterdam: Elsevier/North-Holland.

Pincus, J. (1986). The impact of paid employment on women's community participation. *Proceedings of the New England–St. Lawrence Valley Geographical Society, 16,* 58–62.

Pipkin, J. S. (1981). Cognitive behavioral geography and repetitive travel. In K. R. Cox & R. G. Golledge (Eds.), *Behavioral Problems in Geography Revisited* (pp. 145–181). New York: Methuen.

Plog, S. (1974). Why distination areas rise and fall in popularity. *Cornell HRA Quarterly, February.*

Popper, K. R. (1972). *Objective Knowledge: An Evolutionary Approach.* London: Oxford University Press.

Port Authority of New York and New Jersey. (1986). *The Regional Economy: Review 1985, Outlook 1986.* New York: Port Authority of New York and New Jersey.

Potter, R. B. (1967). Directional bias within the usage and perceptual fields of urban consumers. *Psychological Reports, 38,* 988–990.

Potter, R. B. (1977). Spatial patterns of consumer behavior and perception in relation to the social class variable. *Area, 9,* 153–156.

Potter, R. B. (1978). Aggregate consumer behavior and perception in relation to urban retailing structure: A preliminary investigation. *Tijdschrift voor Economische en Sociale Geografie, 69,* 345–352.

Potter, R. B. (1979). Perception of urban retailing facilities: An analysis of consumer information fields. *Geografiska Annaler B, 61,* 19–29.

Pratt, G. (1986). Against reductionism: The relations of consumption as a model of social structuration. *International Journal of Urban and Regional Research, 10,* 377–400.

Pratt, G., & Hanson, S. (1988). Gender, class, and space. *Environment and Planning D: Society and Space, 6,* 15–35.

Pred, A. (1977). *City Systems in Advanced Economies: Past Growth, Present Processes, and Future Development Options.* London: Hutchinson.

Pred, A. (1981). Of paths and projects: Individual behavior and its societal context. In K. R. Cox & R. G. Golledge (Eds.), *Behavioral Problems in Geography Revisited.* (pp. 131–235). New York: Methuen.

Prelec, D. (1982). Matching, maximizing, and the hyperbolic reinforcement feedback function. *Psychological Review, 89,* 189–230.

Presson, C. C., & Hazelrigg, M. D. (1984). Building spatial representation through primary and secondary learning. *Journal of Experimental Psychology: Learning, Memory, and Cognition, 10,* 716–722.

Quandt, R. E. (1956). A probabilistic theory of consumer behavior. *The Quarterly Journal of Economics, 30,* 507–536.

Quigley, J. (1976). Housing demand in the short run: An analysis of polytomous choice. *Explorations in Economic Research, 3,* 76–102.

Rapoport, A. (1977). *Human Aspects of Urban Form: Towards a Man Environment Approach to Urban Form and Design.* Oxford: Pergamon.

Ravenstein, E. G. (1885). The laws of migration. *Journal of the Royal Statistical Society, 48,* 167–235.

Recker, W. W., McNally, M. G., & Root, G. S. (1982). *An activity-based model of complex travel behavior.* Paper presented at the annual meeting of the Association of American Geographers, San Antonio, TX.

Recker, W. W., McNally, M. G., & Root, G. S. (1986a). A model of complex travel behaviour: Theoretical development. *Transportation Research, 20A,* 307–318.

Recker, W. W., McNally, M. G., & Root, G. S. (1986b). A model of complex travel behaviour: An operational model. *Transportation Research, 20A,* 319–330.

Redford, A. (1926). *Labour Migration in England (1800–1850).* Manchester: University of Manchester Press.

Reilly, W. J. (1931). *The Law of Retail Gravitation.* New York: G. P. Putnam.

Reime, M., & Hawkins, C. (1979). Tourism development: A model for growth. *Cornell HRA Quarterly, May.*

Relph, E. (1976). *Place and Placelessness.* London: Pion.

Rengert, G. F. (1980). Spatial aspects of criminal behavior. In D. Georges-Abeyie & K. Harries (Eds.), *Crime: A Spatial Perspective* (pp. 47–57). New York: Columbia University Press.

Rengert, G. F. (1994, March). *Perceptions of ethnic neighborhoods and safety in the inner city: Geographic issues in police training.* Paper presented to the Association of American Geographers, San Francisco.

Rengert, G. F., & Waselchick, J. (1985). *Suburban Burglary: A Time and Place for Everything.* Springfield, IL: Charles C. Thomas.

Reskin, B., & Hartmann, H. (1986). *Women's Work, Men's Work: Sex Segregation on the Job.* Washington, DC: Academy Press.

Ricardo, D. (1924). *The First Six Chapters of the Principles of Political Economy and Taxation of David Ricardo, 1817.* New York: Macmillan. (Originally published 1817)

Richardson, G. D. (1979). *The appropriateness of using various Minkowskian metrics for representing cognitive maps produced by nonmetric multidimensional scaling.* Unpublished Masters Thesis, Department of Geography, University of California Santa Barbara.

Richardson, H. W. (1989, August). *Urban development issues in the Pacific Rim.* Paper presented at the conference on Urban Development of the Pacific Rim, Los Angeles.

Rieser, J. J., Guth, D. A., & Hill, E. W. (1982). Mental processes mediating independent travel: Implications for orientation and mobility. *Journal of Visual Impairment and Blindness, 76*(6), 213–218.

Rieser, J. J., Guth, D. A., & Hill, E. W. (1986). Sensitivity to perspective structure while walking without vision. *Perception, 15,* 173–188.

Rieser, J. J., Hill, E. W., Rosen, S., Long, R. G., Talor, C., & Zimmerman, G. J. (1985). *The Locomotion and Mobility of Low Vision Persons.* National Institute for Handicapped Research (Grant No. 1231413680A1). Nashville, TN: Vanderbilt University.

Rieser, J. J., Hill, E. W., Taylor, C. R., Bradfield, A., & Rosen, S. (1992). Visual experience, visual field size, and the development of nonvisual sensitivity to the spatial structure of outdoor neighborhoods explored by walking. *Journal of Experimental Psychology: General, 121,* 210–221.

Rieser, J. J., Lockman, J. L., & Pick, H. L. (1980). The role of visual experience in knowledge of a spatial layout. *Perception and Psychophysics, 28,* 185–190.

Rimmer, P. J. (1989, July). The emerging infrastructure arena in Pacific Rim in the early 1990s. Invited paper. In *Proceedings of Japan Society of Engineers. Infrastructure Planning and Management, 432,* iv-15.

Ritchey, P. N. (1976). Explanations of migration. *Annual Review of Sociology, 2,* 363–404.

Rogers, A. (1992). Elderly migration and population redistribution in the United States. In A. Rogers (Ed.), *Elderly Migration and Population Redistribution: A Comparative Study* (pp. 226–248). London: Belhaven.

Rogers, D. S. (1979). Evaluating the business and planning impacts of suburban shopping developments: A proposed framework of analysis. *Regional Studies, 13,* 395–408.

Rogers, D. S. (1970). *The role of search and learning in consumer space behavior: The case of urban in-migrants.* Unpublished Masters Thesis, Department of Geography, University of Wisconsin, Madison.

Rogerson, P., & MacKinnon, R. D. (1981). *A geographical model of job search, migration, and unemployment.* Buffalo: State University of New York.

Root, G. S., & Recker, W. W. (1983). Towards a dynamic model of individual activity pattern formulation. In S. Carpenter & P. M. Jones (Eds.), *Recent Advances in Travel Demand Analysis* (pp. 371–382). Aldershot: Gower.

Roseman, C. C. (1971). Migration as a spatial and temporal process. *Annals, Association of American Geographers, 61,* 589–598.

Rossano, M. J., & Warren, D. H. (1989). The importance of alignment in blind subject's use of tactual maps. *Perception, 18,* 805–816.

Rossi, P. H. (1955). *Why Families Move: A Study in the Social Psychology of Urban Residential Mobility.* Glencoe, IL: Free Press.

Rowland, D. T. (1982). Living arrangements and the later family life cycle in Australia. *Australian Journal of Ageing, 1,* 3–6.

Rowles, G. D. (1978). *Prisoners of Space? Exploring Geographical Experiences of Older People.* Boulder, CO: Westview Press.

Rushton, G. (1965). *The spatial pattern of grocery purchases in Iowa.* Unpublished Doctoral Dissertation, Department of Geography, University of Iowa.

Rushton, G. (1969a). Analysis of spatial behavior by revealed space preference. *Annals of the Association of American Geographers, 59,* 391–400.

Rushton, G. (1969b).The scaling of locational preference. In K. R. Cox & R. G. Golledge (Eds.), *Behavioral Problems in Geography: A Symposium* (pp. 197–227). Evanston, IL: Northwestern University Press.

Rushton, G., Golledge, R. G., & Clark, W. A. V. *(1967). Formulation and test of a normative model for the spatial allocation of grocery expenditures by a dispersed population.* Annals of the Association of American Geographers, 57, 389–400.

Rutherford, B., & Wekerle, G. (1988). Captive rider, captive labor: Spatial constraints on women's employment. *Urban Geography, 9,* 193–193.

Saarinen, T. F. (1966). *Perception of the Drought Hazard on the Great Plains.* University of Chicago, Department of Geography, Research Paper No. 106. Chicago, University of Chicago Press.

Saarinen, T. F. (1973a). Student views of the world. In R. M. Downs & D. Stea (Eds.), *Image and Environment* (pp. 148–161). Chicago: Aldine.

Saarinen, T. F. (1973b). The use of projective techniques in geographic research. In W. H. Ittelson (Ed.), *Environmental Cognition* (pp. 29–52). New York: Seminar Press.

Saarinen, T. F., Seamon, D., & Sell, J. L. (1984). *Environmental perception and behavior: An inventory and prospect.* University of Chicago, Department of Geography Research Paper No. 209. Chicago: University of Chicago Press.

Sabel, C. F. (1989). Flexible specialisation and the re-emergence of regional economies. In P. Hirst & J. Zeitlin (Eds.), *Reversing Industrial Decline* (pp. 17–70). Oxford: Berg.

Sadalla, E. K., & Staplin, L. J. (1980a). The perception of traversed distance: Intersections. *Environment and Behavior, 12,* 167–182.

Sadalla, E. K., & Staplin, L. J. (1980b). An information storage model for distance cognition. *Environment and Behavior, 12,* 183–193.

Sant, M., & Simons, P. (1993). The conceptual basis of counterurbanisation: Critique and development. *Australian Geographical Studies 31*(2), 113–126.

Savage, L.J. (1962) Bayesian statistics. In R. E. Machol & P. Grey (Eds.), *Recent Developments in Information and Decision Processes* (pp. 161–194). New York: Macmillan.

Saxenian, A. (1985). The genesis of Silicon Valley. In P. Hall & A. Markusen (Eds.), *Silicon Landscapes.* Boston: George Allen and Unwin.

Scheckle, G. L. (1949) *Expections in Economics.* Cambridge: Cambridge University Press.

Schoenburg, S. P. (1980). Some trends in the community participation of women in their neighborhoods. *Signs: Journal of Women in Culture and Society, 5*(3), s261–s268.

Schuler, H. J. (1979). A disaggregate store choice model of spatial decision making. *The Professional Geographer, 31,* 146–156.

Scott, A. J. (1985). Location processes, urbanization, and territorial development: An exploratory essay. *Environment and Planning A, 17,* 479–501.

Self, C. M., & Golledge, R. G. (1994). Sex-related differences in spatial ability: What every geography educator should know. *Journal of Geography, 93*(5), 234–243.

Self, C. M., Golledge, R. G., & Montello, D. R. (1995, March). *The gendering of spatial abilities.* Paper presented at the annual Association of American Geographers meeting,Chicago.

Self, C. M., Gopal, S., Golledge, R. G., & Fenstermaker, S. (1992). Gender-related differences in spatial abilities. *Progress in Human Geography 16*(3), 315–342.

Sell, J. L., Taylor, J. G., & Zube, E. H. (1984). Toward a theoretical framework for landscape perception. In T. F. Saarinen, D. Seamon, & J. L. Sell (Eds.), *Environmental Perception and Behavior: An Inventory and Prospect* (pp. 61–83). University of Chicago, Department of Geography Research Paper No. 209. Chicago: University of Chicago Press.

Selye, H. (1956). *The Stress of Life.* New York: McGraw-Hill.

Serow, W. J., & O'Cain, S. M. (1992). The role of region and coastal location in explaining metropolitan population growth differentials during the 1980s. *Review of Regional Studies, 22*(3), 217–226.

Shapcott, M., & Steadman, P. (1978). Rhythms of urban activity. In T. Carlstein, D. N. Parkes, & N. J. Thrift (Eds.), *Timing Space and Spacing Time II: Human Activity and Time Geography* (pp. 49–74). London: Edward Arnold.

Shaw, R. P. (1975). *Migration Theory and Fact: Bibliography Series.* Philadelphia: Regional Science Research Unit.

Shemyakin, F. N. (1962). General problems of orientation in space and space representations. In B. G. Anan'yev, *et al.* (Eds.), *Psychological Science in the USSR* (Vol. 1), NTIS Report No. TT62-11083 (pp. 184–225). Washington, DC: Office of Technical Services.

Shepard, R. N., & Metzler, J. (1971). Mental rotation of three-dimensional objects. *Science, 171,* 701–703.

Shevky, E., & Bell, W. (1955). *Social Area Analysis.* Stanford, CA: Stanford University Press.

Shuval, J. T. (1982). Migration and stress. In L. Goldberger & S. Breznitz (Eds.), *Handbook of Stress: Theoretical and Clinical Aspects* (pp. 677–691). New York: Free Press.

Siegel, A. W. (1981). The externalization of cognitive maps by children and adults: In search of better ways to ask better questions. In L. S. Liben, A. Patterson, & N. Newcombe (Eds.), *Spatial Representation and Behavior across the Life Span: Theory and Application* (pp. 167–194). New York: Academic Press.

Siegel, A. W., & White, S. H. (1975). The development of spatial representations of large-scale environments. In W. H. Reese (Ed.), *Advances in Child Development and Behavior* (pp. 9–55). New York: Academic Press.

Simmons, J. (1974). *Patterns of Residential Movement in Metropolitan Toronto.* Toronto: Department of Geography Research Publications,University of Toronto.

Simms, M. & Malveaux, J. (Eds.) (1986). *Slipping Through the Cracks: The Status of Black Women.* New Bruswick, NJ: Transaction Books.

Simon, H. A. (1957). *Models of Man.* New York: John Wiley.

Simon, H. A. (1990). Invariants of human behaviour. *Annual Review of Psychology, 41,* 1–19.

Singson, R. (1975). Multidimensional scaling analysis of store image and shopping behavior. *Journal of Retailing, Summer,* 38–52.

Sjaastad, L. A. (1972). The costs and returns of human migration. *Journal of Political Economy, 70,* 80–93.

Skinner, B. F. (1938). *The Behavior of Organisms.* New York: Appleton-Century-Crofts.

Skinner, J. L., Rampa, H., & Spriggs, J. (1983). Push and pull factors influencing migration to Queensland: 1961–1981. In *Migration in Australia: Proceedings of Symposium, 1–2 December* (pp. 101–144). Brisbane: Royal Geographical Society of Australia and Australian Population Association.

Slovic, P. (1993). Perceptions of environmental hazards: Psychological perspectives. In T. Gärling & R. Golledge (Eds.), *Behavior and Environment* (pp. 223–248). Amsterdam: Elsevier.

Slovic, P., Fischoff, B., & Lichtenstein, S. (1977). Behavioral decision theory. *Annual Review of Psychology, 28,* 1–39.

Slovic, P., Fischoff, B., & Lichtenstein, S. (19779). Rating the risks. *Environment, 21,* 14–20, 36–39.

Slovic, P., Layman, M., Kraus, N., Flynn, J., Chalmers, J., & Gesell, G. (1991). Perceived risk, stigman, and potential economic impacts of a high-level nuclear waste repository in Nevada. *Risk Analysis, 11,* 683–696.

Smilor, R. W., & Wakelin, M. (1990). Smart infrastructure and economic development: the role of technology and global networks. In *The Technopolis Phenomenon* (pp. 100–125). Austin: IC^2 Institute, The University of Texas at Austin.

Smith, C. J. (1983, April). *Innovation in mental health policy: The political economy of the community mental health movement, 1965–1980.* Paper presented in a special session "Community Mental Health Care in Crisis" hosted by the Medical Geography Specialty Group of the Association of American Geographers, Denver Colorado.

Smith, D. M. (1973). *The Geography of Well-Being in the United States.* New York: McGraw-Hill.

Smith, D. M. (1977). *Human Geography: A Welfare Approach.* London: Edward Arnold.

Smith, D. M. (1981). *Industrial Location: An Industrial–Geographical Analysis* (2nd ed.). New York: John Wiley.

Smith, N. (1984). *Uneven Development.* Oxford: Blackwell.

Smith, R. M. (1979). Some reflections on the evidence for the origins of the European marriage pattern in England. In C. Harris (Ed.), *The Sociology of the Family: New Directions for Britain, Sociological Review No. 28* (pp. 74–112). Keele: University of Keele.

Smith, T. R., & Clark, W. A. V. (1980). Housing market search: Information constraints and efficiency. In W. A. V. Clark & E. G. Moore (Eds.), *Residential Mobility and Public Policy.* Beverly Hills, CA: Sage Publications.

Smith, T. R., Clark, W. A. V., Huff, J. O., & Shapiro, P. (1979). A decision making and search model for intraurban migration. *Geographical Analysis, 11,* 1–22.

Smith, T. R., Pellegrino, J. W., & Golledge, R. G. (1982). Computational process modeling of spatial cognition and behavior. *Geographical Analysis, 14,* 305–325.

Sommer, J., & Hicks, D. A. (Eds.). (1993). *Rediscovering Urban America: Perspectives on the 1980's.* Washington, DC: US Department of Housing and Urban Development, Office of Policy Development Research.

Sommer, R. (1969). *Personal Space: The Behavioral Basis of Design.* Englewood Cliffs, NJ: Prentice-Hall.

Sonnenfeld, J. (1969). Equivalence and distortion of the perceptual environment. *Environment and Behavior, 1,* 83–99.

Sorkin, D. L., Ferris, D. L., & Hudak, J. (1984). *Strategies for Cities and Countries: A Strategic Planning Guide.* Washington, DC: Public Technology, Inc.

Sorre, M. (1955). *Les Migrations des Peuples: Essai sur la Mobilité Geographique.* Paris: Marcel Riviere.

Sorre, M. (1957). *Rencontres de la Geographie et de la Sociologie.* Paris: Marcel Riviere.

Sorre, M. (1958). La geographie phsychologique: L'adoptation au milieu elimatique et biosocial. *Traite de Psychologie Appliqueé, 6,* 1343–1393.

Spector, A. N. (1978). *An analysis of urban spatial imagery.* Unpublished Doctoral Dissertation, Department of Geography, Ohio State University.

Spencer, A. H. (1978). Deriving measures of attractiveness for shopping centres. *Regional Studies, 12,* 713–726.

Spencer, A. H. (1980). Cognition and shopping choice: A multidimensional scaling approach. *Environment and Planning A, 12,* 1235–1251.

Spencer, C., & Blades, M. (1986). Pattern and process: A review essay on the relationship between behavioural geography and environmental psychology. *Progress in Human Geography, 10*(2), 230–248.

Stanback, T. M. (1991). *The New Suburbanization: Challenge to the Central City.* Boulder: The Eisenhower Centre for the Conservation of Human Resources Studies in the New Economy, Westview Press.

Stanley, T., & Sewall, M. (1976). Image inputs to probabilistic model: Predicting retail potential. *Journal of Marketing, 40,* 48–53.

Stansfield, C. (1978). Atlantic City and the resort cycle. *Annals of Tourism Research, 5.*

Stea, D. (1969). The measurement of mental maps: An experimental model for studying conceptual spaces. In K. R. Cox & R. G. Golledge (Eds.), *Behavioral Problems in Geography: A Symposium* (pp. 228–253). Evanston, IL: Northwestern University Press.

Stea, D. (1973). The use of environmental modelling ("toy play") for studying the environmental cognition of children and adults. Proceedings of the Fourteenth Congress, Sociedad Interamericana de Psicologia. Sao Paulo, Brazil.

Stea, D. (1976). Notes on spatial fugue. In G. T. Moore & R. G. Golledge (Eds.), *Environmental Knowing* (pp. 106–120). Stroudsburg, PA: Dowden, Hutchinson and Ross.

Stea, D., & Blaut, J. M. (1973). Some preliminary observations on spatial learning in school children. In R. M. Downs and D. Stea (Eds.), *Image and Environment: Cognitive Mapping and Spatial Behavior* (pp. 226–234). Chicago: Aldine.

Stevens, A., & Coupe, P. (1978). Distortions in dudged spatial relations. *Cognitive Psychology, 10,* 422–437.

Stevens, S. S. (1956). The direct estimation of sensory magnitudes-loudness. *American Journal of Psychology, 69,* 1–25.

Stewart, J. Q. (1948). Demographic gravitation: Evidence and applications. *Sociometry, 11*(1), 31–58.

Stimson, R. J. (1978). *Social space, preference space and residential location behavior: A social geography of Adelaide.* Unpublished doctoral dissertation, Department of Geography, Flinders University, Australia.

Stimson, R. J. (1982). *The Australian City: A Welfare Geography.* Melbourne: Longman Cheshire.

Stimson, R. J. (1985). Evaluating the trade area impact of shopping center developments: Gravity model approaches. In R. J. Stimson & R. Sanderson (Eds.), *Assessing the Economic Impacts of Retail Centres: Issues, Methods and Implications for Government Policy.* Publication no. 122 (pp. 106–123). Canberra: Australian Institute of Urban Studies.

Stimson, R. J. (1992, July). Strategic planning for managing regional growth and development. *Pacific Regional Science Conference Organization,* The Second Summer Institute: Taipei.

Stolz, B. (1985). *Still Struggling: America's Low-Income Working Women Confront the 80s.* Lexington, MA: DC Heath.

Stopher, P. R. (1977). On the application of psychological measurement techniques to travel demand estimation. *Environment and Behavior, 9,* 67–80.

Stopher, P. R., Hartgen, D. T., & Li, Y. J. (1993). *SMART: Simulation Model for Activities, Resources and Travel.* Paper presented at the 73rd Annual Meeting of the Transportation Research Board.

Storper, M., & Christopherson, S. (1987). Flexible specialisation and regional industrial agglomerations: The case of the US motion picture industry. *Annals of the Association of American Geographers, 17,* 104–117.

Stouffer, S. D. (1940). Intervening opportunities: A theory relating to mobility and distance. *American Sociological Review, 5,* 845–867.

Stough, R. R., & Feldman, M. (1982). Tourist attraction development modeling: Public sector and management cases. *Review of Regional Studies, 12,* 22–39.

Stumpf, H. (1993). Performance factors and gender-related differences in spatial ability: Another assessment. *Memory and Cognition, 21*(6), 828–836.

Supernak, J. (1983). Transportation modeling: Lessons from the past and tasks for the future. *Transportation, 12*(1), 79–90.

Szalai, A., Converse, P. E., Feldheim, P., Scheuch, E. K., & Stone, P. J. (1972). *The Use of Time: Daily Activities of Urban and Suburban Populations in Twelve Countries.* The Hague: Mouton.

Talmy, L. (1983). How language structures space. In H. L. Pick & L. P. Acredolo (Eds.), *Spatial Orientation: Theory, Research, and Application* (pp. 225–282). New York: Plenum Press

Tanaka, N. (1993). *Container shipping and container ports of the world in recent times.* Unpublished Paper, Tokyo: Research Division, NYK Line.

Tatham, A. F., & Dodds, A. G. (1988). *Proceedings of the Second International Symposium on Maps and Graphics for Visually Handicapped People.* Nottingham, England: University of Nottingham.

Theodorson, G. A., & Theodorson, A. G. (1969). *A Modern Dictionary of Sociology.* New York: Thomas Coowell.

Thiel, P. (1961). A sequence-experience notation for architectural and urban spaces. *Town Planning Review, 32,* 33–53.

Thomas, D. S. (1938). *Research Memorandum on Migration Differentials.* New York: Social Science Research Council.

Thomlinson, R. (1969). *Urban Structure: The Social and Spatial Characteristics of Cities.* New York: Random House.

Thompson, D. (1963). New concept: Subjective distance–store impressions affect estimates of travel time. *Journal of Retailing, 39,* 1–6.

Thorndyke, P. W. (1981). Distance estimation from cognitive maps. *Cognitive Psychology, 13,* 526–550.

Thrift, N. J. (1970). Time and theory in human geography. In C. Board, R. J. Chorley, P. Haggett, & D. R. Stoddart (Eds.), *Progress in geography* (Vol. 1, pp. 65–101). London: Edward Arnold.

Thurstone, L. L. (1927). A law of comparative judgment. *Psychological Review, 34,* 273–286.

Timmermans, H. J. P. (1979). A spatial preference model of regional shopping behavior. *Tijdschrift Voor Economische en Sociale Geografie, 70,* 45–48.

Timmermans, H. J. P. (1980). Consumer spatial choice strategies: A comparative study of some alternative behavioural spatial shopping models. *Geoforum, 11,* 123–131.

Timmermans, H. J. P. (1982). Consumer choice of shopping center: An information integration approach. *Regional Studies, 16,* 171–182.

Timmermans, H. J. P. (1985). *Variety seeking models and recreational choice behavior.* Paper presented at the Pacific Regional Science Conference, Molokai, Hawaii.

Timmermans, H. J. P. (1991). Retail environments and spatial shopping behavior. Unpublished manuscript Department of Architecture, Building, and Planning, University of Technology, Eindhoven, The Netherlands.

Timmermans, H. J. P., & Golledge, R. G. (1990). Applications of behavioral research on spatial problems II: Preference and choice. *Progress in human Geography, 14,* 311–354.

Timmermans, H. J. P., van der Heijden, R. V., & Westerveld, H. (1982a). Cognition of urban retailing structures: A Dutch case study. *Tijdschrift voor Economische en Sociale Geografie, 73,* 1–12.

Timmermans, H. J. P., van der Heijden, R. V., & Westerveld, H. (1982b). The identification of factors influencing destination choice: An application of the repertory grid methodology. *Transportation, 11,* 189–203.

Tobler, W. R. (1976). The geometry of mental maps. In R. G. Golledge & G. Rushton (Eds.), *Spatial Choice and Spatial Behavior* (pp. 69–82). Columbus, OH: The Ohio State University Press.

Tobler, W. R. (1977). *Bidimensional regression.* Unpublished manuscript, Department of Geography, University of California, Santa Barbara.

Tobler, W. R. (1978). Comparison of plane forms. *Geographical Analysis, 10,*(2), 154–162.

Tolman, E. C. (1948). Cognitive maps in rats and men. *Psychological Review, 55,* 189–208.

Triandis, H. C. (1971). *Attitude and Change.* New York: Wiley.

Trowbridge, C. C. (1913). On fundamental methods of orientation and imaginary maps. *Science, 38,* 888–897.

Tuan, Y-F. (1971). Geography, phenomenology and the study of human nature. *Canadian Geographer, 15,* 181–192.

Tuan, Y-F. (1974). *Topophilia: A Study of Environmental Perception, Attitudes, and Values.* Englewood Cliffs, NJ: Prentice-Hall.

Tuan, Y-F. (1977). *Space and Place: Perspective of Experience.* Minneapolis: University of Minneapolis Press.

Tversky, A. (1972) Elimination by aspects: A theory of choice. *Psychological Review, 79*(4), 281–299.

Tversky, B. (1981). Distortions in memory for maps. *Cognitive Psychology, 13,* 407–433.

Tversky, B. (1992). Distortions in cognitive maps. *Geogorum, 23*(2), 131–138.

Tversky, A., & Kahneman, D. (1974). Judgment under uncertainty: Heuristics and biases. *Science, 185,* 1124–1131.

Tyler, P. R., Loomis, L. M. & Sorce, P. (1989). A life style analysis of healthy retirees and their interests in moving into a retirement community. In L. A. Pastalan (Ed.), *Journal of Housing for the Elderly, The Retirement Community Movement: Contemporary Issues, special issue 5, no. 2,* 19–35. New York: The Hayworth Press.

Ulrich, R. S. (1979). Visual landscapes and psychological well-being. *Landscape Research, 4*(1), 17–23.

Ulrich, R. S. (1983). Aesthetic and affective response to natural environments. In I. Altman & J. F. Wohlwill (Eds.), *Human Behavior and Evironment* (pp. 85–125). New York: Plenum Press.

Ulrich, R. S. (1984). View through a window may influence recovery from surgery. *Science, 224,* 420–421.

Ulrich, R. S., & Simons, R. F. (1986). Recovery from stress during exposure to everyday outdoor environments. *Proceedings EDRA 18,* 115–122.

Ungar, S., Blades, M., Spencer, C., & Morsley, K. (1994). Can visually impaired children use tactile maps to estimate directions? *Journal of Visual Impairment and Blindness, May–June,* 221–233.

United States Architectural and Transportation Barriers Compliance Board. (1982). *A Guidebook to the Minimum Federal Guidelines and Requirements for Accessible Design.* Washington, DC: Author.

United States Government (1979). *Public Resources Code USA, Section 2115.* Washington, DC: Government Printing Office.

Urban Land Institute (1976). *Shopping Center Development Handbook.* Washington, DC: The Urban Land Institute.

Urban Land Institute (1981). *Shopping Center Development Handbook* (pp. 74–76). Washington, DC: The Urban Land Institute.

Vance, J. E. Jr., (1976). Institutional forces that shape the city. In D. T. Herbert & R. J. Johnston (Eds.)., *Social Areas in Cities: Volume 1, Spatial Processes and Form* (pp. 81–110). Chichester, NY: Wiley.

Van Den Berg, L., Drewett, R., Klassen, L., Rossi, A., & Vijverberg, C. (1982). *Urban Europe: A Study of Growth and Decline.* Oxford: Pergamon.

van der Heijden, R. E. C. M., and Timmermans, H. J. P. (1988). The spatial transferability of a decompositional multiatribute preference model. *Environment and Planning A, 20,* 1013–1025.

Vining, D. R., Dobronyi, J. B., Otness, M., & Schwinn, J. (1982). A principal axis shift in the American spatial economy? *Professional Geographer, 34,* 270–278.

Vipond, J. (1992). The demographic dimensions of urban change and implications for housing. *Australian Planner, 30*(2), 81–85.

von Saint Paul, U. (1982) Do geese use path integration for walking home? In F. Papi & H. G. Wallraff (Eds.), *Avian Navigation.* (pp. 298–306). Berlin: Springer–Verlag.

von Senden, S. M. (1932). *Space and Sight: The Perception of Space and Shape by the Congenitally Blind Before and After Operation.* Glencoe, IL: The Free Press.

Walker, R. (1988). The geographical organization of production systems. *Environment and Planning D: Society and Space, 6,* 377–408.

Walmsley, D. J., Saarinen, T. G., & MacCabe, C. L. (1990). Down under or centre stage? The world images of Australian students. *Australian Geographer, 21*(2), 164–173.

Wapner, S., & Werner, H. (1957). *Perceptual Development.* Worcester, MA: Clark University Press.

Warner, K. P. (1983). Demographics and housing. In *Housing for a Maturing Population (pp. 2–23).* Washington, DC: Urban Land Institute.

Warnes, A. M. (1992). Age-related variation and temporal change in elderly migration. In A. Rogers

(Ed.), *Elderly Migration and Population Redistribution: A Comparative Study* (pp. 35–55). London: Belhaven.

Warren, C. A. B. (1988). *Gender Issues in Field Research. Qualitative Research Methods Series, No. 9, A Sage University Paper.* Newbury Park, CA: Sage Publications.

Warren, D. H., Rossano, M. J., & Wear, T. D. (1990). Perception of map-environment correspondence: The roles of features and alignment. *Ecological Psychology, 2,* 131–150.

Warren, D. H., & Scott, T. E. (1993). Map alignment in traveling multisegment routes. *Environment and Behavior, 25*(5), 643–666.

Watson, D. A. (1990). A message from the World Blind Union. *Journal of Visual Impairment and Blindness, 84*(6), 245.

Webber, I., & Osterbind, C. C. (1961). Types of retirement villages. In E. Burgess (Ed.), *Retirement Villages* (pp. 3–10). Ann Arbor: University of Michigan.

Webber, M. J. (1972). *Impact of Uncertainty on Location.* Cambridge: MIT Press.

Welsh, R. L., & Blasch, B. B. (Eds.). (1980). *Foundation of Orientation and Mobility.* New York: American Foundation for the Blind.

Werner, H. (1948). *Comparative psychology of mental development: Revised Edition.* New York: International Universities Press.

Werner, H., & Kaplan, B. (1963). *Symbol Formation: An Organismic–Developmental Approach to Language and the Expression of Thought.* New York: John Wiley.

Wheeler, J. (1972). Trip purposes and urban activity linkages. *Annals of the Association of American Geographers, 62,* 641–654.

Wheeler, J., & Stutz, F. P. (1971). Spatial dimensions of urban social travel. *Annals of the Association of American Geographers, 61,* 371–386.

White, G. F. (1945). *Human Adjustment to Floods: A Geographical Approach to the Flood Problem in the United States.* Department of Geography, Research Paper No. 29. Chicago: University of Chicago.

White, G. F., et al. (1958). *Changes in Urban Occupance of Flood Plains in the United States.* Department of Geography, Research Paper No. 57. Chicago: University of Chicago.

White, G. F. (1986). Formation and role of public attitudes. In R. W. Kates & I. Burton (Eds.), *Geography, Resources, and Environment: Themes from the Work of G. F. White* (pp. 219–245). Chicago: University of Chicago Press. (Original work published 1966)

White, G. F. *et al.* (1958). *Changes in Urban Occupance of Flood Plains in the United States,* Department of Geography Research Paper No. 57. Chicago: University of Chicago Press.

White, S. E., Brown, L. A., Clark, W. A. V., Gober, P., Jones, R., McHugh, K., & Morrill, R. L. (1989). Population geography. In G. L. Gaile & C. J. Willmott (Eds.), *Geography in America.* Melbourne: Merrill.

White, S. K. (1991). *Political Theory and Postmodernism.* Cambridge: Cambridge University Press.

Wiedel, J. (Ed.). (1983). *Proceedings of the First International Conference on Maps and Graphics for the Visually Impaired.* Washington, DC: The Association of American Geographers.

Wilson, A. G. (1969). The use of analogies in geography. *Geographical Analysis, 1,* 225–233.

Wilson, A. G. (1972). Some recent developments in microeconomic approaches to modelling household behavior with special reference to spatiotemporal organization. In A. Wilson (Ed.), *Papers in Urban Regional Analysis.* London: Pion.

Wilson, R. L. (1962). Livability of the city: Attitudes and urban development. In F. S. Chapin & S. F. Weiss (Eds.), *Urban Growth Dynamics* (pp. 359–399). New York: John Wiley.

Wingo, L. (1961). Utilization of urban land. *Papers and Proceedings of the Regional Science Association, 11,* 195.

Winston, G. C. (1982). *The Timing of Economic Activities.* Cambridge: Cambridge University Press.

Winston, G. C. (1987). Activity choice: A new approach to economic behavior. *Journal of Economic Behavior and Organization, 8,* 567–585.

Wiseman, R. J. (1987). Concentration and migration of older Americans. In V. Reignier & J. Pynoos (Eds.), *Housing the Aged: Design Directives and Policy Considerations* (pp. 69–82). New York: Elsevier Press.

Wish, M. (1972). Notes on the variety, appropriateness and choice of proximity measures. Unpublished paper, Bell Laboratories Workshop on Multidimensional Scaling, Murray Hill, NJ.

Wohlwill, J. F. (1968). Amount of stimulus exploration and preference as differential functions of the stimulus complexity. *Perception and Psychophysics, 4,* 307–312.

Wolcott, H. F. (1990). *Writing up Qualitative Research. Qualitative Research Methods Series, No. 20, A Sage University Paper.* Newbury Park, CA: Sage Publications.

Wolpert, J. (1964). The decision process in a spatial context. *Annals of the Association of American Geographers, 54,* 537–558.

Wolpert, J. (1965). Behavioral aspects of the decision to migrate. *Papers and Proceedings of the Regional Science Association, 15,* 159–169.

Wolpert, J. (1966). Migration as an adjustment to environmental stress. *Journal of Social Issues, 22,* 92–102.

Wood, D. (1973). The cartography of reality. Unpublished paper, Graduate School of Geography, Clark University.

Wood, D., & Beck, R. (1990). Tour personality: The independence of environmental orientation in interpersonal behavior. *Journal of Environmental Psychology, 10,* 177–207.

Worchel, P. (1951). Space perception and orientation in the blind. *Psychological Monographs: General and Applied, 65,* 1–27.

Wright, J. K. (1947). Terrae incognitae: The place of the imagination in geography. *Annals of the Association of American Geographers, 37,* 1–15.

Wrigley, N. (1984). *Categorical Data Analysis for Geographers and Environmental Scientists.* New York: Longman, Inc.

Wrigley, N., & Dunn, R. (1984a). Stochastic panel-data models of urban shopping behavior 3: The interaction of store choice and brand choice. *Environment and Planning A, 16,* 1221–1236.

Wrigley, N., & Dunn, R. (1984b). Stochastic panel-data models of urban shopping behavior 2: Multistore purchasing pattern and the Dirichlet model. *Environment and Planning A, 16,* 759–778.

Wrigley, N., & Dunn, R. (1984c) Diagnostics and resistant fits in logit choice models. In D. Pitfield (Ed.), *London Papers in Regional Science, Volume 14: Discrete Choice Models in Regional Science* (pp. 44–66). London: Pion

Wu, T. (1992). Taipei: From industrial centre to international city. In E. J. Blakely & R. J. Stimson (Eds.), *The New City of the Pacific Rim, Monograph No. 43.* Berkeley: Institute for Urban and Regional Development, University of California, Berkeley.

Yin, R. K. (1989). *Case Study Research: Design and Method.* Newbury Park, CA: Sage Publications.

Zannaras, G. (1968). *An empirical analysis of urban neighborhood perception.* Unpublished Masters Thesis, Department of Geography, The Ohio State University.

Zannaras, G. (1973). The cognitive structure of urban areas. In W. S. Preiser (Ed.), *Proceedings of EDRA IV* (vol. 2) (pp. 214–220). Stroudsburg, PA: Dowden, Hutchinson and Ross.

Zannaras, G. (1976). The relation between cognitive structure and urban form. In G. T. Moore & R. G. Golledge (Eds.), *Environmental Knowing* (pp. 336–350). Stroudsburg, PA: Dowden, Hutchinson and Ross.

Zelinsky, W. (1971). The hypothesis of mobility transition. *Geographical Review, 61*(2), 219–249.

Zonabend, F. (1992). The monograph in European ethnology. *Current Sociology, 4,* 49–54.

Zonn, L. E. (1984). Landscape depiction and perception: A transacctional approach. *Landscape Journal, 3,* 144–150.

Zube, E. H., & Pitt, D. G. (1981). Cross-cultural perceptions of scenic and heritage landscapes. *Landscape Planning, 8,* 69–87.m 401–421.

Zube, E. H. (1984). Themes in landscape assessment. *Landscape Journal, 3,* 104–110.

Index

AAMD Adaptive Behavior Scale Test, 531
Abbott, 112
Abelson and Levy, 312
abilities, spatial, 156–158, 500, 505–6, 509, 517, 546
Abrahamson and Sigelman, 543
accessibility, 330, 379, 384, 442, 472–473, 498, 505
acquiring spatial knowledge, 2, 163, 165–167, 169, 171, 173, 234, 500
action, 43, 52, 174, 206, 292, 390
  behavior, 6, 22, 35, 148, 161
  space, 136, 189, 268, 277–9, 28–29, 465, 470
activities, 279, 388, 442
  classification, 306
  household, 100, 289, 317, 332, 339–340, 342–33, 346, 461
  human, 191, 267–268, 288, 291, 296, 314
  in-home, 283, 286, 326, 334
  location, 89, 284, 461
  obligatory and discretionary, 292
  out-of-home, 283–284, 321, 334
  participation rate, 343
  patterns, 52, 112, 135, 167, 288–289, 292–293, 303, 306, 312, 316, 326, 338, 340–343, 550, 556
  planned, 296, 298
  records, 346–347
  sequence, 308
  space, 74, 136, 267–268, 277, 279–284, 288
  spatial, 547
  system, 280, 290
  time and space, 314–315, 319, 339, 342
  time budgets, 304–305
  travel records, 305, 345
  unexpected, 296, 298
activity space, 285
acts
  California Environmental Quality, 381

  disability, 495
  discretionary, 290–291
  National Environmental Policy, 381
  obligatory, 290
ADA (Americans with Disabilities Act, 1990), 490, 495, 497–498
Adams, 159, 283, 481
Adamsons and Taylor, 502
Adler and Ben-Akiva, 312
administration system, 41, 65, 146
affect, 251, 280, 388, 502
affective
  attitude, 204, 332
  components, 32, 224, 387, 390–391, 403
  responses, 391
age structures, 549
aggregate societal approach, 4, 6
aging, 103, 163, 500, 503–504, 536, 549–550
AHURI (Australian Housing and Urban Research Institute), 100
aids, 496, 506
  navigation and wayfinding, 259, 261, 506
airports, 89, 123, 127, 219
  and seaports, 127–8
Aitken, 392
Aitken, Cutter, Foote, and Sell, 239, 392–393
Ajzen, 335–336
allocentric individual, 419
Alonso, 42, 460
Amedeo, 388–391
  and Golledge, 5, 7, 31, 188, 191–192, 199
  and York, 391
analysis, 10, 13–14, 21, 26, 35, 451
  external representation, 511
  hypothetical deductive, 12
  inductive, 11–12, 15, 17
  ordered tree, 254
  of social areas, 136, 138, 140
anchor, 167, 172, 251, 253, 367